Electrical Engineering Essentials

Series Editor
Anantha P. Chandrakasan

For further volumes:
http://www.springer.com/series/10854

Dejan Marković · Robert W. Brodersen

DSP Architecture Design Essentials

Springer

Dejan Marković
Associate Professor
Electrical Engineering Department
University of California, Los Angeles
Los Angeles, CA 90095
USA

Robert W. Brodersen
Professor Emeritus
Berkeley Wireless Research Center
University of California, Berkeley
Berkeley, CA 94704
USA

Please note that additional material for this book can be downloaded from http://extras.springer.com

ISBN 978-1-4899-7778-6   ISBN 978-1-4419-9660-2   (eBook)
DOI 10.1007/978-1-4419-9660-2
Springer New York Heidelberg Dordrecht London

Contents

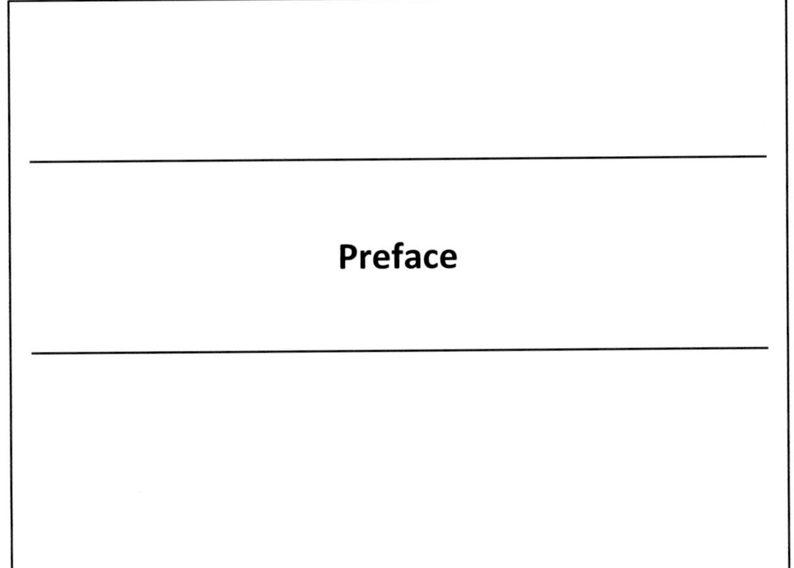

The advancement of semiconductor industry over the past few decades has made significant social and economic impacts by providing inexpensive computing and communication technologies. Our ability to access and process increasing amounts of data has created a major shift in information technology towards parallel data processing. Today's microprocessors are deploying multiple processor cores on a single chip to increase performance; radios are starting to use multiple antennas to transmit data faster and farther; new technologies are needed for processing large records of data in biomedical applications. The fundamental challenge in all these applications is how to map data processing algorithms onto the underlying hardware while meeting application constraints for power, performance, and area. Digital signal processing (DSP) architecture design is the key for successful realization of many diverse applications in hardware.

The tradeoff of various types of architectures to implement DSP algorithms has been a topic of research since the initial development of the theory. Recently, the application of these DSP algorithms to systems that require low cost and the lowest possible energy consumption has placed a new emphasis on defining the most appropriate solutions. The flexibility consideration has become a new dimension in the algorithm/architecture design. Traditional approach to provide flexibility has been through software programming a Von Neumann architecture. This approach was based on technology assumptions that hardware was expensive and the power consumption was not critical so time multiplexing was used to provide maximum sharing of the hardware resources. The situation now for highly integrated system-on-a-chip implementations is fundamentally different: hardware is cheap with potentially 1000's of multipliers and adders on a chip and the energy consumption is a critical design constraint in portable applications. Even in the case of applications that have an unlimited energy source, we have moved into an era of power-constrained performance since heat removal requires the processor to operate at lower clock rates than dictated by the logic delays.

This book, therefore, addresses DSP architecture design and the application of advanced DSP algorithms to heavily power-constrained micro-systems.

Why This Book?

♦ **Goal: to address the need for area/energy-efficient mapping of advanced DSP algorithms to the underlying hardware technology**

♦ **Challenges in digital signal processing (DSP) chip design**
  – Higher computational complexity for advanced DSP algorithms
  – More flexibility (multi-mode, multi-standard) required
  – Algorithm and hardware design are often separate
  – Power-limited performance

♦ **Solution: systematic methodology for algorithm specification, architecture mapping, and hardware optimizations**
  – Outcome 1: hardware-friendly algorithm development
  – Outcome 2: optimized hardware implementation

P.2

Slide P.2

This book addresses the need for DSP architecture design that maps advanced DSP algorithms to the underlying hardware technology in the most area- and energy-efficient way. Architecture design is expensive and architectural changes have not been able to track the pace of technology scaling. The ability to quickly explore many architectural realizations is essential for selecting the architecture that best utilizes the intrinsic computational efficiency of silicon technology.

In addition to tracking the advancements in technology, advanced DSP algorithms greatly increase computational complexity. At the same time, more flexibility to support multiple operation modes and/or multiple standards is needed in portable devices. Traditionally, algorithms and architectures are developed by different engineering teams, who also use different tools to describe their designs. Clearly, there is a pressing need for DSP architecture design that tightly couples into algorithmic and technology parameters, in order to deliver the most effective solution in power-limited regime.

In response to the above challenges, this book provides systematic methodology for algorithm modeling, architecture description and mapping, and various hardware optimizations that take into account algorithm, architecture, and technology layers. This interaction is essential, because algorithmic simplifications can often far outstrip any energy savings possible in the implementation step. The outcomes of the proposed approach, generally speaking, are hardware-aware algorithm development and its optimized hardware implementation.

Highlights

- **A design methodology starting from a high-level description to an implementation optimized for performance, power and area**

- **Unified description of algorithm and hardware parameters**
  - Methodology for automated wordlength reduction
  - Automated exploration of many architectural solutions
  - Design flow for FPGA and custom hardware including chip verification

- **Examples to show wide throughput range (kS/s to GS/s)**
  - Outcomes: energy/area optimal design, technology portability

- **Online resources: examples, references, tutorials etc.**

P.3

Slide P.3

The key feature of this book is a design methodology based on a high-level design model that leads to hardware implementation that is optimized for performance, power, and area. The methodology includes algorithm-level considerations such as automated wordlength reduction and unique data properties that can be leveraged to simplify the arithmetic. Starting point for architectural optimizations is a direct-mapped architecture, because it is well defined. From a high-level data-flow graph (DFG) model for the reference architecture, a methodology based on linear programming is used to create many different architectural solutions, within constraints dictated by the underlying technology. Once architectural solutions are available, any of the architecture design points can be mapped through commercial and semi-custom flows to field-programmable gate array (FPGA) and application-specific integrated circuit (ASIC) hardware platforms. As a final step, FPGA-based logic analysis is used to verify ASIC chips using the same design environment, which greatly simplifies the debugging process.

To exemplify the use of the design methodology described above, many examples will be discussed to demonstrate diverse range of application requirements. Applications ranging from kHz to GHz rates will be illustrated and results from working ASIC chips will be presented.

The slide material provided in the book is supplemented with additional examples, links to reference material, CAD tutorials, and custom software. All the supplements are available online. More detail about the online content is provided in Slide P.11.

Book Development

- **Over 15 years of effort and revisions…**
  - Course material from UC Berkeley (Communication Signal Processing, EE225C), ~1995-2003
    - Profs. Robert W. Brodersen, Jan M. Rabaey, Borivoje Nikolić
  - The concepts were applied and expanded by researchers from the Berkeley Wireless Research Center (BWRC), 2000-2006
    - W. Rhett Davis, Chen Chang, Changchun Shi, Hayden So, Brian Richards, Dejan Marković
  - UCLA course (VLSI Signal Processing, EE216B), 2006-2008
    - Prof. Dejan Marković
  - The concepts expanded by researchers from UCLA, 2006-2010
    - Sarah Gibson, Vaibhav Karkare, Rashmi Nanda, Cheng C. Wang, Chia-Hsiang Yang
- **All of this is integrated into the book**
  - Lots of practical ideas and working examples

P.4

Slide P.4

The material in this book is a result of many years of development and classroom use. It started as a class material (Communications Signal Processing, EE225C) at UC Berkeley, developed by professors Bob Brodersen, Jan Rabaey, and Bora Nikolić in the 1990s and early 2000s. Many concepts were applied and extended in research projects at the Berkeley Wireless Research Center in the early 2000s. These include automated Simulink-to-silicon toolflow (by R. Davis, H. So,

B. Richards), automated wordlength optimization (by C. Shi), the BEE (Berkeley Emulation Engine) FPGA platforms (by C. Chang et al.), and the use of this infrastructure in chip design (by D. Marković and Z. Zhang).

The material was further developed at UCLA as class material by Prof. D. Marković and EE216B (VLSI Signal Processing) students. Additional topics include algorithms and architectures for neural-spike analysis (by S. Gibson and V. Karkare), automated architecture transformations (by R. Nanda), revisions to wordlenght optimization tool (by C. Wang), flexible architectures for multi-mode and multi-band radio DSP (by C.-H. Yang).

All this knowledge is integrated in this book. The material will be illustrated on working hardware examples and supplemented with online resources.

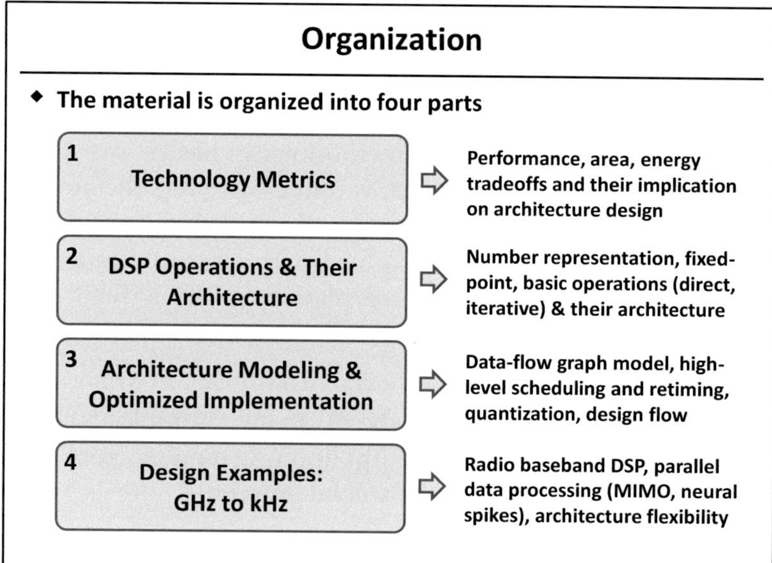

Slide P.5

The material is organized into four parts: (1) technology metrics, (2) DSP operations and their architecture, (3) architecture modeling and optimized implementation, and (4) design examples. The first part introduces technology metrics and their impact on architecture design. Towards implementation, the second part discusses number representation, fixed-point effects, basic direct and recursive DSP operations and their architecture. Putting the technology metrics and architecture concepts together, Part 3 provides data-flow graph based model and discusses automated architecture exploration using linear programming methods. Quantization effects and hardware design flow are also discussed. Finally, Part 4 demonstrates the use of architecture design methodology and hardware mapping flow on several examples to show architecture optimization under different sampling rates and amounts of flexibility. The emphasis is placed on flexibility and parallel data processing. To get a quick grasp of the book content, visual highlights from each of the parts are provided in the next few slides.

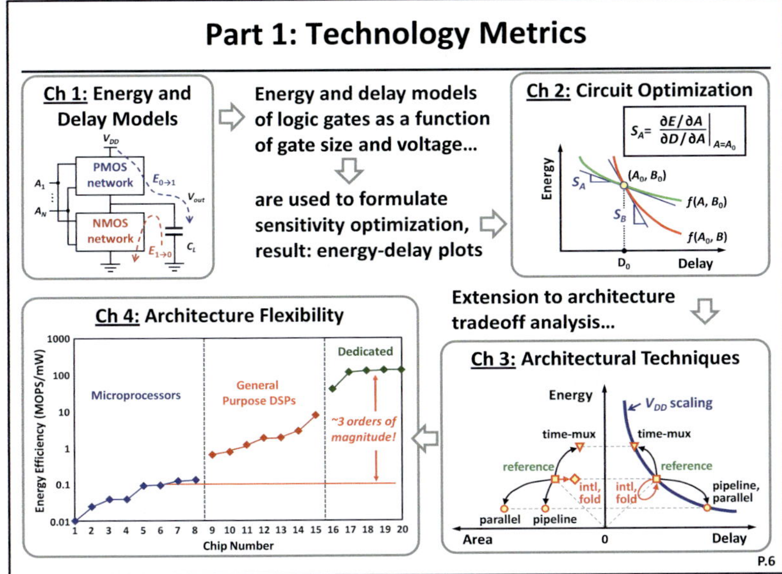

Slide P.6

Part 1 begins with energy and delay models of logic gates, which are discussed in Chap. 1. The models describe energy and delay as a function of circuit design variables: gate size, supply and threshold voltage. With these models, we formulate sensitivity-based circuit optimization in Chap. 2. The output of the sensitivity framework is the plot of energy-delay tradeoffs in digital circuits, which allows for comparing multiple circuit realizations of a function. Since performance range of circuit tuning is limited, the concept is extended to architecture level in Chap. 3. Energy-delay tradeoffs in datapaths are used to navigate architectural transformations such as time-multiplexing, parallelism, pipelining, interleaving and folding. This way, tradeoffs between area and energy for a given performance can be analyzed. To further understand architectural issues, Chap. 4 compares a set of representative chips from various categories: microprocessors, general-purpose DSPs, and dedicated. Energy and area efficiency are analyzed to understand architectural features.

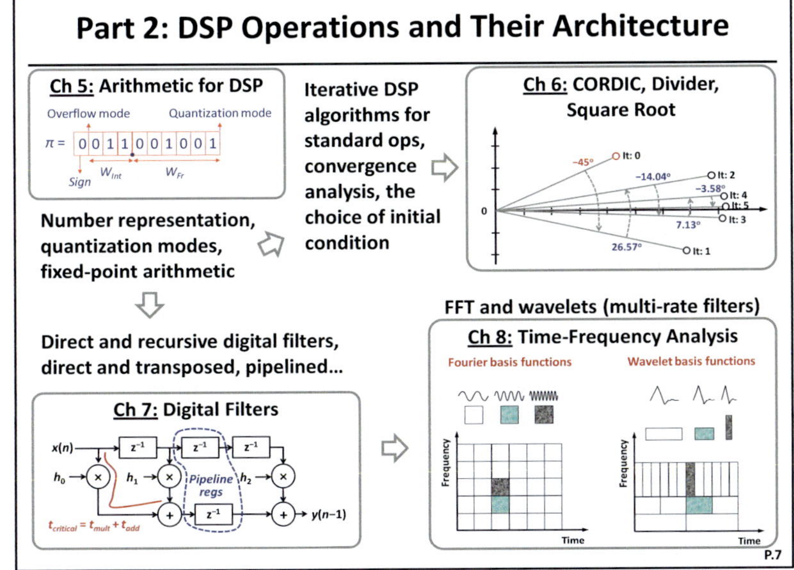

Slide P.7

Part 2 first looks at number representation and quantization modes in Chap. 5, which is required for fixed-point description of DSP algorithms. Chap. 6 then presents commonly used iterative DSP operators such as CORDIC and Newton-Raphson methods for square rooting and division. Convergence analysis with respect to the number of iterations, required quantization accuracy and the choice of initial condition is presented. Chap. 7 continues with algorithms for digital filtering. Direct and recursive filters are considered as well as direct and transposed architectures. The impact of pipelining on performance is also discussed. As a way of frequency analysis, Chap. 8 discusses FFT and wavelet transforms. Baseline architecture for the FFT and wavelet transforms is presented.

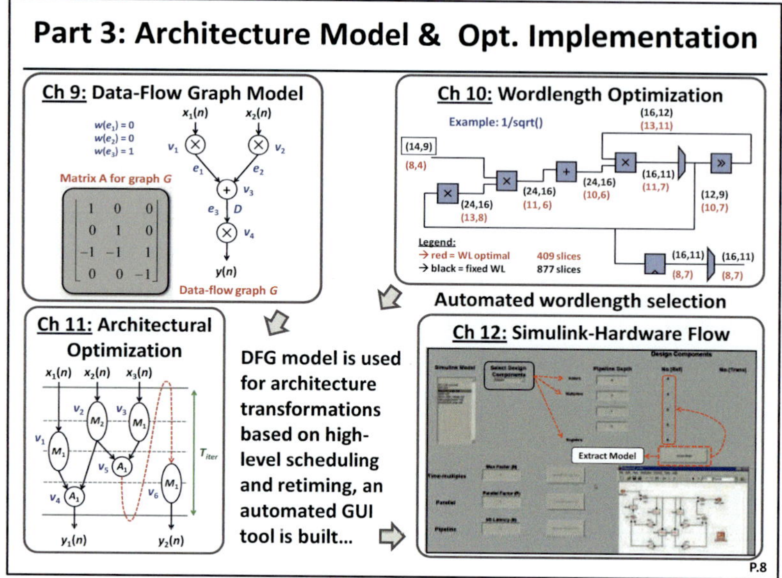

Slide P.8

Having defined technology metrics in Part 1, algorithm and architecture techniques in Part 2, we move on to algorithm models that are convenient for hardware optimization. Modeling approach is based on data-flow graph (DFG) description of a design, presented in Chap. 9, which defines the graph connectivity through incidence and loop matrices. As a first step in hardware optimization, Chap. 10 presents a method based on perturbation theory that is used to minimize wordlengths subject to constrained mean-square error (MSE) degradation due to quantization. Upon wordlength reduction, Chap. 11 discusses high-level scheduling and retiming approach as a basis for automated architecture transformations. A custom tool based on integer linear programming is implemented in a GUI environment (available for download) to demonstrate automated architecture exploration for several common DSP algorithms. Chapter 12 presents Simulink-to-hardware mapping flow that includes FPGA-based chip verification. The infrastructure from Part 3 is applied to a range of examples.

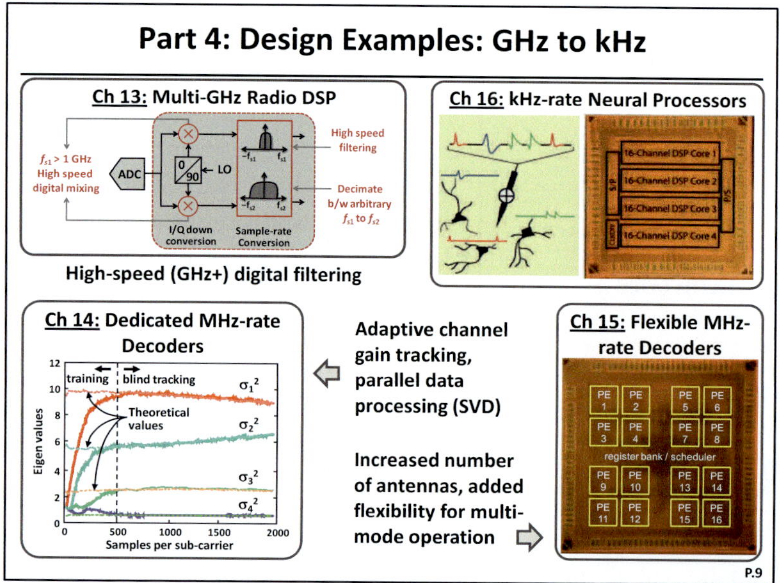

Slide P.9

Several practical design examples are demonstrated in Part 4. Starting with a GHz-rate sampling speeds, Chap. 13 discusses digital front-end architecture for software-defined radio. It illustrates multi-GHz (2.5–3.6 GHz) digital mixing, high-speed filtering, and fractional sample-rate conversion down to the modem frequency. Chap. 14 illustrates multi-antenna (MIMO) DSP processor that estimates channel gains in a 4×4 MIMO system. The algorithm implemented performs singular value decomposition (SVD) on a 4×4 matrix and makes use of iterative Newton-Raphson divider and square root. It also demonstrates adaptive LMS and retiming of multiple nested feedback loops. The SVD design serves as a reference point for energy and area efficiency and studies of design flexibility. Based on the SVD reference, Chap. 15 presents multi-mode sphere decoder that can work with up to 16×16 antennas, involves adaptive QAM modulations from BPSK to 64-QAM, sequential and parallel search methods, and variable number of sub-carriers. It demonstrates multi-core architecture achieving better energy efficiency than the SVD chip with less than 2x area cost to operate the multi-core architecture. Finally, as a demonstration of leakage-limited application, Chap. 16 discusses architecture design for neural spike sorting. The chip dissipates just $130\,\mu$W for simultaneous processing of 64 channels. The chip makes use of architecture folding for area (leakage) reduction and power gating to minimize the leakage of inactive units.

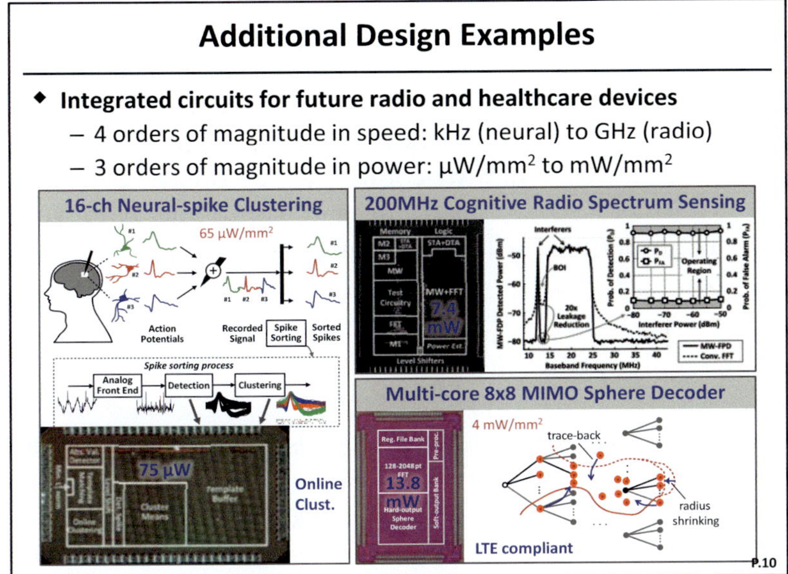

Slide P.10

The framework presented in Part 4 has also been applied to many other chips, which range about 4 orders of magnitude in sampling speed and 3 orders of magnitude in power density. Some of interesting applications include wideband cognitive-radio spectrum sensing that demonstrates sensing of 200 MHz with 200 kHz spectral resolution. The chip dissipates 7.4 mW and achieves probability of detection >0.9, probability of false-alarm <0.1, for −5 dB SNR and adjacent-band interferer of 30 dB. It shows the use of multitap-windowed FFT, adaptive decision threshold and sensing times. Further extending the flexibility of MIMO processors to multiple signal bands, an 8×8 sphere decoder featuring programmable FFT processor with 128–2048 points, multi-core hard decision, and soft output unit is integrated in 13.8 mW for a 20 MHz bandwidth. The chip meets the LTE standard specifications with power consumption of 5.8 mW. Finally, we show online spike clustering algorithm in 75 µW for 16 channels. The clustering chip exemplifies optimization of memory-intensive design. These and many other examples can be effectively optimized for low power and area using the techniques presented in this book.

Online Material

- **Online content**
  - References (papers, books, links, etc.)
  - Design examples (mini projects)
  - Custom software (architecture transformations)
  - CAD tutorials (hardware mapping flow)

- **Web sites**
  - Official public release: http://extras.springer.com
    - Updates will be uploaded as frequently as needed
  - Development wiki: http://icslwebs.ee.ucla.edu/dejan/dspwiki
    - Pre-release material will be developed on the book wiki page
    - Your contributions would be greatly appreciated and acknowledged

P.11

Slide P.11

The book is supplemented with online content that will be regularly updated. This includes references (papers, textbooks, online links, etc.), design examples, CAD tutorials and custom software. There are two places you should check for online material.

The official publisher website will contain release material, the development wiki page will contain pre-release content. Your contributions to the wiki are most welcome. Please contact us for an account and contribute with your own examples and suggestions. Your contributions will be greatly appreciated and also acknowledged in the next edition.

Acknowledgments

- **UC Berkeley / Berkeley Wireless Research Center**
  - Chen Chang, Henry Chen, Rhett Davis, Hayden So, Kimmo Kuusilinna, Borivoje Nikolić, Ada Poon, Brian Richards, Changcun Shi, Dan Wertheimer, EE225C students
- **UCLA**
  - Henry Chen, Jack Judy, Vaibhav Karkare, Sarah Gibson, Rashmi Nanda, Richard Staba, Cheng Wang, Chia-Hsiang Yang, Tsung-Han Yu, EE216B students, DM Group
- **Infrastructure support**
  - FPGA hardware: Xilinx, BWRC, BEEcube
  - Chip fabrication: IBM, ST Microelectronics
  - Software: Cadence, Mathworks, Synopsys, Synplicity

P.12

Slide P.12

Many people contributed to the development of the material. Special thanks go to UC Berkeley/ BWRC researchers for the development of advanced DSP algorithms (A. Poon), hardware platforms (C. Chen, H. Chen, H. So, K. Kuusilinna, B. Richards, D. Wertheimer), hardware flows (R. Davis, H. So, B. Nikolić), wordlength tool (C. Shi). EE225C students at UC Berkeley and EE216B students at UCLA are acknowledged for testing the material and valuable suggestions for improvement. UCLA researchers are acknowledged for the development of algorithms and architectures for neural-spike processing (S. Gibson, V. Karkare, J. Judy) and test data (R. Staba), revisions of the wordlength optimizer (C. Wang), development of architecture optimization tool (R. Nanda), and application of the methodology in chip design (V. Karkare, C.-H. Yang, T.-H. Yu). Students from DM group and ASL group are acknowledged for proofreading the manuscript. We also acknowledge hardware support from Xilinx (chips for BEE boards), BWRC (BEE boards), FPGA mapping tool flow development (BEEcube), chip fabrication by IBM and ST Microelectronics, and software support from Cadence, Mathworks, Synopsys and Synplicity.

Part I

Technology Metrics

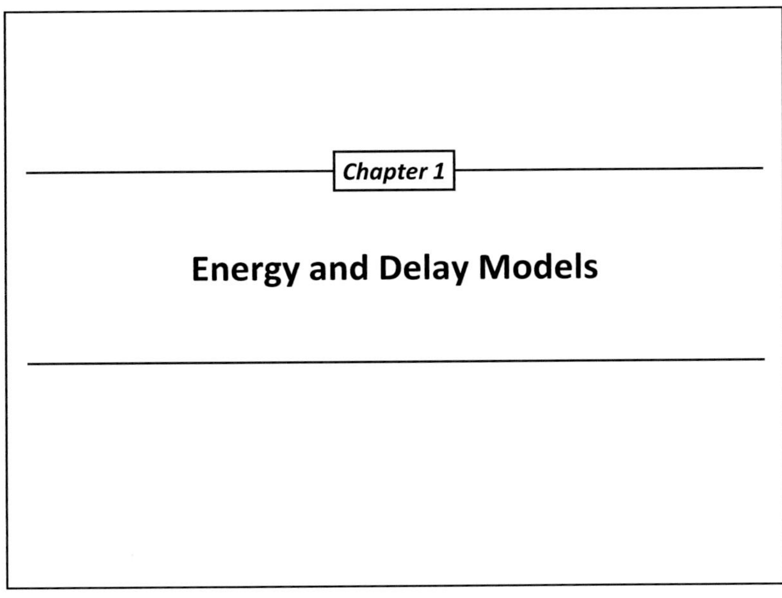

Slide 1.1

This chapter introduces energy and delay metrics of digital circuits used to implement DSP algorithms. The discussion begins with energy and delay definitions for logic gates, including the analysis of various factors that contribute to energy consumption and propagation delay. Design tradeoffs with respect to tuning gate size, supply and threshold voltages are analyzed next, followed by setting up an energy-delay tradeoff analysis for use in circuit-level optimizations. The discussion of energy and delay metrics in this chapter aims to give DSP architecture designers an understanding of hardware cost for implementing their algorithms.

Chapter Overview

- **Goal: tie-in parameters of the underlying implementation technology together with algorithm-level specifications**

- **Strategy**
  - Technology characterization (energy, delay)
  - Circuit-level tuning (gate size, supply voltage)
  - Tradeoff analysis (E-D space, logic depth, activity)

- **Remember**
  - We will go all the way down to these low-level results to match algorithm specs with technology characteristics

1.2

Slide 1.2

The goal is to bring parameters of the underlying technology into the algorithm space – to bring together two areas that are traditionally kept separate.

Technology characterization (energy, delay) is required to gain insight into circuit tuning variables that are used to adjust the delay and energy. It is important to establish technology models that can be used in the algorithms space. This model propagation will allow designers to make tradeoff analyses in the energy-delay-area space as a function of circuit, architecture, and algorithm parameters. Going from the device level and up, circuit-level analysis will mostly consider the effects of logic depth and activity since these properties strongly influence architecture design.

The analysis of circuit-level tradeoffs is important to ensure that the implemented algorithms fully utilize the performance capabilities of the underlying technology.

D. Marković and R.W. Brodersen, *DSP Architecture Design Essentials*, Electrical Engineering Essentials,
DOI 10.1007/978-1-4419-9660-2_1, © Springer Science+Business Media New York 2012

Power and Energy Figures of Merit

- **Power consumption in Watts**
  - Determines battery life in hours
- **Peak power**
  - Determines power ground wiring designs
  - Sets packaging limits
  - Impacts signal noise margin and reliability analysis
- **Energy efficiency in Joules**
  - Rate at which power is consumed over time
- **Energy = Power * Delay**
  - Joules = Watts * seconds
  - Lower energy number means less power to perform a computation at the same frequency

1.3

Slide 1.3

Let's start with power and energy metrics.

Power is the rate at which energy is delivered or exchanged; power dissipation is the rate at which energy is taken from the source (V_{DD}) and converted into heat (electrical energy is converted into heat energy during circuit operation). Power is expressed in watts and determines battery life in hours (instantaneously measures how much energy is taken out of energy source over time). We must design for peak power to ensure proper design robustness.

Energy is the rate over which power is consumed over time. It is also equal to the power-delay product. Lower energy means less power to perform the computation at the same frequency. (We will show in Chap. 4 that power efficiency is the same as energy efficiency.)

Delivering power or energy into a chip is not an easy task. For example, a 30 W mobile processor requires 30 A of current to be provided from a 1 V supply. Careful analysis of power must also include units for power delivery and conversion, not just computations. In Chap. 4, we will focus on the analysis of power required for computation.

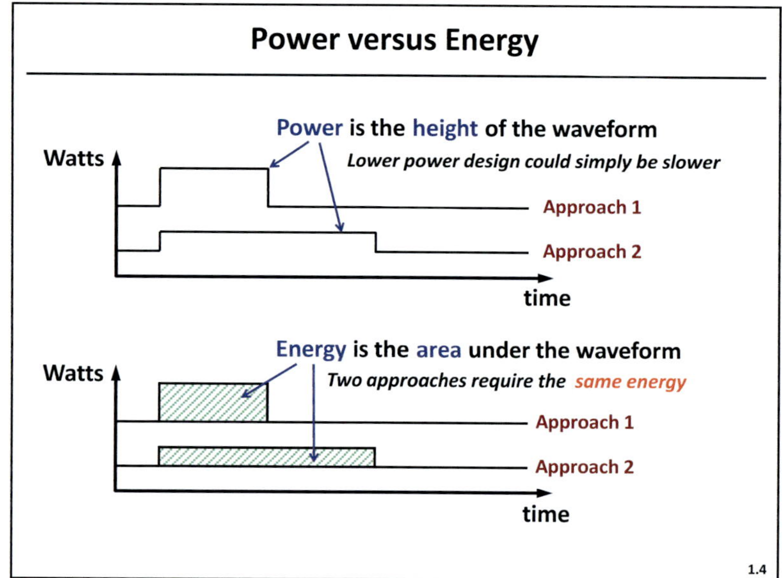

Slide 1.4

Power and energy can be graphically illustrated as shown in this slide.

If we look at the power consumption alone, we are looking at the height of the waveform. Note that Approach 1 and Approach 2 start from the same absolute level on the vertical axis, but are shown separately for improved visual clarity. The Approach 1 requires more power while Approach 2 has lower power, but takes longer to perform a task — it is a performance-power tradeoff.

If we look at the energy consumption for these two examples, we are looking at the product of power and delay (area under the waveforms). The two approaches use the same amount of energy. The same amount of energy can be spent to operate fast with a higher power or operate slow with a lower power. So, changing the operating frequency does not necessarily change the energy consumption. In other words, we consume the same amount of energy to perform a task whether we do it fast or we do it slow.

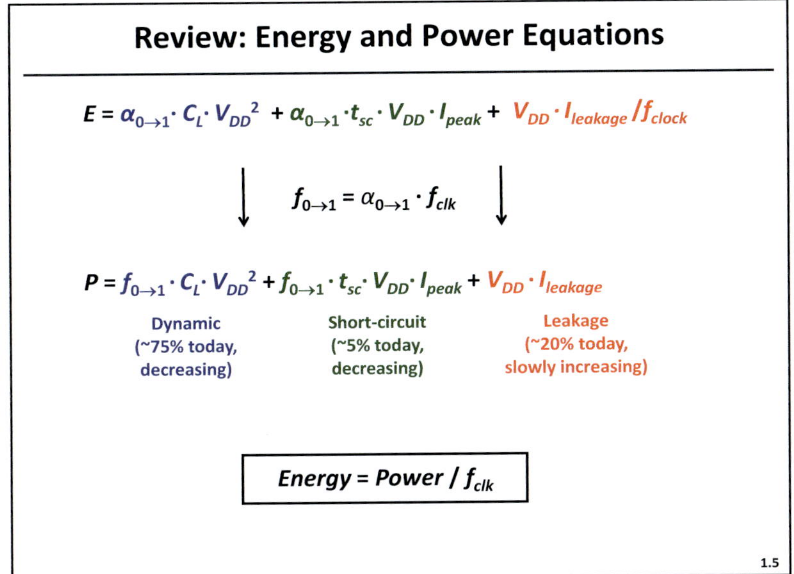

Slide 1.5

Here, we review the components of energy (power). We consider three main components of power: switching, short-circuit, and leakage. The two dominant components are switching and leakage. Switching (dynamic) power can be calculated using the well-known $f_{0\to1} \cdot C_L \cdot V_{DD}^2$ formula, where $f_{0\to1}$ represents the frequency of $0\to1$ transition at the gate output, C_L is the output capacitance of a gate, and V_{DD} is supply voltage. Leakage power is proportional to the leakage current (during idle periods) and V_{DD}. Short-circuit power can be neglected in most circuits – it typically contributes to about 5% of the total power.

As we scale technology, the short-circuit power is decreasing relatively due to supply and threshold scaling (supply scales down faster). However, threshold voltage reduction results in an exponential increase in the leakage component. Therefore, we must balance dynamic and leakage energy in order to minimize the total energy of a design. A simple relationship that allows us to convert between energy and power is *Energy = Power/f_{clk}*, where f_{clk} represents the clock frequency of a chip.

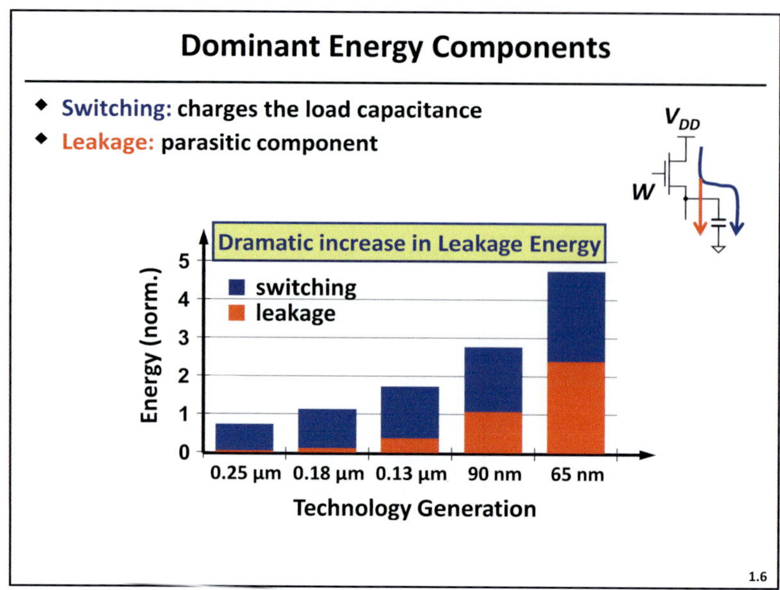

Slide 1.6

We need to model two primary components of energy: switching component, and leakage component. The switching component relates to the energy consumption during active circuit operation, when internal nodes are switched between logic levels "0" and "1". The leakage component is primarily related to the energy consumption during periods of time when the circuit is inactive.

With the scaling of CMOS technology, relative contributions of these two components have been changing over time. If the technology scaling were to continue in the direction of reduced V_{DD} and V_{TH}, the leakage component would have quickly become the dominant component of energy. To keep the energy balanced, general scaling of technology dictates nearly constant V_{TH} and slow V_{DD} reduction in sub-100 nm process nodes. Another reason for the reduced pace in voltage scaling is increase in process variations. Sub-100 nm technologies mainly benefit from increased integration density and reduced gate capacitance, not so much from raw performance improvements.

In the past, when leakage energy was negligible, minimization of energy could be simplified to minimization of switching energy. Today, leakage energy also needs full attention in the overall energy minimization.

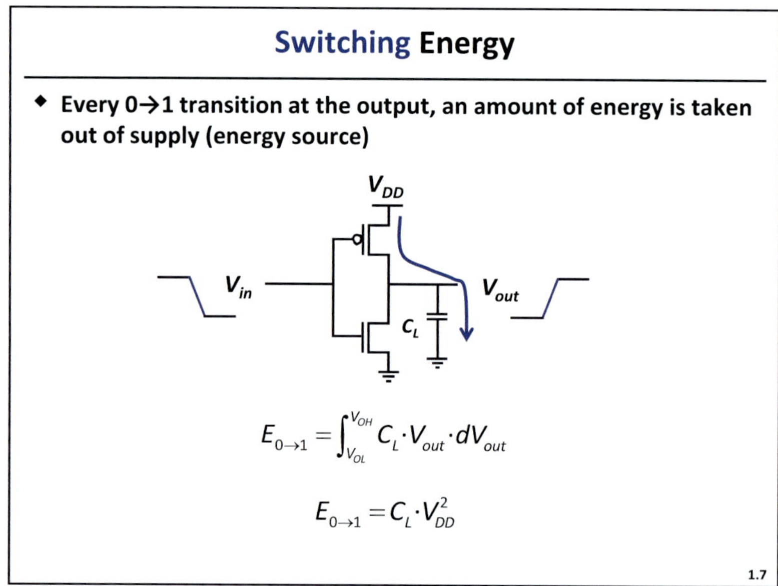

Slide 1.7

The switching energy can be explained by looking at transistor-level representation of a logic gate. For every 0→1 transition at the output we must charge the capacitance C_L and an amount of energy equal to $E_{0 \to 1}$ is taken out of the supply. For CMOS gates ($V_{OL} = 0$, $V_{OH} = V_{DD}$), the energy-per-transition formula evaluates to the well-known result: $E_{0 \to 1} = C_L \cdot V_{DD}^2$. Thus the switching energy is quadratically impacted by the supply voltage, and is proportional to the total capacitance at the output of a gate.

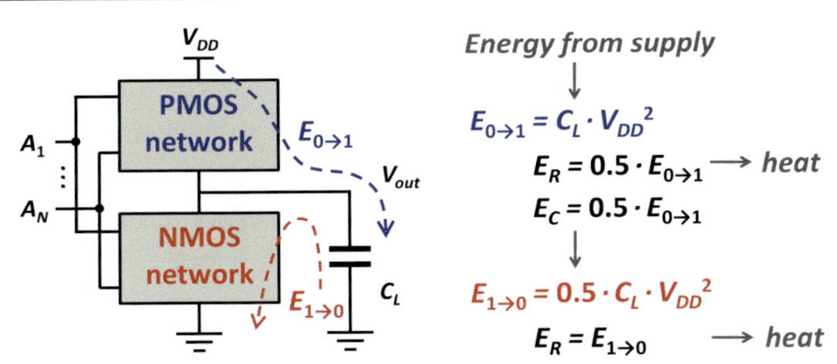

- One half of the energy from supply is consumed in the pull-up network and one half is stored on C_L

- Charge from C_L is discharged to *Gnd* during the 1→0 transition

1.8

Slide 1.8

What happens to the energy taken out of the supply? Where does it go to?

Consider a CMOS gate as shown on this slide. PMOS and NMOS networks are responsible for establishing a logic "1" and "0" at V_{out}, respectively, when inputs $A_1 - A_N$ are applied. During the pull-up, when logic output undergoes a 0→1 transition, $E_{0\rightarrow1}$ is taken out of the supply. One half of the energy is stored on the load capacitance (E_C) and the other half is dissipated as heat by transistors in the PMOS network (E_R). During the pull-down, when logic output makes a 1→0 transition, the charge from C_L is dissipated as heat by transistors in the NMOS network.

Therefore, in a complete charge-discharge cycle at the output, $E_{0\rightarrow1}$ is consumed as heat. Half of the energy is temporarily stored on the output capacitance to hold logic "1". The next question is how often does the output of a gate switch?

<hr>

Node Transition Activity and Energy

- Consider switching a CMOS gate for *N* clock cycles

$$E_N = C_L \cdot V_{DD}^2 \cdot n(N)$$

$E_N:$ the energy consumed for *N* clock cycles
$n(N):$ the number of 0→1 transitions in *N* clock cycles

$$E_{avg} = \lim_{N\to\infty} \frac{E_N}{N} = \left(\lim_{N\to\infty} \frac{n(N)}{N} \right) \cdot C_L \cdot V_{DD}^2$$

$$\alpha_{0\rightarrow1} = \lim_{N\to\infty} \frac{n(N)}{N}$$

$$\boxed{E_{avg} = \alpha_{0\rightarrow1} \cdot C_L \cdot V_{DD}^2}$$

1.9

Slide 1.9

This slide introduces the concept of switching activity. If we observe a logic gate over N switching cycles and count the number of 0→1 transitions $n(N)$, then we can count the average number of transitions by taking the limit for large N. The limit of $n(N)/N$ as N approaches infinity is the switching probability $\alpha_{0\rightarrow1}$, also known as the activity factor. The average energy E_{avg} dissipated over N cycles is directly proportional to the switching activity: $E_{avg} = \alpha_{0\rightarrow1} \cdot E_{0\rightarrow1}$. The switching activity becomes the third factor, in addition to C_L and V_{DD}, impacting the switching energy. The activity factor is best extracted from the input data and helps accurately predict the energy consumption.

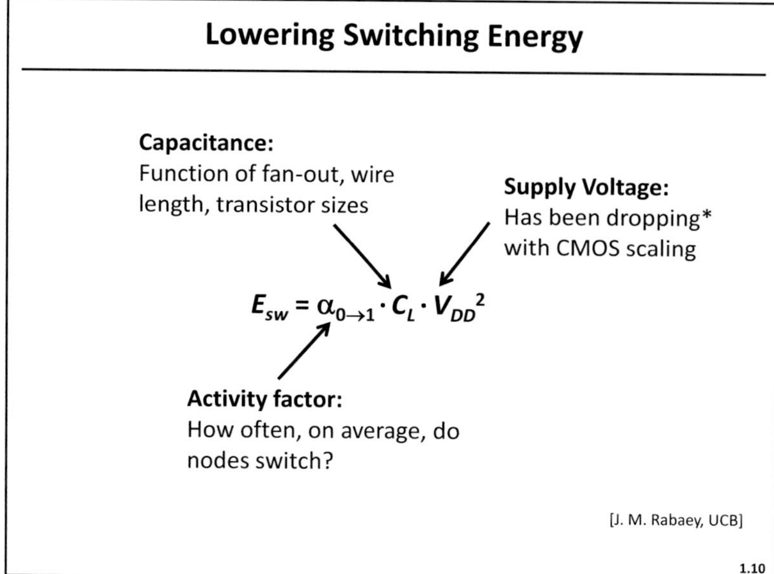

Slide 1.10

How can we control switching energy? The most obvious way is to control each of the terms in the expression. Supply voltage has the largest impact, but it is no longer scaling down with technology at a rate of 0.7x per technology generation, as already discussed. Activity factor depends on how often the data switches and it also depends on the particular realization of a logic function (internal node switching matters). Switching capacitance is a function of gate size and topology. Some design guidelines for energy reduction are outlined below.

Lowering C_L improves performance. C_L is minimized simply by minimizing transistor width (keeps intrinsic capacitance, gate and diffusion, small). We also need to consider performance and the impact of interconnect. A simple heuristic is to upsize transistors only when C_L is dominated by extrinsic capacitance (fanout and wires).

Reducing V_{DD} has a quadratic effect! At the same time, V_{DD} reduction degrades performance especially as V_{DD} approaches $2V_{TH}$. The energy-performance tradeoff, again, has to be carefully analyzed.

Reducing the switching activity, $f_{0\to1} = p_{0\to1} \cdot f_{clk}$, is another way to reduce energy. Switching activity is a function of signal statistics and clock rate. It is impacted by the logic and architecture design decisions.

Focusing only on V_{DD} and C_L (via gate sizing), a good heuristic is to lower the supply voltage as much as possible and to compensate for the loss in performance by increasing the transistor size. This strategy has limited benefit at very low V_{DD}. As we approach the sub-threshold region, leakage energy grows exponentially and there is a well-defined minimum energy point (beyond which further scaling of voltage results in higher energy).

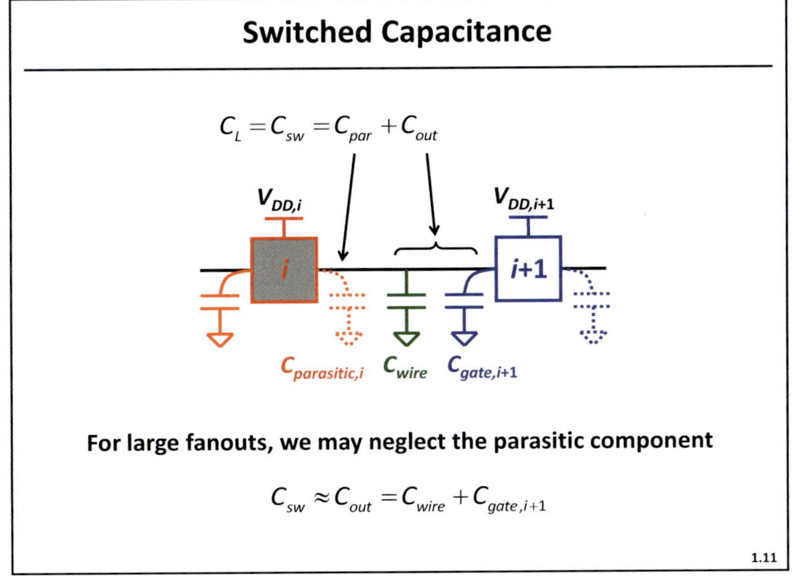

Slide 1.11

Let's further analyze the switched capacitance. Consider two stages of logic, stages i and $i+1$. We assume that each stage has its own supply voltage in ideal case. Each logic gate has two capacitances: intrinsic parasitic capacitance of the drain diffusion ($C_{parasitic}$), and input capacitance of the gate electrode (C_{gate}). Additionally, there is an interconnect connecting the gates that also contributes to capacitive loading.

At the output of the logic gate in the ith stage, there are three capacitances: intrinsic parasitic capacitance $C_{parasitic,i}$ of the gate in stage i, external wire capacitance C_{wire}, and input capacitance of the next stage, $C_{gate,i+1}$. These three components combined are C_L from the previous slides.

Changing the size of the gate in stage i affects only the energy stored on the gate, at its input and parasitic capacitance. Logic gates typically have large fanouts (e.g. 4 or higher), so the total external capacitance $C_{out} = C_{wire} + C_{gate,i+1}$ is a dominant component of C_L. For large fanouts (large C_{out}), we may thus neglect $C_{parasitic,i}$. Let's discuss the origin of $C_{parasitic}$ and C_{gate}.

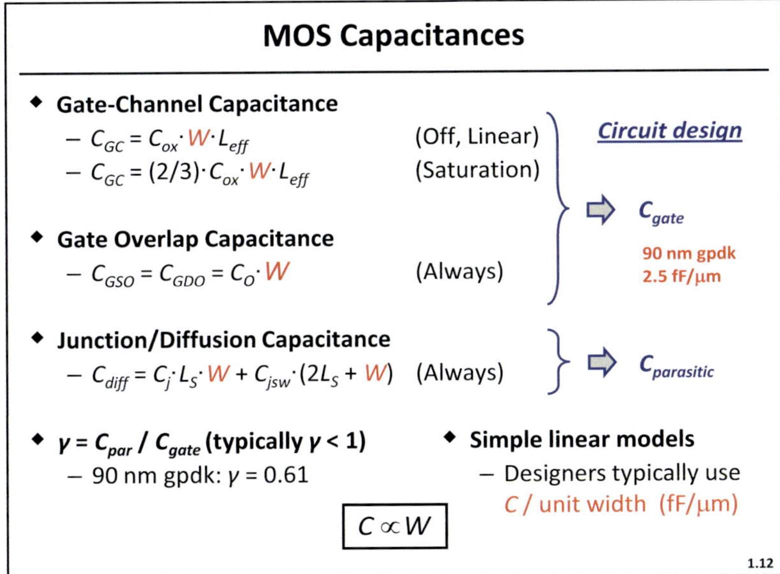

Slide 1.12

Going another level down to device and circuit parameters, we see that all capacitances are geometric capacitances and can be abstracted as some capacitance per unit width. Transistor width (W) is the main design variable in digital design (transistor length L is typically set at L_{min} as given by the process and rarely used as a variable). There are two physical components of the gate capacitance: gate-to-channel and gate-to-source/drain overlap capacitances. For circuit design, we want a lumped model for C_{gate} to be able to look into macroscopic gate and parasitic capacitances as defined on the previous slide. Since components of C_{gate} are proportional to W, we can easily derive a macro model for C_{gate}. A typical value of C_{gate} per unit width is $2.5\,\text{fF}/\mu\text{m}$ in a 90-nm technology. The same W dependency can be observed in the parasitic diffusion capacitance. The C_{par}/C_{gate} ratio is typically less than 1. This ratio is usually labeled as γ. For a 90-nm general process design kit, the value of γ is about 0.6. Starting from the values given here, and using $C(W)$ relationship, we can derive typical capacitance values in scaled technology nodes.

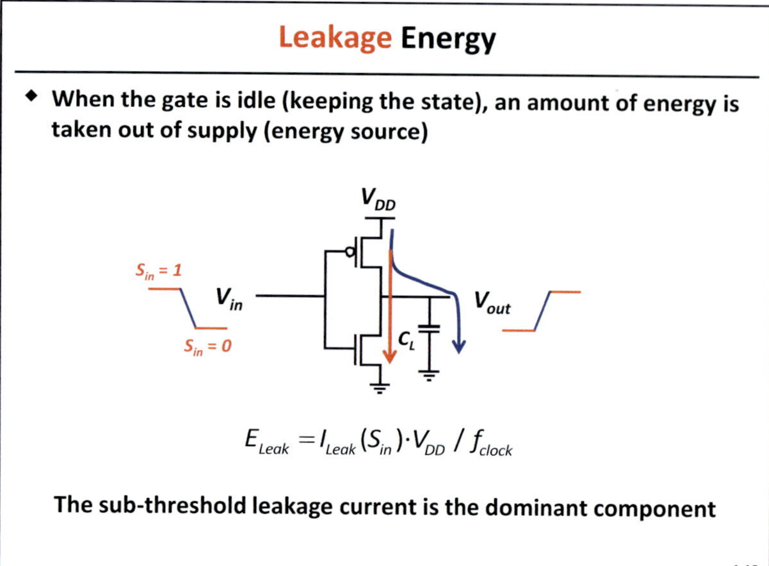

Slide 1.13

Leakage energy is the second most dominant component of energy and is dissipated when the gate is idling (red lines on the slide). There is also active leakage that flows between V_{DD} and Gnd during switching, but we neglect it because in the active mode switching energy dominates. Leakage current/energy is input-state dependent, because states correspond to different transistor topologies for pull-up and pull-down. We have to analyze leakage for all combinations of fixed inputs since the amounts of leakage can easily vary by an order of magnitude across the states. As indicated in Slide 1.5, leakage energy can be calculated as the leakage power divided by the clock frequency.

In today's processes, sub-threshold leakage is the main contributor to leakage current. Gate leakage used to be a threat until high-K gate technology got introduced (at the 45 nm node). So, modeling sub-threshold source-to-drain leakage current is of interest for circuit designers.

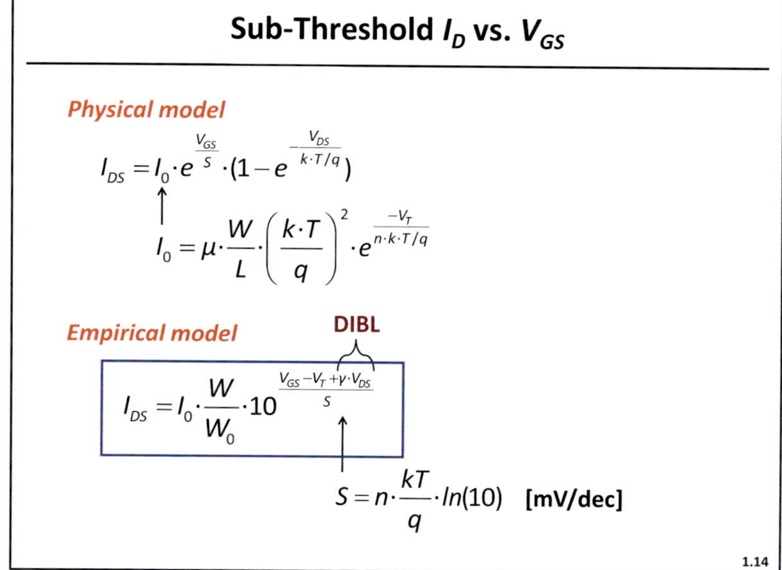

Slide 1.14

We can use two models for sub-threshold leakage current. One approach is physics-based, which is close to what SPICE uses, but is not convenient for hand calculations or quick intuitive reasoning because of the exponent terms. Instead of working with natural logarithm, we prefer to work with decimal system and discuss by how much we have to tune V_{GS} to get an order of magnitude reduction in leakage current. It is much easier to work in decades, so we derive an empirical model. The empirical model shown on the slide also exposes gate size and transistor terminal voltages as variables that designers can control. It is also useful to normalize the number to a unit device. So, in the formula shown on the slide, W/W_0 is the ratio of actual with to a unit width, or the relative size of a device with respect to the unit device. Notice the exponential impact of V_{GS} and V_{TH}, and also DIBL effect γ on V_{DS}.

The parameter S represents the amount of V_{GS} tuning for an order of magnitude change in leakage current. In CMOS, S is typically 80–90 mV/dec. Device research is focused on improving the slope, which would mean better I_{on}/I_{off} ratio and less leakage current. In the ideal case of zero diffusion capacitance, S is lower-bounded to 60 mV ($n = 1$).

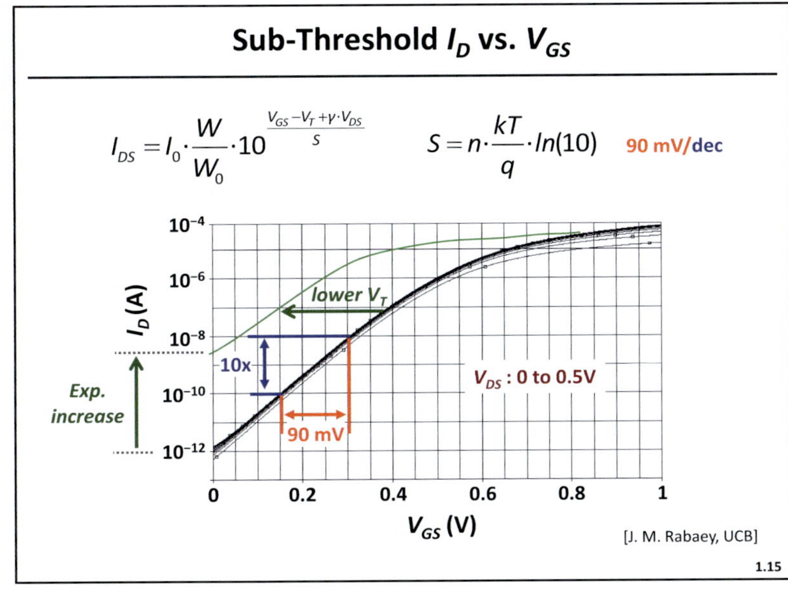

Slide 1.15

It is interesting to see how threshold or V_{GS} impact the sub-threshold leakage current. If we look at the formula, the slope of the $\log(I_{DS})$ versus V_{GS} line simply represents the sub-threshold slope parameter S. In sub-threshold for this 0.25 μm technology, we can see that the current can be reduced by 10x for a 90 mV decrease in V_{GS} ($n = 1.5$). If we scale technology, the leakage current starts to roll down at lower values of V_{GS} (lower V_{TH}), so the amount of current at a particular V_{GS} is exponentially larger in a scaled technology. This exponential dependence presents a problem, because it is hard to control.

An important question is: with scaling of technology and relatively increasing leakage, what is the relative relationship between the switching and leakage energy when the total energy is minimized?

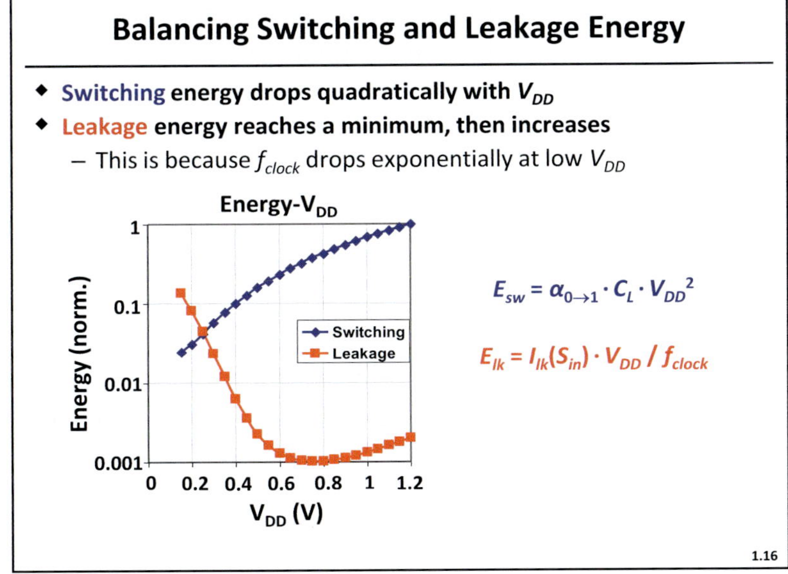

Slide 1.16

The most common way to reduce energy is through supply voltage scaling. V_{DD} has a large impact on energy, especially switching. It is also a global variable, so it affects the whole design (as opposed to tweaking size of individual gates).

The plot on this slide shows simulations from a typical 65-nm technology with nominal V_{DD} of 1.2 V. We can see that the switching energy drops quadratically with V_{DD}. The red line shows leakage energy which is the current multiplied by V_{DD} and divided by f_{clk}. It is interesting that the leakage energy has a minimum around 0.7 V for a 65 nm technology. With further reduction of V_{DD} leakage current is relatively constant, but exponential delay increase (f_{clk} reduction) causes leakage energy to increase in the same exponential manner. How does this affect total energy?

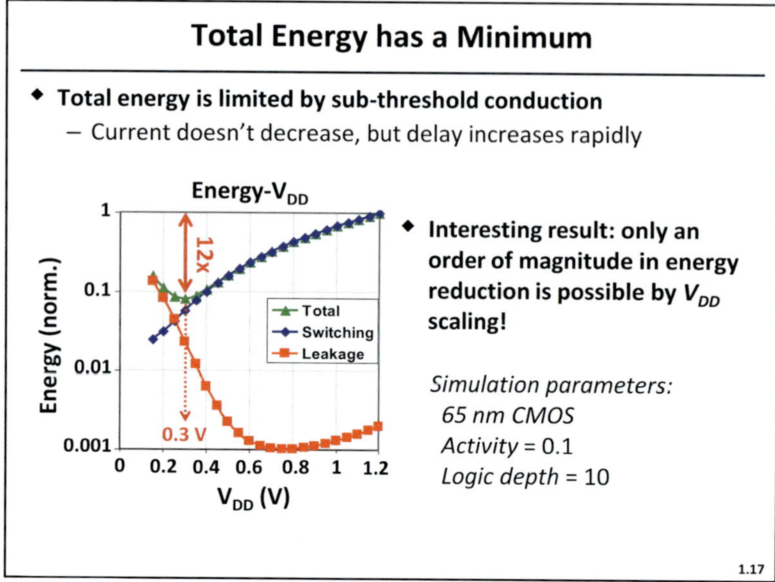

Slide 1.17

The total energy has a minimum and happens to be around 0.3 V, slightly higher than the device threshold. This voltage varies as a function of logic depth (cycle time) and switching activity. The result presented in this slide assumes logic depth of 10 and activity factor of 10% for a chain of CMOS inverters. The total energy is limited by sub-threshold conduction when the increase in leakage energy offsets savings in switching energy. Thus, we can only reduce energy down to the minimum-energy point. A very interesting result is that only about one order of magnitude energy reduction is possible by V_{DD} scaling. This means that at the circuit level there is not much energy to be gained due to this limitation. More improvements can be obtained in finding algorithms that minimize the number of operations in each task.

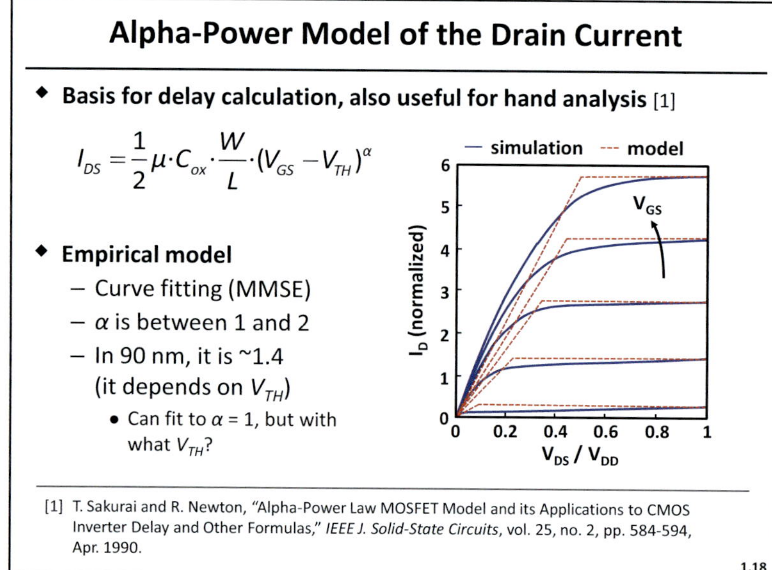

Slide 1.18

As a basis for delay calculation, we consider the alpha-power model of the drain current [1]. This model works quite well in velocity saturated devices (e.g. gate lengths below 100 nm). The expression on the slide departs from the long-channel quadratic model to include parameter α that measures the degree of velocity saturation. Values closer to 1 correspond to a higher degree of velocity saturation. This value is obtained by minimum mean square error (MMSE) curve fitting of simulation data. A typical value for α is around 1.4 in a 90 nm technology and largely depends on fitting accuracy and the value of transistor threshold V_{TH}. Considering V_{TH} as another fitting parameter, one can find an array of

(α, V_{TH}) values for a given accuracy. A good practice is to set an initial value of V_{TH} to V_{T0} from the transistor-level (Spectre, SPICE) model.

Designers often use the alpha-power law model for current to approximate the average current during logic transitions from $\{0, V_{DD}\}$ to $V_{DD}/2$ in order to calculate logic delay.

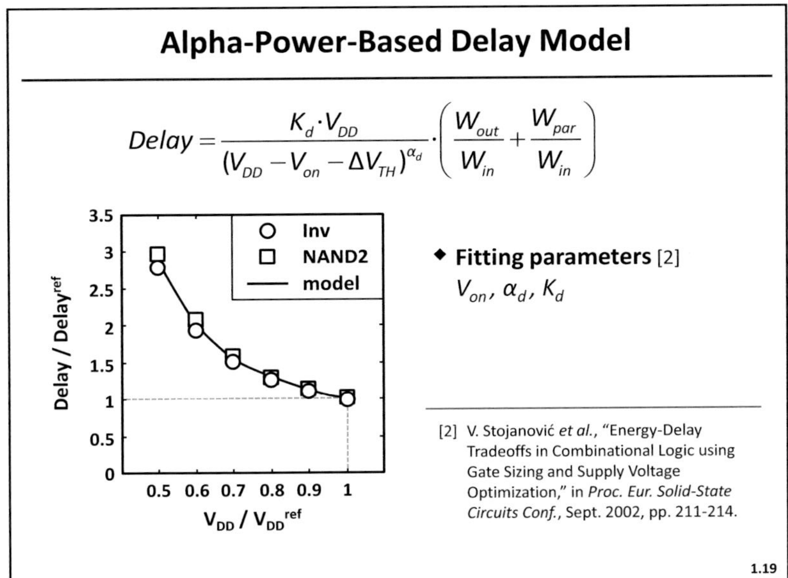

Slide 1.19

A very simple delay model that includes the effects of changes in supply, threshold and transistor sizes is shown in this slide. In the formula, W_{out} is the size of the fanout gates, W_{in} is the size of the input gate, and W_{par} corresponds to the parasitic capacitance of the input gate. The ratio W_{par}/W_{in} is the ratio of parasitic capacitance to gate capacitance.

This is a curve-fitting expression based on the alpha-power law model for the drain current, with parameters V_{on} and α_d related to, but not equal to the transistor threshold and the velocity saturation index from the current model [2]. Due to the curve-fitting approach, ΔV_{TH} is used as a parameter ($\Delta V_{TH} = 0$ for nominal V_{TH}) in order to capture the effect of threshold adjustment on the delay.

As shown on the plot, the delay model fits SPICE simulated data quite nicely, across a range of supply voltages, with a nominal supply voltage of 1.2 V for a 0.13-μm CMOS process. This model will be later used for gate sizing and supply and threshold voltage adjustment in order to obtain the energy-delay tradeoff for logic computations.

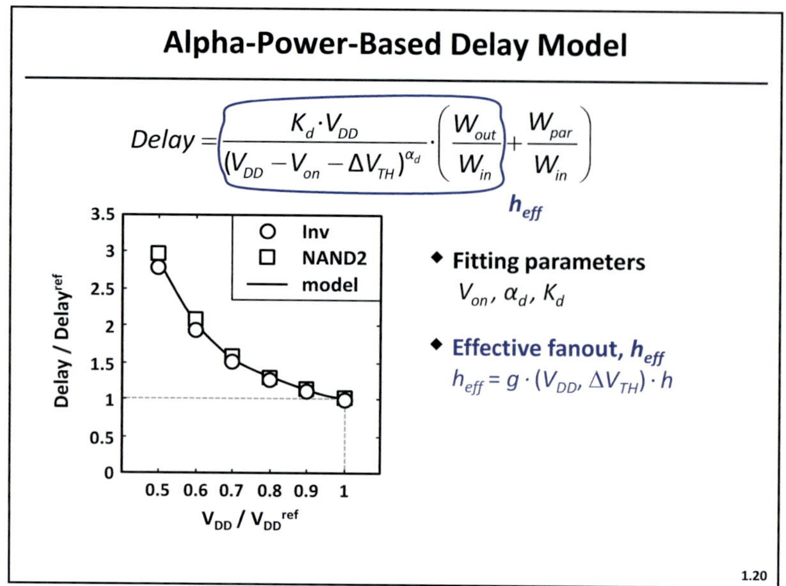

Slide 1.20

The impact of gate size and supply voltage can be lumped together in the effective fanout parameter h_{eff} as defined on this slide. In logical effort terms, this parameter is $g(V_{DD},\ \Delta V_{TH}) \cdot h$, where g is the logical effort and h is the electrical effort (fanout). Such a formulation allows for consideration of voltage effects together with the sizing problem typically considered in the logical effort delay model. Chapter 2 will make use of h_{eff} in energy-delay sensitivity analysis.

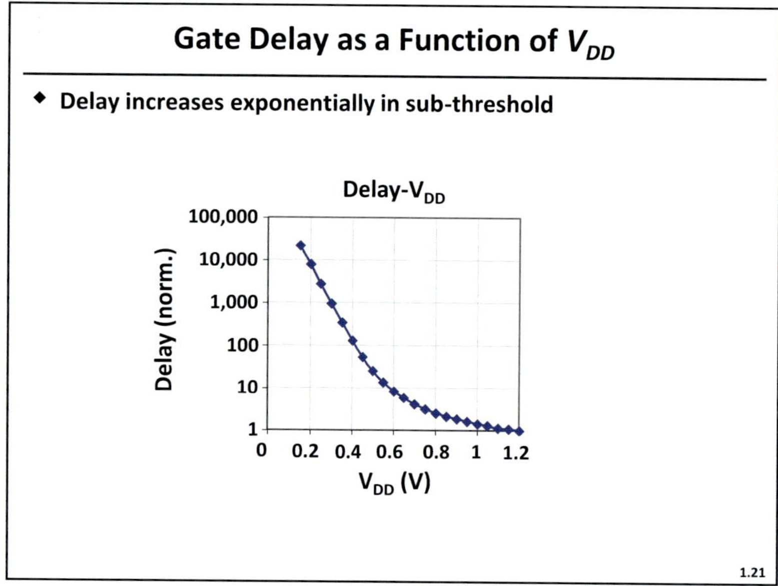

Slide 1.21

As discussed in Slide 1.10, supply voltage reduction is the most attractive approach for energy reduction. A plot of delay as a function of V_{DD} is shown in this slide for a 90-nm technology. Starting from the nominal voltage of 1.2V the delay increases by about an order of magnitude when the supply is reduced by half. Further V_{DD} reduction towards V_{TH} results in exponential delay increase, as predicted by the alpha-power model. It is important to remember this result as we move on to energy-delay optimization in Chap. 2.

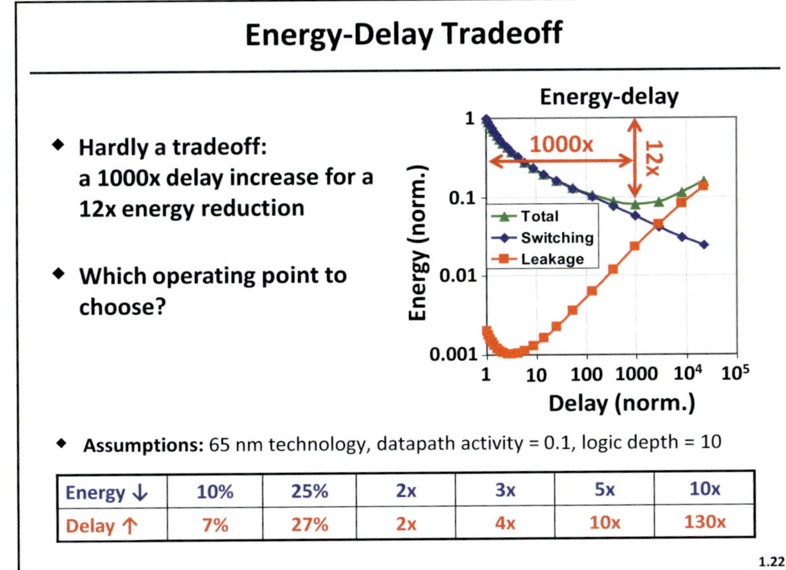

Energy ↓	10%	25%	2x	3x	5x	10x
Delay ↑	7%	27%	2x	4x	10x	130x

Slide 1.22

Putting the energy and delay models together, we can plot the energy-delay tradeoff due to voltage scaling. There is a minimum energy point, when taking into account the tradeoff between switching and leakage currents. A simple way to minimize energy in a digital system is to perform all operations at the minimum-energy point. However, the delay penalty at the minimum-energy point is enormous: a 1000x increase in delay is needed for around a 10x reduction in energy as compared to the minimum-delay point obtained for design operating at nominal supply voltage. This is hardly a tradeoff considering the delay cost.

The question is which operating point to choose from the energy-delay line? The table shows energy reduction and corresponding delay increase for several points along the green line for a 65-nm technology assuming logic activity of 0.1 and 10 logic stages between the pipeline registers. The data shows a good energy-delay tradeoff, up to about 3–5x increase in delay. For delays longer than 5x, the relative delay increase is much higher than the relative energy reduction, making the tradeoff less favorable. The energy-delay tradeoff is the essence of design optimization and will be discussed in more detail in Chap. 2.

The important message from this slide is the significance of algorithm-level optimizations. Once an algorithm is fixed, we only have an order of magnitude margin for further energy reduction (not considering special effects such as stacking or power gating, but even with these, the energy reduction will be limited). Therefore, it is very important to consider algorithmic simplifications together with gate-level design.

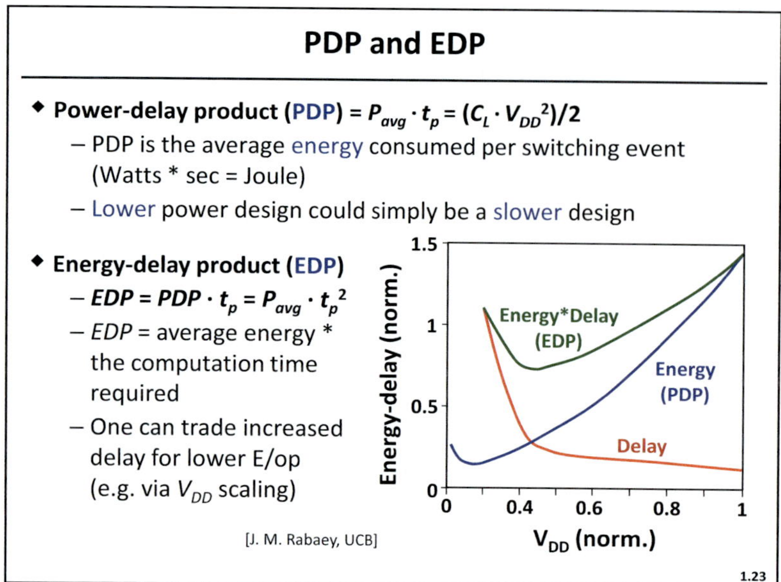

Slide 1.23

To measure the compromise between energy and delay, several synthetic metrics are used. Power-delay product (PDP) and energy-delay product (EDP) are the most common of such metrics. Power-delay product is the average energy consumed per switching event and it evaluates to $C_L \cdot V_{DD}^2 / 2$. Each switching cycle contains a $0 \to 1$ and a $1 \to 0$ transition, so E_{avg} is twice the PDP. PDP is the energy metric and does not tell us about the speed of computation. For a given circuit, PDP may be made arbitrarily low by reducing the supply voltage that comes at the expense of performance.

Energy-delay product, or power-delay-squared product, is the average energy multiplied by the time it takes to do the computation. EDP, thus, takes performance into account and is the preferred metric. The graph on this slide plots energy and delay metrics on the vertical axis versus supply voltage (normalized to the reference V_{DD} for the technology) on the horizontal axis. As V_{DD} is scaled down, EDP (the green line) reaches a minimum before PDP (energy) reaches its own minimum. Thus, EDP is minimized somewhere between the minimum-delay and minimum-energy points and generally represents a good energy-delay tradeoff.

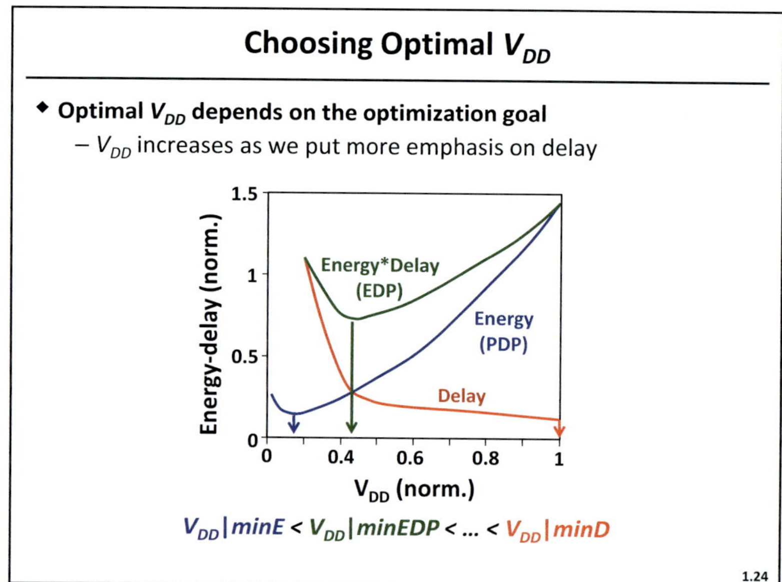

Slide 1.24

The supply voltage corresponding to minimum EDP is roughly $V_{DD}(\text{min-EDP}) = 3/2\ V_{TE}$ where $V_{TE} = V_{TH} + V_{DSAT}/2$. This value of V_{DD} optimizes both performance and energy simultaneously and is roughly 0.4 to 0.5 V_{DD} and corresponds to the onset of strong inversion ($V_{DSAT}/2$ above V_{TH}). This value is between V_{DD} corresponding to minimum energy (lowest) and minimum delay (highest). Minimum EDP, to a first order, is independent of supply voltage and could be used for architecture comparison across technology. EDP, however, is just one of the points on the energy-delay tradeoff line and, as such, is rarely optimal (for actual designs). The optimal point in the energy-delay space depends on the required level of performance, as well as design architecture.

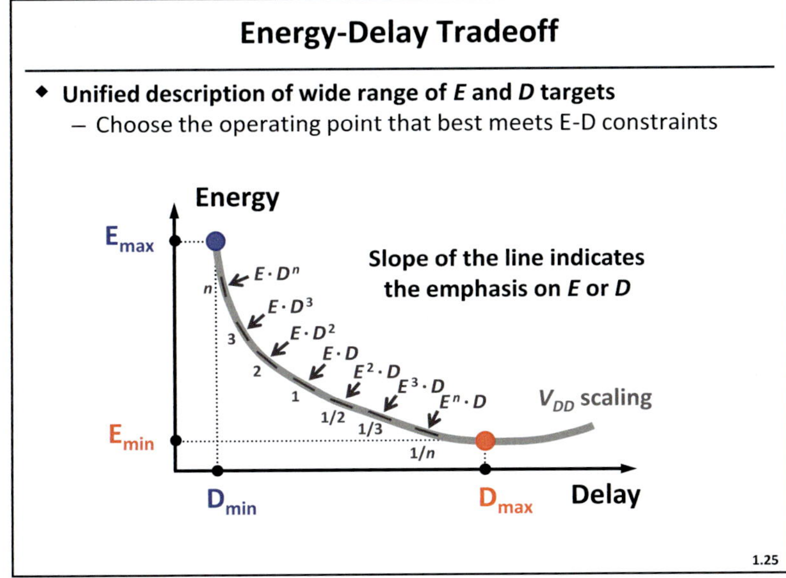

Slide 1.25

The energy-delay tradeoff plot is essential for evaluating energy efficiency of a design. It gives a unified description of all achievable energy and delay targets. As such, the tradeoff curve can also be viewed as a continuous set of (α, β) values that model a synthetic metric $E^{\alpha} \cdot D^{\beta}$. This includes minimum energy-delay product ($\alpha = \beta = 1$) as well as points that put more emphasis on delay ($\alpha = 1$, $\beta > 1$) or energy ($\alpha > 1$, $\beta = 1$). The two boundary points are the point of minimum delay and the point of minimum energy. The separation between these two points is about three orders of magnitude in energy and about one order of magnitude in delay. These points define the range of energy and delay tuning at the circuit level. It also allows us to formulate optimization problems under energy or delay constraints.

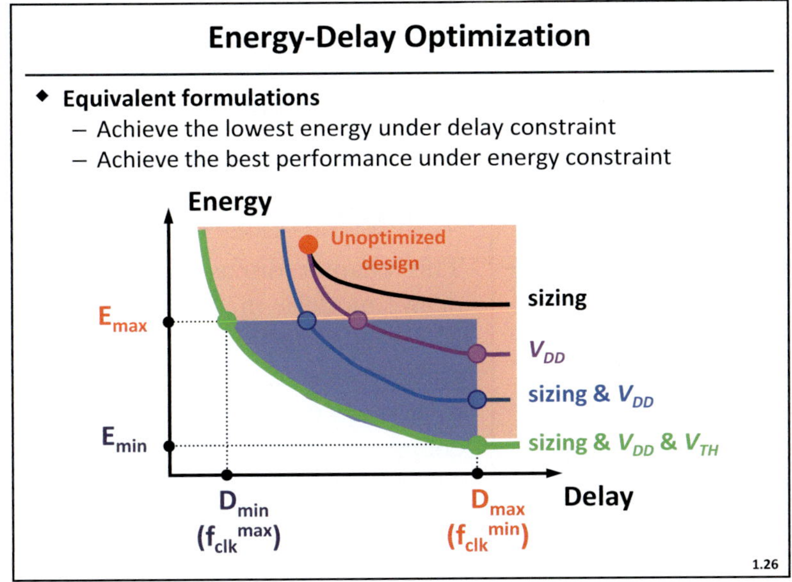

Slide 1.26

The goal of energy-delay optimization is to find the best energy-delay tradeoff by tuning various design variables. The plot shows the energy-delay tradeoff in CMOS circuits obtained by the adjustment of gate size and threshold and supply voltages. When limited by energy (E_{max}) or delay (D_{max}), designers have to be able to quickly find solutions that meet the requirements of their designs.

Local optimizations focusing on one variable are the easiest to perform, but may not meet the specs (e.g. tuning gate sizing as shown on the plot). Some variables are more effective. Tuning V_{DD}, for example, results in a design that meets the energy and delay targets. However, we can do better. A combined effort from sizing and V_{DD} gives a better solution than what is achievable by individual variables. Clearly, the best energy-delay tradeoff is obtained by jointly tuning all variables in the design. The green curve represents this global optimum, because all other points have longer delay for the same energy or higher energy for the same delay. The next step is to formulate an optimization problem, based on energy and delay models presented in this chapter, in order to quickly generate this optimal tradeoff.

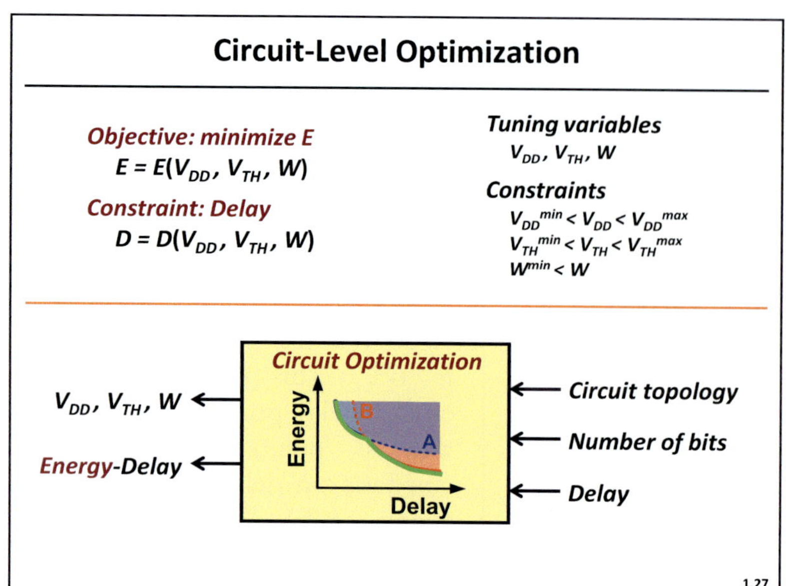

Slide 1.27

Generating the optimal energy-delay tradeoff is done in the circuit optimization routine that minimizes energy subject to a delay constraint. Notice that energy minimization subject to a delay constraint is dual to delay minimization subject to an energy constraint, since both optimizations result in the same tradeoff curve.

We choose to perform a delay-constrained energy minimization, because the delay constraint can be simply derived from the required cycle time. The optimization is performed using gate size (W), supply (V_{DD}) and threshold (V_{TH}) voltages. Design variables can be tuned within various operational constraints that define the lower and upper bounds. The circuit optimization problem can be defined on a fixed circuit topology, for a selected number of bits, and for a range of delay constraints that allow us to generate the entire tradeoff curve as defined in Slide 1.25. The output of the circuit optimization is the optimal energy-delay tradeoff and the values of tuning variables at each point along the curve. The optimal curve can then be used in a macro model to assist in the architectural selection (topic of Chap. 3). The ability to quickly generate energy-delay tradeoff curves for various circuit topologies allows us to compare many possible circuit topologies (A and B shown on the slide) used to implement a logic function.

Slide 1.28

This chapter discussed energy and delay models in CMOS circuits. There is a well-defined minimum-energy point in CMOS circuits due to leakage currents. This minimum energy places a lower limit on energy-per-operation (E/op) that can be practically achieved. Limited energy reduction in circuits underscores the importance of algorithm-level reduction in the number of operations required to perform a task. Energy and delay models as a function of gate size, supply and threshold voltage will be used for circuit optimizations in Chap. 2.

References

- T. Sakurai and A.R. Newton, "Alpha-Power Law MOSFET Model and its Applications to CMOS Inverter Delay and Other Formulas," *IEEE J. Solid-State Circuits,* vol. 25, no.2, pp. 584-594, Apr. 1990.

- V. Stojanović *et al.*, "Energy-Delay Tradeoffs in Combinational Logic using Gate Sizing and Supply Voltage Optimization," in *Proc. Eur. Solid-State Circuits Conf.,* Sept. 2002, pp. 211-214.

Additional References

- Predictive Technology Models from ASU, [Online]. Available: http://www.eas.asu.edu/~ptm

- J. Rabaey, A. Chandrakasan, and B. Nikolić, Digital Integrated Circuits: A Design Perspective, (2nd Ed), Prentice Hall, 2003.

Circuit Optimization

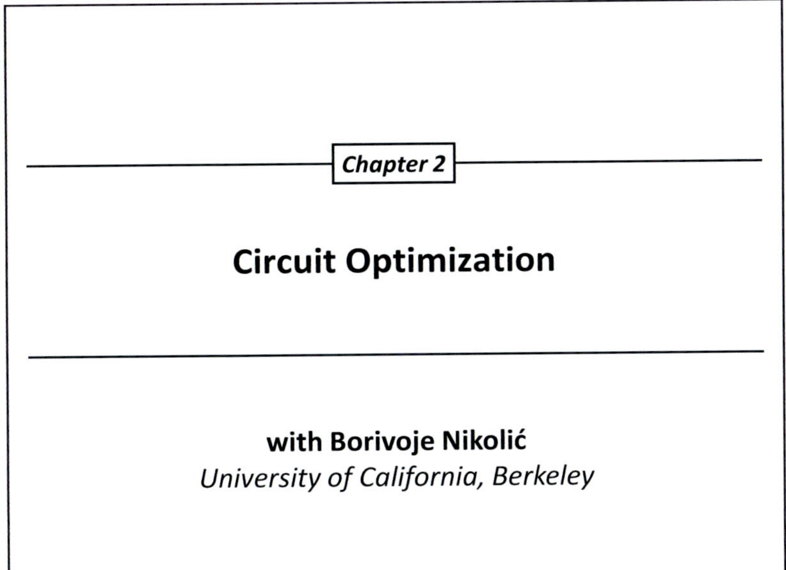

Slide 2.1

This chapter discusses methods for circuit-level optimization. We discuss a methodology for generating the optimal energy-delay tradeoff by tuning gate size and supply and threshold voltages. The methodology is based on the sensitivity approach to measure and balance the benefits of all the tuning variables. The analysis will be illustrated on datapath logic, and the results will serve as a guideline for architecture-level design in later chapters.

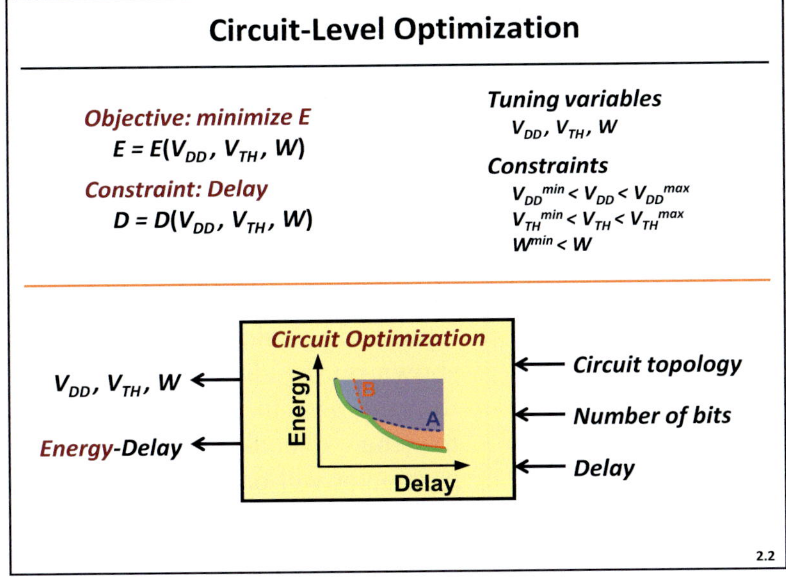

Slide 2.2

As introduced in Chap. 1, circuit-level optimization can be viewed as an energy minimization problem subject to a delay constraint. Key variables at the circuit level are supply (V_{DD}) and threshold (V_{TH}) voltages, and gate size (W). Gate size is typically normalized to a unit inverter, according to the logical effort theory. Variables are bounded by technology and operation constraints. For example, V_{DD} cannot exceed $V_{DD}{}^{max}$ as dictated by the oxide reliability limit and cannot be lower than $V_{DD}{}^{min}$ as mandated by noise margins or the minimum energy point. The threshold voltage cannot be lower than $V_{TH}{}^{min}$ due to leakage and variability constraints, and cannot be larger than $V_{TH}{}^{max}$ for performance reasons. Gate size W is limited by the minimum gate size W^{min} as defined by manufacturing constraints or noise, while the upper limit is derived from fanout and signal integrity constraints (increasing W to be arbitrarily large would result in self-loading and the effects of fanout would be negligible – W increase has to stop well before that point).

Circuit optimization is defined as a problem of finding the optimal energy-delay tradeoff curve for a given circuit topology and a given number of bits. This is accomplished by varying delay constraint and minimizing energy at each point, starting from a design sized for minimum delay at nominal supply and threshold voltages. Energy minimization is accomplished by tuning V_{DD}, V_{TH}, and W. The result of the optimization is the optimal energy-delay tradeoff line, and the values of the tuning variables at each point along the tradeoff line.

D. Marković and R.W. Brodersen, *DSP Architecture Design Essentials*, Electrical Engineering Essentials,
DOI 10.1007/978-1-4419-9660-2_2, © Springer Science+Business Media New York 2012

Energy Model for Circuit Optimization

- **Switching Energy**

$$E_{sw} = \alpha_{0 \to 1} \cdot (C(W_{out}) + C(W_{par})) \cdot V_{DD}^2$$

- **Leakage Energy**

$$E_{lk} = \frac{W_{in}}{W_0} \cdot I_0(S_{in}) \cdot 10^{-\frac{V_{TH} - \gamma \cdot V_{DD}}{S}} \cdot V_{DD} \cdot Delay$$

2.3

Slide 2.3

Based on energy and delay models from Chap. 1, we model energy as shown in this slide. Switching and leakage components are considered as they dominate energy consumption, while the short-circuit component is safely ignored. The switching energy model depends on the switching probability $\alpha_{0 \to 1}$, parasitic and output load capacitances, and supply voltage. The leakage energy is modeled using the standard input-state-dependent exponential leakage current model with the DIBL effect. Delay represents cycle time. Circuit variables (W, V_{DD}, V_{TH}) are made explicit for the purpose of providing insight into tuning variables and for energy-delay sensitivity analysis.

Switching Component of Energy

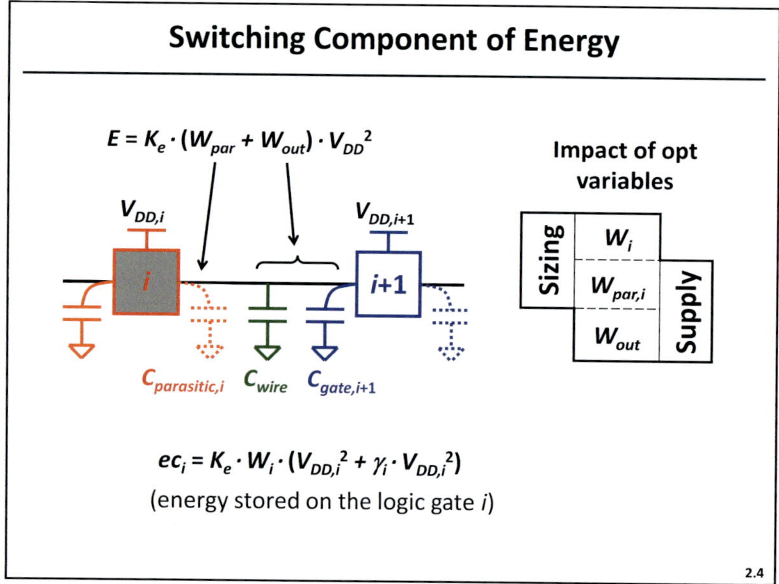

2.4

Slide 2.4

The switching component of energy, for example, affects only the energy stored on the gate, at its input and parasitic capacitances. As shown on the slide, ec_i is the energy that the gate in stage i contributes to the overall energy. This parameter will be used in sensitivity analysis. Supply voltage affects the energy due to total load at the output, including wire and gate loads, and the self-loading of the gate. The total energy stored on these three capacitances is the energy taken out of the supply voltage in stage i.

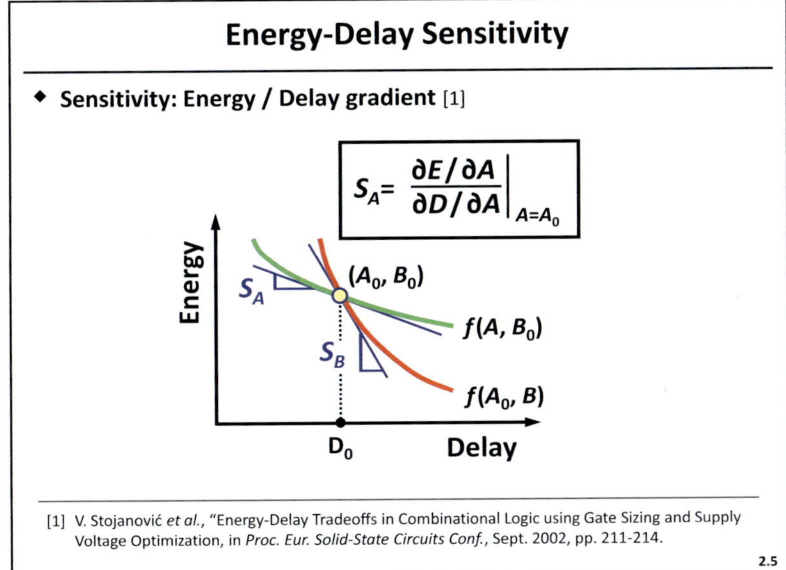

$$S_A = \left. \frac{\partial E / \partial A}{\partial D / \partial A} \right|_{A=A_0}$$

Slide 2.5

Energy-delay sensitivity is a formal way to evaluate the effectiveness of various variables in the design [1]. This is the core of circuit-level optimization infrastructure. It relies on simple gradient expressions that describe the rate of change in energy and delay by tuning a design variable; for example by tuning design variable A at point A_0. At point (A_0, B_0) illustrated in the graph, the sensitivity to the variable is simply the slope of the curve with respect to the variable. Observe that sensitivities are negative due to the nature of energy-delay tradeoff. We will compare their absolute values, where the larger absolute values indicate higher potential for energy reduction. For example, variable B has higher energy-delay sensitivity ($|S_B| > |S_A|$) at point (A_0, B_0) than variable A.

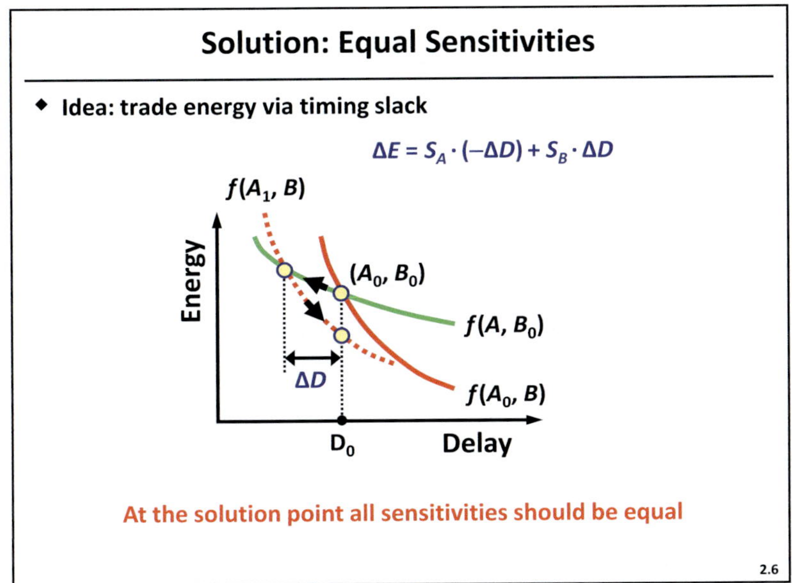

Slide 2.6

The key concept is that at the solution point, the sensitivities for all design variables should be equal. If the sensitivities are not equal, we can utilize a low-energy cost variable (variable A) to create timing slack ΔD and increase energy by ΔE, proportional to sensitivity S_A. Now we are at point (A_1, B_0), so we can use a higher-energy-cost variable B and reduce energy by $S_B \cdot \Delta D$. Since $|S_B| > |S_A|$, the overall energy is reduced by $\Delta E = (|S_B| - |S_A|) \cdot \Delta D$. A fixed point in the optimization is reached, therefore, when all sensitivities are equal.

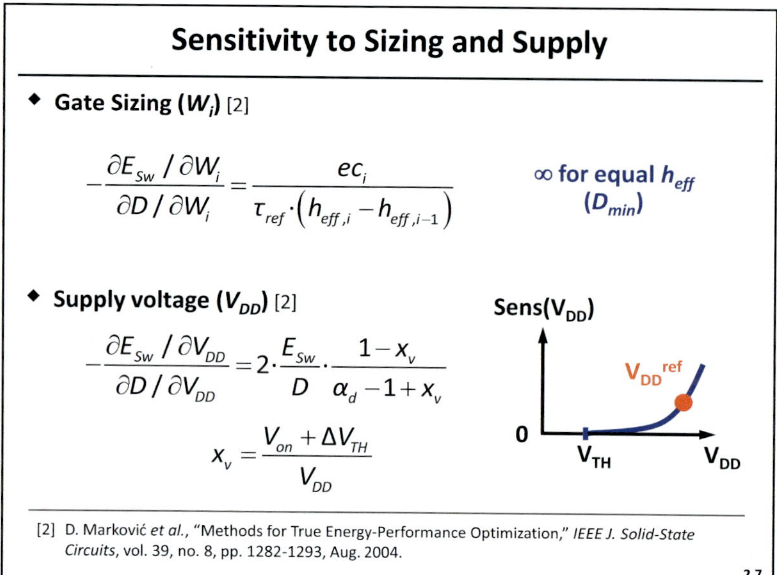

The equations on the slide:

$$-\frac{\partial E_{Sw}/\partial W_i}{\partial D/\partial W_i} = \frac{ec_i}{\tau_{ref}\cdot\left(h_{eff,i}-h_{eff,i-1}\right)}$$

$$-\frac{\partial E_{Sw}/\partial V_{DD}}{\partial D/\partial V_{DD}} = 2\cdot\frac{E_{Sw}}{D}\cdot\frac{1-x_v}{\alpha_d-1+x_v}$$

$$x_v = \frac{V_{on}+\Delta V_{TH}}{V_{DD}}$$

Slide 2.7

Based on the presented energy and delay models we can define the sensitivity of each of the tuning variables [2].

The formulas for sizing indicate that the largest potential for energy savings is at the point where the design is optimized for minimum delay. The design that is sized for minimum delay has equal effective fanouts, which means infinite sensitivity to sizing. This makes sense because at minimum delay no amount of energy added through sizing can further improve the delay.

The power supply sensitivity is finite at the nominal point but decreases to zero when V_{DD} approaches V_{TH}, because the delay approaches infinity.

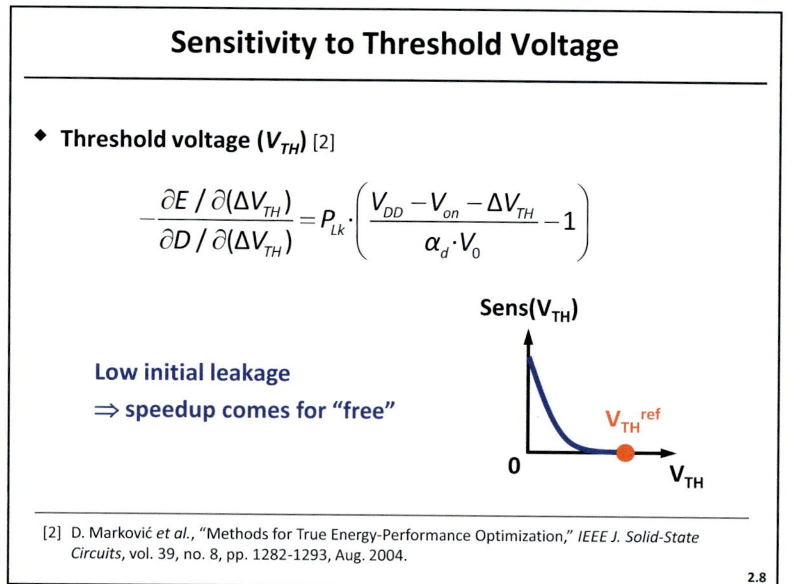

The equation on the slide:

$$-\frac{\partial E/\partial(\Delta V_{TH})}{\partial D/\partial(\Delta V_{TH})} = P_{Lk}\cdot\left(\frac{V_{DD}-V_{on}-\Delta V_{TH}}{\alpha_d\cdot V_0}-1\right)$$

Slide 2.8

The last parameter we would like to explore is threshold voltage. Here, the sensitivity is opposite to that of supply voltage. At the reference point it starts off low with very low sensitivity and increases exponentially as the threshold gets reduced to zero.

The performance improves as we decrease V_{TH}, but the leakage increases. If the leakage power is nominally very small, we get the speedup almost for "free". The problem is that the leakage power is exponential in threshold and after a while decreasing threshold becomes very expensive in energy.

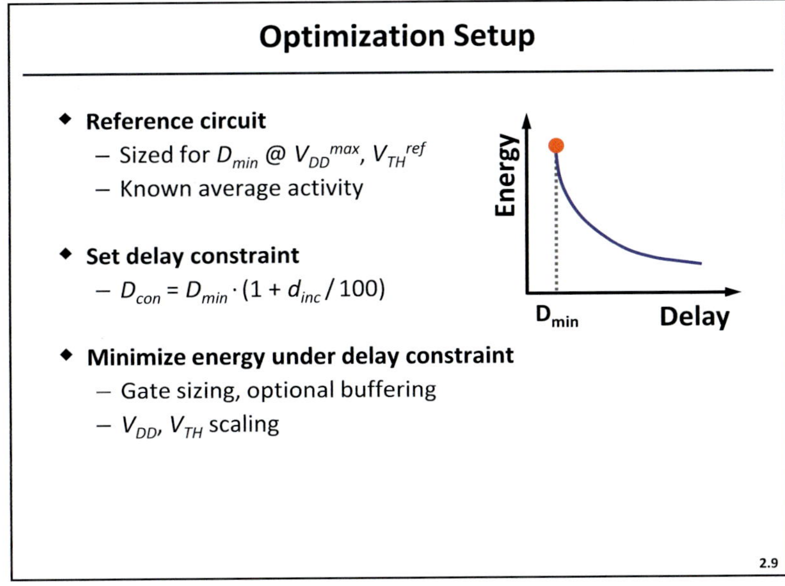

Slide 2.9

We initialize our optimization starting from a design which, for a given delay, is the most energy-efficient, and then trade off energy and delay from that point. The minimum delay point is one of the points on the energy-efficient curve and is convenient because it is well defined.

We start from the minimum delay achieved at the reference supply and threshold voltages provided by the technology. Then we define a delay constraint, D_{con}, by specifying some incremental increase in delay d_{inc} with respect to the minimum delay D_{min}. Under this delay constraint, the minimum energy is found using supply and threshold voltages, gate sizing, and optional buffering.

Supply optimizations we investigate include global supply reduction, two discrete supplies, and per-stage supplies. We limit supply voltage to only decrease from input to output of a block assuming that low-to-high level conversion is done in registers. Sizing is allowed to change continuously, and buffering preserves the logic polarity.

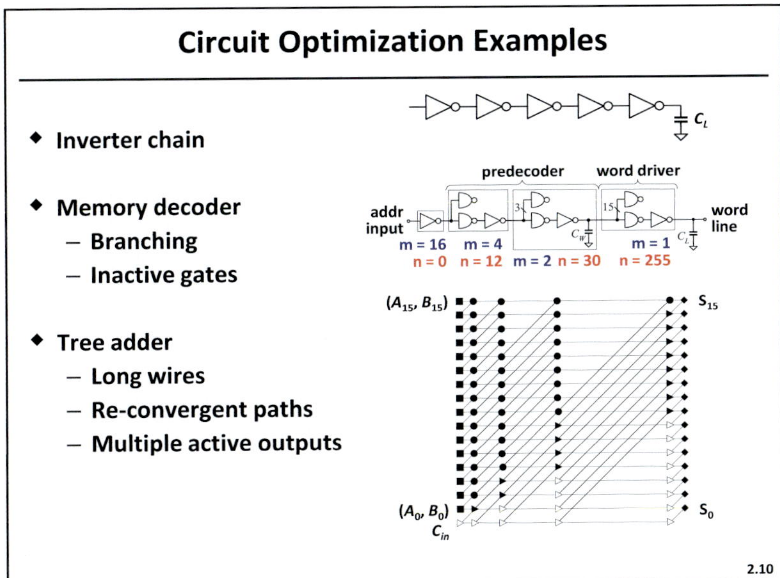

Slide 2.10

To illustrate circuit optimization, we look at a few examples. In re-examining circuit examples representing common topologies, we realize that they differ in the amount of off-path loading and path reconvergence. By analyzing how these properties affect a circuit energy profile, we can better define principles for energy reduction relating to logic blocks. We analyze examples of an inverter chain, memory decoder and tree adder that illustrate all of these properties.

The inverter chain is a simple topology with single path and geometrically increasing energy profile. The memory decoder is another topology where the total number of gates increases geometrically. The memory decoder has branching and inactive gates toward the output, which results in an energy peak in the middle of the structure. Finally, we analyze the tree adder that has long wires, reconvergent fanout and multiple active outputs qualified by paths of various logic depth.

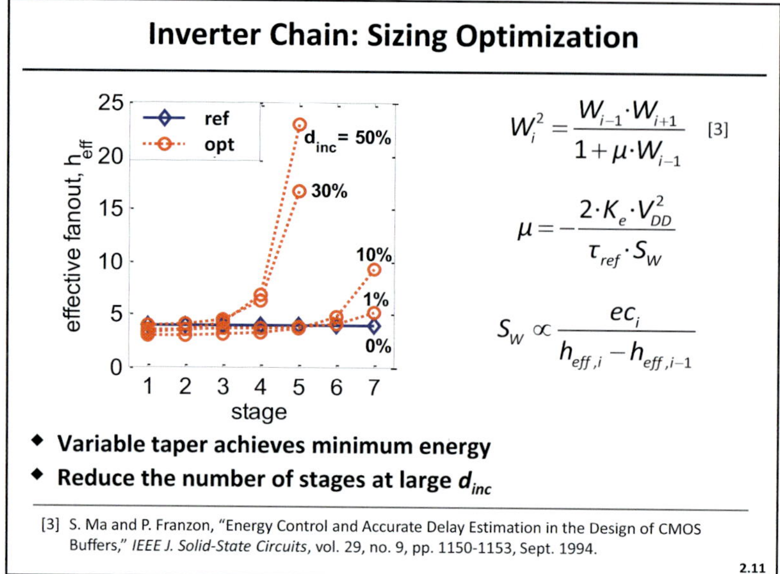

- **Variable taper achieves minimum energy**
- **Reduce the number of stages at large d_{inc}**

[3] S. Ma and P. Franzon, "Energy Control and Accurate Delay Estimation in the Design of CMOS Buffers," *IEEE J. Solid-State Circuits*, vol. 29, no. 9, pp. 1150-1153, Sept. 1994.

2.11

Slide 2.11

The inverter chain is the most commonly used example in sizing optimizations. Because of its widespread use the inverter chain has been the focus of many studies. When sized for minimum delay, the inverter chain's topology dictates geometrically increasing energy towards the output. Most of the energy is stored in the last few stages, with the largest energy in the final load.

The plot shows the effective fanout going over various stages through the chain, for a family of curves that correspond to various delay increments (0–50%). General result for the optimal stage size, derived by Ma and Franzon [3], will be explained here by using the sensitivity analysis. Recall the result from Slide 2.7: the sensitivity to gate sizing is proportional to the energy stored on the gate, and is inversely proportional to the difference in effective fanouts. What this means is that, for equal sensitivity in all stages, the difference in the effective fanouts must increase in proportion to the energy of the gate. This indicates that the difference in the effective fanouts ends in an exponential increase towards the output.

An energy-efficient solution may sometimes require a reduced number of stages. In this example, the reduction in the number of stages is beneficial at large delay increments.

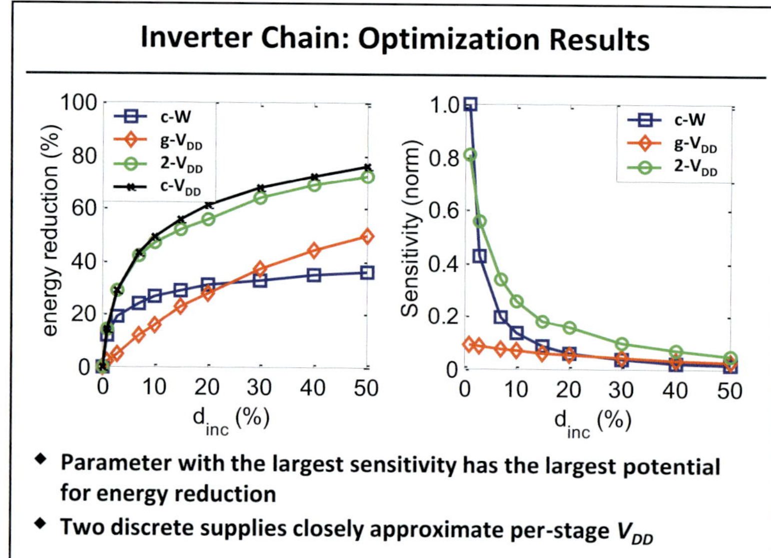

- **Parameter with the largest sensitivity has the largest potential for energy reduction**
- **Two discrete supplies closely approximate per-stage V_{DD}**

2.12

Slide 2.12

To further analyze energy-delay optimization, this slide shows the result of various optimizations performed on the inverter chain: sizing, global V_{DD}, two discrete V_{DD}'s, and per-stage V_{DD} ("c-V_{DD}" in the slide). The graphs show energy reduction and sensitivity versus delay increment. The key concept to realize is that the parameter with the largest sensitivity has the largest potential for energy reduction. For example, at small delay increments sizing has the largest sensitivity, so it offers the largest energy reduction, but the potential for energy reduction from sizing quickly falls off. At large delay increments, it pays to scale the supply voltage of the entire circuit, achieving the sensitivity equal to that of sizing at around 25% excess delay.

We also see from the graph on the left that dual supply voltage closely approximates optimal per-stage supply reduction, meaning that there is almost no additional benefit of having more than two discrete supplies for improving energy in this topology.

The inverter chain has a particularly simple energy distribution, which grows geometrically until the final stage. This type of profile drives the optimization over sizing and V_{DD} to focus on the final stages first. However, most practical circuits like adders have a more complex energy profile.

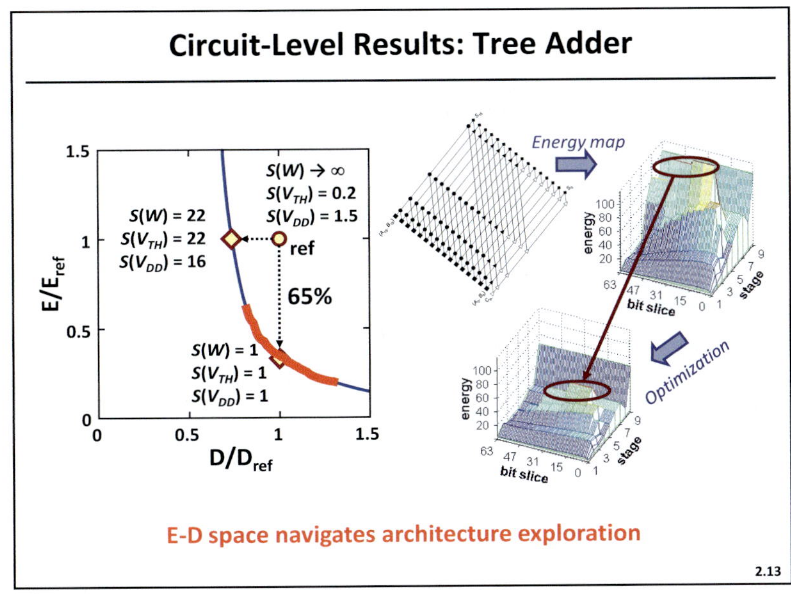

Slide 2.13

The adder is an interesting arithmetic block for DSP applications, so let's take a closer look at this example. This slide shows energy-delay tradeoff in a 64-bit Kogge-Stone tree adder. Energy and delay are normalized to the reference point, which is the design sized for minimum delay at nominal supply and threshold voltage. Starting from the reference point, by equalizing sensitivity to W, V_{DD}, and V_{TH} we can move down vertically and achieve a 65% energy reduction without any performance penalty. Equivalently, we can improve speed about by about 25% without loss in energy.

To gain further insights into energy reduction, the energy map for the adder is shown on the right for the reference and optimized designs. This adder is convenient for 3-D representation, where the horizontal axes correspond to individual bit slices and logic stages. For example, a 64-bit design requires ($1 + \log_2 64 + 1 = 8$ stages). The adder has propagate/generate blocks at the input (first stage), followed by carry-merge operators (six stages), and finally XORs for the final sum generation (last stage). The plot shows the impact of sizing optimization on reducing the dominant energy peak in the middle of the adder in order to balance sensitivity of W with V_{DD} and V_{TH}.

The performance range due to tuning of circuit variables is limited to about ±30% around the reference point. Otherwise, optimization becomes costly in terms of delay or energy. This can also be explained by looking into values of the optimization variables.

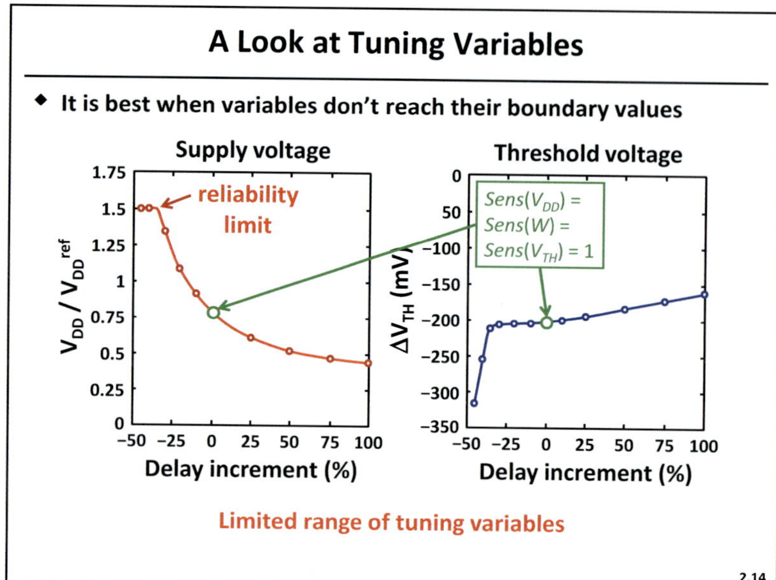

Slide 2.14

Physical limitations put constraints on the values of tuning variables. While all the variables are within their bounds, their sensitivities are the same. The reference case, optimized for nominal delay, has equal sensitivities. Here, the sensitivity of 1 means "reference" sensitivity. In case any of the variables hits a constraint, its sensitivity cannot be made equal to that of the rest. For example, on the left, speeding up the design requires the supply to increase up to the device failure limit. Near that point supply has to be bounded and further speed-ups can be achieved most effectively through threshold scaling, but with higher energy cost.

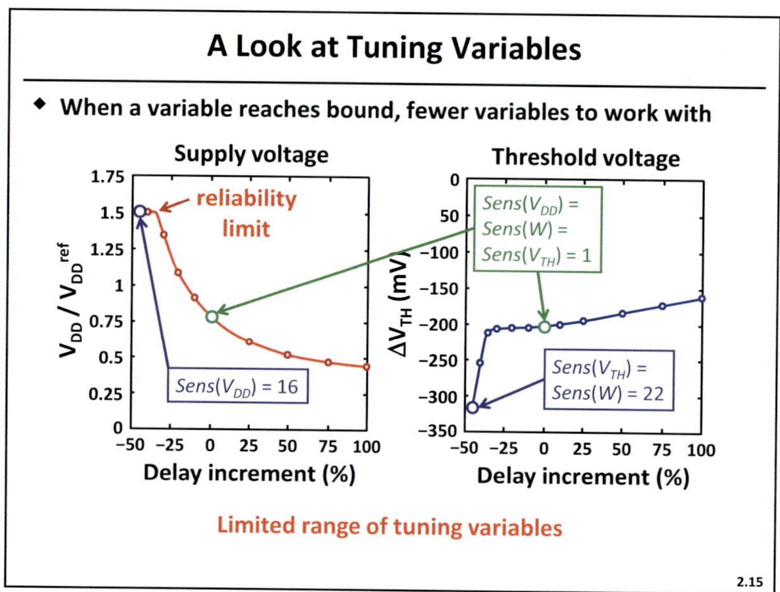

Slide 2.15

After V_{DD} reaches its upper bound, only the threshold and sizing sensitivities are equalized in further optimization. They are equal to 22 in the case of minimum achievable delay. At the minimum-delay point, sensitivity of V_{DD} is 16, because no additional benefit is available from V_{DD} for speed improvement. The slope of the threshold vs. delay increment line clearly indicates that V_{TH} has to work "extra hard" together with W to further reduce the delay.

Many people just choose a subset of tuning variables to optimize. Choosing a variable with the largest sensitivity is the right approach for single-variable optimization. A more refined heuristic is to use two variables – the variable with the highest sensitivity and the variable with the lowest sensitivity – and exploit the sensitivity gap by trading off timing slack as described in Slide 2.6 to minimize energy. For example, in the tree adder we analyzed, one would exploit W and V_{TH} in a two-variable approach.

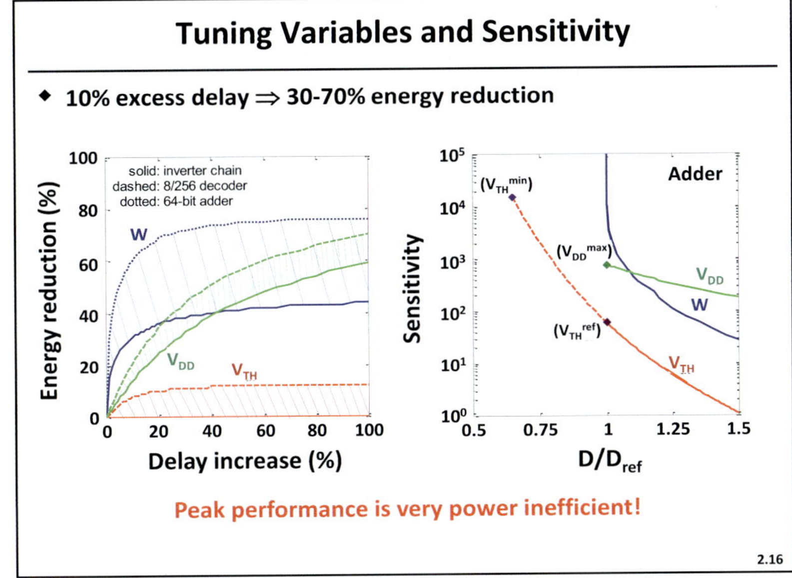

Slide 2.16

Looking at the three examples (inverter chain, memory decoder, and adder) that represent a wide variety of circuit topologies, we can make general conclusions about energy-delay optimization. The left graph shows energy reduction versus delay increment for these three different circuit topologies, which loosely define bounds on energy reduction. We take the best-case and worst-case energy reduction across the three examples, so we can observe general trends.

Sizing is the most efficient at small incremental delays. To understand this, we can also take a look at sensitivities on the right and observe that at the minimum delay point sizing sensitivity is infinite. This makes sense because at minimum delay there is no amount of energy that can be spent to improve the delay. This is consistent with the result from Slide 2.7. The benefits of sizing get quickly utilized at about 20% excess delay. For larger incremental delays, supply voltage becomes the most dominant variable. Threshold voltage has the smallest impact on energy because this technology was not leaky enough, and this is reflected in its sensitivity. If we balance all the sensitivities, we can achieve significant energy savings. A 10% delay slack allows us to achieve 30–70% reduction in energy at the circuit level. So, peak performance is very power-inefficient and should be avoided unless architectural techniques are not available for performance improvement.

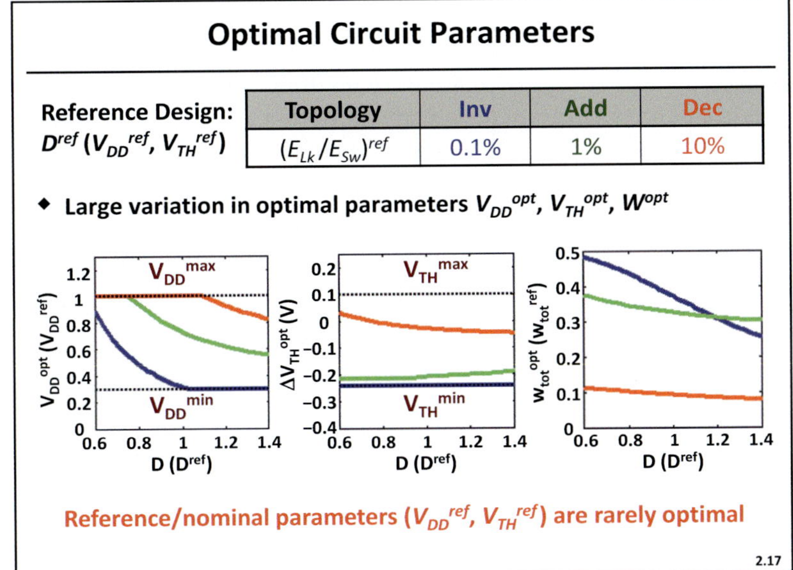

Topology	Inv	Add	Dec
$(E_{Lk}/E_{Sw})^{ref}$	0.1%	1%	10%

Slide 2.17

It is interesting to analyze the values of circuit variables for the examples from Slide 2.10. The reference design in all optimizations is sized for minimum delay at nominal supply and threshold voltages of the technology. In this example, for a 0.13-μm technology, $V_{DD}^{ref} = 1.2$ V, $V_{TH}^{ref} = 0.35$ V. Tuning of circuit variables is related to balancing of the leakage and switching components of energy.

The table represents the ratio of the total leakage to switching energy in the three circuit examples. We see about two orders of magnitude difference in nominal leakage-to-switching ratio. This clearly suggests that nominal technology parameters are not optimal for all circuit topologies. To explore that further, let's now look at the optimal values of V_{DD}, V_{TH}, and W, normalized to the reference case, in these circuits, as shown in this slide. The delay is normalized to minimum delay (at the nominal supply and threshold voltages) in its respective circuit.

Plots on the left and middle graph show that, at nominal delay, supply and threshold are close to optimal only in the memory decoder (*red line*), which has the highest initial leakage energy. The adder has smaller relative leakage, so its optimal threshold has to be reduced to balance switching and leakage energy, while the inverter has the smallest relative leakage, leading to the lowest threshold voltage. To maintain circuit speed, the supply voltage has to decrease with reduction of the threshold voltage.

The plot on the right shows the relative total gate size versus target delay. The memory decoder has the largest reduction in size due to a large number of inactive gates. The size of inactive gates also decreases by the same factor as that in active paths.

In the past, designers used to tune only one parameter, most commonly being supply voltage scaling, in order to minimize energy for a given throughput. However, the only way to truly minimize energy is to utilize all variables. This may seem like a complicated task, but it is well worth the effort, when we consider potential energy savings.

<table>
<tr><td>

Lessons from Circuit Optimization

- **Sensitivity-based optimization framework**
  - Equal marginal costs ⇔ Energy-efficient design

- **Effectiveness of tuning variables**
  - Sizing is the most effective for small delay increments
  - Vdd is better for large delay increments

- **Peak performance is VERY power inefficient**
  - About 70% energy reduction for 20% delay penalty

- **Limited performance range of tuning variables**
  - Additional variables for higher energy-efficiency

2.18

</td></tr>
</table>

Slide 2.18

This chapter has so far presented sensitivity-based optimization framework that equalizes marginal costs for the most energy-efficient design. Below are key results we have been able to derive from this framework.

First, sizing is most effective for small delay increments while supply voltage is better at large incremental delays relative to the minimum-delay design. We are going to extensively use this result in the synthesis based environment by performing incremental compilations to utilize delay slack.

We also learned that peak performance is very power inefficient: about 70% energy reduction is possible with only 20% relaxation in timing. We also learned that there is a limited performance range of tuning variables so we need to consider other layers in the design abstraction to further increase delay and energy efficiency.

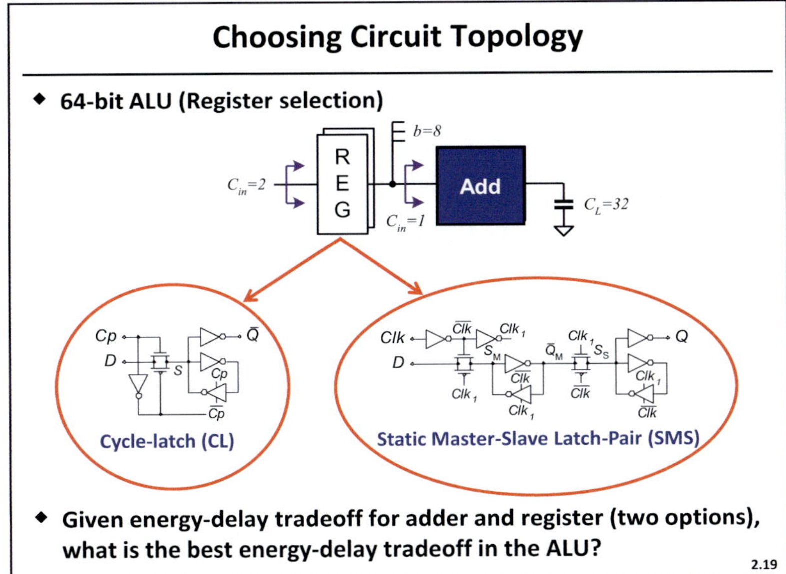

Slide 2.19

Energy-delay tradeoff in logic blocks can extended hierarchically. This slide shows an example of a 64-bit ALU. This is a simplified bit-slice model of the ALU that uses the Kogge-Stone tree adder from Slide 2.13. Now we need to implement input registers. This can be done using a high-speed cycle-latch based design or using a low-power master-slave based design. What is the optimal energy-delay tradeoff in the ALU given the energy-delay tradeoff in each of the circuit blocks? Individual circuit examples can be misleading because the overall energy cost of the system is what really matters.

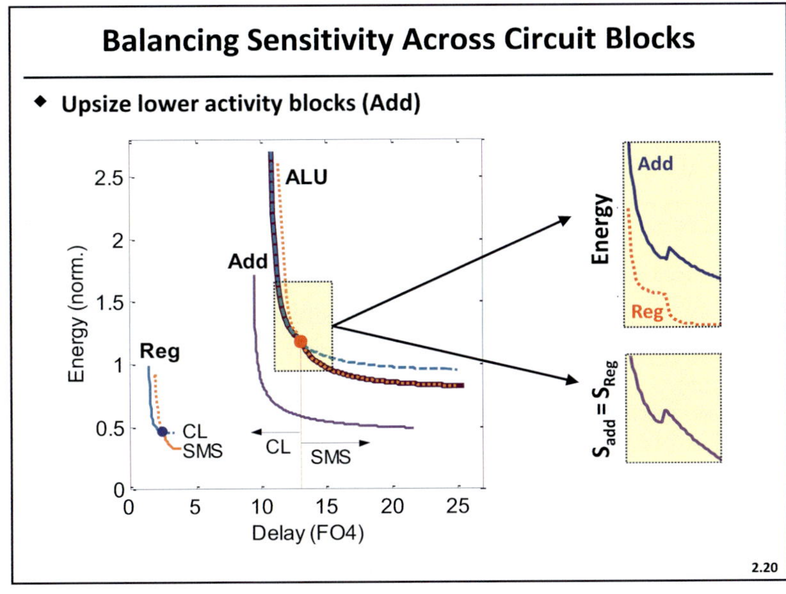

Slide 2.20

In this case it actually pays off to upsize lower activity blocks such as adders and downsize flip-flops so that we can more effectively utilize the energy that is available to us. Globally optimal curves for the register and adder combine to define the energy-delay curve for the ALU. We find that the cycle latch is best used for high performance while the master-slave design is more suitable for low power.

What happens at the boundary is that the adder energy increases to create a delay slack that can be much more efficiently utilized by downsizing higher activity blocks such as registers (this tradeoff was explained in Slide 2.6). As a result, sensitivity increases at the boundary delay point. In other words, the slope of optimal energy-delay line in the ALU increases. After re-optimizing the adder and register, their sensitivities become equal as shown on the bottom-right plot. The concept of balancing sensitivities at the gate level also holds at the block level. This is a nice property which allows us to formulate a hierarchical system optimization approach.

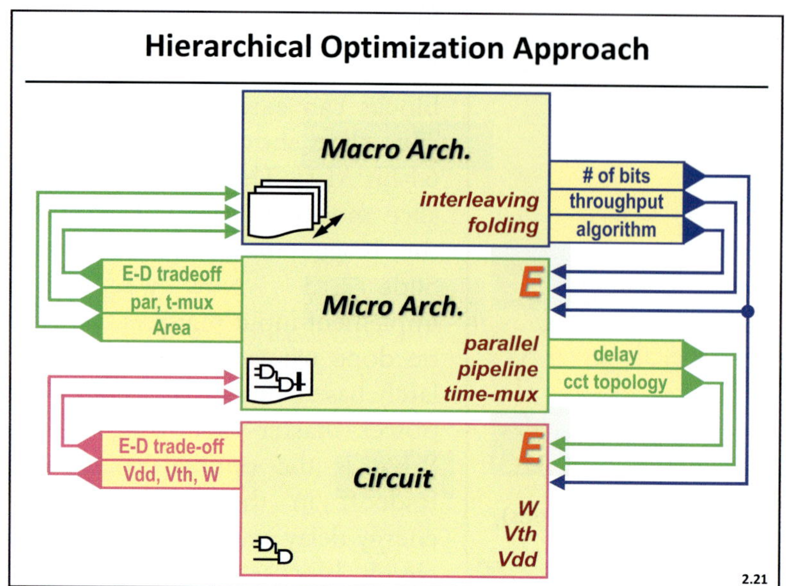

Slide 2.21

To achieve a globally optimal result, we therefore expand circuit optimization hierarchically by decomposing the problem into several abstraction layers. Within each layer, we try to identify independent sub-spaces so we can exchange design tradeoffs between these various layers. At the circuit level, we minimize energy subject to a delay constraint using gate size, supply and threshold voltage. The result is the optimal value of these tuning parameters as well as the energy-delay tradeoff. At the micro-architectural level we have more degrees of freedom: we can choose circuit topology, we can use parallelism/pipelining or time-multiplexing, and for a given algorithm, number of bits and throughput. Finally, at the macro-architecture level, we may also introduce interleaving and folding to deal with recursive and multi-dimensional problems. Since architectural techniques affect design area, we also have to take into account implementation area.

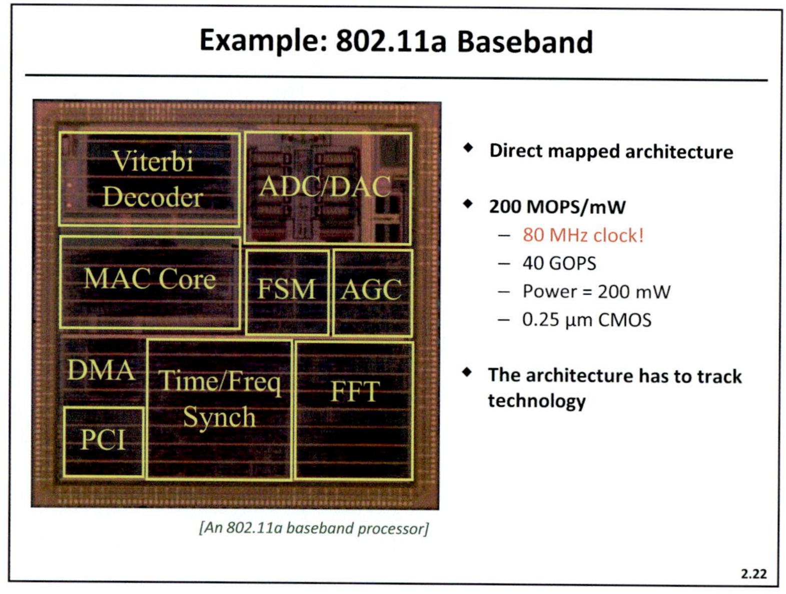

[An 802.11a baseband processor]

Slide 2.22

To provide motivation for architectural optimization, let's look at an example of an 802.11a baseband chip. The DSP blocks operate with an 80 MHz clock frequency. The chip performs 40 GOPS (GOPS = Giga Operations per Second) and consumes 200 mW, which is typical power consumption for a baseband DSP. The chip was implemented in a 0.25-µm CMOS process. Since the application is fixed, the performance requirement would not scale with technology. However, as technology scales, the speed of technology itself gets faster. Thus, this architecture would be sub-optimal if ported to a faster technology by simply shrinking transistor dimensions. The architecture would also need to change in response to technology changes, in order to minimize the chip cost. However, making architectural changes incurs significant design time and increases non-recurring engineering (NRE) costs. We must, therefore, find a way to quickly explore various architectural realizations and make use of the available area savings in scaled technologies.

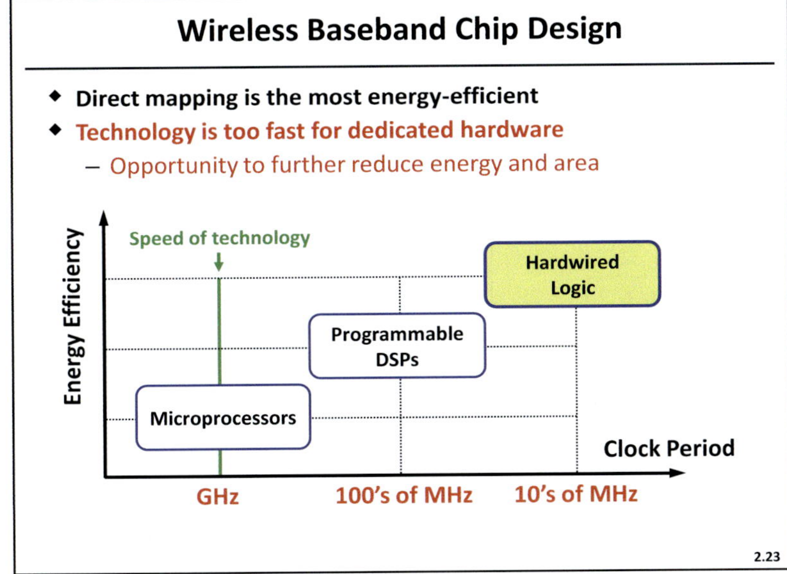

Slide 2.23

To further illustrate the issue of architecture design, let's examine several architectures for digital baseband in a radio. This slide shows energy efficiency (to be defined in Chap. 4) versus clock period in three digital processing architectures: microprocessors, programmable DSPs, and hardwired logic. Digital baseband functionality for a radio is typically provided with direct-mapped, hardwired logic. This approach offers the highest energy-efficiency, which is a prime concern in battery operated devices. Another extreme is the heavily time-multiplexed microprocessor architecture, which has the highest flexibility, but it is also the least energy efficient because of extra overhead to support the time multiplexing.

If we take another look at these architectural choices along the horizontal axis, we see that the clock speed required from direct-mapped architectures is significantly lower than the speed of technology tailored for microprocessor designs. We can take this as an opportunity to further reduce energy and area of chip implementations using architecture design.

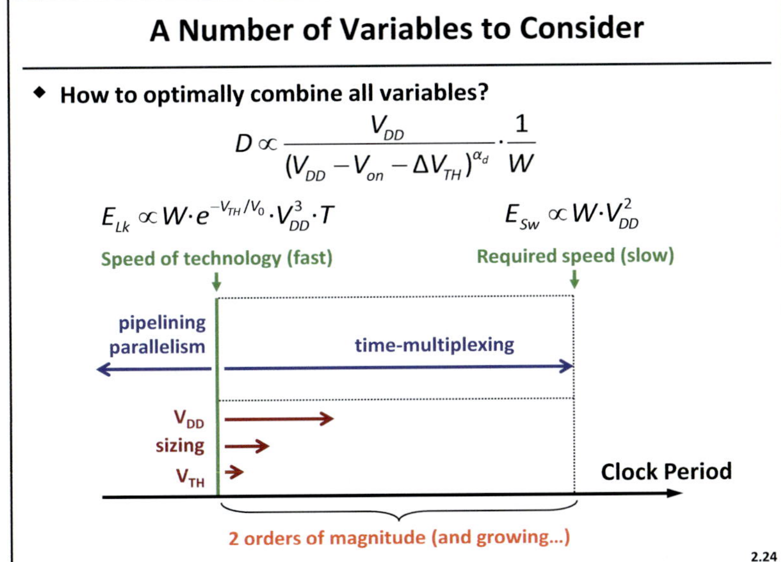

$$D \propto \frac{V_{DD}}{(V_{DD} - V_{on} - \Delta V_{TH})^{\alpha_d}} \cdot \frac{1}{W}$$

$$E_{Lk} \propto W \cdot e^{-V_{TH}/V_0} \cdot V_{DD}^3 \cdot T \qquad E_{Sw} \propto W \cdot V_{DD}^2$$

Slide 2.24

As seen in the previous slide, there is a large disparity in the required clock speed for an application and the speed of operation available by the technology. In order to take advantage of this disparity, we need to tune circuit and architecture variables. The challenge is to come up with an optimal combination of these variables for a given application. Supply voltage scaling, sizing and threshold voltage adjustment can be used at the circuit level, as discussed earlier in this chapter. At the architectural level, we have pipelining, parallelism etc. that can be used to exploit the available performance margin. The goal is to reduce the performance gap in order to minimize energy per operation. The impact of circuit-level variables on energy and delay can be clearly understood by looking at the simple expressions for delay, leakage and switching components of energy. The next challenge is to figure out how to work-in architectural variables and perform global optimization across circuit and architecture levels. Also, how do we make the optimizations work within existing design flows?

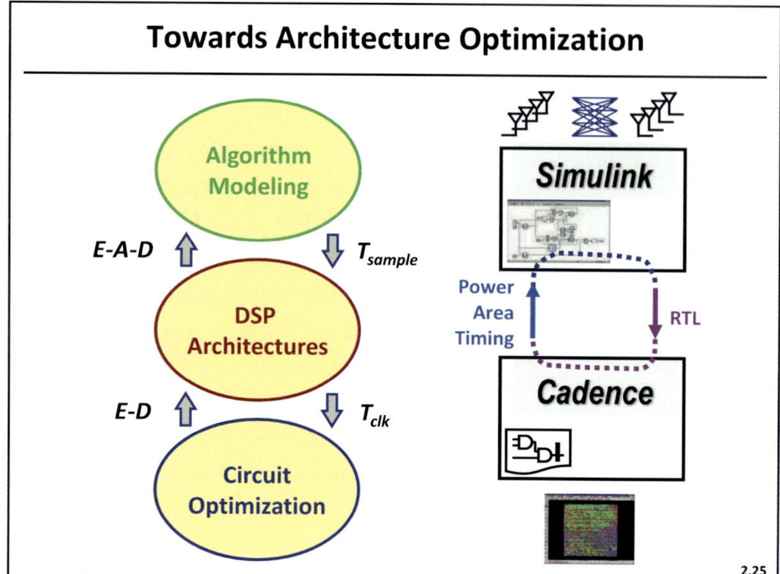

Slide 2.25

This diagram illustrates algorithm, architecture, and circuit design levels. In a typical design flow, we start with a high-level algorithm model, followed by architectural optimization, and further optimizations at the circuit level. The design starts off in MATLAB/Simulink and converts into a layout with the help of Cadence and Synopsys CAD tools. However, in order to ensure that the final design is energy- and area-efficient, algorithm designers need to have feedback about power, area and timing from the circuit level. Essentially, architecture optimizations ensure that algorithm specifications meet technology constraints. Algorithm sample time and architectural clock rate are the timing constraints used to navigate architecture and circuit optimizations, respectively. It is, therefore, important to include hardware parameters in the algorithm model early in the design process. This will be made possible with the hierarchical approach illustrated on Slide 2.19.

Architectural Feedback from Technology

- Simulink hardware library implicitly carries information only about latency and wordlength (we can later choose sample period when targeting an FPGA)

- For ASIC flow, block characterization also has to include technology features such as speed, power, and area

- But, technology parameters scale each generation
  - Need a general and quick characterization methodology
  - Propagate results back to Simulink to avoid iterations

2.26

Slide 2.26

In high-level descriptions such as MATLAB/Simulink, algorithms can be modeled with realistic latency and wordlength information to provide a bit-true cycle-accurate representation. The sample rate can be initially chosen to map the design onto FPGA for hardware emulation. Further down the implementation path, we have to know speed, power and area characteristics of the building blocks in order to perform top-level optimization.

As technology parameters change with scaling, we need a way to provide technology feedback in a systematic manner that avoids design iterations. Simulink modeling (as will be described in later chapters) provides a convenient way for this architecture-circuit co-design.

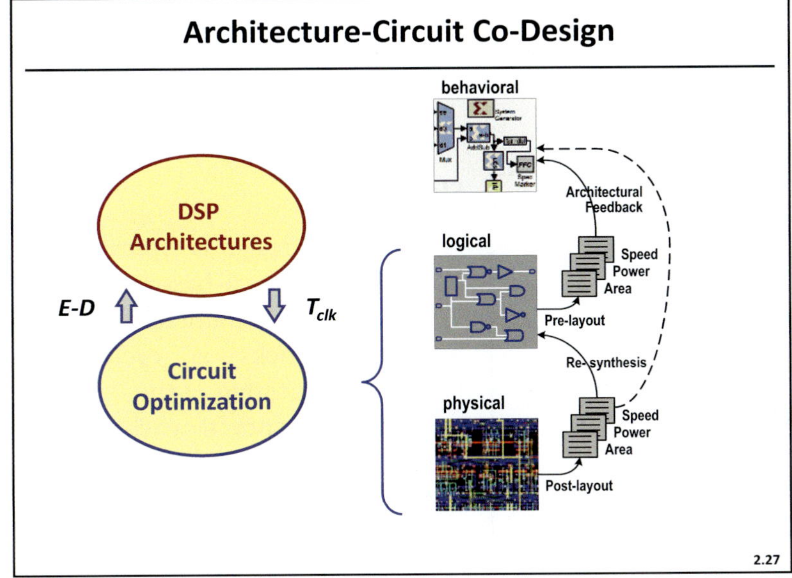

Slide 2.27

To facilitate architecture-circuit co-design, it is crucial to obtain speed, power, and area estimates from the circuit layer and introduce them to Simulink to navigate architecture-level changes. The estimates can be obtained from the logic-level or physical-level synthesis, depending on the desired level of accuracy and available simulation time. Characterization of many blocks may sound like a big effort, but the characterization can be greatly simplified by decomposing the problem into simple datapaths at the circuit level and choosing an adequate level of pipelining at the micro-architectural level. Further, designers don't need to make extensive library characterizations, but only characterize blocks used in their designs.

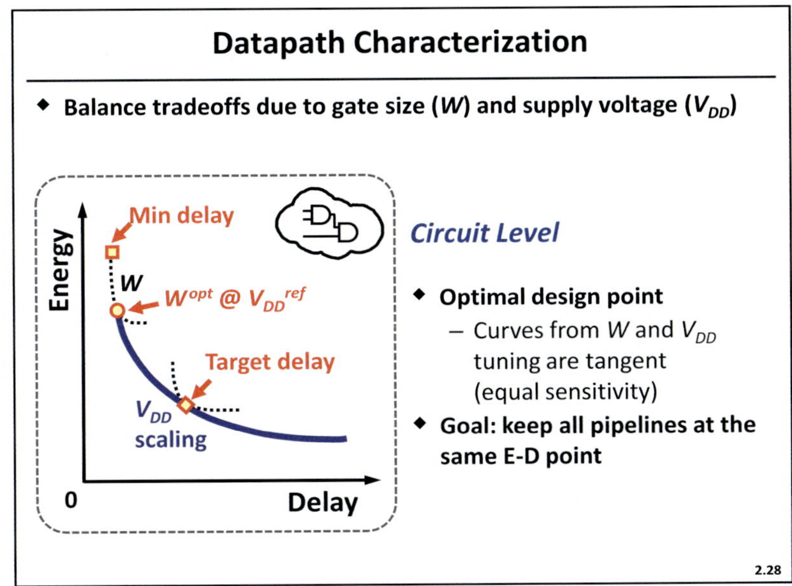

Slide 2.28

The first step in technology characterization is to obtain simple energy-delay (E-D) relationships with respect to voltage scaling. This assumes that pipeline logic has uniform logic depth and equal sizing sensitivity so that supply voltage scaling can be done globally to match the sizing sensitivity. The E-D tradeoff curve shown on this slide can be generated for a fanout-of-four inverter, a typical benchmark for speed of logic, because scaling of other CMOS gates follows a similar trend. The E-D curve serves as a guideline for the range of energy and delay adjustment that can be made by voltage scaling. It also tells us about the energy-delay sensitivity of the design. At the optimal E-D point, sizing and supply voltage scaling tradeoff curves have to be tangent, representing equal sensitivity. As shown on the plot, the min-delay point has highest sensitivity to sizing and requires downsizing to match supply sensitivity at nominal voltage. Lowering supply will thus require further downsizing to balance the design.

Simple E-D characterizations provide great insights for logic and architectural adjustments. For example, assume a datapath with L_D logic stages where each stage has a delay T_1 and let T_{FF} be the register delay. Cycle time T_{clk} can be expressed simply as: $T_{clk} = L_D \cdot T_1 + T_{FF}$. We can satisfy the equation with many combinations of T_1 and L_D by changing the pipeline depth and the supply voltage to reach a desired E-D point. By automating the exploration, we can quickly search among many feasible solutions in order to best minimize area and energy.

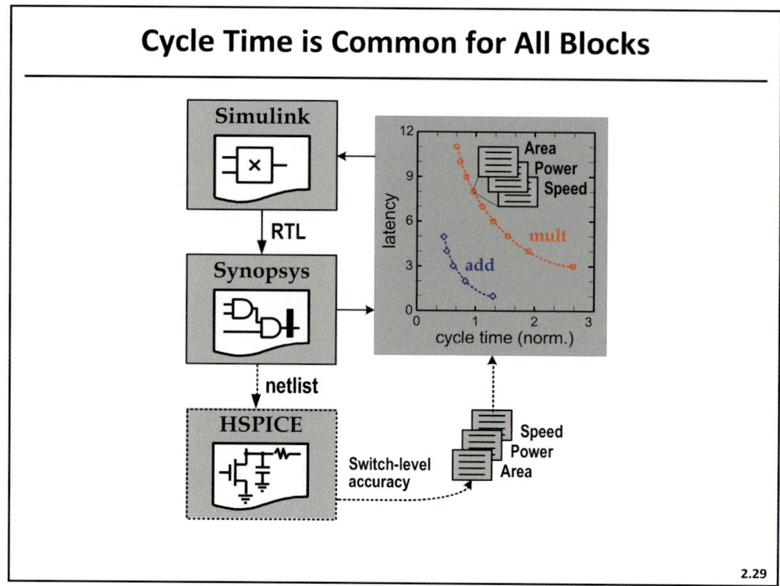

Slide 2.29

Architecture design is typically done in a modular block-based approach. The common parameter for DSP hardware blocks is cycle time. Hence, allocating the proper amount of latency to each block is crucial to achieve overall design optimality. For example, an N-by-N multiplier is about N times more complex than an N-bit adder. This complexity difference is reflected in multiplier latency as shown on the slide. With this characterization flow, we can quickly obtain latency versus cycle time for library blocks.

This way, we augment high-level Simulink blocks with library cards for area, power, and speed. This characterization approach greatly simplifies top-level retiming (to be discussed in Chaps. 3, 9 and 11).

Slide 2.30

Energy-delay optimization at circuit level was discussed. The optimization uses sensitivity-based approach to balance marginal returns with respect to tuning variables. Energy and delay models from Chap. 1 allow for convex formulation of delay-constrained energy minimization. As a result, optimal energy-delay tradeoff is obtained by tuning gate size, supply and threshold voltage. Sensitivity theory suggests that gate sizing is the most effective at small delays (relative to the minimum-delay point), supply voltage reduction is the most effective variable for medium delays and threshold voltage is the most effective for long delays (around minimum-energy point). Circuit-level optimization is limited to about $\pm 30\%$ around the minimum delay; outside of this region optimization becomes too costly either in energy or delay. To expand the optimization across broader range of performance, more degrees of freedom are needed. Next chapter discusses the use of architectural-level variables for area-energy-delay optimization.

References

- V. Stojanović *et al.* "Energy-Delay Tradeoffs in Combinational Logic using Gate Sizing and Supply Voltage Optimization," in *Proc. Eur. Solid-State Circuits Conf.,* Sept. 2002, pp. 211-214.

- D. Marković *et al.,* "Methods for True Energy-Performance Optimization," *IEEE J. Solid-State Circuits,* vol. 39, no. 8, pp. 1282-1293, Aug. 2004.

- S. Ma and P. Franzon, "Energy Control and Accurate Delay Estimation in the Design of CMOS Buffers," *IEEE J. Solid-State Circuits,* vol. 29, no. 9, pp. 1150-1153, Sept. 1994.

Additional References

- R. Gonzalez, B. Gordon, and M.A. Horowitz, "Supply and Threshold Voltage Scaling for Low Power CMOS," *IEEE J. Solid-State Circuits,* vol. 32, no. 8, pp. 1210-1216, Aug. 1997.

- V. Zyuban *et al.,* "Integrated Analysis of Power and Performance for Pipelined Microprocessors," *IEEE Trans. Computers,* vol. 53, no. 8, pp. 1004-1016, Aug. 2004.

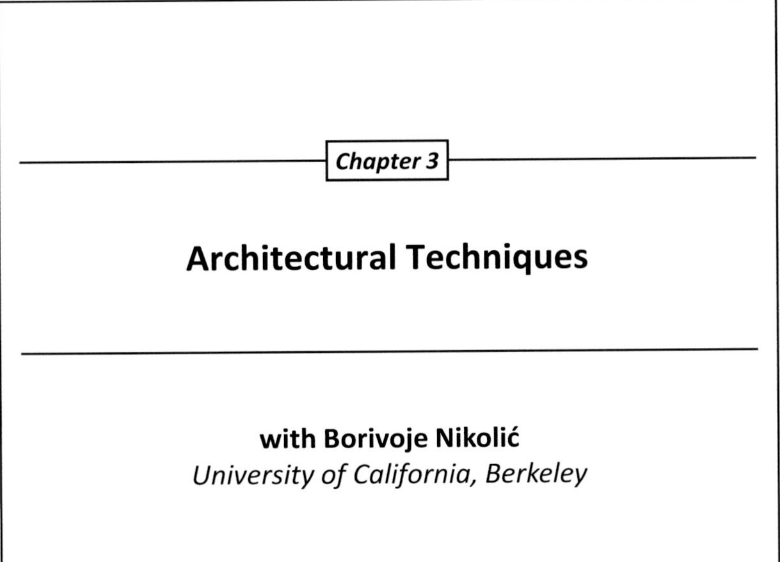

Slide 3.1

This chapter discusses architectural techniques for area and energy reduction in chips for digital signal processing. Parallelism, time-multiplexing, pipelining, interleaving and folding are compared in the energy-delay space of pipeline logic as a systematic way to evaluate different architectural options. The energy-delay analysis is extended to include area comparison and quantify time-space tradeoffs. So, the energy-area-performance representation will serve as a basis for evaluating multiple architectural techniques. It will also give insight into which architectural transformations need to be made to track scaling of the underlying technology for the most cost- and energy-efficient solutions.

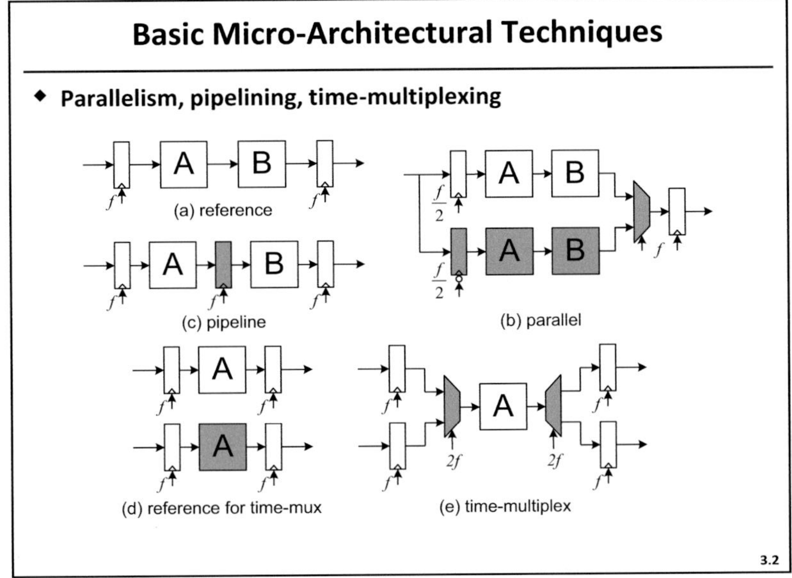

Slide 3.2

Three basic architectural techniques are parallelism, pipelining, and time-multiplexing. Figure (a) shows the reference datapath with logic blocks A and B between pipeline registers. Starting from the reference architecture, we can make transformations into parallel, pipelined, or time-multiplexed designs without affecting the data throughput.

Parallelism and pipelining are used for power reduction. We could introduce parallelism as shown in Figure (b). This is accomplished by replicating the input register and the datapath, and adding a multiplexer before the output register. Parallelism trades increased area (blocks shaded in *gray*) for reduced speed of the pipeline logic (A and B are running at half the original speed), which allows for supply voltage reduction to decrease power. Another option is to pipeline the blocks, as shown in Figure (c). An extra pipeline register is inserted between logic blocks A and B. This lets blocks A and B run at half the speed and also allows for supply voltage reduction, which leads to a decrease in power. The logic depth is reduced at the cost of increased latency.

Time-multiplexing is used for area reduction, as shown in Figures (d) and (e). The reference case shown in Figure (d) has two blocks to execute two operations of the same kind. An alternative is to

D. Marković and R.W. Brodersen, *DSP Architecture Design Essentials*, Electrical Engineering Essentials, DOI 10.1007/978-1-4419-9660-2_3, © Springer Science+Business Media New York 2012

do the same by using a single hardware module, but at twice the reference speed, to process the two data streams as shown in Figure (e). If the area of the multiplexer and demultiplexer is less than the area of block A, time-multiplexing results in an area reduction. This is most commonly the case since datapath logic is much more complex than a de/multiplexing operation.

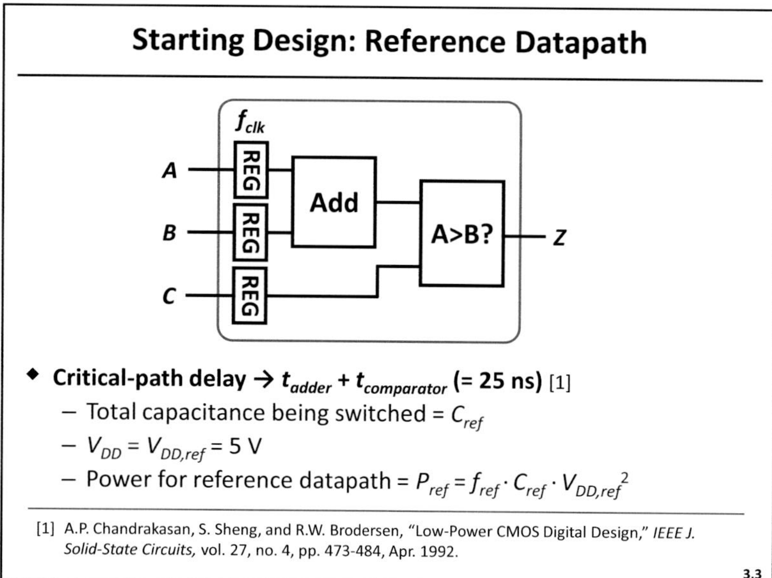

Slide 3.3

To establish a baseline for architecture study, let's consider a reference datapath which includes an adder that adds two operands A and B, followed by a comparator that compares the sum to a third operand C [1]. The critical-path delay here is $t_{adder} + t_{comparator}$, which is equal to 25 ns, for example (this is a delay for the 0.6-μm technology used in [1]). Let C_{ref} be the total capacitance switched at a reference voltage $V_{DD,ref} = 5V$. The switching power for the reference datapath is $f_{ref} \cdot C_{ref} \cdot V_{DD,ref}^2$. Now, let's see what kind of tradeoffs can be made by restructuring the datapath by using parallelism.

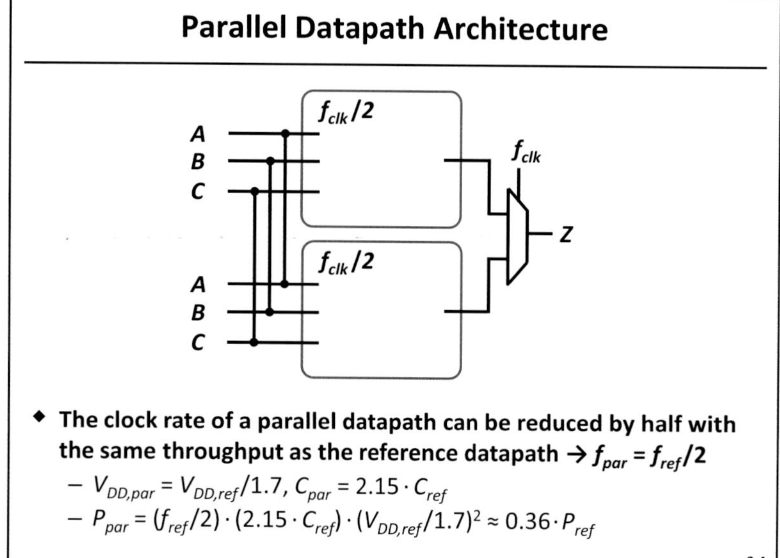

Slide 3.4

Parallelism is now employed on the previous design. The two modules now process odd and even samples of the incoming data, which are then combined by the output multiplexer. The datapath is replicated and routed to the output multiplexer. The clock rate for each of the blocks is reduced by half to maintain the same throughput. Since the cycle time is reduced, supply voltage can also be reduced by a factor of 1.7. The total capacitance is now $2.15C_{ref}$ including the overhead capacitance of $0.15C_{ref}$. The total switching power, however, is $(f_{ref}/2) \cdot 2.15C_{ref} \cdot (V_{DD,ref}/1.7)^2 \approx 0.36 \ P_{ref}$. Therefore, over 60% of power reduction is achieved without performance penalty by using a parallel-2 architecture.

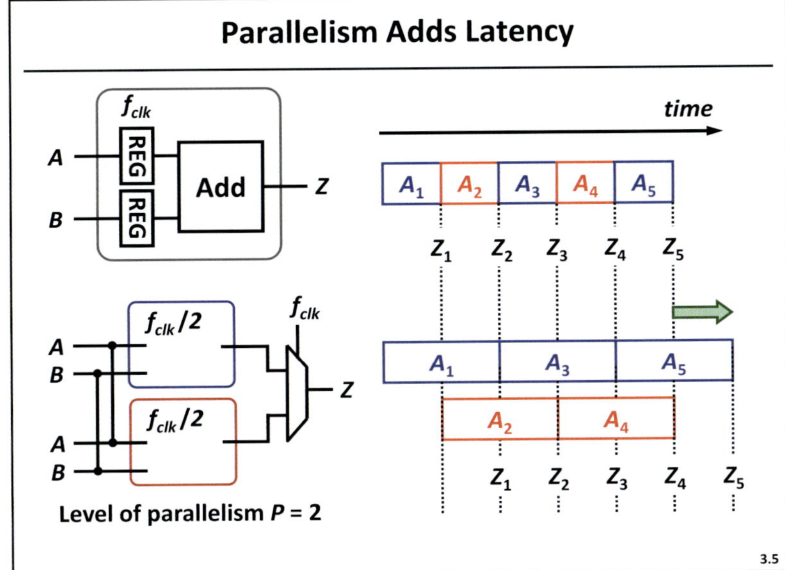

Slide 3.5

It should, however, be kept in mind that parallelism adds latency, as illustrated in this slide. The top half of the slide shows the incoming data stream A and the corresponding output Z. The design has a single-cycle latency, so samples of Z are delayed by one clock cycle with respect to samples of A. The parallel-2 design takes samples of A and propagates resulting samples of Z after two clock cycles. Therefore, one extra cycle of latency is introduced. Parallelism is, therefore, possible if extra latency is allowed. This is the case in most low-power applications. Besides energy reduction, parallelism can also be used to improve performance; if the clock rate for each of the blocks were kept at the original rate, for example, then the throughput would increase by twofold.

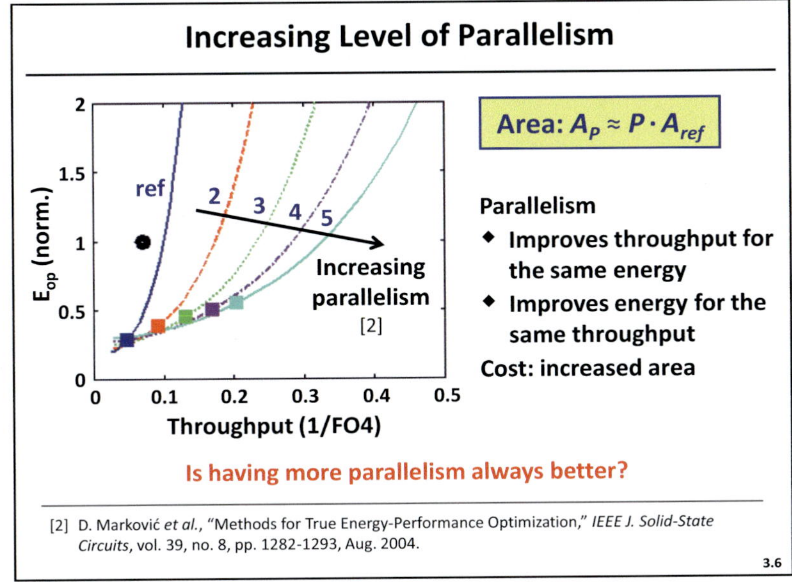

[2] D. Marković *et al.*, "Methods for True Energy-Performance Optimization," *IEEE J. Solid-State Circuits*, vol. 39, no. 8, pp. 1282-1293, Aug. 2004.

Slide 3.6

The plot in this slide shows energy per operation (E_{op}) versus throughput for architectures with varying degrees of parallelism [2]. We see that parallelism can be used to reduce power at the same frequency or increase the frequency for the same energy. Squares show the minimum energy-delay product (EDP) point for each of the architectures. The results indicate that as the amount of parallelism increases, the minimum EDP point corresponds to a higher throughput and also a higher energy. With an increasing amount of parallelism, more energy is needed to support the overhead. Also, leakage energy has more impact with increasing parallelism.

The plot also shows the increased performance range of micro-architectural tuning as compared to circuit-level optimization. The reference datapath (min-delay sizing at reference voltage) is shown as the black dot in the energy-delay space. This point is also used as reference for energy. The reference design shows the tunability at the circuit level, which is limited in performance to $\pm 30\%$ around the reference point as discussed in Chap. 2. Clearly, parallelism goes much beyond $\pm 30\%$ in extending performance while keeping the energy consumption low. However, the area and performance benefits of parallelism come at the cost of increased area, which imposes a practical (cost) limit on the amount of parallelism.

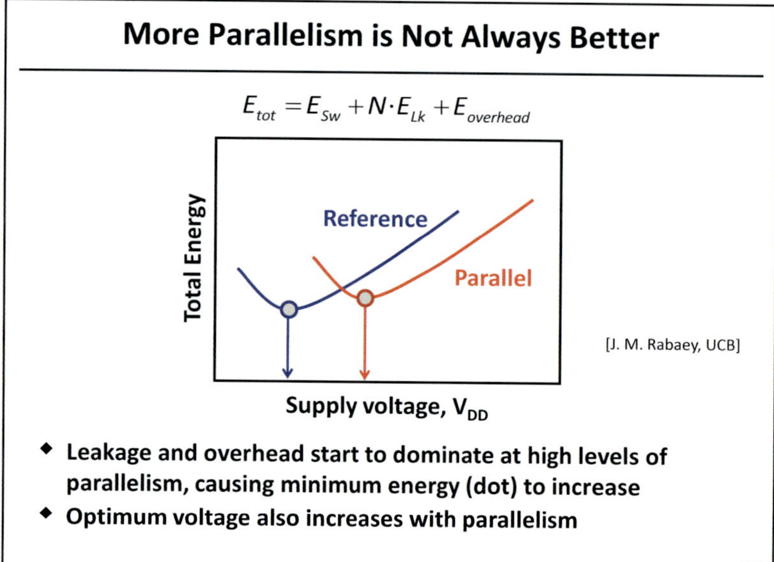

Slide 3.7

Area is not the only limiting factor to increasing the level of parallelism. Total energy also increases at some point due to the large impact of leakage. Let's look at the E-D tradeoff curve to explain this effect. The total energy we expend in a parallel design is the sum of switching energy of the active block, the leakage energy of all the parallel blocks and the energy incurred in the multiplexing. Even if we neglect the multiplexing overhead, (which is a fair assumption for complex blocks), the increase in leakage energy at high levels of parallelism would imply that the minimum-energy point now shifts to a higher voltage. The total energy to perform one operation with higher levels of parallelism is thus higher, since the increased leakage and cost of overhead pull the energy per operation to a higher value.

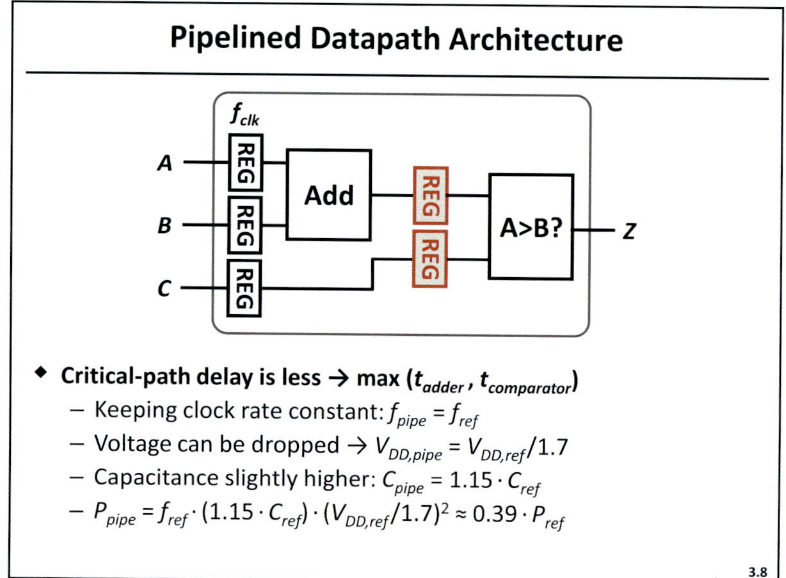

Slide 3.8

Let us now use the same example to look into pipelining. We introduce latency via a set of pipeline registers to cut down the critical path of $t_{adder} + t_{comparator}$ to the maximum of (t_{adder}, $t_{comparator}$). The clock frequency is the same as the original design. Given the reduced critical path, the voltage can now be reduced by a factor of 1.7 (same as in the parallelism example on Slide 3.4). The capacitance is slightly higher than the reference capacitance to account for the switching energy dissipated in the pipeline registers. The switching power of the pipelined design is thus given by $f_{ref} \cdot (1.15 C_{ref}) \cdot (V_{DD,ref}/1.7)^2 \approx 0.39 P_{ref}$. Pipelining, therefore, can also be used to reduce the overall power consumption.

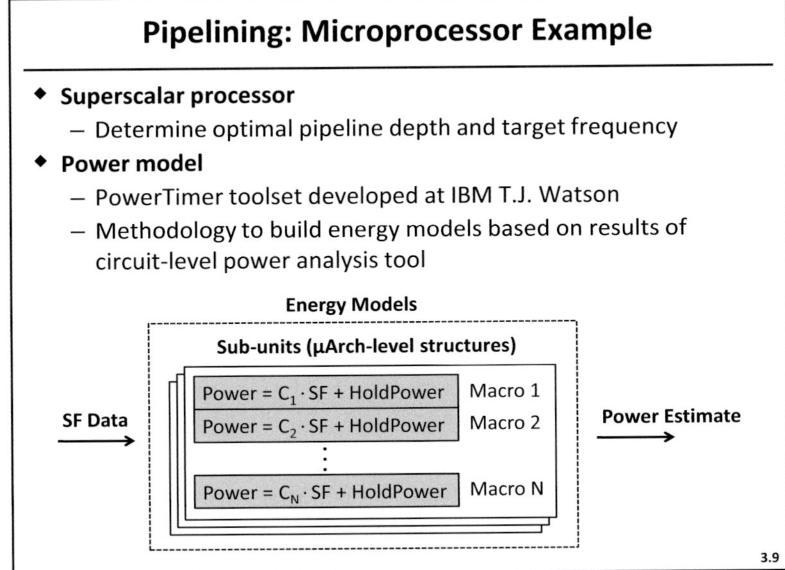

Slide 3.9

Pipelining is one of the key techniques used in microprocessor design. This slide shows a practical design: a superscalar processor from IBM, which illustrates the effect of pipelining on power and performance. This example is taken from the paper by Srinivasan *et. al.* (see Slide 3.13 for complete reference). A latch-level accurate power model in the PowerTimer tool is developed for a power-performance analysis. The analysis accounts for power of several processor units in hold and switching modes. The approach to model energy is illustrated on the slide. The processor can be analyzed as a set of sub-units (micro-architectural-level structures), where each sub-unit consists of a number of macros. Each macro can execute an operation on input SF Data or be in the Hold mode. Power consumption of all such macros in the processor gives an estimate of the total power.

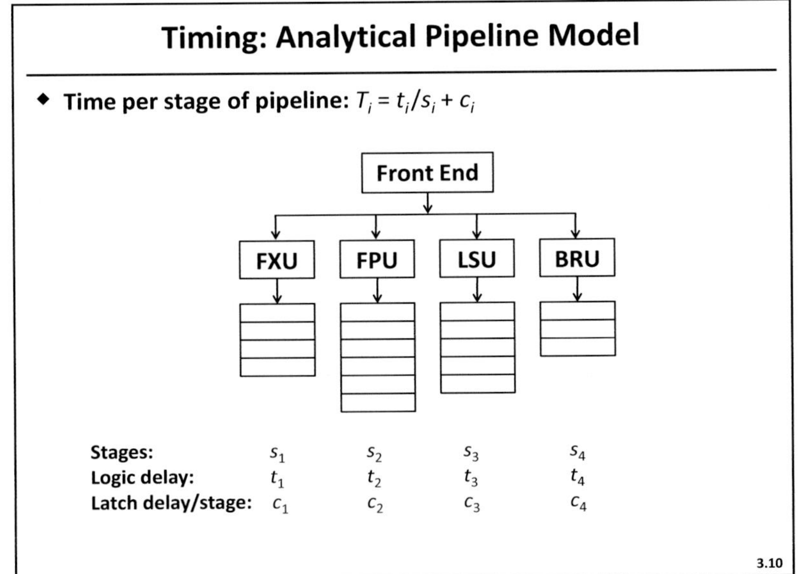

Slide 3.10

The timing model for the processor is based on the delay per stage of the pipeline including register (latch) delay. The processor front end dispatches operations into four units: fixed-point (FXU), floating-point (FPU), load/shift (LSU), and branching unit (BRU). Each of these units is modeled at the datapath level by the number of stages (s), logic delay (t), and latch delay per stage (c). The time per stage of pipeline is $T_i = t_i/s_i + c_i$. This gate-level model is used to calculate the time it takes to perform various operations by the processor units.

Timing: Analytical Pipeline Model

- **Time to complete FXU operation in presence of stalls**

$$T_{fxu} = T_1 + Stall_{fxu\text{-}fxu} \cdot T_1 + Stall_{fxu\text{-}fpu} \cdot T_2 + \ldots + Stall_{fxu\text{-}bru} \cdot T_4$$

$$Stall_{fxu\text{-}fxu} = f_1 \cdot (s_1 - 1) + f_2 \cdot (s_1 - 2) + \ldots$$

f_i is conditional probability that an FXU instruction m depends on FXU instruction $(m - i)$

$$\text{Throughput} = u_1/T_{fxu} + u_2/T_{fpu} + u_3/T_{lsu} + u_4/T_{bru} \qquad [3]$$

u_i fraction of time pipe i has instructions arriving from FE of the machine $u_i = 0$ unutilized pipe, $u_i = 1$ fully utilized

[3] V. Srinivasan et al., "Optimizing Pipelines for Power and Performance," in *Proc. IEEE/ACM Int. Symp. on Microarchitecture*, Nov. 2002, pp. 333-344.

3.11

Slide 3.11

Using the model from the previous slide, we can now estimate the time for completing instructions and taking into account instruction dependence. For instance, the fixed-point unit may be stalled if an instruction m depends on instruction $(m - i)$. Hence, the time to complete a fixed-point operation is the sum of the time to complete the operation without stalls, T_1, time to clear a stall within FXU, $Stall_{fxu\text{-}fxu} \cdot T_1$, time to clear a stall in FPU, $Stall_{fxu\text{-}fpu} \cdot T_2$, and so on for all functional units which may have co-dependent instructions. Suppose that parameter u is the fraction of time that a pipeline has instructions arriving from the front end (0 indicates no utilization, 1 indicates full utilization). Based on the time it takes to complete an instruction and the pipeline utilization, we can calculate throughput of a processor as highlighted on the slide [3]. Lower pipeline utilization naturally results in a lower throughput and vice versa. Fewer stalls reduce the time per instruction and increase throughput.

Simulation Results

- **Optimal pipeline depth was determined for two applications (SPEC 2000, TPC-C) under different optimization metrics**
  - Performance-aware optimization: maximize BIPS
  - Power-aware optimization: maximize BIPS³/W

- **More pipeline stages are needed for low power (BIPS³/W)**

Application	Max BIPS	Max BIPS³/W
Spec 2000	10 FO4	18 FO4
TPC-C	10 FO4	25 FO4

- **Choice of pipeline register also impacts BIPS**
  - Overall BIPS performance improved by 20% by using a register with 2 FO4 delay as compared to a register with 5 FO4 delay

3.12

Slide 3.12

Simulation results from the study are summarized here. Relative performance versus the number of logic stages (logic depth) for a variety of metrics was investigated. The metrics include performance in billion instructions per second (BIPS), various degrees of power-performance tradeoffs, and power-aware optimization (that maximizes BIPS³/W). The results show that performance is maximized with shallow logic depth ($L_D = 10$). This is in agreement with earlier discussion about pipelining (equivalent to shortening logic depth), when used for performance improvement. The situation in this processor is a bit more complex with the consideration of stalls, but general trends still hold. Increasing the logic depth reduces the power, due to the decrease in register power overhead. Power is minimized for $L_D = 18$. Therefore, more logic stages are needed for lower power. Other applications such as TPC-C require 23 logic stages for power minimization while performance is maximized for $L_D = 10$.

The performance also depends on the underlying datapath implementation. A 20% performance variation is observed for designs that use latches with delay from 2 FO4 to 5 FO4 delays. Although the performance increases with faster latches, the study showed that the logic depth at which performance is maximized increases with increasing latch delay. This makes sense, because more stages of logic are needed to compensate for longer latch delay (reduce latch overhead).

This study of pipelining showed several important results. Performance-centric designs require shorter pipelines than power-centric designs. A ballpark number for logic depth is about 10 for high-performance designs and about 20 for low-power designs.

Architecture Summary (Simple Datapath)

- Pipelining and parallelism relax performance of a datapath, which allows voltage reduction and results in power savings
- Pipelining has less are overhead than parallelism, but is generally harder to implement (involves finding convenient logic cut-sets)

Results from [1]

Architecture type	Voltage	Area	Power
Reference datapath (no pipelining of parallelism)	5 V	1	1
Pipelined datapath	2.9 V	1.3	0.39
Parallel datapath	2.9 V	3.4	0.36
Pipeline-Parallel	2.0 V	3.7	0.2

[1] A.P. Chandrakasan, S. Sheng, and R.W. Brodersen, "Low-Power CMOS Digital Design," *IEEE J. Solid-State Circuits,* vol. 27, no. 4, pp. 473-484, Apr. 1992.

3.13

Slide 3.13

This slide summarizes the benefits of pipelining and parallelism for power reduction for the two examples presented in Slides 3.3–3.9 [1]. Adding one level of parallelism or pipelining to the reference design has a similar impact on power consumption (about 60% power reduction). This assumes that the reference design operates at its maximum frequency. Power savings achieved are consistent with energy-delay analysis from Chap. 2. Also, since the delay increase is 100%, supply reduction is the most effective for power reduction. The gains summarized in this slide could have been even larger had gate sizing been used together with V_{DD} reduction. The inclusion of sizing, however, would complicate the design since layout would need to be modified to include new sizing of the gates. When parallelism and pipelining are combined, even larger energy/power savings can be achieved: the slide shows 80% power reduction when both parallelism and pipelining are employed.

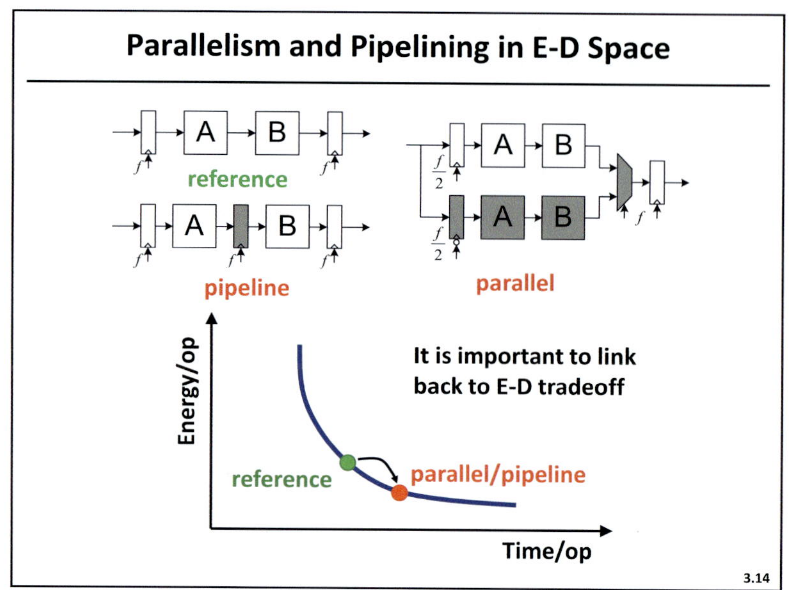

Slide 3.14

Let us further analyze parallelism and pipelining using the energy-delay space for pipeline logic. The block diagrams show reference, parallel and pipeline implementations of a logic function. Parallelism simply slows down the computation by introducing redundancy in area. It directs operands to parallel branches in an interleaved fashion and later combines them at the output. Pipelining introduces an extra pipeline registers to relax timing constraints on blocks A and B so that we can then scale the voltage to reduce energy.

The energy-delay behavior of logic blocks during these transformations is very important to understand, because pipeline logic is a fundamental micro-architectural component. From a datapath logic standpoint, parallelism and pipelining are equivalent to moving the reference energy-delay point toward reduced *Energy/Op* and increased *Time/Op* on the energy-delay curve. Clearly, the energy saving would be the largest when the reference design is in the steep-slope region of the energy-delay curve. This corresponds to the minimum-delay design (which explains why parallelism and pipelining were so effective in previous examples). It is also possible to reduce energy when the performance requirement on the reference design is beyond its minimum achievable delay as illustrated on Slide 3.6. In that case, parallelism would be needed to meet the performance first, followed by the use of voltage scaling to reduce energy. How is the energy minimized?

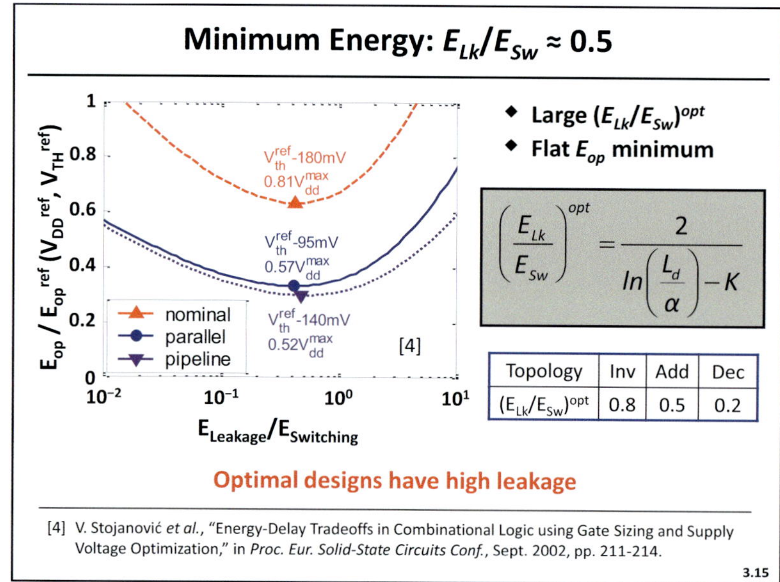

Topology	Inv	Add	Dec
$(E_{Lk}/E_{Sw})^{opt}$	0.8	0.5	0.2

[4] V. Stojanović *et al.*, "Energy-Delay Tradeoffs in Combinational Logic using Gate Sizing and Supply Voltage Optimization," in *Proc. Eur. Solid-State Circuits Conf.*, Sept. 2002, pp. 211-214.

3.15

Slide 3.15

Minimum energy is achieved when the leakage energy is about one half of the switching energy, $E_{Lk}/E_{Sw} \approx$ 0.5. This ratio can be analytically derived, as shown in the box [4]. The equation states that the optimal E_{Lk}/E_{Sw} depends on the logic depth (L_D), switching activity (α), and the technology process (parameter K). The minimum is quite broad, due to the logarithmic formula, with total energy being within about 10% of the minimum for E_{Lk}/E_{Sw} from 0.1 to 10.

These diagrams show reference, parallel and pipeline implementations of an ALU. In this experiment, V_{TH} was swept in increments of 5 mV and for each V_{TH}, the optimal V_{DD} and sizing were found to minimize energy. Each point, hence, has an E_{Lk}/E_{Sw} ratio that corresponds to minimum energy. The plot illustrates several points. First, the energy minimum is very flat and roughly corresponds to equalizing the leakage and switching components of energy. This is the case in all three architectures. The result coincides with the three examples from the circuit level: inverter chain, adder, and memory decoder, as shown in the table. Second, parallel and pipeline designs are more energy efficient than the reference design, because their logic runs slower and voltage scaling is possible. Third, the supply voltage at the minimum-energy point is lower in the pipeline design than in the parallel design, because the pipeline design has smaller area and less leakage. This is also evidenced by a lower V_{TH} in the pipeline design.

Optimal designs, therefore, have high leakage when the total energy is minimized. This makes sense for high-activity designs where we can balance the E_{Lk}/E_{Sw} ratio to minimize energy. In the idle mode $E_{Sw} = 0$, so the energy minimization problem reduces to leakage minimization.

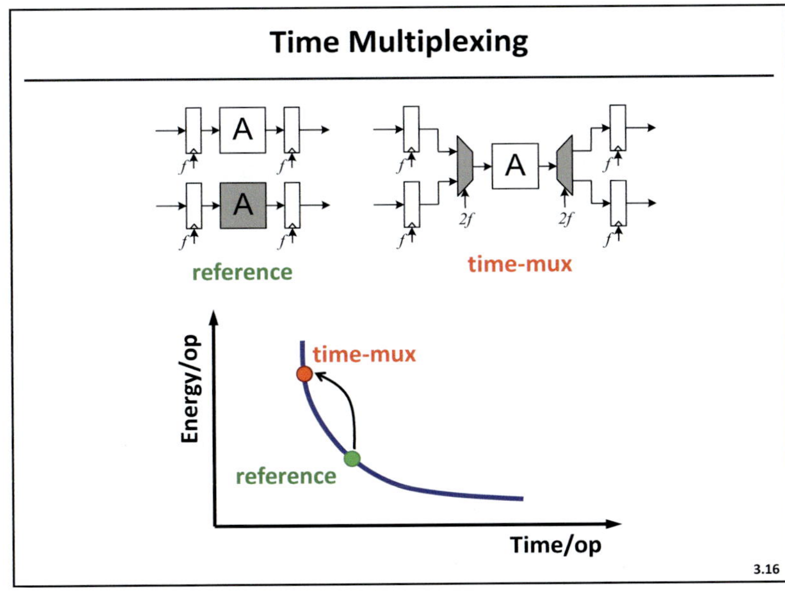

Slide 3.16

Time multiplexing does just the opposite of pipelining/parallelism. It executes a number of parallel data streams on the same processing element (A). Due to the parallel-to-serial conversion at the input of block A, logic block A works with up-sampled data ($2 \cdot f$) to maintain external throughput. This imposes a more aggressive timing constraint on logic block A. The reference design, hence, moves toward higher energy and lower delay in the E-D space. Since the hardware for logic is shared, the area of the time-multiplexed design is reduced.

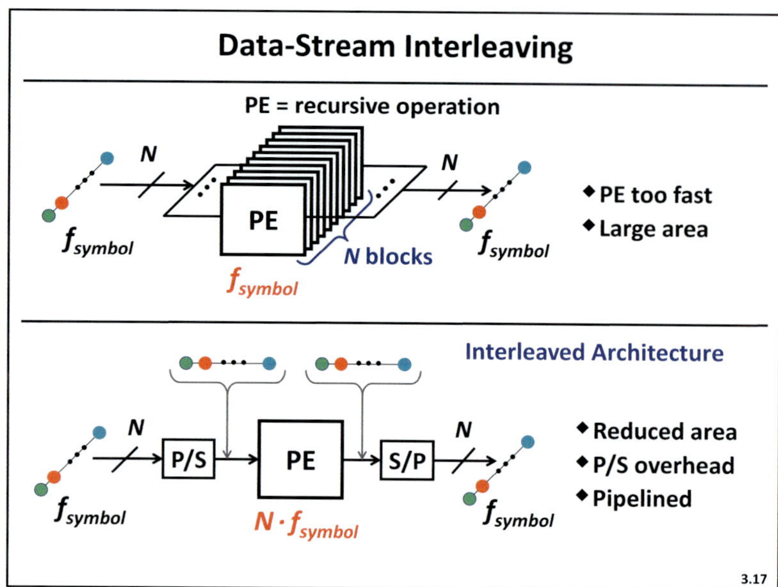

Slide 3.17

We frequently encounter parallel data in signal processing algorithms such as those found in multi-carrier communication systems or medical applications. The example shown in the upper figure shows N processing elements (PEs) that process N independent streams of data. Given the speed of nano-scale technologies, the processing element can work much faster than the application requirement. Under such conditions we can time-interleave parallel data streams on the same hardware in order to take advantage of the difference in application speed requirements and the speed of the technology. The interleaving approach is used to save hardware area.

Parallel-to-serial (P/S) conversion is applied to the incoming data stream to time-interleave the samples. The processing element in the interleaved architecture now runs at a rate N-times faster to process all incoming data streams. A serial-to-parallel (S/P) converter then splits the output data stream back into N parallel channels. It is important to note that extra pipeline registers need to be introduced for the interleaved processing element.

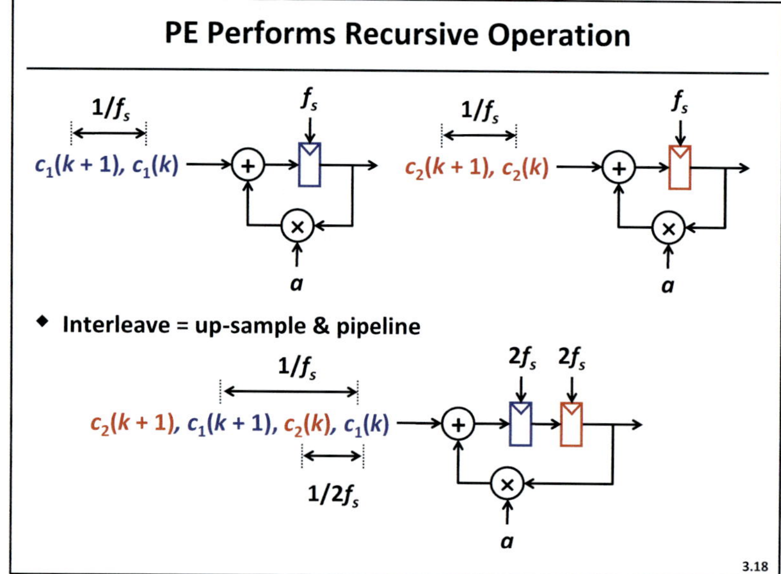

Slide 3.18

Let us consider a simple example of a processing element to illustrate interleaving. We assume a simple add-multiply operation and two inputs C_1 and C_2.

The processing element shown in this slide is a two-stream interleaved implementation of the function $1/(1 - a \cdot z^{-1})$. Two input streams at a rate f_s are interleaved at a rate $1/2f_s$. The key point of interleaving is to add an extra pipeline stage (*red*) as a memory element for the stream C_2. This extra pipeline register can be pushed into the multiplier, which is convenient because the multiplier now needs to operate at a higher speed to support interleaving. Interleaving, thus, means up-sampling and pipelining of the computation. It saves area by sharing datapath logic (add, mult). In recursive systems, the total loop latency has to be equal to the number of input sub-channels to ensure correct functionality.

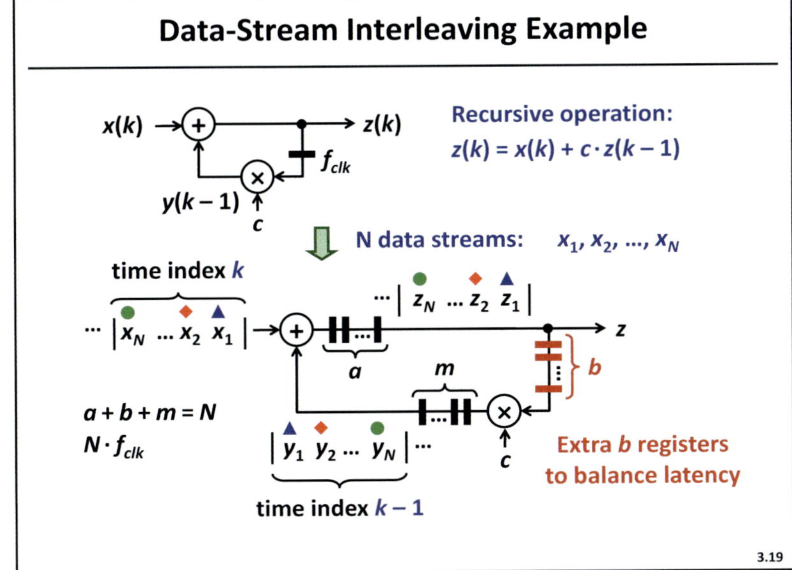

Slide 3.19

The example in this slide is an N data stream generalization of the previous example.

The initial PE that implements $H(z) = 1/(1 - c \cdot z^{-1})$ is now interleaved to operate on N data streams. This implies that $N-1$ additional registers are added to the original single-channel design. Since the design now needs to run N-times faster, the extra registers can be used to pipeline the adder and multiplier. This could result in further performance improvement or energy reduction. The number of pipeline stages pushed into the adder/multiplier is determined in such a way as to keep uniform logic depth for all paths. If, for a given cycle time, N is greater than the number of registers needed to pipeline the adder and multiplier, $b = N - a - m$ extra registers are needed to balance the latency of the design. Data-stream interleaving is applicable for parallel data processing.

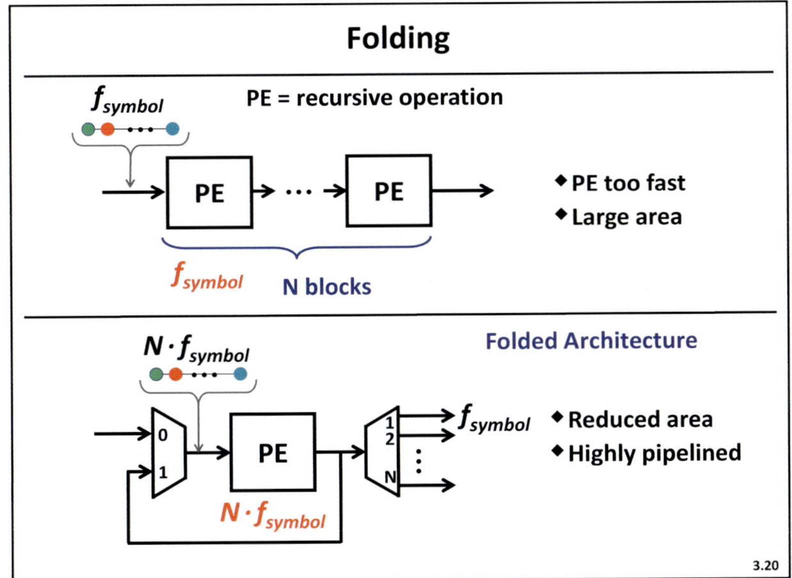

Slide 3.20

Folding, which is similar to data-stream interleaving in principle, is used for time-serial data processing. It involves up-sampling and pipelining of recursive loops, like in data-stream interleaving. Folding also needs an additional data ordering step (the mux in the folded architecture) to support time-serial execution. Like interleaving, folding also reduces area. As shown in the slide, a reference design with N PEs in series is transformed to a folded design with a single PE. This is possible if the PE can run faster than the application requires, so the excess speed can be traded for reduced area. The folded architecture operates at a rate N-times higher than the reference architecture. The input to the PE is multiplexed between the external input and PE's output.

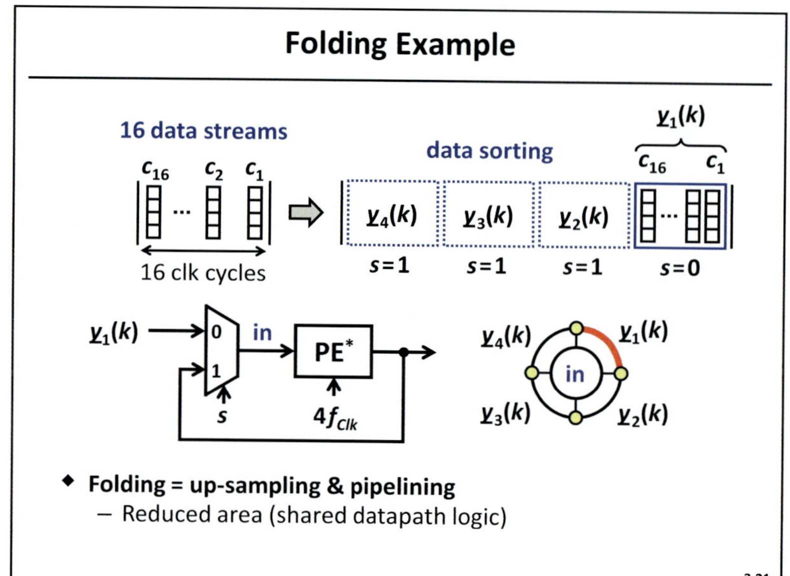

Slide 3.21

This example illustrates folding of 16 data streams representing frequency sub-carriers in a multi-input multi-output (MIMO) baseband DSP. Each stream is a vector of dimensionality four sampled at rate f_{clk}. The PE performs recursive operations on the sub-carriers (PE* indicates extra pipelining inside the PE in order to accommodate all sub-carriers). We can take the output of the PE block and fold it over in time back to its input or select the incoming data stream y_1 by using the life-chart on the right. The 16 sub-carriers, each carrying a vector of real and imaginary data, are sorted in time and space, occupying 16 consecutive clock cycles to allow folding by 4. This amount of folding corresponds to four antennas in a MIMO system.

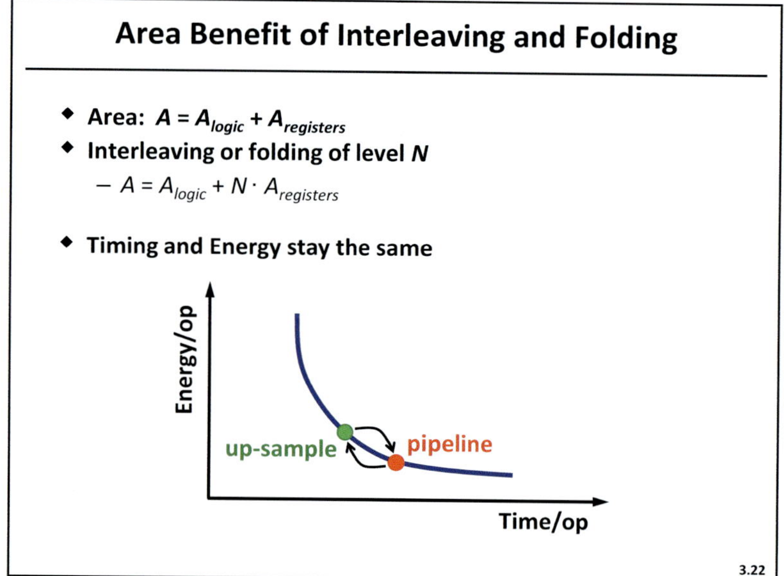

Slide 3.22

Both interleaving and folding introduce pipeline registers to store internal states, but share pipeline logic to save overall area. We can use this simple area model to illustrate area savings. Both techniques introduce more area corresponding to the states, but share the logic area. Timing and energy stay the same because in both cases we do pipelining and up-sampling, which basically brings us back to the starting point. Adding new pipeline registers raises the question of how to optimally balance the pipeline stages (retiming). Retiming will be discussed at the end of this chapter.

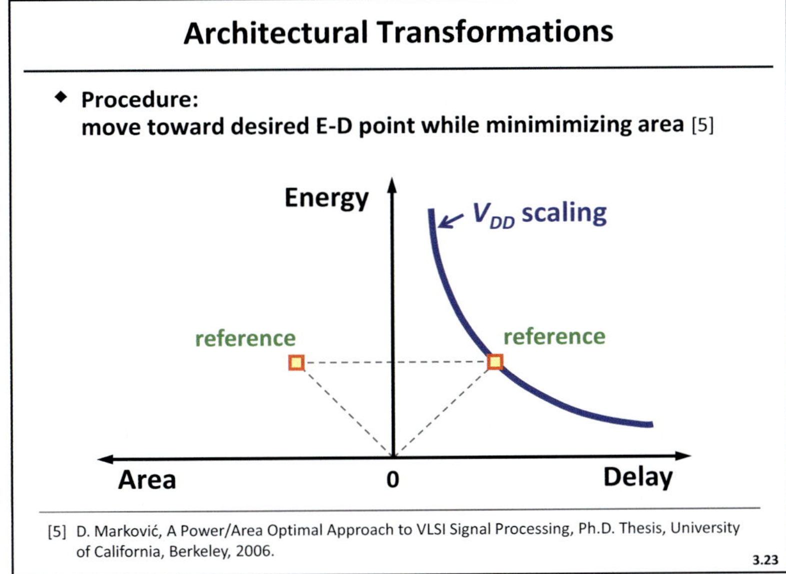

Slide 3.23

The energy-delay tradeoff in datapath logic helps explain architectural transformations. To include area, we plot the energy-area tradeoff beside the energy-delay tradeoff [5]. The energy axis is shared between the energy-area and energy-delay curves. This way, we can analyze energy-delay at the datapath level and energy-area at the micro-architecture level. The energy-delay tradeoff from the datapath logic drives energy-area plots on the left. The E-D tradeoff can be obtained by simple V_{DD} scaling, as shown in this slide. The reference point indicates a starting design for architecture transformations. The energy-area-delay framework shown in this slide is a simple guideline for architectural transformations, which will be described next.

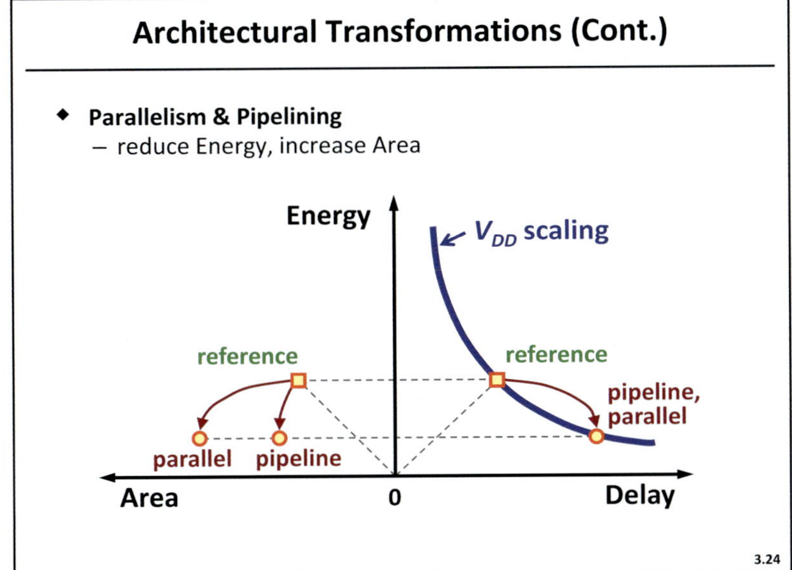

Slide 3.24

Pipelining and parallelism both relax the timing constraint on the pipeline logic and they map roughly to the same point on the energy-delay line. Voltage is scaled down to reduce energy per operation, but the area increases as shown on the left. Area increases more in the parallel design than in the pipeline design. The energy-area tradeoff is very important design consideration the system, because energy relates to the battery life and area relates to the cost of silicon.

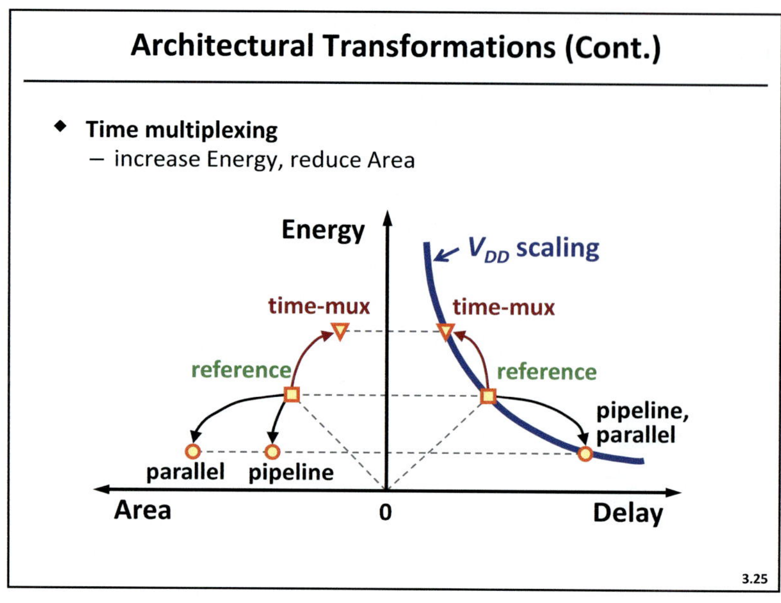

Slide 3.25

In contrast to pipelining and parallelism, time multiplexing requires datapath logic to run faster in order to process many streams sequentially. Starting from the reference design, this means shorter time available for computation. Voltage and/or sizing need to increase in order to achieve faster delay. As a result of resource sharing, the area of the time-multiplexed design is reduced as compared to the reference design. The area reduction comes at the expense of increased energy-per-operation.

Slide 3.26

Interleaving and folding reduce the area for the same energy by sharing the logic gates. Both techniques involve up-sampling and interleaving, so there is no time slack available to utilize and, hence, the supply voltage remains constant. That is why interleaving and folding map back to the reference point in the energy-delay space, but move toward reduced area in the energy-area space.

We can use these architectural techniques to reach a desired energy-delay-area point and this slide shows a systematic way of how to do it.

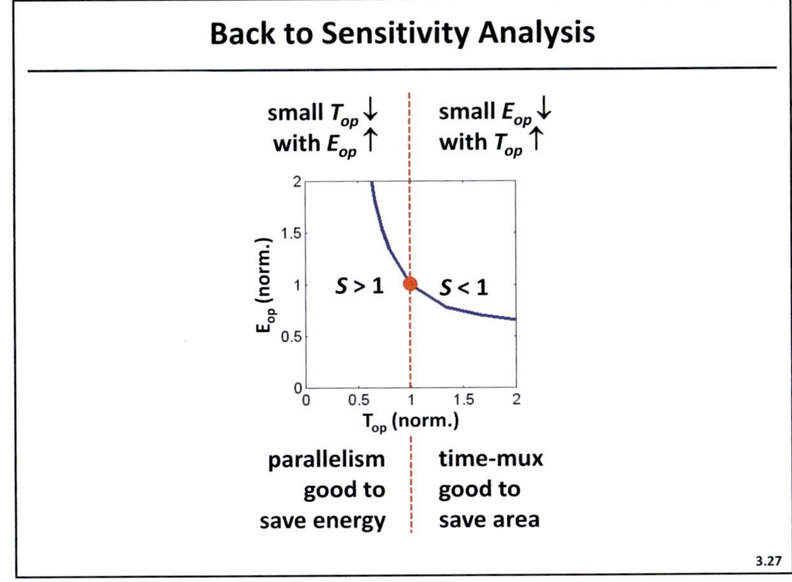

Slide 3.27

We can also use the sensitivity curve to decide on which architectural transformation to use. The plot shows energy-per-operation (E_{op}) versus time-per-operation (T_{op}) for an ALU design. The reference design is made such that energy-delay sensitivities are balanced so that roughly 1% of energy increase corresponds to 1% of delay reduction (and vice versa). This point has sensitivity $S = 1$.

In the $S > 1$ region, a small decrease in time-per-operation costs a significant amount of energy. Hence, it would be better to move back to longer T_{op} by using parallelism/pipelining and save energy.

In the $S < 1$ region, a small reduction in energy corresponds to a large increase in delay. Thus we can use time multiplexing to move to a slightly higher energy point, but save on area.

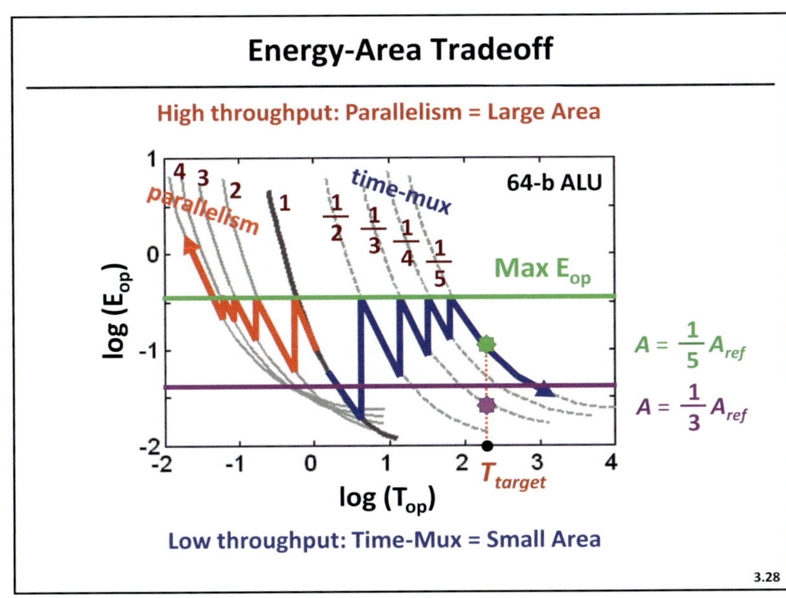

Slide 3.28

Here is another way to look into area-energy-performance tradeoff. This graphs plots energy-per-operation versus time-per-operation for various implementations of a 64-bit ALU. Reference, parallel and time-multiplexed designs are shown. The numbers indicate the relative area of each design (2–4 for parallelism, 1/5-1/2 for time multiplexing) as compared to the reference design ($A_{ref} = 1$).

Suppose that energy is limited to E_{op} (*green line*). How fast can we operate? We can see that parallelism is a good technique for improving throughput while time multiplexing is a good solution for low throughput when the required delay per operation is long.

When a target performance (T_{target}) is given, we choose the architecture with the lowest area that satisfies the energy budget. For the max E_{op} given by the green line, we can use time multiplexing level 5 that has $1/5$ A_{ref}. However, if the maximum E_{op} is limited to the purple line, then we can only use time multiplexing by level 3, so the area increases to $1/3$ A_{ref}.

Therefore, parallelism is good for high-throughput applications (requires large area) and time-multiplexing is good for low-throughput applications (requires small area). The "high" and "low" are relative to the speed of technology.

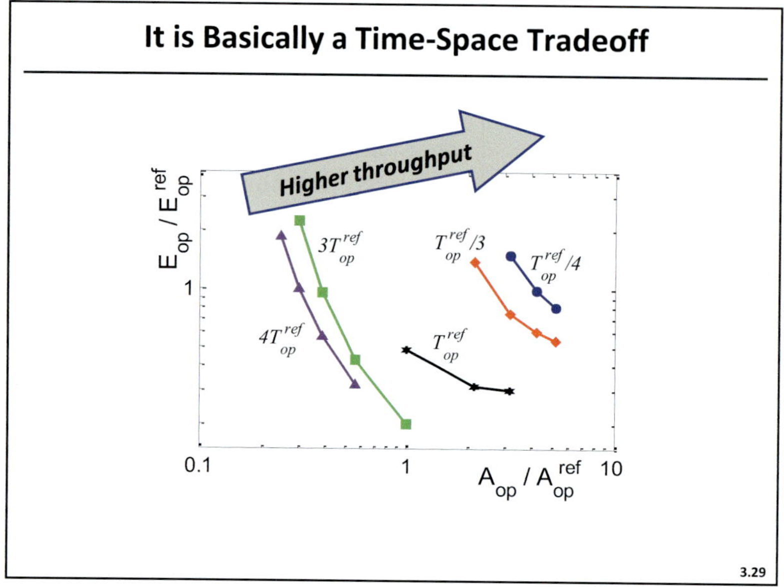

Slide 3.29

Essentially, by employing time-multiplexing and parallelism, we are trading off energy for area. This plot shows the energy-area tradeoff for different throughput requirements. The lines show points of fixed throughput ranging from $T_{op}^{ref}/4$ to $4T_{op}^{ref}$. Higher throughput means larger area (the use of parallelism). For any given throughput, an energy-area tradeoff can be made by employing various levels of time-multiplexing (low area, high energy) or parallelism (low energy, high area).

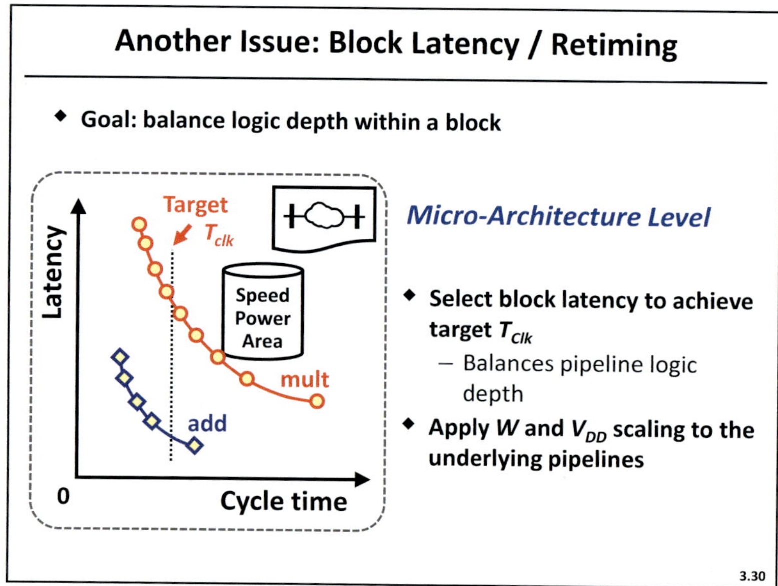

Slide 3.30

An important issue to consider with pipelining-based transformations is how many cycles of latency to allocate to each DSP block. Remember, pipelining helps performance, but we also need to balance logic depth within a block to maximize performance out of our design. This step is called retiming and involves moving existing registers around. This becomes particularly challenging in recursive designs when registers need to be shifted around loops. To simplify the problem, it is very important to assign the correct amount of latency to each block and retime feed-forward blocks.

DSP block characterization for latency vs. cycle time is shown on the slide. In a given design, we have blocks of varying complexity which take different number of cycles to propagate input to output. The increased latency of the complex block comes from the motive to balance pipeline depths for all paths in the block. Once all the logic depths are balanced, then we can globally scale supply voltage on locally sized logic pipelines. In balanced datapaths, design optimization consists of optimizing a chain of gates (i.e. circuit optimization, which we have learned how to do in Chap. 2).

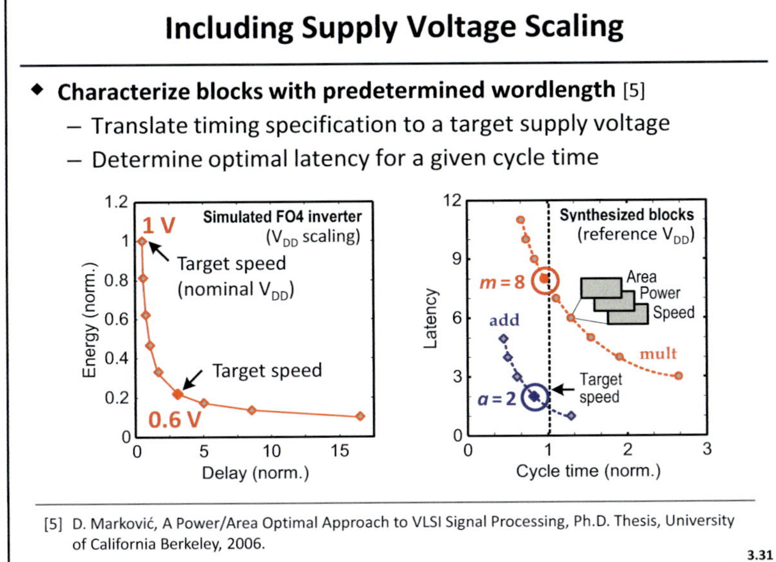

Slide 3.31

Supply voltage scaling is an important variable for energy minimization, so we need to include it in our block characterization flow. This is done by translating timing specifications for logic synthesis to a more aggressive target, so that the design can operate with a reduced voltage. The amount of delay vs. voltage scaling is estimated from gate-level simulations [5].

On the left, we have a graph that plots an E-D curve for a fanout-4 inverter for a target technology. From synthesis, we have the cycle time and latency as shown on the right. Given a cycle time target, we select the latency for all building blocks. For instance, in the plot shown on the slide, for a unit cycle time (normalized), the multiplier has a latency of 8 and the adder has a latency of 2 in order to meet the delay specification. Now, suppose that the target delay needs to be met at supply voltage of 0.6 V. Since libraries are characterized at a fixed (reference) voltage, we need to account for the delay margin in synthesis. For example, if the cycle time of adder at 0.6 V is found to be four units, and if the delay at 0.6 V is 5x the delay at 1 V, then the target cycle time for synthesis at 1 V would be assigned as $4/5 = 0.8$ units. This way, we can incorporate supply voltage optimization within constraints of chip synthesis tools.

Summary

- **Architecture parallelism and pipelining can be used to reduce power (improve energy efficiency) by voltage scaling**
  - Equivalently, performance can be improved for the same energy per operation
- **Performance-centric designs (that maximize BIPS) require shorter (fewer FO4 stages) logic pipelines**
- **Energy-performance optimal designs have about equal leakage and switching components of energy**
  - Otherwise, one can be traded for another for further energy reduction
- **Architecture techniques for direct (parallelism, pipelining) and recursive (interleaving, folding) systems can be analyzed in area-energy-performance plane for compact comparison**
  - Latency (number of pipeline stages) is dictated by cycle time

3.32

Slide 3.32

Architecture techniques for direct and recursive algorithms are presented. Techniques of parallelism and pipelining can be used to reduce energy for the same performance or, equivalently, improve performance for the same level of energy per operation. Study of a practical processor showed that performance-centric designs require shallower pipelines while power-centric designs require deeper pipelines. Optimization of energy subject to a delay constraint showed that optimal designs have balanced leakage and switching components. This is intuitively clear because otherwise one could trade one type of energy for another to achieve further energy reduction. Techniques of interleaving and folding involve up-sampling and pipelining of recursive loops to share logic area without impacting power and performance. Architectural examples emphasizing the use of parallelism and time-multiplexing will be studied in the next chapter to show energy and even area benefits of parallelism.

References

- A.P. Chandrakasan, S. Sheng, and R.W. Brodersen, "Low-Power CMOS Digital Design," *IEEE J. Solid-State Circuits,* vol. 27, no. 4, pp. 473-484, Apr. 1992.

- D. Marković *et al.,* "Methods for True Energy-Performance Optimization," *IEEE J. Solid-State Circuits,* vol. 39, no. 8, pp. 1282-1293, Aug. 2004.

- V. Srinivasan *et al.,* "Optimizing Pipelines for Power and Performance," in *Proc. IEEE/ACM Int. Symp. Microarchitecture,* Nov. 2002, pp. 333-344.

- R.W. Brodersen *et al.,* "Methods for True Power Minimization," in *Proc. Int. Conf. on Computer Aided Design,* Nov. 2002, pp. 35-42.

- D. Marković, A Power/Area Optimal Approach to VLSI Signal Processing, Ph.D. Thesis, University of California, Berkeley, 2006.

Additional References

- K. Nose and T. Sakurai, "Optimization of V_{DD} and V_{TH} for Low-Power and High-Speed Applications," in *Proc. Asia South Pacific Design Automation Conf.,* (ASP-DAC) Jan. 2000, pp. 469-474.

- V. Zyuban and P. Strenski, "Unified Methodology for Resolving Power-Performance Tradeoffs at the Microarchitectural and Circuit Levels," in *Proc. Int. Symp. Low Power Electronics and Design,* Aug. 2002, pp. 166-171.

- H.P. Hofstee, "Power-Constrained Microprocessor Design," in *Proc. Int. Conf. Computer Design,* Sept. 2002, pp 14-16.

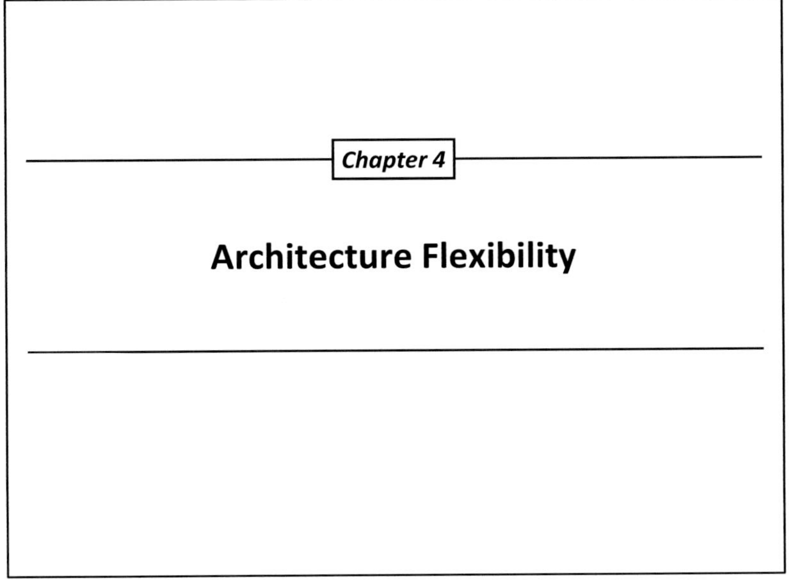

Slide 4.1

This chapter studies architecture flexibility and its implication on energy and area efficiency. Having a flexible architecture would be nice. It would be convenient if we could design a chip and program it to do whatever it needs to do. There would be no need for optimizations prior to any design decisions. What is the cost of flexibility? What are we giving up? How much more area, power, etc?

This chapter provides answers to those questions. First, energy- and area-efficiency metrics for DSP computations will be defined, and then we will study how architecture affects energy and area efficiency. Examples of general-purpose microprocessors, programmable DSPs, and dedicated hardware will be compared. The comparison for a number of chips will show that architecture parallelism is the key for maximizing energy (and even area) efficiency.

Another Requirement: Architecture Flexibility

- **Determining how much to include and how to do it in the most efficient way possible** [1]

- **Claims (to be shown)**
  - There are good reasons for flexibility
  - The "cost" of flexibility is orders of magnitude of inefficiency over an optimized solution
  - There are many different ways to provide flexibility

[1] R.W. Brodersen, "Technology, Architecture, and Applications," in *Proc. Int. Solid-State Circuits Conf.*, Special Topic Evening Session: Low Voltage Design for Portable Systems, Feb. 2002.

4.2

Slide 4.2

The main issue is determining how much flexibility to include, how to do it in the most efficient way and how much it would cost [1]. There are good reasons to make flexible designs, and we will discuss the cost of doing it in this chapter. However there are different ways to provide flexibility. Flexibility has in some sense become equal to software programmability. There are lots of ways to do flexibility, and software programmability is only one of those.

D. Marković and R.W. Brodersen, *DSP Architecture Design Essentials*, Electrical Engineering Essentials,
DOI 10.1007/978-1-4419-9660-2_4, © Springer Science+Business Media New York 2012

Good Reasons for Flexibility

- **One design for a number of SoC customers – more sales volume**
- **Customers able to provide added value and uniqueness**
- **Unsure of specification or can't make a decision**
- **Backwards compatibility with debugged software**
- **Risk, cost and time of implementing hardwired solutions**

Important to note: these are business, not technical reasons

4.3

Slide 4.3

There are several good reasons for flexibility:

A flexible design can serve multiple applications. By selling one design in large volume, the cost of the layout masks can be greatly reduced. This applies to Intel's and TI's processors, for instance.

Another reason is that we are unsure of specifications or can't make a decision. This is the reason flexibility is one of the key factors for cell phone companies, for example. Flexibility is so important because they have to delay the decisions until the last possible minute. People who are making the decisions on what the features, standards, etc. will be won't make a decision until they absolutely have to. If the decision can be delayed until software design stage, then the most up-to-date version would be available and that will have the best sales.

Despite the fact that dedicated design could provide orders of magnitude better energy efficiency than DSP processors, the DSP parts still sell in large volumes (on the order of a million chips a day). For the DSPs the main reason flexibility is so important is backwards compatibility. Software is so difficult to do, so difficult to verify, that once it is done, people don't want to change it. It actually turns out that software is not flexible and easy to change. Customers are attracted by the idea of getting new hardware that is compatible with legacy code.

Building dedicated chips is very expensive. For a 90-nm technology, the cost for a mask set is over $1 million and even more for advanced technologies. If designers do something that may potentially be wrong, that scares people.

So, What is the Cost of Flexibility?

- **We need technical metrics that we can use to compare flexible and non-flexible implementations**
    - A power metric because of thermal limitations
    - An energy metric for portable operation
    - A cost metric related to the area of the chip
    - Performance (computational throughput)

Let's use metrics normalized to the amount of computation being performed – so now lets define computation

4.4

Slide 4.4

We need some metrics to determine the cost of flexibility.

Let's use a power metric, which is for thermal limitations. This is how many computations (number of atomic operations) we do per mW.

An energy metric tells us how much a set of operations cost in energy. In other words, how much we can do with one battery. The energy is measured in the number of operations per Joule. Energy issue is a battery issue, whereas operations/mW is a power issue.

We also need a cost metric, which is equal to the area of a chip.

Finally, performance requirements need to be met. We will assume that designs meet performance metrics and we look at the other metrics.

<table>
<tr><td>

Definitions

Computation
- Operation = OP =algorithmically interesting computation (i.e. multiply, add, delay)
- MOPS = Millions of OP's per Second
- N_{op} = Number of parallel OP's in **each** clock cycle

Power
- P_{chip} = Total power of chip = $A_{chip} \cdot C_{sw} \cdot V_{DD}^2 \cdot f_{clk}$
- C_{sw} = Switched Capacitance / mm^2
  = $P_{chip} / (A_{chip} \cdot V_{DD}^2 \cdot f_{clk})$

Area
- A_{chip} = Total area of chip
- A_{op} = Average area of each operation = A_{chip}/N_{op}

4.5

</td><td>

Slide 4.5

The metrics are going to be defined around operations. An operation is an algorithmically interesting computation, like a multiply, add or delay. It is related to the function an algorithm performs. This is different from an instruction. Often we talk about MIPS (millions of instructions per second) and we categorize complexity by how many instructions it takes. Instructions are tied to a particular implementation, a particular architecture. If we talk about operations, we want to get back to

</td></tr>
</table>

what we are trying to do (algorithm), not to the particular implementation. So, in general it takes several instructions to do one operation. We are going to base ourselves around operations.

MOPS is millions of OP/sec (OP = operation). This is a rate at which operations are being performed.

N_{op} is number of parallel operations per clock cycle. This is an important number because it tells us the amount of parallelism in our hardware. For each clock cycle, if we have 100 multipliers and they each contribute to an algorithm, we have an N_{op} of 100 in that clock cycle, or that chip has a parallelism of 100.

P_{chip} is the total power of the chip, the area of the chip * C_{sw} (C_{sw} is switched capacitance, some average value of capacitance and its activity that's being switched each cycle per unit area). Some parts of the chip will be changing rapidly, so all that capacitance will be linearly factored in. Other parts will be changing slowly, so that capacitance will be weighted down, because the activity factor is lower. C_{sw} = activity factor * Capacitance of gates, as explained in Chap. 1.

C_{sw} is equal to switched capacitance/mm^2. Solving for C_{sw} yields power of the chip divided by area of the chip, divided by V_{DD}^2, divided by f_{clk}. C_{sw} is the average capacitance over the chip. To find the power of the chip, C_{sw} needs to multiply area of the chip, the clock rate, and V_{DD}^2. The question is how C_{sw} changes between different variations of the design?

A_{chip} is the total area of the chip.

A_{op} is the average area of each operation. So if we take N_{op} (the number of parallel operations in each clock cycle), divide that by the area of the chip, this tells us how much area each parallel operation takes.

The idea is to figure out what the basic metrics are so we can see how good a design is. How do we compare different implementations and see which ones are good?

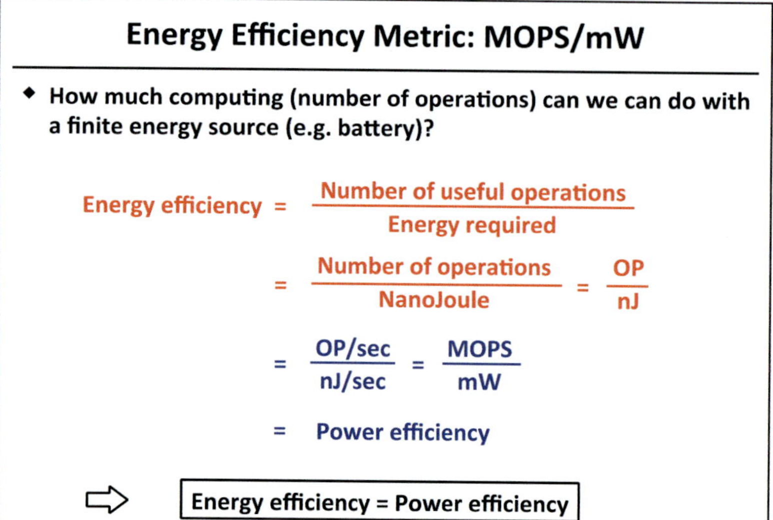

Slide 4.6

Energy efficiency is the number of useful OPs divided by the energy required to do them. So if we have 1000 operations, we know how many nJ it takes to do them, and therefore we can calculate the energy efficiency. Energy efficiency is the OP/nJ, the average number of Joules per operation. Joules are power * time (Watt * s). OP/sec divided by nJ/sec gives MOPS/mW. Therefore, if we calculate OP/nJ, it is exactly the same as MOPS/mW. This implies that the energy metric is exactly the same as the power metric. Therefore, power efficiency and energy efficiency are actually the same.

Slide 4.7

Now we can compare different designs and answer the question of how many mW does it take to do one million operations or how many nJ per operation.

ISSCC Chips (0.18μm – 0.25μm)

- Chips published at ISSCC over a 5-year span

Chip	Year	Paper	Description
1	1997	10.3	S/390
2	2000	5.2	PPC
3	1999	5.2	G5
4	2000	5.1	Alpha
6	1998	15.4	P6
7	1998	18.4	Alpha
8	1999	5.6	PPC

Chip	Year	Paper	Description
9	1998	18.6	Strong-Arm
10	2000	4.2	Comm.
11	1998	18.1	Graphics
12	1998	18.2	Multimedia
13	2000	14.6	Multimedia
14	2002	22.1	MPEG Dec.
15	1998	18.3	Multimedia
16	2001	21.2	Encryption
17	2000	14.5	Hearing Aid
18	2000	4.7	FIR
19	1998	2.1	MPEG Dec.
20	2002	7.2	802.11a

Chip type:
Microprocessor
General purpose DSP
Dedicated design

Slide 4.8

This slide summarizes a number of chips from ISSCC, top international conference in chip design, over a 5-year period (1996–2001). Only chips for 0.18 μm to 0.25 μm were analyzed (because technology will have a big effect on these numbers). In this study, there were 20 different chips. The chips which had all the numbers needed for calculating C_{sw}, power, area and OPs were chosen. Eight of the chips were microprocessors; the DSPs are another set of chips – software programmable with extra hardware to support the particular domain they were interested in (multimedia, graphics, etc); the final group is dedicated chips – hard-wired, do-one-function chips. How does MOPS/mW change over these three classes?

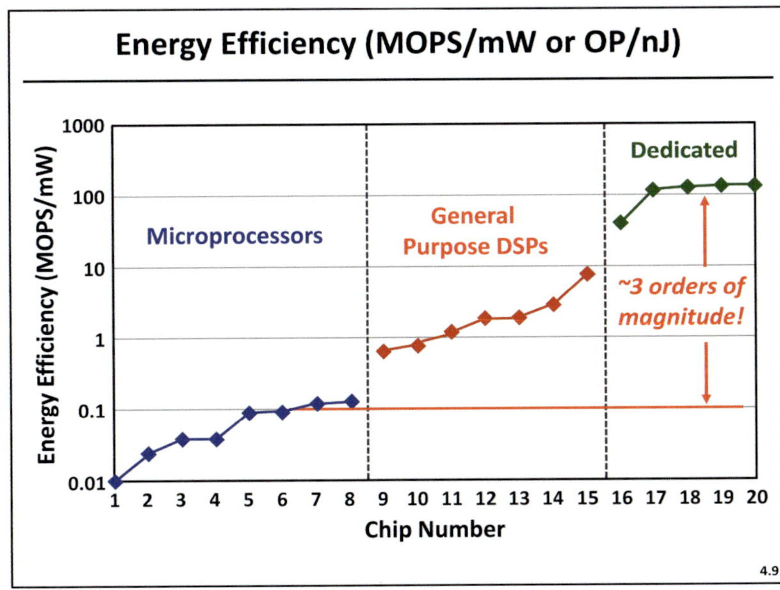

Slide 4.9

The y-axis is energy efficiency, in MOPS/mW or nJ/op on a log scale, and the x-axis indicates the chip analyzed. The numbers range from 0.01 to about 0.1 MOPS/mW for microprocessors. The general purpose DSPs (still software programmable, so quite flexible), are about 10 to 100 times more efficient than the general purpose microprocessors, so this parallelism is having quite an effect – a factor of 10. The dedicated chips are going up another couple orders of magnitude. Overall, we can observe a factor of 1000 between dedicated chips and software-programmable chips. This number is sometimes intuitively clear when people say there are just some things you don't do in software. For a really fast bit manipulation, for example, you would not think to do it in software.

We had many good reasons for software programmability, but at a cost higher by a factor of 1000. We can't be trading off things that are so different. There are other reasons for doing high levels of flexibility, but it is not going to come from engineers optimizing the area or energy efficiency. The numbers are too different.

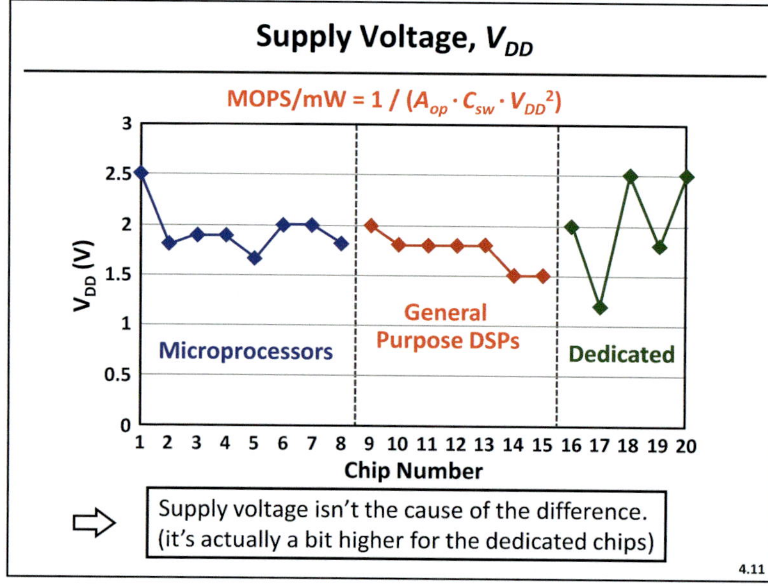

Slide 4.10

Let's look at the components of MOPS/mW. Operations per second: N_{op} is the number of parallel operations per clock cycle. If we want to look at MOPS, we take the clock rate times the number of parallel operations. Parallelism is the key to understanding energy and area efficiency metrics.

Power of the chip is equal to the same formula we had in Slide 4.5.

If we put in the values for MOPS and take that ratio (MOPS/P_{chip}), we end up with 1/(Area per operation per clock cycle * C_{sw} * V_{DD}^2). (A_{op} * C_{sw}) is the amount of switched capacitance per operation. For example, consider having 100 multipliers on a chip. The chip area divided by 100 will give A_{op}. Take C_{sw} (area cap/unit area for that design), that gives the average switched cap per op, times V_{DD}^2 would be the average power per op. 1 over that is the MOPS/mW.

Let's look at the three components, V_{DD}, C_{sw}, and A_{op}, and see how they change between these three architectures.

Slide 4.11

Supply voltage could be lower for the general purpose microprocessor than for the dedicated chips. The actual reason is that parts run faster if they are run at higher voltages. If microprocessors run at higher voltages, the chips would burn up. Microprocessors run below the voltages they can be run at. Dedicated chips use all the voltage they can get because they don't have a power problem in terms of heat sinking on a chip. Therefore, it is not V_{DD}^2 that is causing the different designs to be so different.

From Chap. 3, you might think it is voltage scaling that is making the difference. It is not.

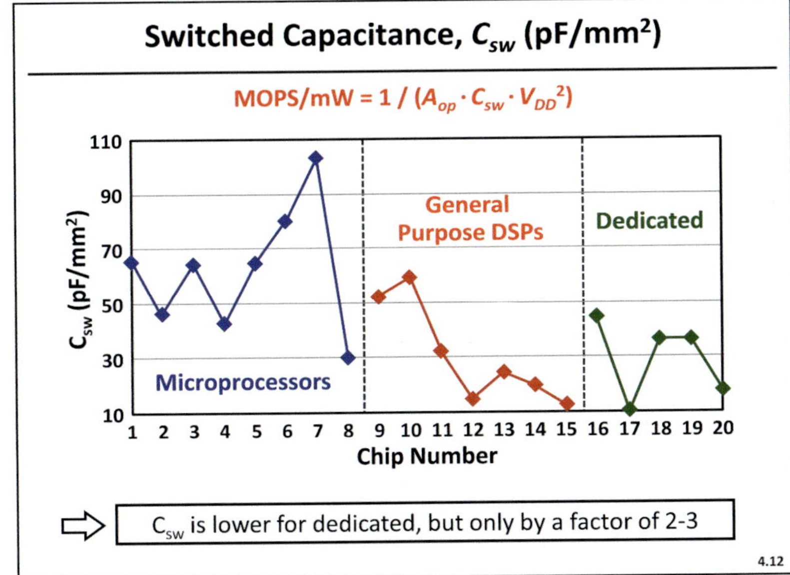

Slide 4.12

What about C_{sw}? It is the capacitance per unit area * activity factor. The CPU is running very fast, which gets hot, but the rest of the chip is memory. The activity factor of that memory would be very low, since you are not accessing all that memory. However, C_{sw} for the microprocessor is actually higher, despite the fact that most of the chip is not doing much. Dedicated chips have the lowest C_{sw}, despite the fact that all the chips are executing every cycle. Why? Essentially, a microprocessor is time multiplexing to do lots of tasks and there is big overhead associated with time multiplexing. To run logic at 2 GHz, there are clocks that are buffered, big drivers (the size of the drivers is in meters), and huge transistors. That leads to big capacitance numbers. With time multiplexing, high-speed operation, and long busses, as well as the necessity to drive and synchronize the data, a lot of power is required.

What about fanout, where you drive buses that end up not using that data? When we look at microprocessor designs of today that try to get more parallelism, we see that they do speculative execution, which is a bad thing for power. You spawn out 3 or 4 guesses for what the next operation should be, perform those operations, figure out which one is going to be used, and throw away the answer to the other 3. In an attempt to get more parallelism, the designers traded off power. When power limited, however, designers have to rethink some of these techniques. One idea is the use of hardware accelerators. If we have to do MPEG decoding, for example, then it is possible to just use an MPEG decoder in hardwired logic instead of using a microprocessor to do that.

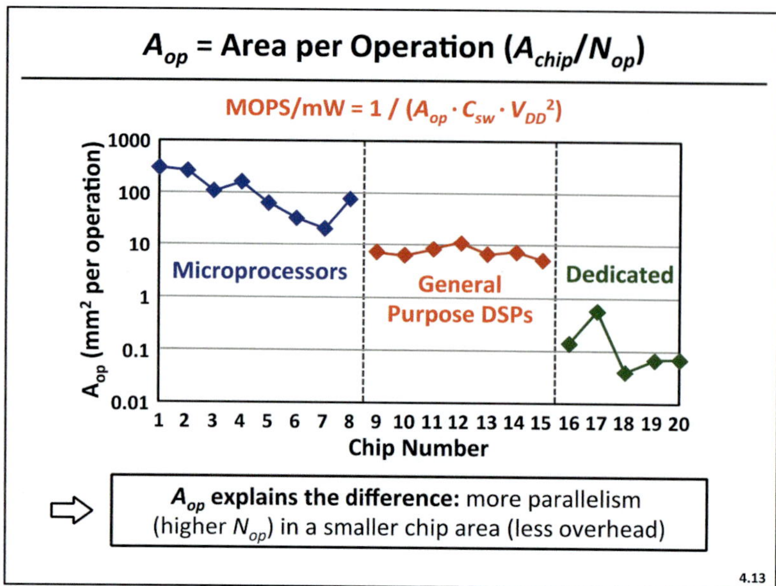

Slide 4.13

So if it is not V_{DD} or C_{sw}, it has to be the A_{op}. A microprocessor can only do three or four operations in parallel. We took the best numbers – assuming every one of the parallel units is being executed. A_{op} is equal to 100's of mm^2 per operation for a microprocessor, because you only get one operation per clock cycle. Dedicated chips give 10's or 100's of operations per clock cycle, which brings A_{op} down from hundreds of mm^2 per op down to 1/10 of mm^2 per op. It is the parallelism achievable in a dedicated design that

makes them so efficient.

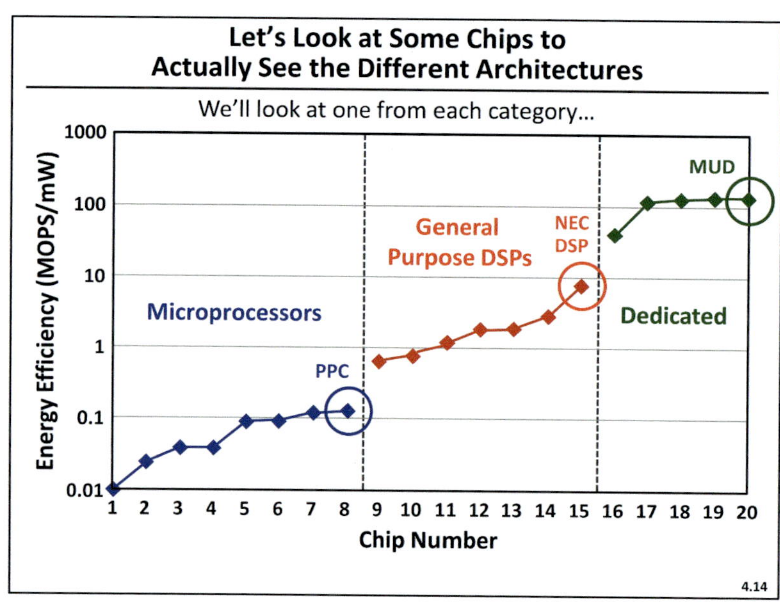

Slide 4.14

Energy is not everything. The other part is cost. Cost is equal to area. How much area does it take to do a function? What is the best way to reduce area? It may seem that the best way to reduce area is to take a bit of area and time multiplex it, like in the von Neumann model. When von Neumann built his architecture over 50 years ago, his model of hardware was that an adder consumed a rack. A register was another rack. If that is a model of hardware, you don't think about parallelism. Instead, you think about time sharing each piece of hardware. The first von Neumann computer was a basement, and that was just one ALU. Chips are cost/mm^2. Cost is more often more important than energy. So wouldn't the best way to reduce cost be to time multiplex?

Let's look at Power PC, the NEC DSP chip, and the MUD chip (multi-user detection chip done at Berkeley) to gain some insight into the three architectures.

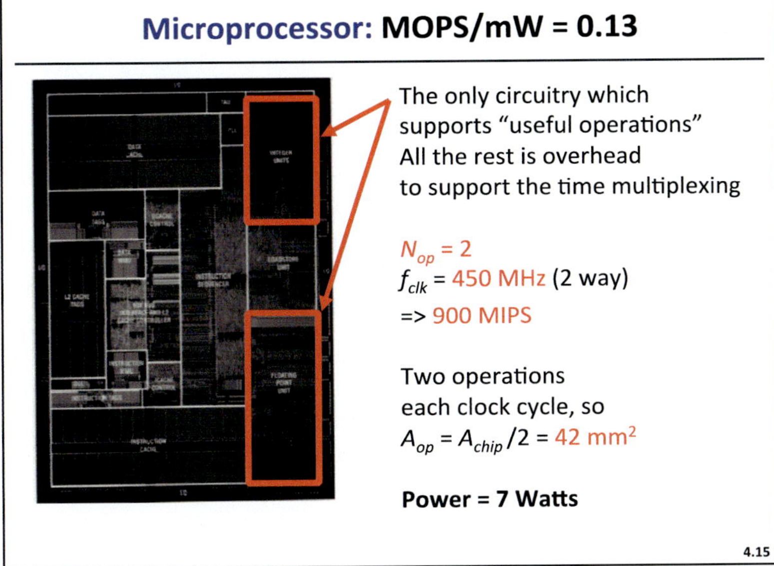

Slide 4.15

Let's look at the Power PC. MOPS/mW = 0.13. The processor is a 2-way superscalar consisting of an integer unit and a floating-point unit, so it executes 2 operations each cycle. The clock rate is 450MHz, so the number of real-time instructions is 900 MIPS. Here, we have blurred the line between instructions and operations. Let's say instruction = operation for this microprocessor. A_{op} = 42 mm^2 per operation.

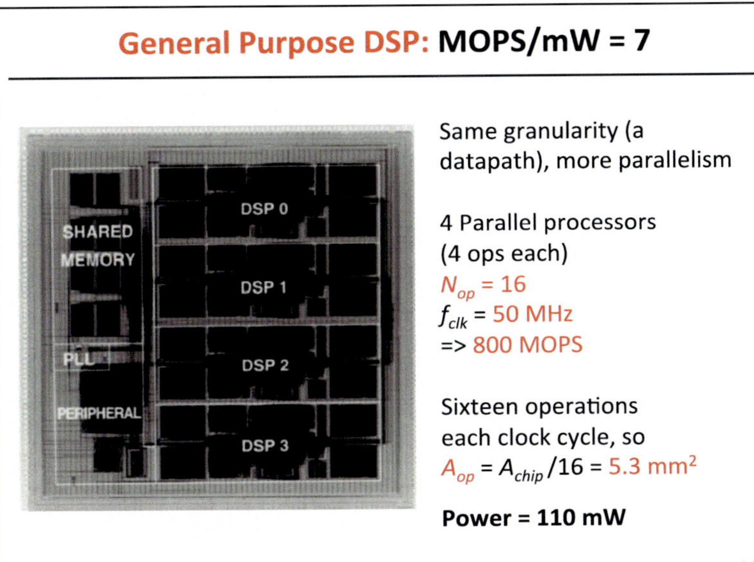

Slide 4.16

Let's take the NEC DSP chip, which has MOPS/mW of 7 (70 times that of the Power PC). It has 4 DSP units, each of which can do 4 parallel operations. Therefore, it can do 16 ops/clock cycle, which is equivalent to 8 times more parallelism than the PPC. The NEC DEC chip has a 50 MHz clock, but it can do 800 MOPS, whereas for the PPC with a 450 MHz clock, it can do 900 MOPS. That is the power of parallelism. A_{op} is 5.3 mm^2. Area efficiency and energy efficiency are a lot higher for this chip. We see some shared memory on the chip. This is a wasted area, just passing data back and forth between these parallel units. Memories are not very efficient. 80% of the chip is operational, whereas in the microprocessor only a fraction of the chip building blocks (see previous slide) are doing operations.

Dedicated Design: MOPS/mW = 200

Slide 4.17

For the dedicated chip, there are 96 ops per clock cycle. At 25 MHz, this chip computes 2400 MOPS. We have dropped the clock rate but ended up with three times more throughput, so clock rate and throughput have nothing to do with each other. Reducing clock rate actually improves throughput because more area can be spent to do operations. A_{op} is 0.15 mm^2, a much smaller chip. Power is 12 mW. The key to lower power is parallelism.

The Basic Problem is Time Multiplexing

- Processor architectures obtain performance by increasing the clock rate, because the parallelism is low [2]
- Results in ever increasing memory on the chip, high control overhead and fast area consuming logic

But doesn't time multiplexing give better area efficiency?

[2] T. A.C.M. Claasen, "High Speed: Not the Only Way to Exploit the Intrinsic Computational Power of Silicon," in *Proc. IEEE Int. Solid-State Circuits Conf.*, Feb. 1999, pp. 22-25.

4.18

Slide 4.18

Parallelism is probably the foremost architectural issue – how to provide flexibility and still retain the area/energy efficiency of dedicated designs. A company called Chameleon systems began working on this problem in the early 2000's and won many awards for their architecture. FPGAs are optimized for random logic, so why not optimize for higher logic – adders, etc? They were targeting the TI DSP chips and their chips didn't improve fast enough. They had just the right tools to get flexibility for this DSP area that works better than an FPGA or TI DSP – they still did not make it. Companies that do the general purpose processors like TI know their solution isn't the right path, and they're beginning to modify their architecture, make dedicated parts, move towards dedicated design, etc. There should probably be a fundamental architecture like an FPGA that is reconfigurable that somehow addresses the DSP domain – that is flexible yet efficient. It is a big time opportunity.

We would argue that the basic problem is time multiplexing. To try to use the same architecture over and over again is at the root of the problem [2]. It is not that it shouldn't be used in some places, but to use it as the main basic architectural strategy seems to be overdoing it.

Area Efficiency

- SOC based devices are often very cost sensitive
- So we need a $ cost metric => for SOC's that is equivalent to the efficiency of area utilization
- Area-efficiency metric:
  Computation per unit area = MOPS/mm²

How much of a $ cost (area) penalty will we have if we put down many parallel hardware units and have limited time multiplexing?

4.19

Slide 4.19

Let's look at MOPS/mm^2, area efficiency. This metric is equivalent to the chip cost. You may expect that parallelism would increase cost, but that is not the case. Let's look at the same examples and compare area efficiency for processors, DSP chips, and dedicated chips.

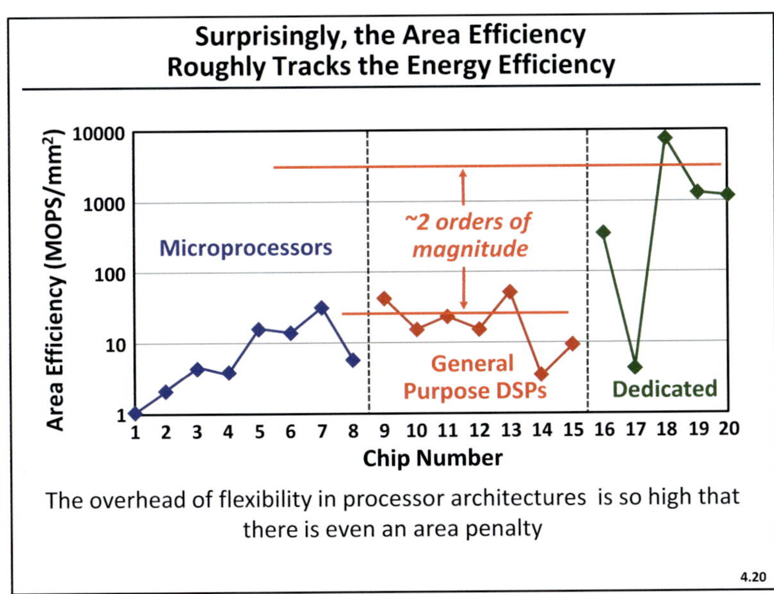

Slide 4.20

You would think that with all that time multiplexing you would get more MOPS/mm^2. The plot shows area efficiency on a log scale for the example chips. Microprocessors achieve around 10 MOPS/mm^2. DSPs are getting a little better, but not much. Dedicated, some designs do a lot better, some don't. The low point is a hearing aid that uses a very low voltage, so this gets very low area efficiency due to reduced processing speed. In general, it is orders of magnitude more area efficient to use parallelism and time multiplex very little. That is also very surprising. You would think the ALU, overdesigned to be 32 or 64 bits, while the dedicated chips are 12 or 16 bits, would be the most area efficient. What causes the area inefficiency in microprocessors? To be more flexible you have to pay an area cost. Signals from the ALU need to get to different places so busses are ran all over the place whereas in dedicated you know where everything is. In flexible designs, you also have to store intermediate values. The CPU is running very fast and you have to feed data constantly into it. There is all sorts of memory all over that chip. Memory is not as fast as the logic, so we put in caches and cache controllers. A major problem is that technology improves the speed of logic much more rapidly than it does memory. All the time multiplexing and all the controllers create lots of area overhead.

The overhead of flexibility in processor architectures is so high that there are about two orders of magnitude of area penalty as compared to dedicated chips.

Hardware / Software

Conclusion:

There is no software/hardware tradeoff.

- **The difference between hardware and software in performance, power and area is so large that there is no "tradeoff".**
- **It is reasons other than power, energy, performance or cost that drives a software solution (e.g. business, legacy, …).**
- **The "Cost of Flexibility" is extremely high, so the other reasons better be good!**

4.21

Slide 4.21

There is no hardware/software tradeoff. The cost of flexibility is extremely high. What we want is something more flexible that can do a lot of different things that is somehow near the efficiency of these highly optimized parallel solutions. Part of the solution has to do with high levels of parallelism.

References

- R.W. Brodersen, "Technology, Architecture, and Applications," in *Proc. IEEE Int. Solid-State Circuits Conf.,* Special Topic Evening Session: Low Voltage Design for Portable Systems, Feb. 2002.

- T. A.C.M. Claasen, "High Speed: Not the Only Way to Exploit the Intrinsic Computational Power of Silicon," in *Proc. IEEE Int. Solid-State Circuits Conf.,* Feb. 1999, pp. 22-25.

Part II

DSP Operations and Their Architecture

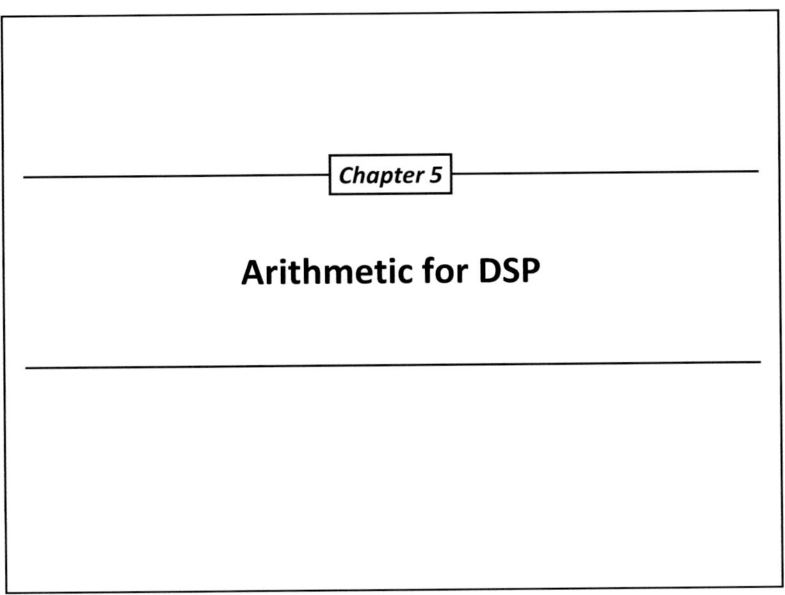

Slide 5.1

This chapter reviews number representation and DSP arithmetic. It starts with floating-point number representation. Fixed-point representations are then introduced, as well as related topics such as overflow and quantization modes. Basic implementations of add and multiply operations are shown as a baseline for studying the impact of micro-architecture on switching activity and power.

Chapter Overview

- **Number systems**
- **Quantization effects**
- **Data dependencies**
- **Implications on power**
- **Adders and multipliers**

5.2

Slide 5.2

The chapter topics include quantization effects, floating-point and fixed-point arithmetic. Data dependencies will be exploited for power reduction, which will be illustrated on adder and multiplier examples. Adders and multipliers are core operators of DSP algorithms. We will look into gate-level implementation of these operations and analyze their performance and energy.

D. Marković and R.W. Brodersen, *DSP Architecture Design Essentials*, Electrical Engineering Essentials,
DOI 10.1007/978-1-4419-9660-2_5, © Springer Science+Business Media New York 2012

Number Systems: Algebraic

Algebraic Number

e.g. $a = \pi + b$

[1]

- **High-level abstraction**
- **Infinite precision**
- **Often easier to understand**
- **Good for theory/algorithm development**
- **Hard to implement**

[1] C. Shi, Floating-point to Fixed-point Conversion, Ph.D. Thesis, University of California, Berkeley, 2004.

5.3

Slide 5.3

Let's start the discussion with a high-level symbolic abstraction of the variables [1]. For example, variable a is the sum of the constant π and variable b, $a = \pi + b$. This is a notation with infinite precision. It is convenient for symbolic calculus, and it is often very easy to understand. Although such a representation is suitable for algorithm development, it is impractical for physical realization. The hardware cost to implement the calculation of a can vary greatly depending on the desired level of accuracy. Hence, we need to study various options and their practical feasibility.

Number Systems: Floating Point

- **Widely used in CPUs**
- **Floating precision**
- **Good for algorithm study and validation**

$$Value = (-1)^{Sign} \times Fraction \times 2^{(Exponent - Bias)} \qquad [2]$$

IEEE 754 standard	Sign	Exponent	Fraction	Bias
Single precision [31:0]	1 [31]	8 [30:23]	23 [22:0]	127
Double precision [63:0]	1 [63]	11 [62:52]	52 [51:00]	1023

[2] J.L. Hennesy and D.A. Paterson, Computer Architecture: A Quantitative Approach, (2nd Ed), Morgan Kaufmann, 1996.

5.4

Slide 5.4

Floating point is a commonly used representation in general processors such as CPUs. It has very high precision and serves well for algorithm study and validation. There are several standards for floating-point numbers; the table on this slide [2] shows the IEEE 754 standard. The value is represented using a sign, fraction, exponent, and bias bits. The single-precision format uses 32 bits, out of which 1 is used for the sign, 8 for the exponent and the remaining 23 for the fraction. A bias of 127 is applied to the exponent. The double-precision format requires 64 bits and has greater accuracy.

Example of Floating-Point Representation

$$Value = (-1)^{Sign} \times Fraction \times 2^{(Exponent - Bias)}$$

A non-IEEE-standard floating point

$$\pi = 0\,1\,1\,0\,0\,1\,0\,1\,0\,1$$

Sign   Frac   Exp   Bias = 3

- **Calculate π**

$$\pi = (-1)^0 \times (1\times 2^{-1} + 1\times 2^{-2} + 0\times 2^{-3} + 0\times 2^{-4} + 1\times 2^{-5} + 0\times 2^{-6})$$
$$\times\, 2^{(1\times 2^2 + 0\times 2^1 + 1\times 2^0 - 3)} = 3.125$$

- **Very few bits are used in this representation, which results in low accuracy (compare to actual value π = 3.141592654…)**

5.5

Slide 5.5

As an example of floating-point representation, we look at the 10-bit representation of the constant π. The sign bit is zero, since π is positive. The fractional and exponent parts are computed using weighted bit-sums, as shown in the first bullet. The truncated value for π in this representation is 3.125. Since a small number of bits are used in this representation, the number deviates from the accurate value (3.141592654…) even at the second decimal point. The IEEE 754 standard formats with 32 or 64 bits would provide a much higher precision.

Floating-Point Standard: IEEE 754

- **Property #1**
  - Rounding a "half-way" result to the nearest float (picks even)

  Example:
  6.1 × 0.5 = 3.05   (base 10, 2 digits)

  even 3.0   3.1   (base 10, 1 digit)

- **Property #2**
  - Includes special values (NaN, ∞, –∞)

  Examples:
  sqrt(–0.5) = NaN, f(NaN) = NaN [*check this in MATLAB*]
  1/0 = ∞, 1/∞ = 0
  arctan(x) → $\pi/2$ as x → ∞ ⇒ arctan(∞) = $\pi/2$

5.6

Slide 5.6

Algorithm developers use floating-point representation, which also serves as a reference for algorithm performance. After the algorithm is developed, it is refined for fixed-point accuracy. Let's look into the properties of the floating-point representation for the IEEE 754 standard.

The first property is the rounding of the "half-way" result to the nearest available even number. For example, 3.05 rounds to 3.0 (even number). Otherwise, rounding goes to the nearest number (round-up if the last digit that is being discarded is greater than 5; round-down if the last digit that is being discarded is less than 5).

The second property concerns the representation of special values such as NaN (not a number), ∞ (infinity), and −∞. These numbers may occur as a result of arithmetic operations. For instance, taking the square root of a negative number will result in a NaN, and any function performed on a NaN result will also produce a NaN. Readers may be familiar with this notation in MATLAB, for example. Division by zero results in ∞, while division by ∞ yields 0. We can, however, take ∞ as a function argument. For example, arctan(∞) gives $\pi/2$.

Floating-Point Standard: IEEE 754 (Cont.)

- ◆ **Property #3**
  - Uses denormals to represent the result $< 1.0 \times e^{Emin}$

  Emin = min exponent $\Big\langle$ Flush to 0
  Use significand < 1.0 and *Emin* ("gradual underflow")

  Example:
  base 10, 4 significant digits, $x = 1.234 \times 10^{Emin}$
  denormals: $x/10 \to 0.123 \times 10^{Emin}$
  $x/1{,}000 \to 0.001 \times 10^{Emin}$
  $x/10{,}000 \to 0$
  $x = y \Leftrightarrow x - y = 0$
  flush-to-0: $x = 1.256 \times 10^{Emin}, y = 1.234 \times 10^{Emin}$
  $x - y = \underbrace{0.022 \times 10^{Emin}}_{\text{denormal number (exact computation)}} = 0$ (although $x \neq y$)

5.7

Slide 5.7

A very convenient technique that aids in the representation of out-of-range numbers is the use of "denormals." Denormals allow representation of numbers smaller than 10^{Emin}, where E_{min} is the minimum exponent. Instead of flushing the result to 0, a significand < 1.0 is used. For instance, consider a decimal system with 4 significant digits and the number $x = 1.234 \cdot 10^{Emin}$. Denormals are numbers where the result is less than 10^{Emin}. The use of denormals guarantees that if two numbers, x and y, are equal then the result of their subtraction will be zero. The reverse also holds. A flush-to-0 system does not satisfy the reverse condition. For example, if $x = 1.256 \cdot 10^{Emin}$ and $y = 1.234 \cdot 10^{Emin}$, the result of $x - y$ in the flush-to-0 sysetm will be 0 although x and y are not equal. The use of denormals effectively allows us to approach 0 more gradually.

Floating-Point Standard: IEEE 754 (Cont.)

- ◆ **Property #4**
  - Rounding modes
    - Nearest (default)
    - Toward 0
    - Toward ∞
    - Toward $-\infty$

5.8

Slide 5.8

There are four different rounding modes that can be used. The default mode is to round to the nearest number; otherwise user can select rounding toward 0, ∞, or $-\infty$.

Representation of Floating-Point Numbers

- **Single precision: 32 bits**
  - Sign: 1 bit
  - Exponent: 8 bits
  - Fraction: 23 bits
    - *Fraction < 1 $\Rightarrow$ Significand = 1 + Fraction*
  - Bias = 127

  Example:

  1 10000001 0100...0

  sign exponent fraction

  129 – 127 $0.01_2 = 0.25$ (significand = 1.25)

  $-1.25 \times 2^2 = -5$

5.9

Slide 5.9

This slide gives an example of a single-precision 32-bit floating-point number. According to the standard, 1 bit is reserved for the sign, 8 bits for the exponent, and 23 bits for the fraction, and the bias is 127. In the case that the fraction is < 1, the significand is calculated as 1 + fraction. The numerical example in the slide shows the use of these techniques. The bit-wise organization of the sign, exponent, and fraction fields is shown. A sign bit of 1 indicates a negative number, the exponent of 129 is offset by the bias of 127, and the fractional part evaluates to 0.25. Since the fractional part is < 1, the significand is 1.25. We calculate the result as $-1.25 \cdot 2^2 = -5$.

Fixed Point: 2's Complement Representation

Overflow mode   Quantization mode

$\pi =$ | 0 | 0 | 1 | 1 | 0 | 0 | 1 | 0 | 0 | 1 |

Sign W_{Int} W_{Fr} fractional

$= 0 \times 2^3 + 0 \times 2^2 + 1 \times 2^1 + 1 \times 2^0 + 0 \times 2^{-1} + 0 \times 2^{-2} + 1 \times 2^{-3} + 0 \times 2^{-4} + 0 \times 2^{-5} + 1 \times 2^{-6}$

$= 3.140625$

- **W_{Int} and W_{Fr} suitable for predictable dynamic range**
  - o-mode (overflow, wrap-around)
  - q-mode (trunc, roundoff)

- **Economic for implementation**

5.10

Slide 5.10

Now, let's consider fixed-point 2's complement representation. This is the most common fixed-point representation used in DSP arithmetic. The slide shows a bit-wise organization of π in the 2's complement notation. The sign bit = 0 indicates a positive number. Besides the sign bit, there are integer and fractional bits that correspond to data range and accuracy. This representation is often expressed in the (W_{Tot}, W_{Fr}) format, where W_{Tot} and W_{Fr} are the total and fractional number of bits, respectively. A simple weighted bit-sum is used to calculate the value as illustrated on the slide.

Fixed-point arithmetic is very convenient for implementation due to a reduced bit count as compared to floating-point systems. The fixed-point representation has a much narrower dynamic range than floating-point representation, however. The implementation also has to consider overflow (large positive or negative numbers) and quantization (fine-precision numbers) effects. 2's complement is commonly used in hardware due to its simplicity and low numerical complexity.

Fixed Point: Unsigned Magnitude Representation

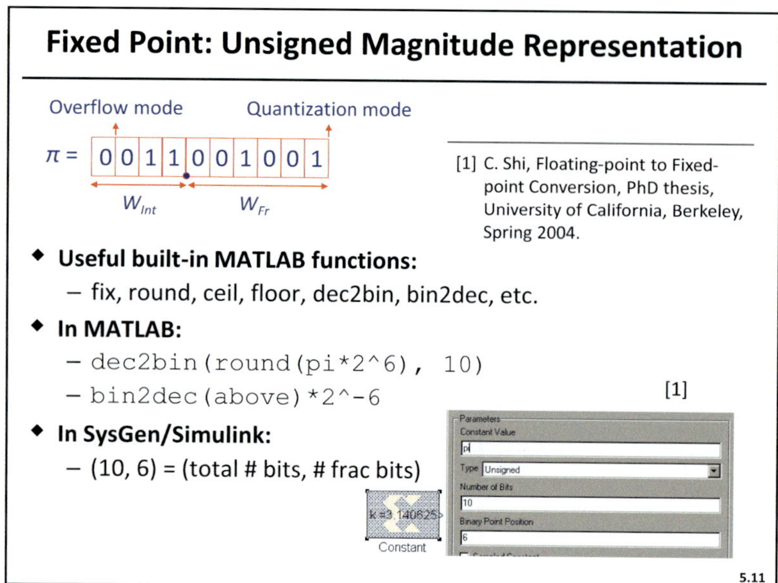

Slide 5.11

Another fixed-point representation is unsigned magnitude. When the overflow bit is "1", it indicates negative values. For instance, in 8-bit representation, $129_{10} = 10000001_2$, which indicates -1. No additional bit-wise arithmetic is required for negative numbers like in 1's or 2's complement.

MATLAB has a number of built-in functions for number conversion. The reader is encouraged to read the tool help pages. Some of these functions are listed on this slide. Additionally, Simulink has a graphical interface for fixed-point types. The example shows an unsigned format with a total of 10 bits of which 6 bits are fractional.

Fixed-Point Representations

- **Sign magnitude**

- **2's complement**
  - $x + (-x) = 2^n$ (complement each bit, add 1)
  - Most widely used (signed arithmetic easy to do)

- **1's complement**
  - $x + (-x) = 2^n - 1$ (complement each bit)

- **Biased** → **add bias, encode as ordinary unsigned number**
  - $k + \text{bias} \geq 0$, $\text{bias} = 2^{n-1}$ (typically)

5.12

Slide 5.12

Commonly used fixed-point representations are summarized here. Sign-magnitude is the most straightforward approach based on a sign bit and magnitude bits. 2's complement and 1's complement numbers require bit-wise inversion to convert between positive and negative numbers. 2's complement, which is most commonly used for DSP applications, also requires adding a 1 to the least significant bit position. Finally, there is a biased representation in which a bias is applied to negative numbers k such that $k + bias$ is always non-negative. Typically, $bias = 2^{n-1}$, where n is the number of bits.

Fixed-Point Representations: Example

Example: n = 4 bits, k = 3, $-k$ = ?

- **Sign magnitude:** $k = 0011_2$ → $-k = 1011_2$

- **2's complement:** $k + 1011 = 2^n$ 0011

  Procedure:
  - Bit-wise inversion
  - Add "1"

  $-k = 1100$ $+1101$
  $+\ \ \ 1$ 10000
  1101_2

- **1's complement:** $-k = 1100_2$ $k + (-k) = 2^n - 1$

- **Biased:** $k + bias = 1011_2$ $-k + bias = 0101_2 = 5 \geq 0$
  $2^{n-1} = 8 = 1000_2$

5.13

Slide 5.13

Here are a few examples. Suppose $n = 4$ (number of bits) and $k = 3$. Given k, we want to come up with a representation of $-k$.

In sign-magnitude, $k = 0011_2$ with MSB being reserved for the sign. Therefore, the negative value is obtained by simply inverting the sign bit, $-k = 1011_2$.

In 2's complement, $k + (-k) = 2^n$ has to hold. The negative value is obtained in two steps. In the first step, we do a bit-wise inversion of k and obtain 1100_2. In the second step, we add 1 to the LSB and obtain $-k = 1101_2$. Now, we can go back and verify the assumption: $k + (-k) = 0011_2 + 1101_2 = 10000_2 = 2^4$. The carry out of 1 is ignored, so the resultant 4-bit number is 0.

In 1's complement, the concept is similar to that of 2's complement except that we stop after the first step (bit-wise inversion). Therefore, $-k = 1100_2$. As a way of verification, $k + (-k) = 0011_2 + 1100_2 = 1111_2 = 2^4 - 1$.

In biased representation, $bias = 2^{n-1} = 8$. $k + bias = 1011_2$, $-k + bias = 0101_2 = 5 \geq 0$.

2's Complement Arithmetic

- **Most widely used representation, simple arithmetic**

- **Example:** 5 + − 2
                              0010 (−2)
      0101_2 (5)   1101
     $+1110_2$ (−2) $+\ \ \ 1$
     $\boxed{1}0011$ = +3 1110_2

  Discard the sign bit (if there is no overflow)

- **Overflow** occurs when Carry into MSB ≠ Carry out of MSB

  MSB
  $0|1\ 0\ 1_2$ (5)
  $+1|1\ 1\ 0_2$ (−2)
  $\overline{1|0\ 0\ 1\ 1}$ (3)
  Carry: $\underline{1}\ 1\ 0\ 0$

  Carry out of MSB = Carry into MSB → **No overflow!**

5.14

Slide 5.14

The 2's complement representation is the most widely used in practice due to its simplicity. Arithmetic operations are preformed regardless of the sign (as long as the overflow condition is being tracked). For example, the addition of 5 and -2 is performed by simple binary addition, as shown on the slide. Since 4 bits are sufficient to represent the result, the sign bit is discarded and we have the correct result, $0011_2 = 3_{10}$.

To keep track of the overflow, we have to look at carries at MSB. If the carry into and out of MSB differ, then we have an overflow. In the previous example, repeated here with annotation of the carry bit, we see that carries into and out of the MSB match, hence no overflow is detected.

Overflow

♦ <u>**Example:**</u> **unsigned 4-bit addition**

$$6 = 0110_2$$
$$+11 = 1011_2$$
$$= 17 = 10001_2 \text{ (5 bits!)}$$

extra bit

♦ **Property of 2's complement**
  – **Negation = bit-by-bit complement + 1 $\rightarrow$ C_{in} = 1, result: $a - b$**

5.15

Slide 5.15

An example of overflow is provided in this slide. If we assume 4-bit numbers and perform the addition of 6 and 11, the result is $17 = 10001_2$. Since the result is out of range, an extra bit is required to represent the result. In this example, overflow occurs.

A nice feature of 2's complement is that by simply bringing a carry into the LSB, addition turns into subtraction. Mathematically, $a + b$ $(C_{in} = 0)$ becomes $a - b$ $(C_{in} = 1)$. Such property is very convenient for hardware realization.

Quantization Effects

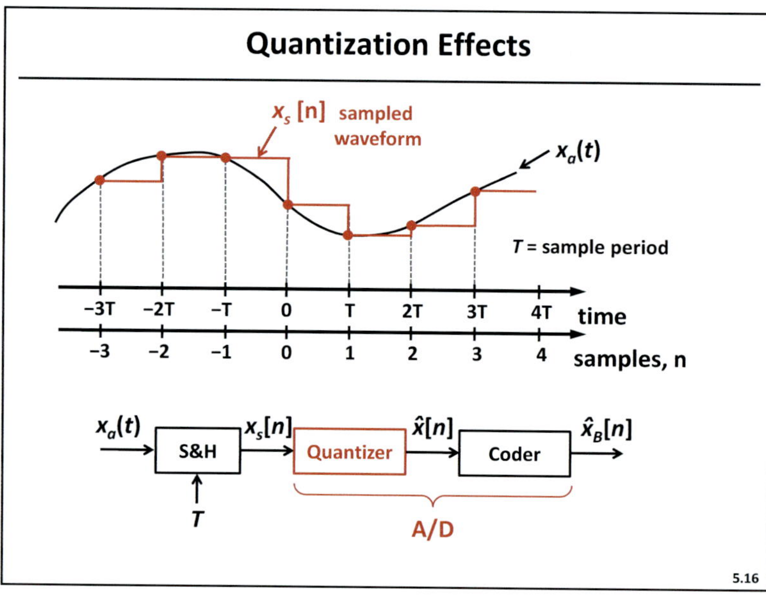

5.16

Slide 5.16

To study quantization, this slide shows a continuous-time analog waveform $x_a(t)$ and its sampled version $x_s[n]$. Sampling occurs at discrete time intervals with period T. The dots indicate the sampled values. The sampled waveform $x_s[n]$ is quantized with a finite number of bits, which are coded into a fixed-point representation. The quantizer and coder implement analog-to-digital conversion, which works at a rate of $1/T$.

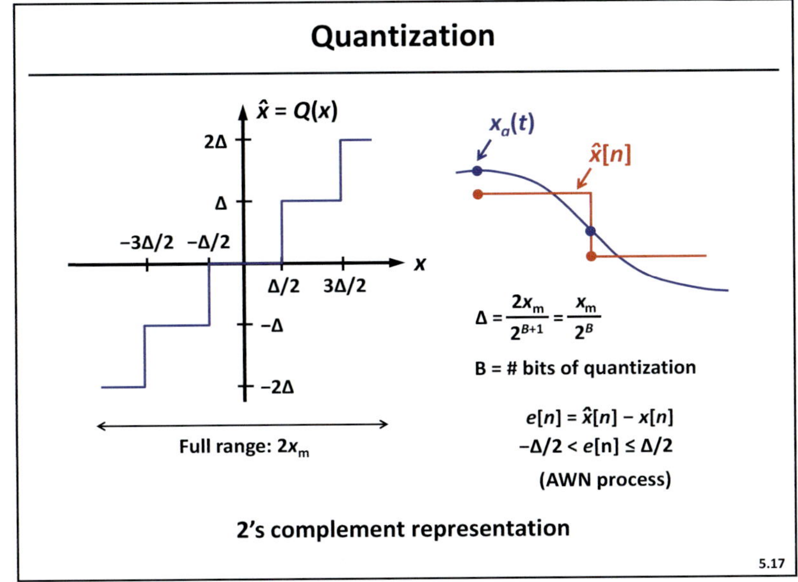

Slide 5.17

To maintain an accurate representation of $x_a(t)$ over the sampling period, quantized values $Q(x)$ are computed as the average of $x_a(t)$ at the beginning and end of the sampling period, as illustrated on the right. The quantization step Δ is chosen to cover the maximum absolute value x_m of input x with B bits. $B+1$ bits are required to cover the full range of $2x_m$, which also includes both positive and negative values. The corresponding quantization characteristic is shown on the left.

The quantization error $e[n]$ can be computed from the quantization characteristic. Due to the nature of quantization, the absolute value of $e[n]$ cannot exceed $\Delta/2$. The error can be modeled as Additive White Noise (AWN).

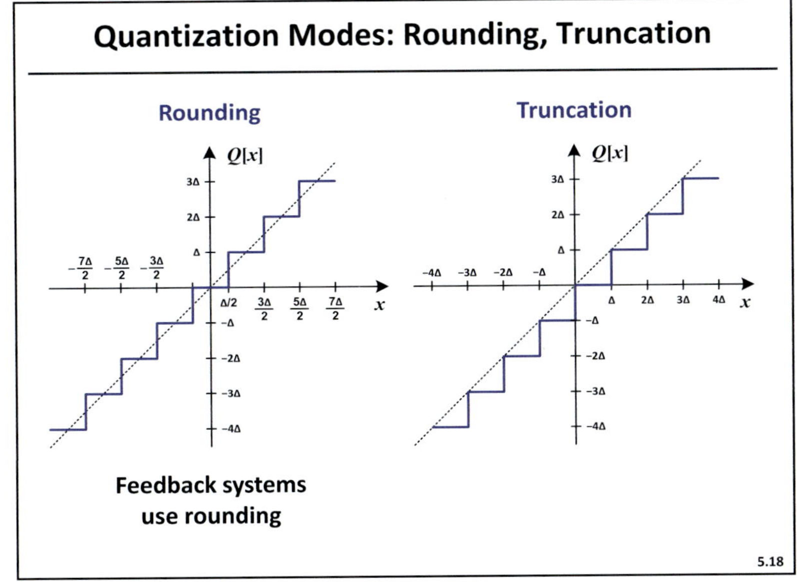

Slide 5.18

In practical systems, we can implement quantization with rounding or truncation. The rounding quantization characteristic is shown on the left and is the same as that on the previous slide. The solid stair-case $Q[x]$ intersects with the dashed 45-degree line $Q[x] = x$, thereby providing a reasonably accurate representation of x.

Truncation can be implemented simply by just discarding the least significant bit, but it introduces two-times larger error than rounding. The quantization characteristic $Q[x]$ is always below the 45-degree line. This representation may work in some feed-forward designs due to its simplicity, but can result in significant error accumulation in recursive algorithms. Feedback systems, for this reason, use rounding as the preferred quantization mode.

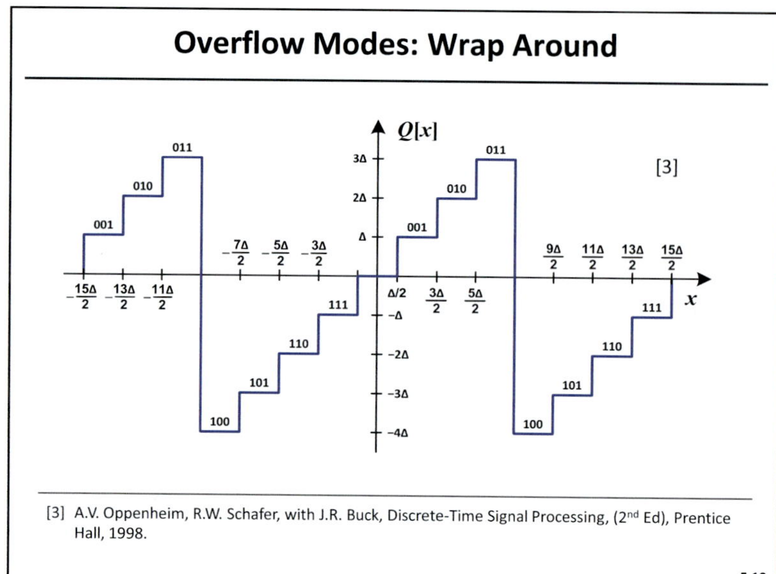

Slide 5.19

Quantization affects least significant bits. We also need to consider accuracy issues related to overflow, which affects the most significant bits. One idea is to use a wrap-around scheme, as shown in this slide [3]. We simply discard the most significant bit when the number goes out of range and keep the remaining bits. This is simple to implement, but could be very inaccurate since large positive values can be represented as negative and vice versa.

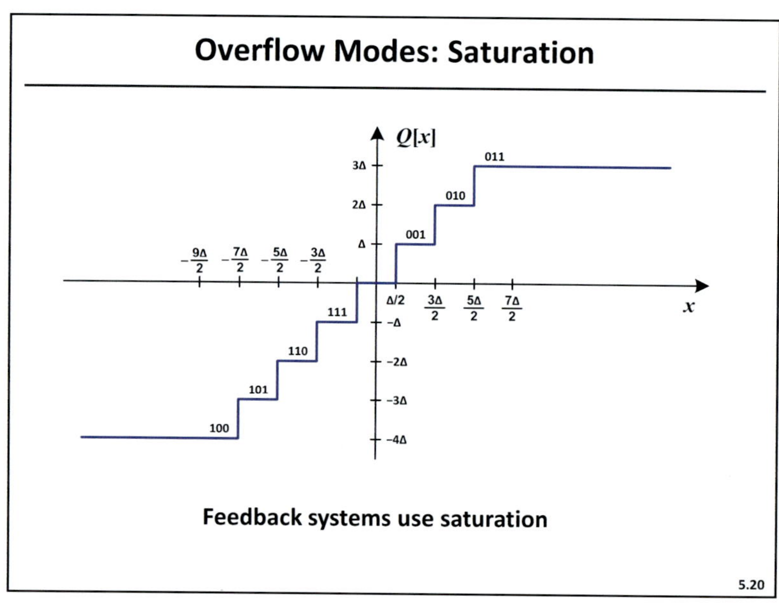

Slide 5.20

Another way to deal with overflow is to use saturation, as shown on this slide. In this case, we also keep the same number of bits, but saturate the result to the largest positive (negative) value. Saturation requires extra logic to detect the overflow and freeze the result, but provides more accurate representation than wrap-around. Saturation is particularly used in recursive systems that are sensitive to large signal dynamics.

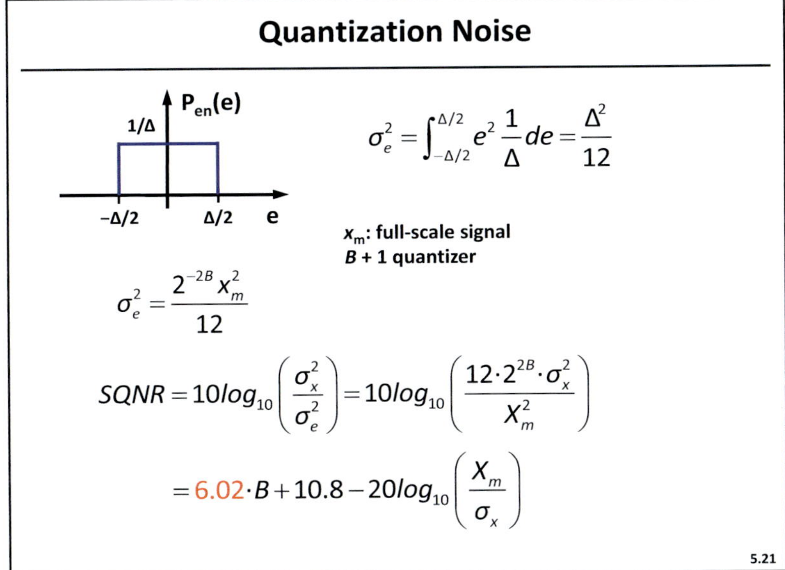

$$\sigma_e^2 = \int_{-\Delta/2}^{\Delta/2} e^2 \frac{1}{\Delta} de = \frac{\Delta^2}{12}$$

$$\sigma_e^2 = \frac{2^{-2B} x_m^2}{12}$$

$$SQNR = 10 log_{10}\left(\frac{\sigma_x^2}{\sigma_e^2}\right) = 10 log_{10}\left(\frac{12 \cdot 2^{2B} \cdot \sigma_x^2}{X_m^2}\right)$$

$$= 6.02 \cdot B + 10.8 - 20 log_{10}\left(\frac{X_m}{\sigma_x}\right)$$

Slide 5.21

Per the discussion in Slide 5.17, quantization noise can be modeled as the Additive White Noise (AWN) process illustrated here. Given the flat noise characteristic, we can calculate the noise variance (corresponding to the noise power) by solving the integral of $e^2 \cdot (1/\Delta)$ when e varies between $-\Delta/2$ and $\Delta/2$. The variance equals $\Delta^2/12$. For $B+1$ bits used to quantize x_m, $\sigma_e = 2^{-2B} \cdot x_m^2/12$.

The noise power can now be used to calculate the signal-to-quantization-noise ratio (SQNR) due to quantization as given by the formula in this slide. Each extra bit of quantization improves the SQNR by 6 dB.

Binary Multiplication

• **Arguments: X, Y**

$$X = \sum_{i=0}^{M-1} X_i \cdot 2^i \qquad Y = \sum_{j=0}^{N-1} Y_j \cdot 2^j$$

• **Product: Z**

$$Z = X \cdot Y = \sum_{k=0}^{M+N-1} z_k \cdot 2^k = \left(\sum_{i=0}^{M-1} X_i \cdot 2^i\right) \cdot \left(\sum_{j=0}^{N-1} Y_j \cdot 2^j\right)$$

$$Z = \sum_{i=0}^{M-1}\left(\sum_{j=0}^{N-1} X_i \cdot Y_j \cdot 2^{i+j}\right) \qquad [4]$$

[4] J. Rabaey, A. Chandrakasan, B. Nikolić, Digital Integrated Circuits: A Design Perspective, (2nd Ed), Prentice Hall, 2003.

5.22

Slide 5.22

As an example of fixed-point operations, let us look at a multiplier. The multiplier takes input arguments X and Y, which have M and N bits, respectively. To perform multiplication without loss of accuracy, the product Z requires $M+N$ bits [4]. If we now take the product as input to the next stage, the required number of bits will further increase. After a few stages, this strategy will result in impractical wordlength. Statistically, not all output bit combinations will occur with the same probability. In fact, the number of values at the output will often be smaller than the total number of distinct values that $M+N$ bits can represent. This gives rise to opportunities for wordlength reduction.

One approach to reduce the number of bits is to encode numbers in a way so as to increase the probability of occurrence of certain output states. Another technique is to reduce the number of bits based on input statistics and desired signal-to-noise specification at the nodes of interest. More details on wordlength optimization will be discussed in Chap. 10.

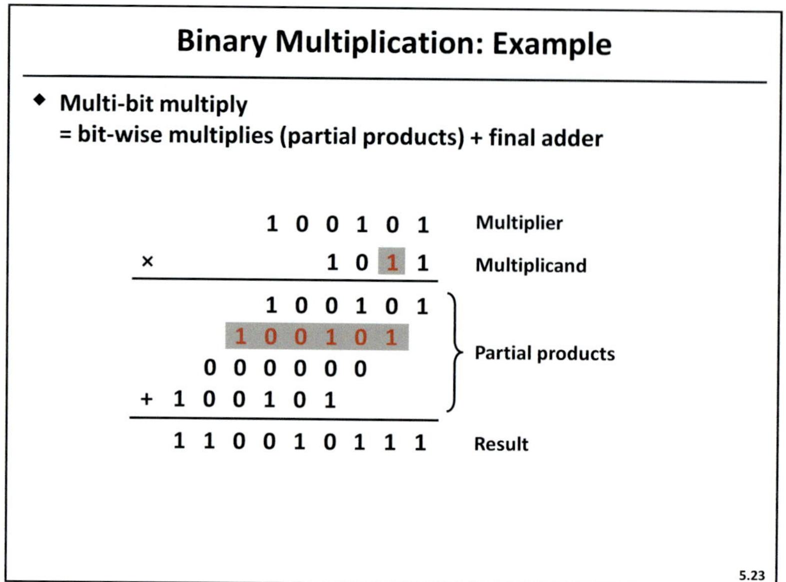

Slide 5.23

Even without considering wordlength reduction, multiplier cost can be reduced by using circuit implementation techniques. Consider a 6-bit multiplier and a 4-bit multiplicand as shown in the slide. Binary multiplication is performed in the same way as decimal multiplication. Each bit of the multiplicand is multiplied with the multiplier to create partial products. The partial products are summed to create the result, which is 10 bits in this example. How to implement the multiplier?

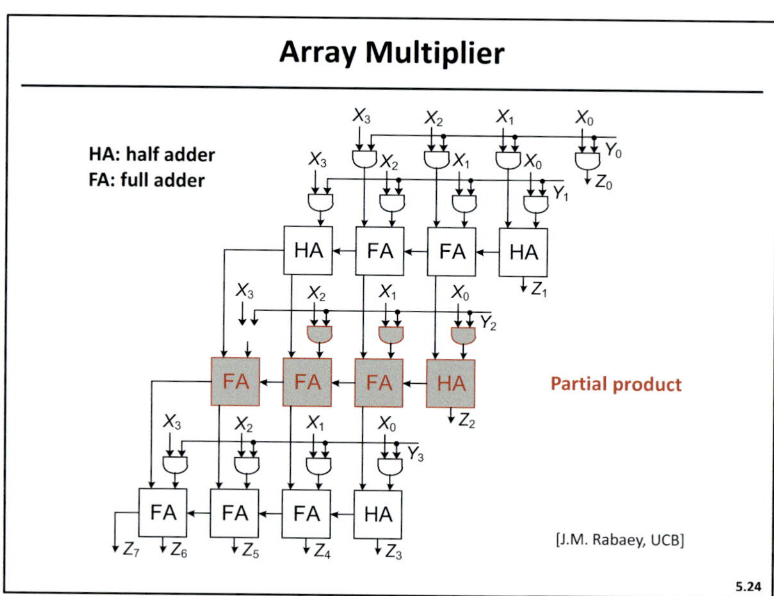

Slide 5.24

A straightforward implementation is shown in Slide 5.23. Single-bit multiplies are implemented with an array of AND gates to realize the partial products. The partial products are then summed with three stages of additions. A simple ripple-carry adder is used as an illustration. Each adder block starts with a half-adder at the least significant bit and propagates the carry to the most significant bit. Intermediate bits are realized with full-adder blocks since they have carry inputs. In all adder stages but the last, the MSBs are also realized using half-adders. This realization is a good reference point for further analysis and optimization.

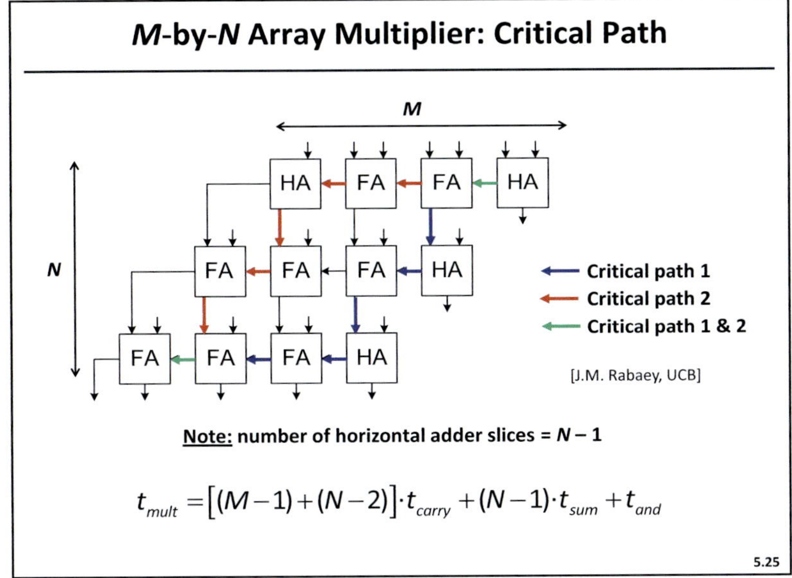

$$t_{mult} = \left[(M-1)+(N-2)\right]\cdot t_{carry} + (N-1)\cdot t_{sum} + t_{and}$$

Slide 5.25

The speed of the multiplier can be improved by restructuring the logic blocks. This slide shows a critical-path analysis of the multiplier architecture from the previous slide. For an M-by-N bit multiplier, there are several critical paths that propagate through the same number of logic blocks. Two such paths are illustrated by the *blue* (critical path 1) and *red* (critical path 2) *arrows*. The *green* arrow indicates shared portion of the two paths. As noted on Slide 5.23, the number of adder stages needed for N partial products is equal to $N-1$.

Tracing the paths through the adder stages, we can calculate the critical path for the multiplier as:

$$t_{mult} = [(M-1)+(N-2)]\cdot t_{carry} + (N-1)\cdot t_{sum} + t_{and},$$

where t_{and} is the delay through the AND gates that compute the partial products, $(N-1)\cdot t_{sum}$ is the delay of the sum bits (vertical arrows), and $[(M-1)+(N-2)]\cdot t_{carry}$ is the delay of the carry bits (horizontal arrows). The multiplier delay, therefore, increases linearly with M and N. Since N has more weight in the t_{mult} formula than M, it is of interest to investigate techniques that would reduce the number of adder stages or make them faster.

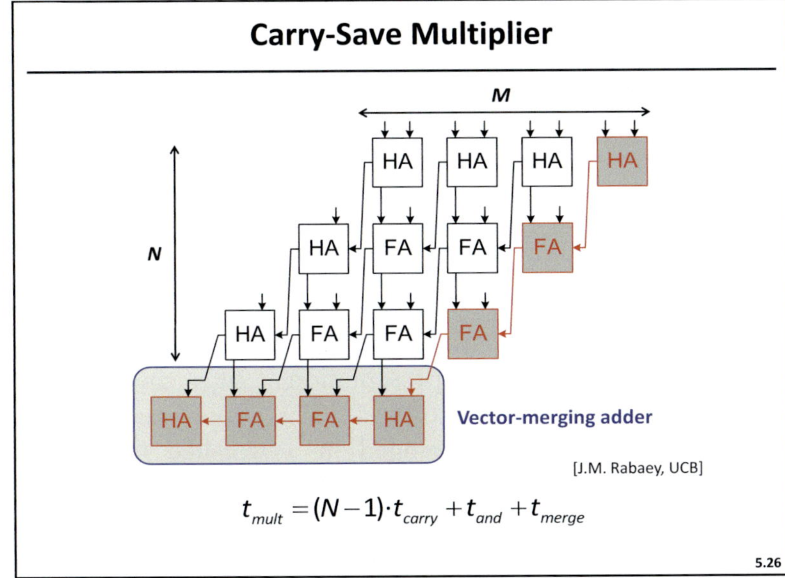

$$t_{mult} = (N-1)\cdot t_{carry} + t_{and} + t_{merge}$$

Slide 5.26

One way to speed up the multiplication is to use carry-save arithmetic, as shown on this slide. The idea is to route the carry signals vertically down to the next stage instead of horizontally within the same adder stage. This is possible, because the addition of the carry out bits in each stage is deferred to final stage, which uses a vector-merge adder. The computation is faster, because the result is propagated down as soon as it becomes available, as opposed to propagating further within the stage. This scheme is, hence, called "carry-save" and is commonly used in practice.

The carry-save architecture requires a final vector-merging adder, which adds to the delay, but the overall delay is still greatly reduced. The critical path of this multiplier architecture is:

$$t_{mult} = (N-1) \cdot t_{carry} + t_{and} + t_{merge},$$

where t_{merge} is the delay of the vector-merging adder, t_{and} is the delay of AND gates in the partial products, and $(N-1) \cdot t_{carry}$ is the delay of carry bits propagating downward. The delay formula thus eliminates the dependency on M and reduces the weight of N.

Assuming the simplistic case where $t_{carry} = t_{and} = t_{sum} = t_{merge} = t_{unit}$, the delay of the adder in Slide 5.25 would be equal to $[(M-1) + 2 \cdot (N-1)] \cdot t_{unit}$ as compared to $(N+1) \cdot t_{unit}$. For $M = N = 4$, the vector-merging adder architecture has a total delay of $5 \cdot t_{unit}$, as compared to $9 \cdot t_{unit}$ for the array-multiplier architecture. The speed gains are even more pronounced for a larger number of bits.

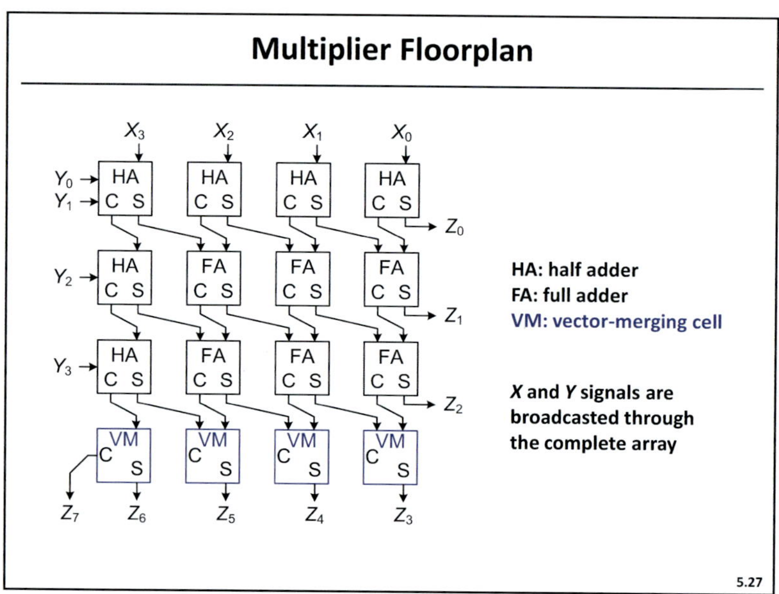

Slide 5.27

The next step is to implement the multiplier and to maximize the area utilization of the chip layout. A possible floorplan is shown in this slide. The blocks organized in a diagonal structure have to fit in a square floorplan as shown on this slide. At this point, routing also becomes important. For example, X and Y need to be routed across all stages, which poses additional constraints on floorplanning. The placement and routing of blocks has to be done in such a way as to balance the wire length (delay) of input signals.

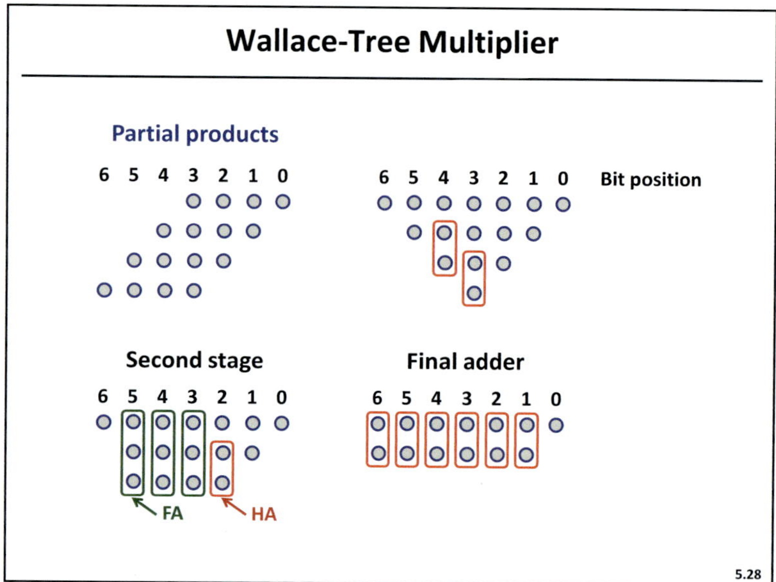

Slide 5.28

The partial-product summation can also be improved at the architectural level by using a Wallace-Tree multiplier. The partial product bits are represented by the blue dots. At the top-left is a direct representation of the partial products. The dots can be re-arranged in a tree-like fashion as shown at the top-right. This representation is more convenient for the analysis of bit-wise add operations. Half adders (HA) take two input bits and produce two output bits. Full adders (FA) take three input bits and produce two output bits. Applying this principle from the root of the tree allows for a systematic computation of the partial-product summation.

The bottom-left of the figure shows bit-wise partial-product organization after executing the circled add operations from the top-right. There are now three stages. Applying the HA at bit position 2 and propagating the result towards more significant bits yields the structure in the bottom-right. The last step is to execute final adder.

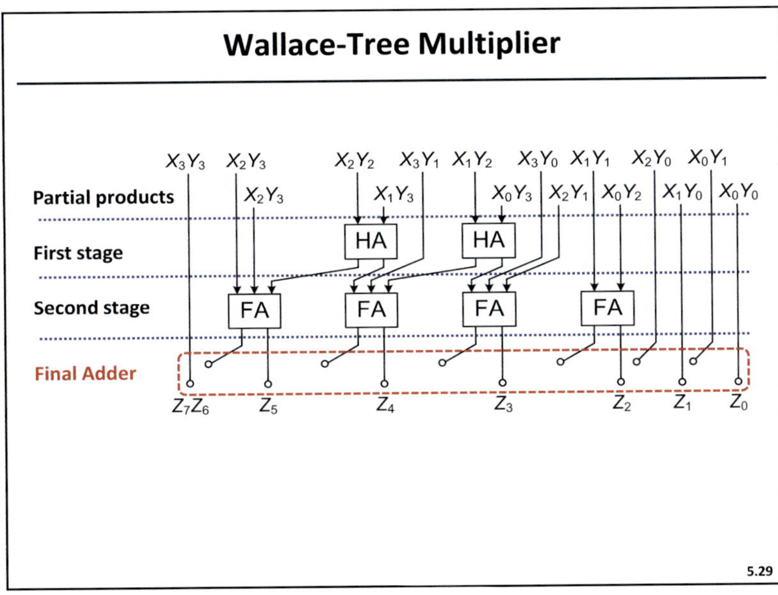

Slide 5.29

The partial-product summation strategy described in the previous slide is graphically shown here. Full-adder blocks are also called 3:2 compressors because they take 3 input bits and compress them to 2 output bits. Effective organization of partial adds, as done in the Wallace Tree, results in a reduced critical-path delay, which allows for voltage scaling and power reduction. The Wallace-Tree multiplier is a commonly used implementation technique.

Multipliers: Summary

- **Optimization goals different than in binary adder**

- **Once again: Identify critical path**

- **Other possible techniques**
  - Logarithmic versus linear (Wallace-Tree multiplier)
  - Data encoding (Booth)
  - Pipelining

Slide 5.30

The multiplier can be viewed as hierarchical extension of the adder.

Other techniques to consider include the investigation of logarithmic adders vs. linear adders in the adder tree. Logarithmic adders require fewer stages, but generally also consume more power. Data encoding as opposed to simple 2's complement can be employed to simplify arithmetic. And pipelining can reduce the critical-path delay at the expense of increased latency.

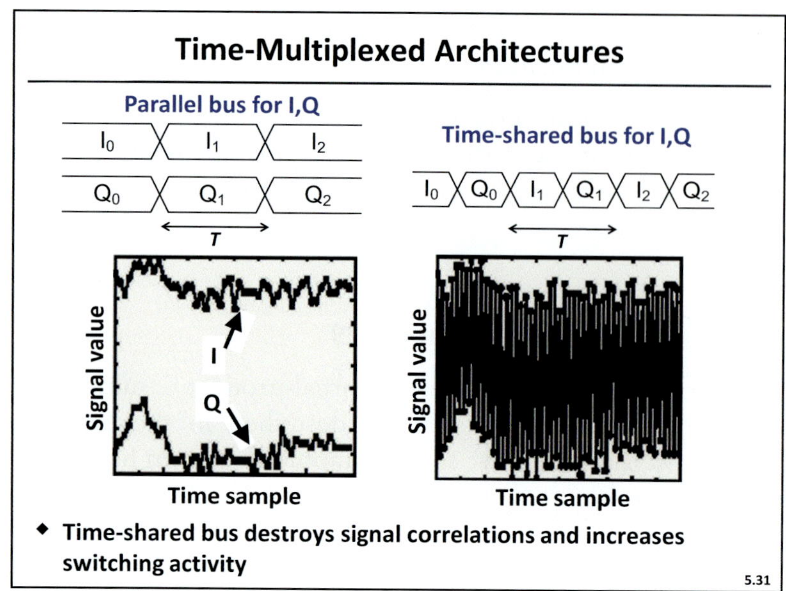

Slide 5.31

Apart from critical-path analysis, we also look at power consumption. Switching activity is an important factor in power minimization. Datapath switching profile is largely affected by implementation. This slide shows parallel and time-multiplexed bus architectures for two streams of data. The parallel-bus design assumes dedicated buses for the I and Q channels with signaling at a symbol rate of $1/T$. The time-shared approach requires just one bus and increases the signaling speed to $2/T$ to accommodate both channels.

Suppose that each of the channels consists of time samples with a high degree of correlation between consecutive samples in time as shown on signal value vs. time sample plot on the left. This implies low switching activity on each of the buses and, hence, low switching power. The time-shared approach would result in very large signal variations due to interleaving. It also results in excess power consumption. This example shows that lower area implementation may not be better from a power standpoint and suggests that arithmetic issues have to be considered both at algorithm and implementation levels.

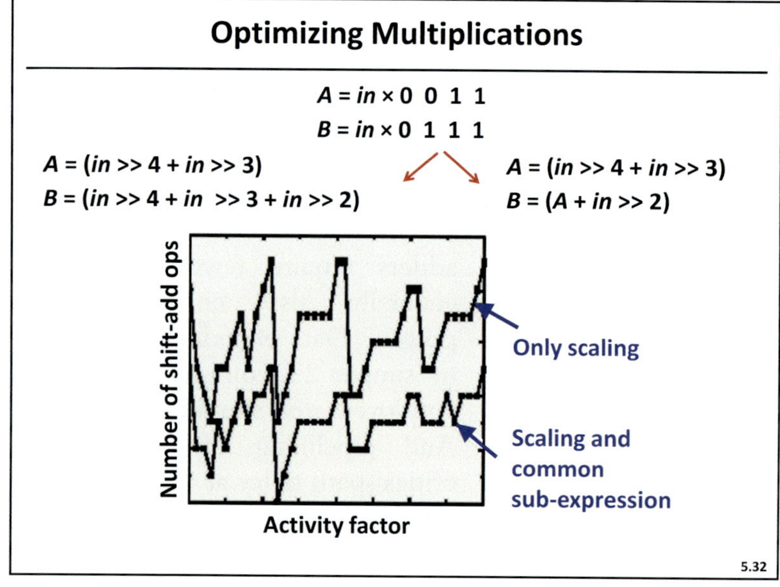

Slide 5.32

Signal activity can be leveraged for power reduction. This slide shows an example of how to exploit signal correlation for a reduction in switching. Consider an unknown input In that has to be multiplied by 0011 to calculate A, and by 0111 to calculate B.

Binary multiplication can be done by using add-shift operations as shown on the left. To calculate A, In is shifted to the right by 3 and 4 bit positions and the two partial results are summed up. The same principle is applied to the calculation of B. When A and B are calculated independently, shifting by 4 and shifting by 3 are repeated, which increases the overall switching activity. This gives rise to the idea of sharing common sub-expressions to calculate both results.

Since the shifts by 3 and by 4 are common, we can then use A as a partial result to calculate B. By adding A and In shifted by 2 bit positions, we obtain B. The plot on this slide shows the number of shift-add operations as a function of the input activity factor for the two implementations. The results show that the use of common sub-expressions greatly reduces the number of operations and, hence, power consumption. Again, we can see how the implementation greatly affects power efficiency.

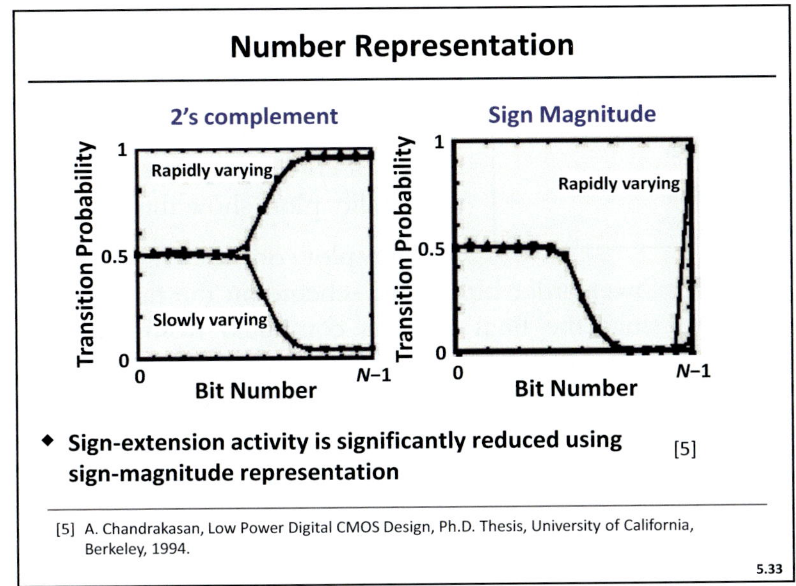

Slide 5.33

In addition to the implementation, the representation of numbers can also play a role in switching activity. This slide shows a comparison of the transition probabilities as a function of bit position for 2's complement and sign-magnitude number representations [5]. Cases of slowly and rapidly varying input are considered. Lower bit positions will always have switching probability around 0.5 in both cases, while more significant bits will vary less frequently in case of the slowly varying input. In 2's complement, the rapidly varying input will cause higher-order bits to toggle more often, while in the sign-magnitude representation only the sign bit would be affected. This makes sense, because many bits will need to flip from positive to negative numbers and vice versa in 2's complement. Sign-magnitude representation is, therefore, more naturally suited for rapidly varying input.

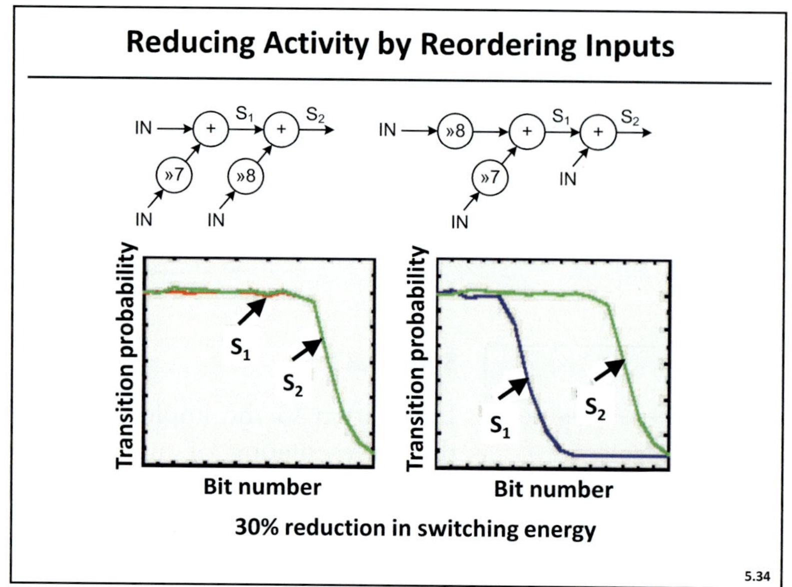

Slide 5.34

Input reordering can be used to maintain signal correlations and reduce switching activity. The sum S_2 can be computed with different orders of the inputs. On the left, S_1 is a sum of inputs that are 7 bits apart, which implies very little correlation between the operands. S_1 is then added to the input shifted by 8 bits. On the right, S_1 is a sum of inputs that differ by only one bit and, hence, have a much higher degree of correlation. The transition probability plots show this result.

The plot on the right clearly shows a reduced switching probability of S_1 for lower-order bits in the scheme on the right. S_2 has the same transition probability for both cases since the final result is computed from the same primary inputs. Input reordering can lead to a 30% reduction in switching energy. This example, once again, emphasizes the need to have joint consideration of the algorithm and implementation issues.

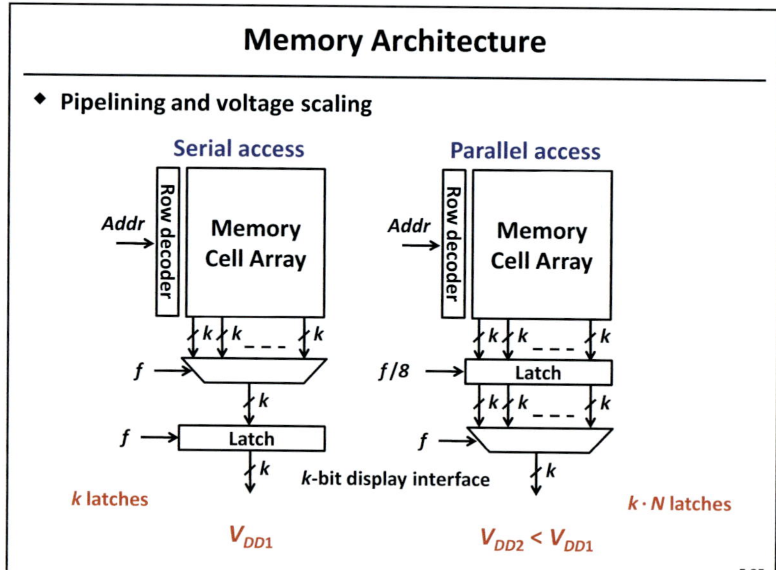

Slide 5.35

The concept of switching activity also applies to the memory architecture. For example, this slide compares serial- and parallel-access memories used to implement a k-bit display interface. On the left, a single k-bit word is multiplexed from 8 words available in memory at a rate of f. This value is then latched at the same rate f. On the right, 8 words are latched in at rate $f/8$ and the multiplexer selects one word at rate f.

Looking at the switching activity of the two schemes, k latches are switched at rate f on the left, while the scheme on the right requires the switching of $k \cdot N$ latches ($N = 8$ in our example) at rate $f/8$. Since the critical-path delay of the serial scheme is longer, the parallel scheme can operate at lower supply voltage ($V_{DD2} < V_{DD1}$) to save power. The parallel-access scheme needs more area for pipeline registers, but the overall switched capacitance is the same for both schemes. Power reduction enabled by pipelining could outweigh the area penalty.

Summary

- **Algorithms are developed in algebraic form and verified using floating-point precision**
- **Fixed-point representation degrades algorithm performance due to quantization noise arising from finite precision**
  - Of particular importance are quantization (rounding, truncation) and overflow (saturation, wrap-around) modes
    - Rounding has zero-mean error and is suitable for recursive systems
    - Saturation is typically used in feedback systems
- **Implementation of a fixed-point adder (key building block in DSP algorithms) needs to focus on the carry-path delay minimization**
  - Carry-save addition can be used to speed up multiplication
- **Data correlation can impact power consumption**
  - Time-shared buses increase the activity (reduce correlation)
  - 2's complement has higher switching than sign-magnitude

5.36

Slide 5.36

DSP arithmetic was discussed in this chapter. Algorithms developed in algebraic form are validated in floating-point simulations, after which a fixed-point representation is typically used for hardware implementation. Translation to fixed point involves determining the number of bits as well as quantization and overflow modes. Rounding and saturation are most often used in feedback systems. Key building element of DSP systems, an adder, requires the reduction in carry-path delay for improved performance (or energy efficiency). The use of carry-save addition was discussed as a way of speeding multiplication. Data activity plays an important role in power consumption. In particular, time-shared buses may increase the activity due to reduced signal correlation and hence increase power. Designers should also consider number representation when it comes to reduced switching activity: it is know that 2's complement has higher switching activity than sign-magnitude. Fixed-point realization of iterative algorithms is discussed in the next chapter.

References

- C. Shi, Floating-point to Fixed-point Conversion, Ph.D. Thesis, University of California, Berkeley, 2004.

- J.L. Hennesy and D.A. Paterson, Computer Architecture: A Quantitative Approach, (2nd Ed), Morgan Kaufmann, 1996.

- A.V. Oppenheim, R.W. Schafer, with J.R. Buck, Discrete-Time Signal Processing, (2nd Ed), Prentice Hall, 1998.

- J. Rabaey, A. Chandrakasan, B. Nikolić, Digital Integrated Circuits: A Design Perspective, (2nd Ed), Prentice Hall, 2003.

- A.P. Chandrakasan, Low Power Digital CMOS Design, Ph.D. Thesis, University of California, Berkeley, 1994.

Additional References

- Appendix A of: D.A. Patterson, and J.L. Hennessy. Computer Organization & Design: The Hardware/Software Interface, (2nd Ed), Morgan Kaufmann, 1996.

- Simulink Help: Filter Design Toolbox -> Getting Started -> Quantization and Quantized Filtering -> Fixed-point Arithmetic/Floating-point Arithmetic, Mathworks Inc.

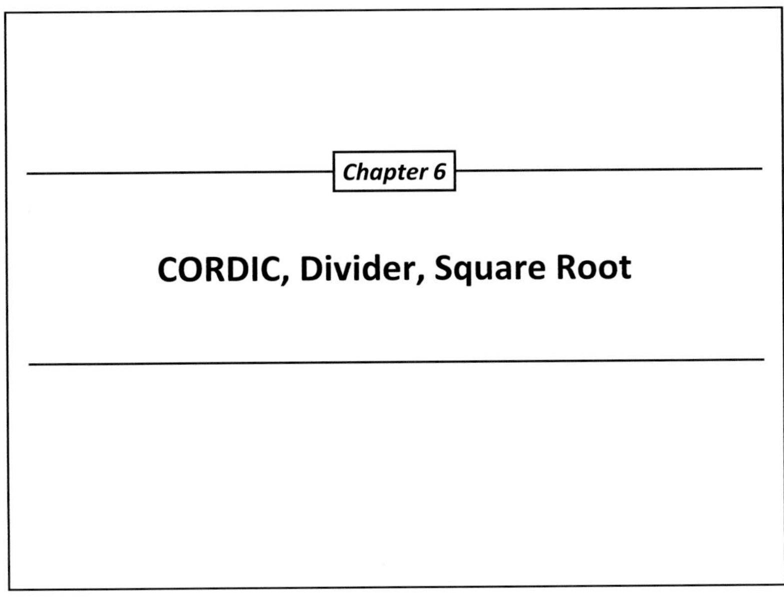

Slide 6.1

This chapter studies iterative algorithms for division, square rooting, trigonometric and hyperbolic functions and their baseline architecture. Iterative approaches are suitable for implementing adaptive signal processing algorithms such as those found in wireless communications.

Chapter Overview

- **The chapter focuses on several important iterative algorithms**
  - CORDIC
  - Division
  - Square root

- **Topics covered include**
  - Algorithms and their implementation
  - Convergence analysis
  - Speed of convergence
  - The choice of initial condition

Slide 6.2

Many DSP algorithms are iterative in nature. This chapter analyzes three common iterative operators: CORDIC (COordinate Rotation DIgital Computer), division, and square root. CORDIC is a widely used block, because it can compute a large number of non-linear functions in a compact form. Additionally, an analysis of architecture and convergence features will be presented. Newton-Raphson formulas for speeding up the convergence of division and square root will be discussed. Examples will illustrate convergence time and block-level architecture design.

D. Marković and R.W. Brodersen, *DSP Architecture Design Essentials*, Electrical Engineering Essentials,
DOI 10.1007/978-1-4419-9660-2_6, © Springer Science+Business Media New York 2012

CORDIC

◆ **To perform the following transformation**

$$y(t) = y_R + j \cdot y_I \rightarrow |y| \cdot e^{j\varphi}$$

and the inverse, we use the CORDIC algorithm

CORDIC - COordinate Rotation DIgital Computer

6.3

Slide 6.3

The CORDIC algorithm is most commonly used in communication systems to translate between Cartesian and polar coordinates. For instance, y_R and y_I, representing the I and Q components of a modulated symbol, can be transformed into their respective magnitude $(|y|)$ and phase (φ) components using CORDIC. The CORDIC algorithm can also accomplish the reverse transformation, as well as an array of other functions.

CORDIC: Idea

◆ **Use rotations to implement a variety of functions**

Examples:

$$x + j \cdot y \Leftrightarrow |\sqrt{x^2 + y^2}| e^{j \cdot tan^{-1}(y/x)}$$

$$z = \sqrt{x^2 + y^2} \qquad z = \cos(y/x) \qquad z = \tan(y/x)$$

$$z = x/y \qquad z = \sin(y/x) \qquad z = \sinh(y/x)$$

$$z = tan^{-1}(y/x) \qquad z = cos^{-1}(y)$$

6.4

Slide 6.4

CORDIC uses rotation as an atomic recursive operation to implement a variety of functions. These functions include: Cartesian-to-polar coordinate translation; square root; division; sine, cosine, and their inverses; tangent and its inverse; as well as hyperbolic sine and its inverse. These functions are implemented by configuring the algorithm into one of the several modes of operation.

CORDIC (Cont.)

♦ **How to do it?**

♦ **Start with general rotation by φ**

$x' = x \cdot \cos(\varphi) - y \cdot \sin(\varphi)$
$y' = y \cdot \cos(\varphi) + x \cdot \sin(\varphi)$

$x' = \cos(\varphi) \cdot [x - y \cdot \tan(\varphi)]$
$y' = \cos(\varphi) \cdot [y + x \cdot \tan(\varphi)]$

♦ **The trick is to only do rotations by values of tan(φ) which are powers of 2**

6.5

Slide 6.5

Let's analyze a CORDIC rotation. We start from (x, y) and rotate by an arbitrary angle φ to calculate (x', y') as described by the equations on this slide. By rewriting $\sin(\varphi)$ and $\cos(\varphi)$ in terms of $\cos(\varphi)$ and $\tan(\varphi)$, we get a new set of expressions. To simplify the math, we do rotations only by values of $\tan(\varphi)$ that are powers of 2. In order to rotate by an arbitrary angle φ, several iterations of the algorithm are needed.

CORDIC (Cont.)

Rotation Number

φ	$\tan(\varphi)$	k	i
45°	1	1	0
26.565°	2^{-1}	2	1
14.036°	2^{-2}	3	2
7.125°	2^{-3}	4	3
3.576°	2^{-4}	5	4
1.790°	2^{-5}	6	5
0.895°	2^{-6}	7	6

♦ **To rotate to any arbitrary angle, we do a sequence of rotations to get to that value**

6.6

Slide 6.6

The number of iterations required by the algorithm depends on the intended precision of the computation. The table shows angles φ corresponding to $\tan(\varphi) = 2^{-i}$. We can see that after 7 rotations, the angular error is less than $1°$. A sequence of progressively smaller rotations is used to rotate to an arbitrary angle.

Basic CORDIC Iteration

$$x_{i+1} = (K_i) \cdot [x_i - y_i \cdot d_i \cdot 2^{-i}]$$
$$y_{i+1} = (K_i) \cdot [y_i + x_i \cdot d_i \cdot 2^{-i}]$$

$$K_i = \cos(\tan^{-1}(2^{-i})) = 1/(1 + 2^{-2i})^{0.5}$$
$$d_i = \pm 1$$

The d_i is chosen to rotate by $\pm\varphi$

- If we don't multiply (x_{i+1}, y_{i+1}) by K_i we get a gain error which is independent of the direction of the rotation
- The error converges to 0.61 - May not need to compensate for it
- We also can accumulate the rotation angle: $z_{i+1} = z_i - d_i \cdot \tan^{-1}(2^{-i})$

6.7

Slide 6.7

The basic CORDIC iteration (from step i to step $i+1$) is described by the formulas highlighted in gray. K_i is the gain factor for iteration i that is used to compensate for the attenuation caused by $\cos(\varphi_i)$. If we don't multiply (x_{i+1}, y_{i+1}) by K_i, we get gain error that converges to 0.61. In some applications, this error does not have to be compensated for.

An important parameter of the CORDIC algorithm is the direction of each rotation. The direction is captured in parameter $d_i = \pm 1$, which depends on the residual error. To decide d_i, we also accumulate the rotation angle $z_{i+1} = z_i - d_i \cdot \tan^{-1}(2^{-i})$ and calculate d_i according to the angle polarity. The gain error is independent of the direction of the rotation.

Example

- Initial vector is described by x_0 and y_0 coordinates

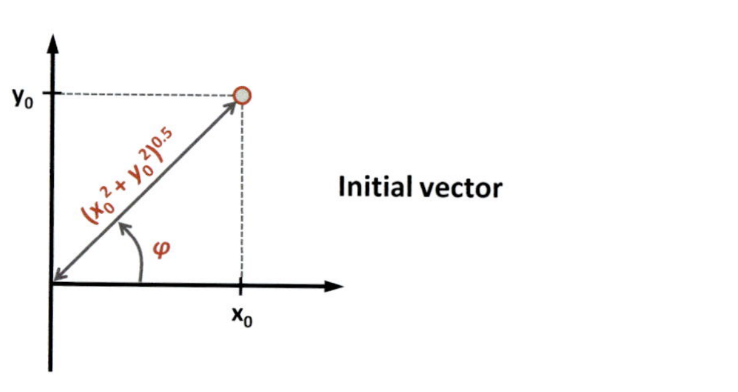

- We want to find φ and $(x_0^2 + y_0^2)^{0.5}$

6.8

Slide 6.8

Let's do an example of a Cartesian-to-polar coordinate translation. The initial vector is described by the coordinates (x_0, y_0). A sequence of rotations will align the vector with the x-axis in order to calculate $(x_0^2 + y_0^2)^{0.5}$ and φ.

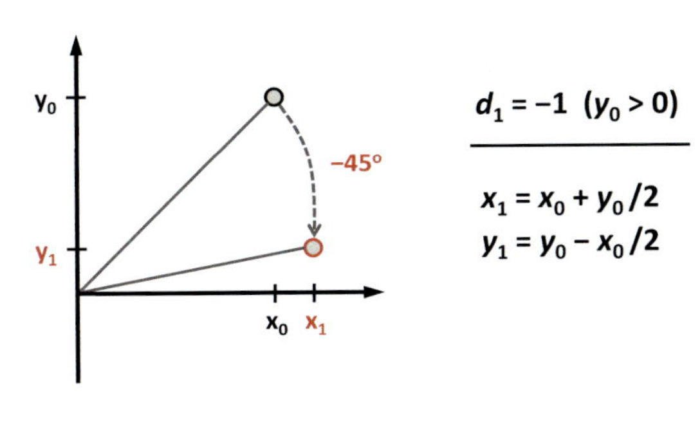

Slide 6.9

The first step of the algorithm is to check the sign of the initial y-coordinate, y_0, to determine the direction d_1 of the first rotation. In the first iteration, the initial vector is rotated by 45° (for $y_0 < 0$) or by −45° (for $y_0 > 0$). In our example, $y_0 > 0$, so we rotate by −45°. The new vector is calculated as $x_1 = x_0 + y_0/2$, $y_1 = y_0 - x_0/2$.

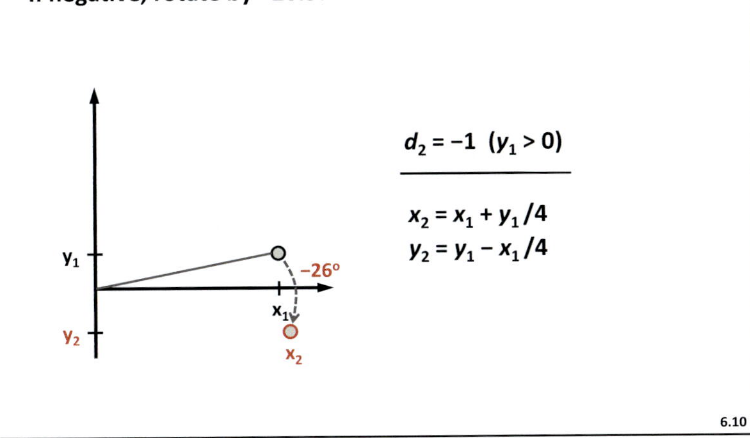

Slide 6.10

The next step is to apply the same procedure to (x_1, y_1) with an updated angle. The angle of the next rotation is $\pm 26.57°$ ($\tan^{-1}(2^{-1}) = 26.57°$), depending upon the sign of y_1. Since $y_1 > 0$, we continue the clockwise rotation to reach (x_2, y_2) as shown on the slide. With each rotation, we are closer to the y-axis.

Repeat Step 2 for Each Rotation k

- Until $y_n = 0$

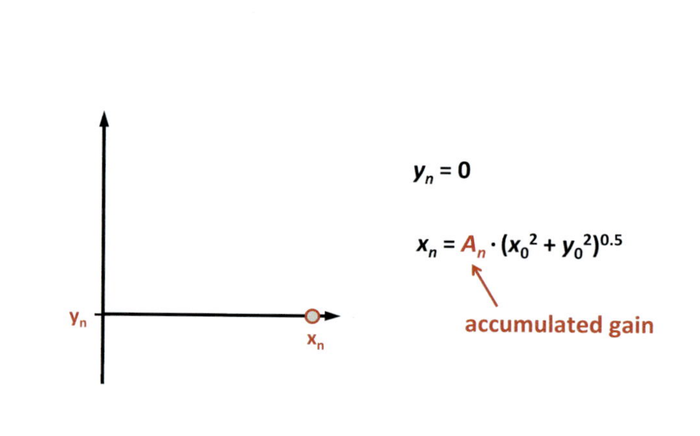

6.11

Slide 6.11

The procedure is repeated until $y_n = 0$. The number of iterations is a function of the desired accuracy (number of bits). When the algorithm converges to $y_n = 0$, we have $x_n = A_n \cdot (x_0^2 + y_0^2)^{0.5}$, where A_n represents the accumulated gain error.

The Gain Factor

- Gain accumulation:
  $G_0 = 1$
  $G_0 G_1 = 1.414$
  $G_0 G_1 G_2 = 1.581$
  $G_0 G_1 G_2 G_3 = 1.630$
  $G_0 G_1 G_2 G_3 G_4 = 1.642$

- So, start with x_0, y_0; end up with:

  Shift & adds of x_0, y_0

  $z_3 = 71°$
  $(x_0^2 + y_0^2)^{0.5} = 1.642 \, (...)$

- We did the rectangular-to-polar coordinate conversion

6.12

Slide 6.12

The accumulation of the gain error is illustrated on this slide. The algorithm has implicit gain unless we multiply by the product of the cosine of the rotation angles. The gain value converges to 1.642 after 4 iterations. Starting from (x_0, y_0), we end up with an angle of $z_3 = 71°$ and magnitude with an accumulated gain error of 1.642. That was the conversion from rectangular to polar coordinates.

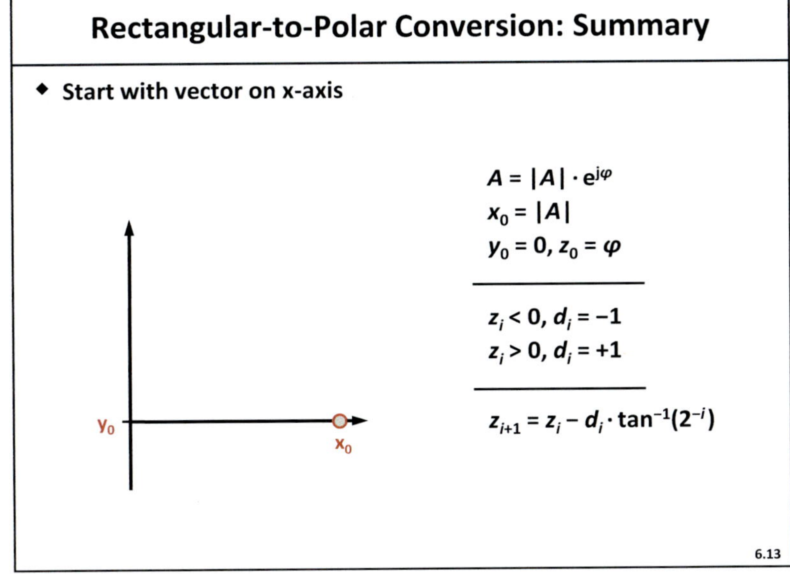

Slide 6.13

We can also do the reverse transformation: conversion from polar to rectangular coordinates. Given $A=|A|\cdot e^{j\varphi}$, we start from $x_0=|A|$ and $y_0=0$, and a residual angle of $z_0=\varphi$. The algorithm starts with a vector aligned to the x-axis. Next, we rotate by successively smaller angles in directions dependent on the sign of z_i. If $z_i > 0$, positive (counterclockwise) rotation is performed. In each iteration we keep track of the residual angle z_{i+1} and terminate the algorithm when $z_{i+1}=0$. The polar-to-rectangular translation is also known as the rotation mode.

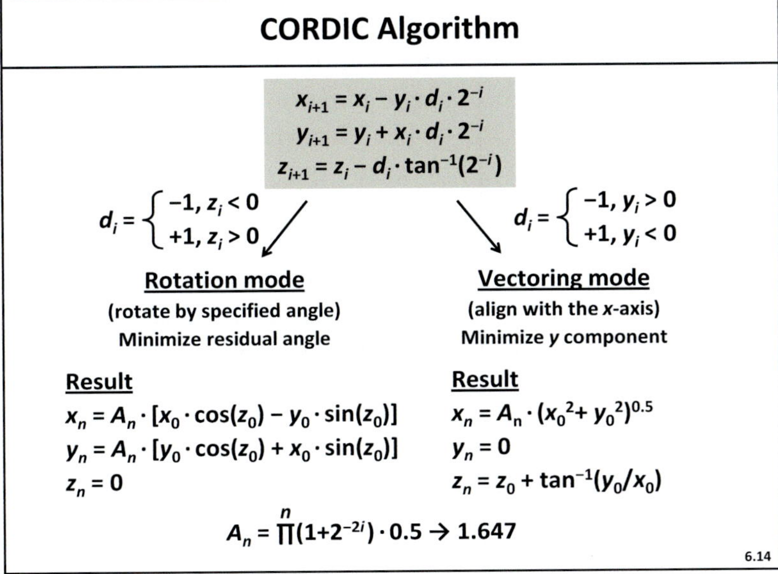

Slide 6.14

The CORDIC algorithm analyzed so far can be summarized with the formulas highlighted in the gray box. We can choose between the vectoring and rotation modes. In the vectoring mode, we start with a vector having a non-zero angle and try to minimize the y component. As a result, the algorithm computes vector magnitude and angle. In the rotation mode, we start with a vector aligned with the x-axis and try to minimize the z-component. As a result, the algorithm computes the rectangular coordinates of the original vector. In both cases, the built-in gain A_n converges to 1.647 and, in cases where scaling factors affect final results, has to be compensated.

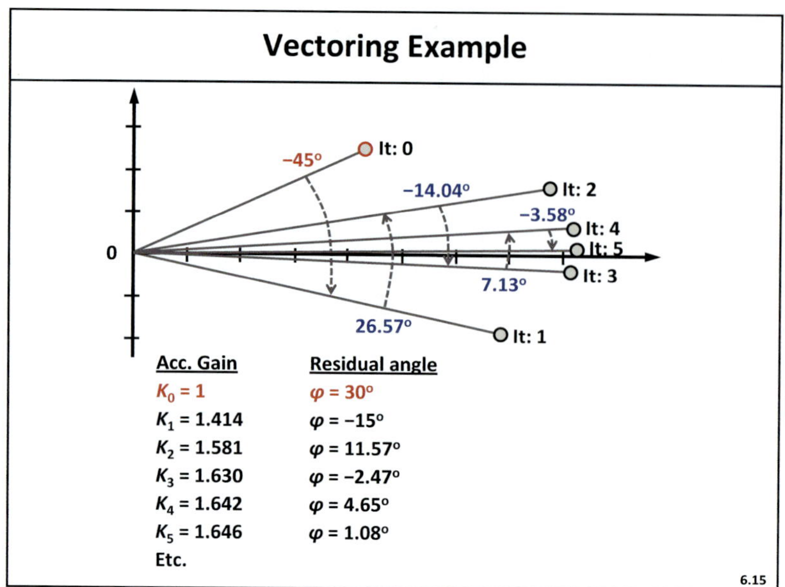

Slide 6.15

This slide graphically illustrates algorithm convergence for vectoring mode, starting with a vector having an initial angle of 30°. We track the accumulated gain and residual angle, starting from the red dot (Iteration: 0). Rotations are performed according to sign of y_i with the goal of minimizing the y-component. The first rotation by −45° produces a residual angle of −15° and a gain of 1.414. The second rotation is by +26.57°, resulting in an accumulated gain of 1.581. This process continues, as shown on the slide, toward smaller residual angles and an accumulated gain of 1.647.

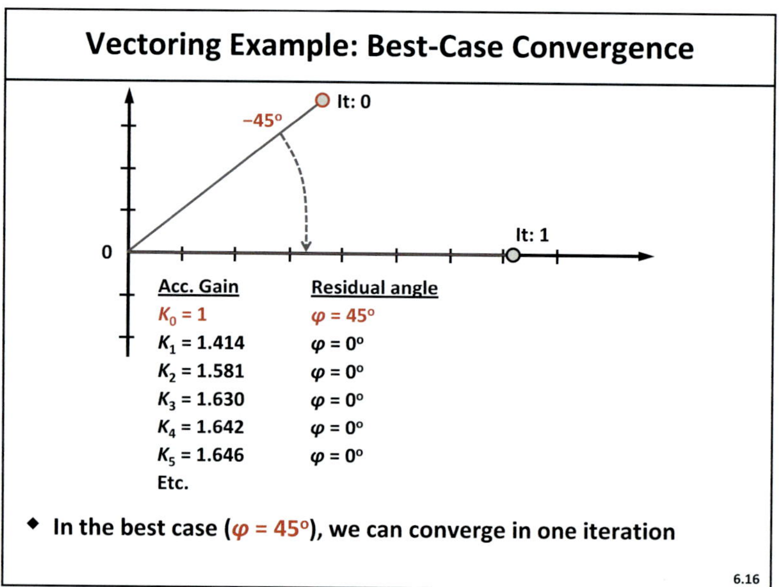

Slide 6.16

The algorithm can converge in a single iteration for the case when φ = 45°. This trivial case is interesting, because it suggests that the convergence accuracy of a recursive algorithm greatly depends upon the initial conditions and the granularity of the rotation angle in each step. We will use this idea later in the chapter.

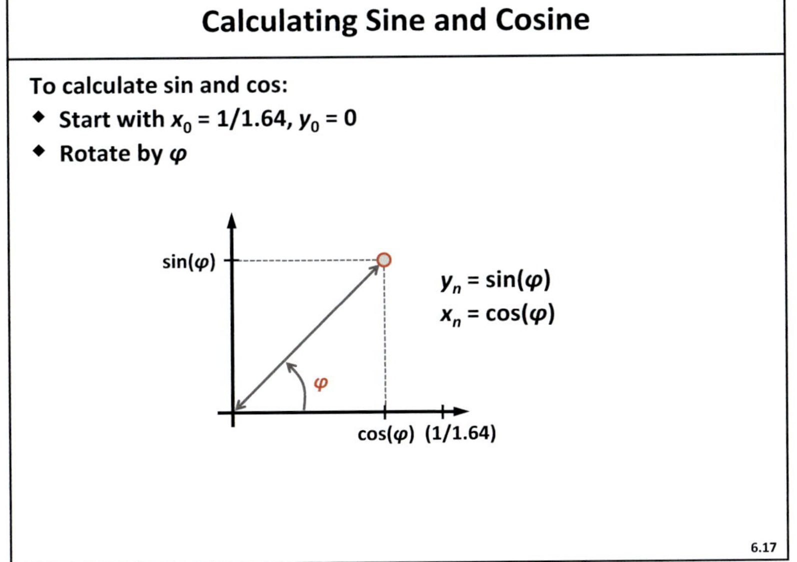

Slide 6.17

CORDIC can also be used to calculate trigonometric functions. This slide shows $\sin(\varphi)$ and $\cos(\varphi)$. We start from a scaled version of x_0 to account for the gain that will be accumulated and use CORDIC in the rotation mode to calculate $\sin(\varphi)$ and $\cos(\varphi)$.

Functions

Rotation mode

sin/cos
z_0 = angle
$y_0 = 0$, $x_0 = 1/A_n$
$x_n = \boxed{A_n \cdot x_0} \cdot \cos(z_0)$
$y_n = \boxed{A_n \cdot x_0} \cdot \sin(z_0)$
(=1)

Polar → Rectangular

$x_n = r \cdot \cos(\varphi)$
$y_n = r \cdot \sin(\varphi)$

$x_0 = r$
$z_0 = \varphi$
$y_0 = 0$

Vectoring mode

$\tan^{-1}$
$z_0 = 0$
$z_n = z_0 + \tan^{-1}(y_0/x_0)$

Vector/Magnitude

$x_n = A_n \cdot (x_0^2 + y_0^2)^{0.5}$

Rectangular → Polar

$r = (x_0^2 + y_0^2)^{0.5}$
$\varphi = \tan^{-1}(y_0/x_0)$

6.18

Slide 6.18

With only small modifications to the algorithm, it can be used to calculate many other functions. So far, we have seen the translation between polar and rectangular coordinates and the calculation of sine and cosine functions. We can also calculate $\tan^{-1}$ and the vector magnitude in the vectoring mode, as described by the formulas on the slide.

CORDIC Divider

- **To do a divide, change CORDIC rotations to a linear function calculator**

$$x_{i+1} = x_i - \boxed{0} \cdot y_i \cdot d_i \cdot 2^{-i} = x_i$$

$$y_{i+1} = y_i + x_i \cdot d_i \cdot 2^{-i}$$

$$z_{i+1} = z_i - d_i \cdot \boxed{(2^{-i})}$$

6.19

Slide 6.19

CORDIC can be used to calculate linear operations such as division. CORDIC can be easily re-configured to support linear functions by introducing modifications to the x and z components as highlighted in this slide. The x component is trivial, $x_{i+1} = x_i$. The z component uses 2^{-i} instead of $\tan^{-1}(2^{-i})$. The modifications can be generalized to other linear functions.

Generalized CORDIC

$$x_{i+1} = x_i - \boxed{m} \cdot y_i \cdot d_i \cdot 2^{-i}$$

$$y_{i+1} = y_i + x_i \cdot d_i \cdot 2^{-i}$$

$$z_{i+1} = z_i - d_i \cdot \boxed{e_i}$$

d_i	
Rotation	Vectoring
$d_i = -1, z_i < 0$	$d_i = -1, y_i > 0$
$d_i = +1, z_i > 0$	$d_i = +1, y_i < 0$
$\text{sign}(z_i)$	$-\text{sign}(y_i)$

Mode	m	e_i
Circular	+1	$\tan^{-1}(2^{-i})$
Linear	0	2^{-i}
Hyperbolic	−1	$\tanh^{-1}(2^{-i})$

6.20

Slide 6.20

The generalized algorithm is described in this slide. Based on m and e_i, the algorithm can be configured into one of three modes: circular, linear, or hyperbolic. For each mode, there are two sub-modes: rotation and vectoring. Overall, the algorithm can be programmed into one of the six modes to calculate a wide variety of functions.

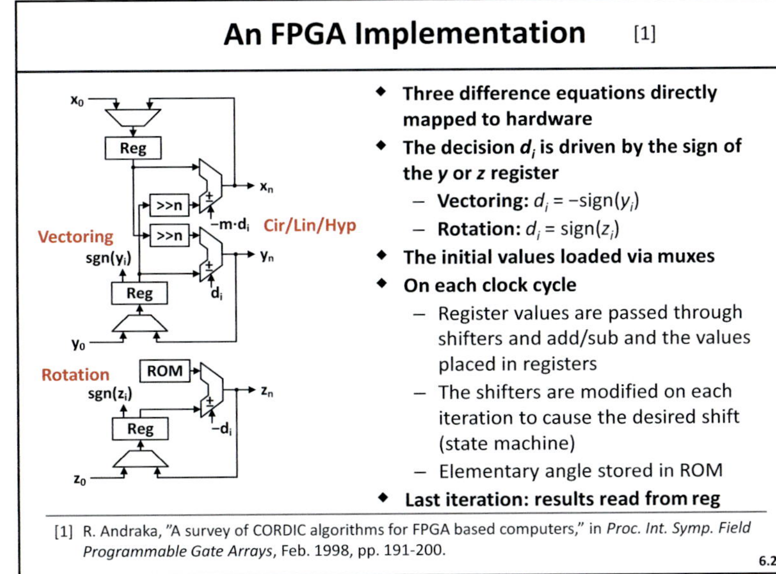

Slide 6.21

This slide shows a direct-mapped realization of CORDIC [1]. Algorithm operations from the tree difference equations are mapped to hardware blocks as shown on the left. The registers can take either the initial conditions (x_0, y_0, z_0) or the result from previous iteration (x_n, y_n, z_n) to compute the next iteration. The shift by n blocks ($>> n$) indicate multiplication by 2^{-n} (in iteration n). The parameter m controls the operating mode (circular, linear, hyperbolic). Rotational mode is determined based on the sign of y_i (vectoring) or the sign of z_i (rotation). Elementary rotation angles are stored in ROM memory. After each clock cycle CORDIC resolves one bit of accuracy.

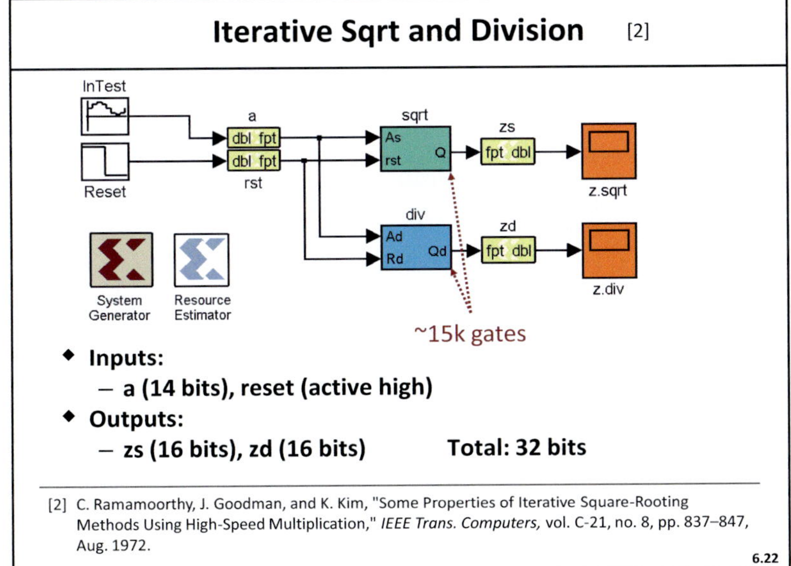

Slide 6.22

The block diagram from the previous slide can be modeled as the sub-system shown here. Inputs from MATLAB are converted to fixed-point precision via the input ports (*yellow blocks*). The outputs of the hardware blocks are also converted to floating-point representations (zs, zd blocks) for waveform analysis in MATLAB. Blocks between the *yellow blocks* represent fixed-point hardware. The resource estimation block estimates the amount of hardware used to implement each block. The system generator block produces a hardware description for implementing the blocks on an FPGA. The complexity of the iterative sqrt and div algorithms in this example is about 15 k gates for 14-bit inputs and 16-bit outputs. The block-based model and hardware estimation tools allow for the exploration of multiple different hardware realizations of the same algorithm. Next, we will look into the implementation of iterative square rooting and division algorithms [2].

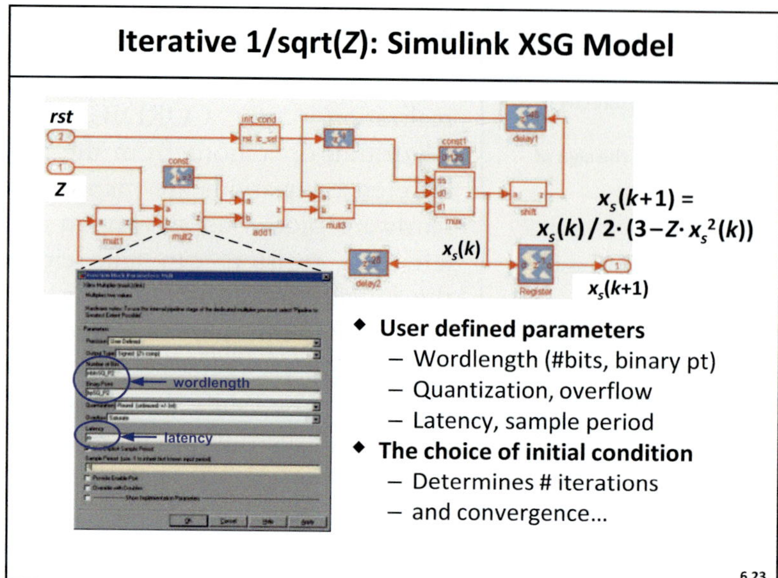

Slide 6.23

The hardware model of an algorithm can be constructed using the Xilinx System Generator (XSG) library. Each block (or hierarchical sub-system) has a notion of wordlength and latency. The wordlength is specified as the total and fractional number of bits; the latency is specified as the number of clock cycles. Additional details such as quantization, overflow modes, and sample period can be defined as well.

Suppose we want to implement $1/\text{sqrt}()$. This can be done with CORDIC in n iterations for n bits of accuracy or by using the Newton-Raphson algorithm in $n/2$ iterations. The Newton-Raphson method is illustrated here. The block diagram shows the implementation of iterative formula $x_s(k+1) = x_s(k)/2 \cdot (3 - Z \cdot x_s^2(k))$ to calculate $1/\text{sqrt}(Z)$. It is important to realize that the convergence speed greatly depends not just on the choice of the algorithm, but also on the choice of initial conditions. The initial condition block (init_cond) computes the initial condition that guarantees convergence in a fixed number of iterations.

Quadratic Convergence: 1/sqrt(N)

$$x_s(k+1) = \frac{x_s(k)}{2} \cdot (3 - N \cdot x_s(k)^2) \qquad x_s(k) \to \left.\frac{1}{\sqrt{N}}\right|_{k \to \infty}$$

$$y_s(k+1) = \frac{y_s(k)}{2} \cdot (3 - y_s(k)^2) \qquad y_s(k) \to 1 \big|_{k \to \infty}$$

$$e_s(k) = y_s(k) - 1$$

$$e_s(k+1) = -\frac{1}{2} \cdot e_s(k)^2 \cdot (3 + e_s(k))$$

Slide 6.24

The convergence of the Newton-Raphson algorithm for $1/\text{sqrt}(N)$ is analyzed here. The difference equation $x_s(k)$ converges to $1/\text{sqrt}(N)$ for large k. Since N is unknown, it is better to analyze the normalized system $y_s(k) = \text{sqrt}(N) \cdot x_s(k)$ where $y_s(k)$ converges to 1. The error formula for $e_s(k) = y_s(k) - 1$ reflects relative error. The algebra yields a quadratic expression for the error. This means that each iteration resolves two bits of accuracy. The number of iterations depends on the desired accuracy as well as the choice of the initial condition.

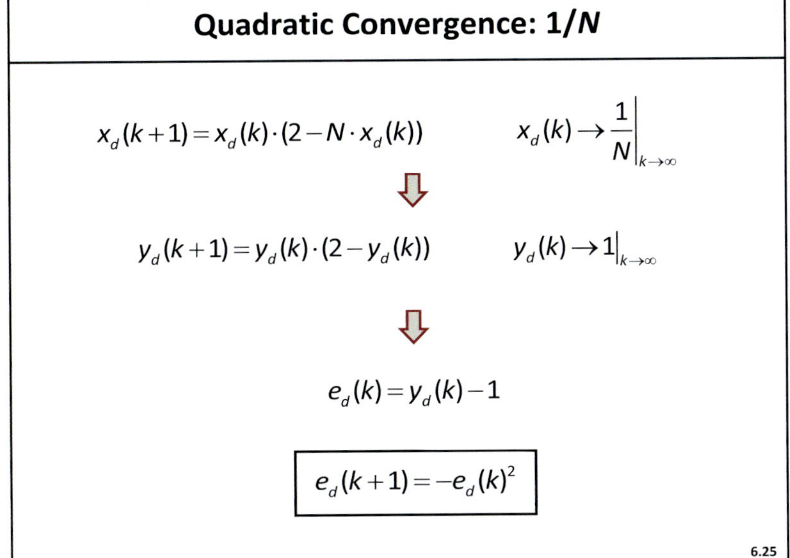

$$x_d(k+1) = x_d(k) \cdot (2 - N \cdot x_d(k)) \qquad x_d(k) \to \left.\frac{1}{N}\right|_{k \to \infty}$$

$$y_d(k+1) = y_d(k) \cdot (2 - y_d(k)) \qquad y_d(k) \to \left.1\right|_{k \to \infty}$$

$$e_d(k) = y_d(k) - 1$$

$$e_d(k+1) = -e_d(k)^2$$

Slide 6.25

Following the analysis from the previous slide, the convergence of the Newton-Raphson algorithm for $1/N$ is shown here. The $x_d(k)$ converges to $1/N$ for large k. We also analyze the normalized system $y_d(k) = N \cdot x_d(k)$ where $y_d(k)$ converges to 1. The relative error formula for $e_d(k) = y_d(k) - 1$ reflect also exhibits quadratic dependence.

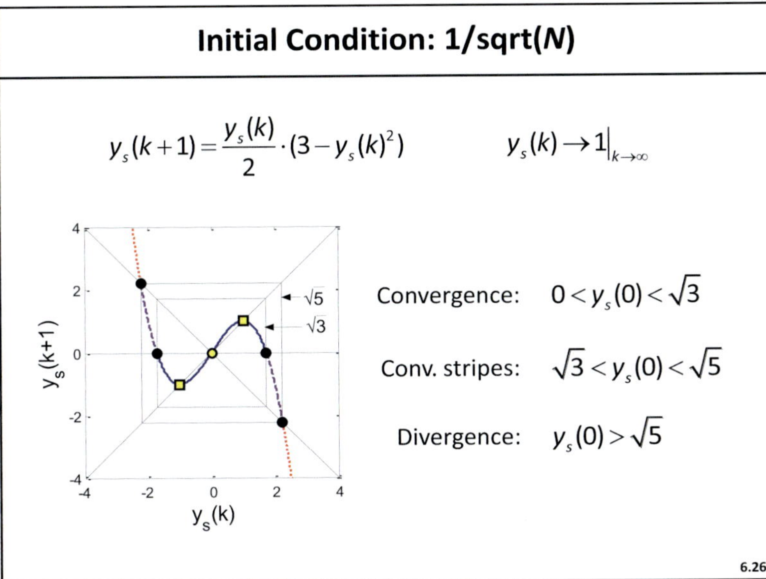

$$y_s(k+1) = \frac{y_s(k)}{2} \cdot (3 - y_s(k)^2) \qquad y_s(k) \to \left.1\right|_{k \to \infty}$$

Convergence: $\quad 0 < y_s(0) < \sqrt{3}$

Conv. stripes: $\quad \sqrt{3} < y_s(0) < \sqrt{5}$

Divergence: $\quad y_s(0) > \sqrt{5}$

Slide 6.26

The choice of the initial condition greatly affects convergence speed. The plot on this slide is a transfer function corresponding to one iteration of the $1/\text{sqrt}(N)$ algorithm. By analyzing the mapping of $y_s(k)$ into $y_s(k+1)$, we can gain insight into how to select the initial condition. The squares indicate convergence points for the algorithms. The round *yellow dot* is another fixed point that represents the trivial solution. The goal is thus to select the initial condition that is closest to the *yellow squares*. Points on the *blue lines* guarantee convergence. Points along the dashed *pink line* would also result in convergence, but with a larger number of iterations. Points on the *red dotted lines* will diverge.

Initial Condition: 1/*N*

$$y_d(k+1) = y_d(k) \cdot (2 - y_d(k)) \qquad \left. y_d(k) \to 1 \right|_{k \to \infty}$$

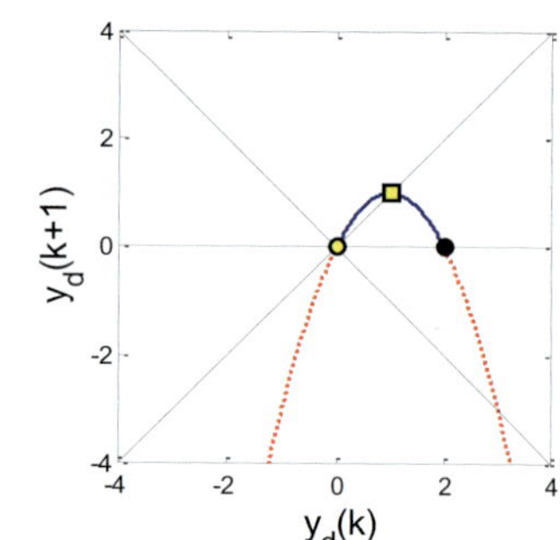

Convergence: $0 < y_d(0) < 2$

Divergence: otherwise

6.27

Slide 6.27

Similarly, we look at the transfer function $y_d(k+1) = f(y_d(k))$ for the division algorithm. The algorithm converges for $0 < y_d(0) < 2$; otherwise it diverges.

1/sqrt(*N*): Convergence Analysis

$$x_{n+1} = \frac{x_n}{2} \cdot \left(3 - N \cdot x_n^2\right) \qquad \left. x_n \to \frac{1}{\sqrt{N}} \right|_{n \to \infty}$$

$$x_n = \frac{y_n}{\sqrt{N}}$$

$$y_{n+1} = \frac{y_n}{2} \cdot \left(3 - y_n^2\right) \quad [3]$$

Error: $e_n = 1 - y_n$

$$e_{n+1} = \frac{3}{2} \cdot e_n^2 - \frac{1}{2} \cdot e_n^3$$

[3] D. Marković, A Power/Area Optimal Approach to VLSI Signal Processing, Ph.D. Thesis, University of California, Berkeley, 2006.

6.28

Slide 6.28

With some insight about convergence, further analysis is needed to better understand the choice of the initial condition. This slide repeats the convergence analysis from Slide 6.24 to illustrate error dynamics. The error formula $e_{n+1} = 1.5 \cdot e_n^2 - 0.5 \cdot e_n^3$ indicates quadratic convergence, but does not explicitly account for the initial condition (included in e_0) [3].

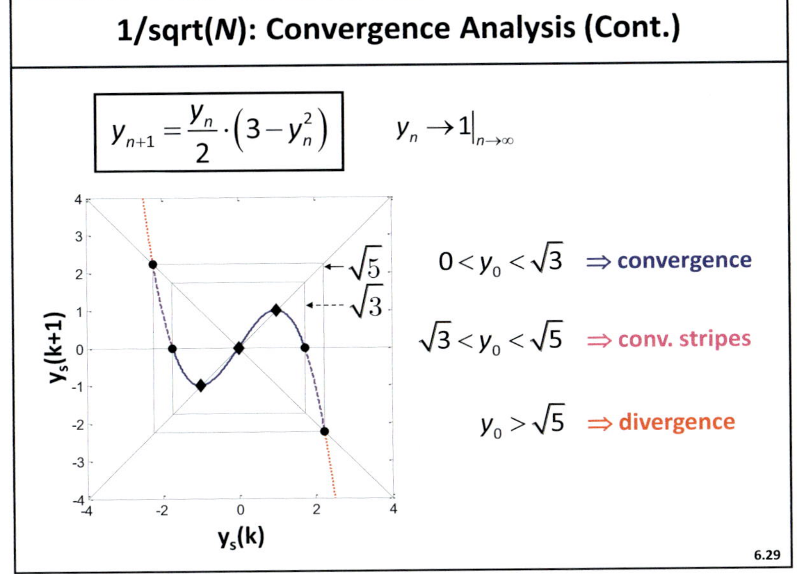

Slide 6.29

To gain more insight into convergence error, let's revisit the transfer function for the $1/\mathrm{sqrt}()$ algorithm. The colored segments indicate three regions of convergence. Since the transfer plot is symmetric around 0, we look into positive values. For $0 < y_0 < \mathrm{sqrt}(3)$, the algorithm converges to 1. For $\mathrm{sqrt}(3) < y_0 < \mathrm{sqrt}(5)$, the algorithm still converges, but it may take on negative values and require a larger number of iterations before it settles. For $y_0 > \mathrm{sqrt}(5)$, the algorithm diverges.

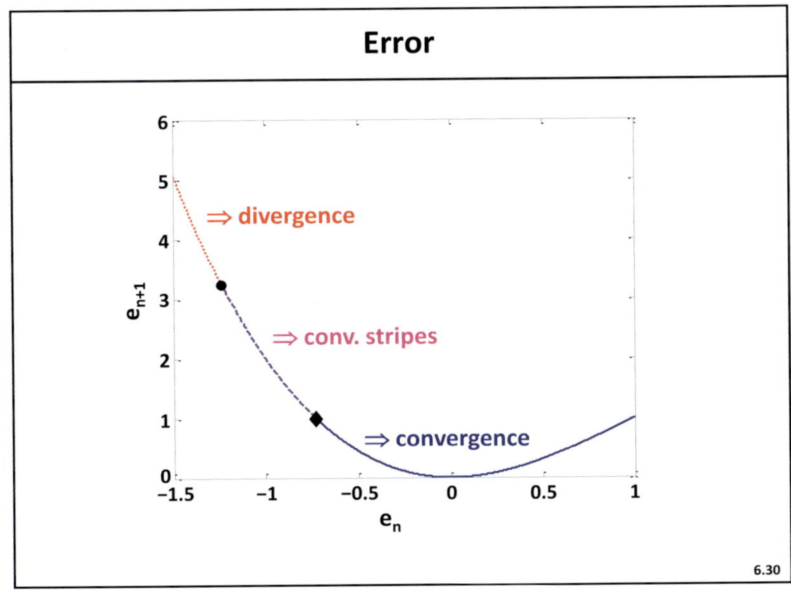

Slide 6.30

The error transfer function is shown on this plot. The convergence bands discussed in the previous slide are also shown. In order to see what this practically means, we will next look into the time response of y_n.

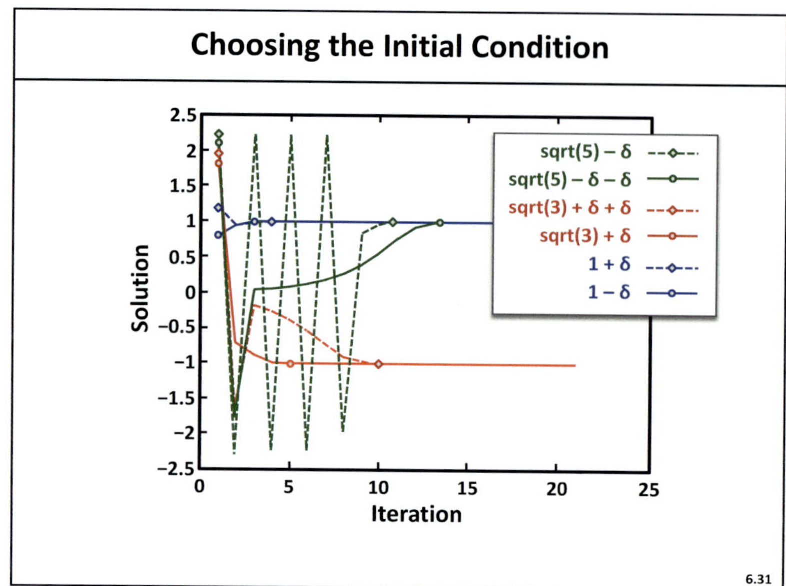

Slide 6.31

This plot shows what happens when we select the initial conditions in each of the convergence bands as previously analyzed. If we start from close neighborhood of the solution and begin with $1+\delta$ or $1-\delta$, the algorithm converges in less than 3 or 4 iterations. If we start from the initial value of sqrt(3)$+\delta$, the algorithm will converge in 5 iterations, but the solution might have the reverse polarity (-1 instead of 1). Getting further away from sqrt(3) will take longer (10 iterations for 2δ away) and the solution may be of either polarity. Starting from sqrt(5) $- 2\delta$, we see a sign change initially and a convergence back to 1. Getting closer to sqrt(5) will incur more sign changes and a slowly decreasing absolute value until the algorithm settles to 1 (or -1).

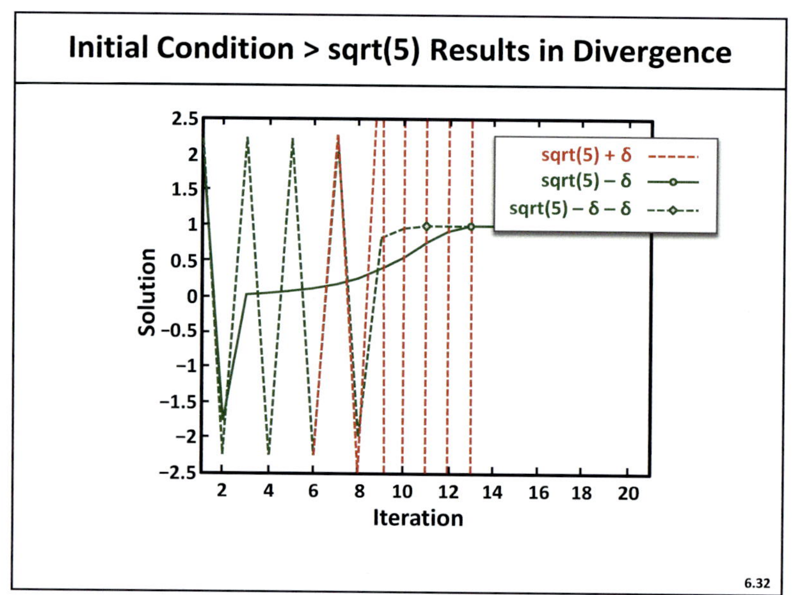

Slide 6.32

For initial conditions greater than sqrt(5), the algorithm diverges as shown on this plot. The key, therefore, is to pick a suitable initial condition that will converge within the specified number of iterations. This has to be properly analyzed and implemented in hardware.

1/sqrt(): Picking Initial Condition

$$x_{n+1} = \frac{x_n}{2} \cdot \left(3 - N \cdot x_n^2\right) \qquad x_n \to \frac{1}{\sqrt{N}}\Big|_{n \to \infty}$$

* **Equilibriums** $\quad 0, \quad \pm \frac{1}{\sqrt{N}}$

* **Initial condition**
  - Take: $\quad V(x_n) = \left(x_n - \frac{1}{\sqrt{N}}\right)^2$ **(6.1)**
  - Find x_0 such that:
    $$V(x_{n+1}) - V(x_n) < 0 \quad \forall n \quad (n=0,1,2,\dots) \qquad \textbf{(6.2)}$$
  - Solution: $\quad S = \{x_0 : V(x_0) < a\}$ **(6.3)**

    "Level set" $V(x_0) = a$ is a convergence bound
    → Local convergence (3 equilibriums)

6.33

Slide 6.33

The idea is to determine the initial condition to guarantee convergence in a bounded number of iterations. Moreover, it is desirable to have decreasing error with every iteration. To do this, we can define a function $V(x_n)$ and set convergence bound by specifying $V(x_0) < a$. To account for symmetry around 0, we define $V(x_n)$ to be positive, as given by (6.1). This function defines the squared absolute error. To satisfy decreasing error requirement, we set $V(x_{n+1}) - V(x_n) < 0$ as in (6.2). Finally, to guarantee convergence in a fixed number of iterations, we need to choose an x_0 that satisfies (6.3).

Descending Absolute Error

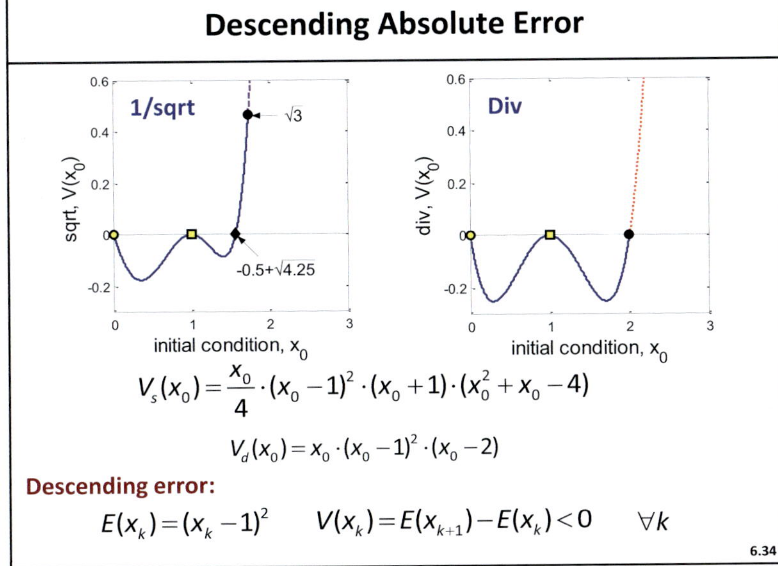

$$V_s(x_0) = \frac{x_0}{4} \cdot (x_0 - 1)^2 \cdot (x_0 + 1) \cdot (x_0^2 + x_0 - 4)$$

$$V_d(x_0) = x_0 \cdot (x_0 - 1)^2 \cdot (x_0 - 2)$$

Descending error:

$$E(x_k) = (x_k - 1)^2 \qquad V(x_k) = E(x_{k+1}) - E(x_k) < 0 \qquad \forall k$$

6.34

Slide 6.34

The initial condition for which the absolute error decreases after each iteration is calculated here. For the square rooting algorithm the initial condition is below sqrt(4.25) − 0.5, which is more restrictive than the simple convergence constraint given by sqrt(3). For the division algorithm, initial conditions below 2 guarantees descending error. Next, we need to implement the calculation of x_0.

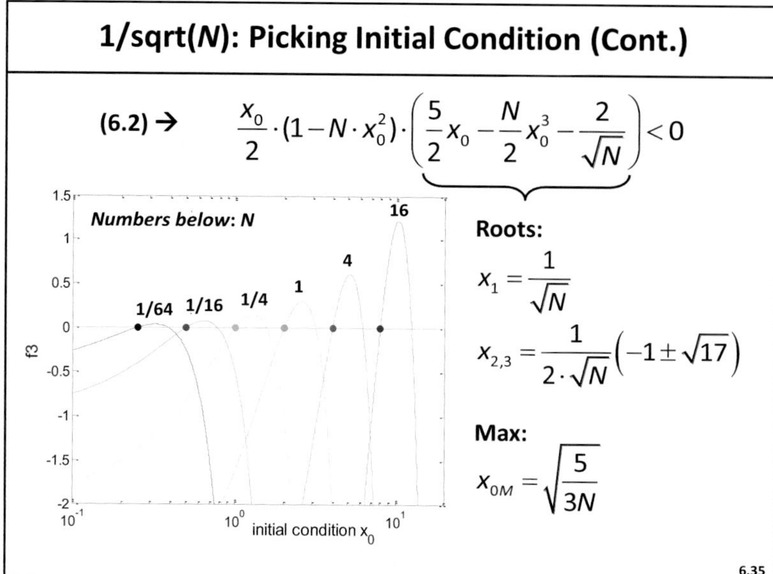

$$(6.2) \rightarrow \quad \frac{x_0}{2} \cdot (1 - N \cdot x_0^2) \cdot \left(\frac{5}{2} x_0 - \frac{N}{2} x_0^3 - \frac{2}{\sqrt{N}} \right) < 0$$

Roots:

$$x_1 = \frac{1}{\sqrt{N}}$$

$$x_{2,3} = \frac{1}{2 \cdot \sqrt{N}} \left(-1 \pm \sqrt{17} \right)$$

Max:

$$x_{0M} = \sqrt{\frac{5}{3N}}$$

Slide 6.35

The solution to (6.2) yields the inequality shown here. The roots are positioned at x_1 and $x_{2,3}$, as given by the formulas on the slide. The plot on the left shows the term $f_3 = (2.5 \cdot x_0 - N/2 \cdot x_0^3 - 2/\text{sqrt}(N))$ as a function of x_0 for N ranging from 1/64 to 16. Zero crossings are indicated with dots. The initial condition has to be chosen from positive values of f_3. The function f_3 has a maximum at $\text{sqrt}(5/3N)$, which could be used as an initial condition.

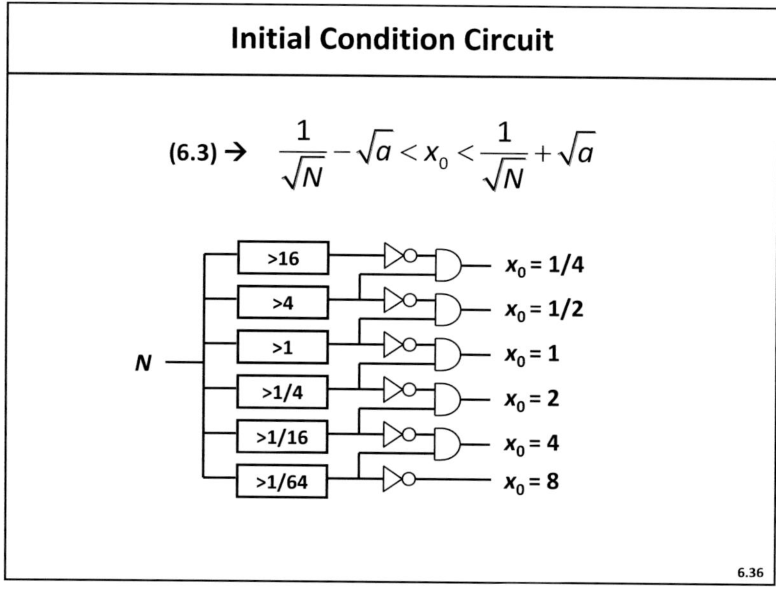

$$(6.3) \rightarrow \quad \frac{1}{\sqrt{N}} - \sqrt{a} < x_0 < \frac{1}{\sqrt{N}} + \sqrt{a}$$

Slide 6.36

Here is an implementation of the initial condition discussed in the previous slide. According to (6.3), $1/\text{sqrt}(N) - \text{sqrt}(a) < x_0 < 1/\text{sqrt}(N) + \text{sqrt}(a)$. For N ranging from 1/64 to 16, the initial circuit calculates 6 possible values for x_0. The threshold values in the first stage are compared to N to select a proper value of x_0. This circuit can be implemented using inexpensive digital logic.

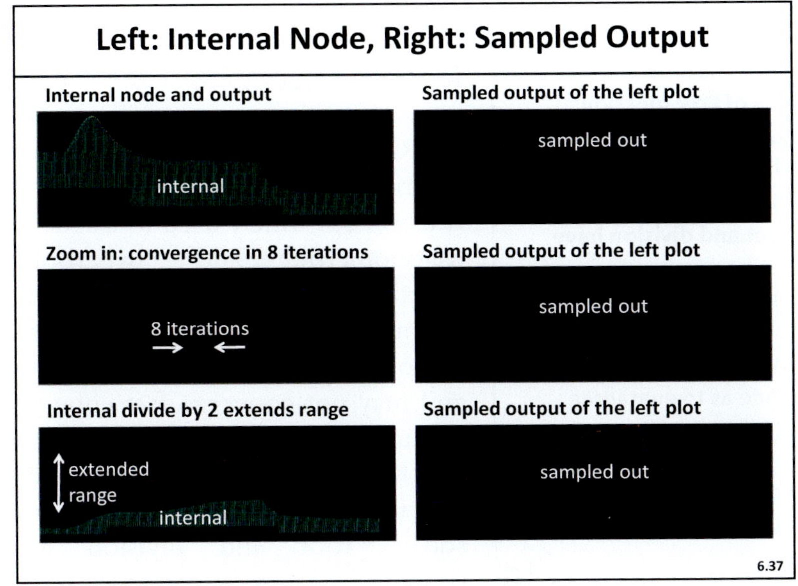

Slide 6.37

This slide shows the use of the initial conditions circuit in the algorithm implementation. On the left, we see the internal node $x_s(k)$ from Slide 6.23, and on the right we have the sampled output $x_s(k+1)$. The middle plot on the left shows a magnified section of the waveform on the top. Convergence within 8 iterations is illustrated. Therefore, it takes 8 clock cycles to process one input. This means that the clock frequency is 8-times higher than the input sample rate. The ideal (*pink*) and calculated (*yellow*) values are compared to show good matching between the calculated and expected results.

Another idea to consider in the implementation is to shift the internal computation ($x_s(k)$ from Slide 6.23) to effectively extend the range of the input argument. Division by 2 is illustrated on the bottom plot to present this concept. The output $x_s(k+1)$ has to be shifted in the opposite direction to compensate for the input scaling. The output on the bottom right shows correct tracking of the expected result.

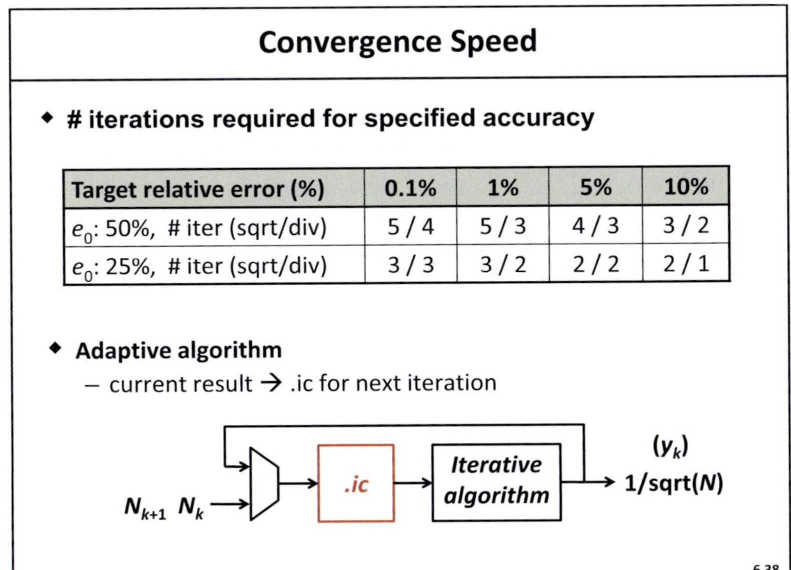

Target relative error (%)	0.1%	1%	5%	10%
e_0: 50%, # iter (sqrt/div)	5 / 4	5 / 3	4 / 3	3 / 2
e_0: 25%, # iter (sqrt/div)	3 / 3	3 / 2	2 / 2	2 / 1

Slide 6.38

The convergence speed is analyzed here as a function of the desired accuracy and of the initial error (initial condition). Both the square-root and the division algorithms converge in at most 5 iterations for an accuracy of 0.1% with initial conditions up to 50% away from the solution. The required number of iterations is less for lower accuracy or lower initial error.

An idea that can be considered for adaptive algorithms with slowly varying input is to use the solution of the algorithm from the previous iteration as the initial condition to the next iteration. This is particularly useful for flat-fading channel estimation in wireless communications or for adaptive filters with slowly varying coefficients. For slowly varying inputs, we can even converge in a single iteration. This technique will be applied in later chapters in the realization of adaptive LMS filters.

Summary

- ◆ Iterative algorithms can be use for a variety of DSP functions
- ◆ CORDIC uses angular rotations to compute trigonometric and hyperbolic functions as well as divide and other operations
  - One bit of resolution is resolved in each iteration
- ◆ Netwon-Raphson algorithms for square root and division have faster convergence than CORDIC
  - Two bits of resolution are resolved in each iteration (the algorithm has quadratic error convergence)
- ◆ Convergence speed greatly depends on the initial condition
  - The choice of initial condition can be made as to guarantee decreasing absolute error in each iteration
  - For slowly varying inputs, adaptive algorithms can use the result of current iteration as the initial condition
  - Hardware latency depends on the initial condition and accuracy

6.39

Slide 6.39

Iterative algorithms and their implementation were discussed. CORDIC, a popular iterative algorithm based on angular rotations, is widely used to calculate trigonometric and hyperbolic functions as well as divide operation. CORDIC resolves one bit of precision in each iteration and may not converge fast enough. As an alternative, algorithms with faster convergence such as Newton-Raphson methods for square root and division are proposed. These algorithms have quadratic convergence, which means that two bits of resolution are resolved in every iteration. Convergence speed also depends on the choice of initial condition, which can be set such that absolute error decreases in each iteration. Another technique to consider is to initialize the algorithm with the output from previous iteration. This is applicable to systems with slowly varying inputs. Ultimately, real-time latency of recursive algorithms depends on the choice of initial condition and desired resolution in the output.

References

- ◆ R. Andraka, "A Survey of CORDIC Algorithms for FPGA based Computers," in *Proc. Int. Symp. Field Programmable Gate Arrays,* Feb. 1998, pp. 191-200.

- ◆ C. Ramamoorthy, J. Goodman, and K. Kim, "Some Properties of Iterative Square-Rooting Methods Using High-Speed Multiplication," *IEEE Trans. Computers,* vol. C-21, no. 8, pp. 837–847, Aug. 1972.

- ◆ D. Marković, A Power/Area Optimal Approach to VLSI Signal Processing, Ph.D. Thesis, University of California, Berkeley, 2006.

Additional References

- ◆ S. Oberman and M. Flynn, "Division Algorithms and Implementations," *IEEE Trans. Computers,* vol. 47, no. 8, pp. 833–854, Aug. 1997.

- ◆ T. Lang and E. Antelo, "CORDIC Vectoring with Arbitrary Target Value," *IEEE Trans. Computers,* vol. 47, no. 7, pp. 736–749, July 1998.

- ◆ J. Valls, M. Kuhlmann, and K.K. Parhi, "Evaluation of CORDIC Algorithms for FPGA Design," *J. VLSI Signal Processing 32 (Kluwer Academic Publishers),* pp. 207–222, Nov. 2002.

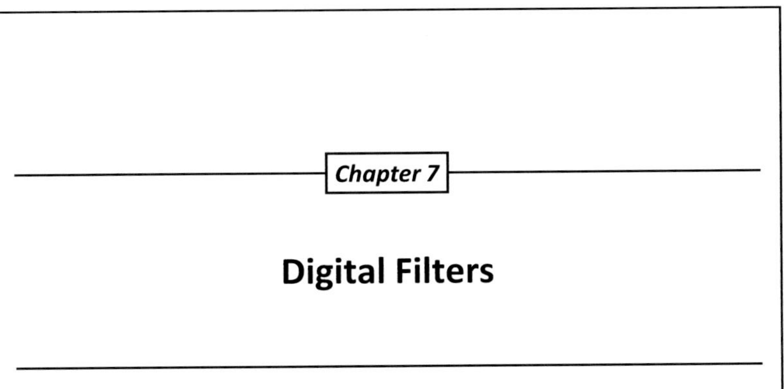

Slide 7.1

This chapter introduces the fundamentals of digital filter design, with emphasis on their usage in radio applications. Radios by definition are expected to transmit and receive signals at certain frequencies, while also ensuring that the transmission does not exceed a specified bandwidth. We will, therefore, discuss the usage of digital filters in radios, and then study specific implementations of these filters, along with pros and cons of the available architectures.

Chapter Outline

Topics discussed:

♦ **Direct Filters**

♦ **Recursive Filters**

♦ **Multi-rate Filters**

Example applications:

♦ **Radio systems** [1]
  – Band-select filters
  – Adaptive equalization
  – Decimation / interpolation

♦ **Implementation techniques**
  – Parallel, pipeline, retimed
  – Distributed arithmetic

[1] J. Proakis, Digital Communications, (3rd Ed), McGraw Hill, 2000.
[2] A.V. Oppenheim and R.W. Schafer, Discrete Time Signal Processing, (3rd Ed), Prentice Hall, 2009.
[3] J.G. Proakis and D.K. Manolakis, Digital Signal Processing, (4th Ed), Prentice Hall, 2006.

7.2

Slide 7.2

Filters are ideally suited to execute the frequency selective tasks in a radio system [1]. This chapter focuses on three classes of filters: feed-forward FIR (finite impulse response), feedback IIR (infinite impulse response), and multi-rate filters [2, 3]. A discussion on implementation examines various architectures for each of these classes, including parallelism, pipelining, and retiming. We also discuss an area-efficient implementation approach based on distributed arithmetic. The presentation of each of the three filter categories will be motivated by relevant application examples.

D. Marković and R.W. Brodersen, *DSP Architecture Design Essentials*, Electrical Engineering Essentials,
DOI 10.1007/978-1-4419-9660-2_7, © Springer Science+Business Media New York 2012

Filter Design for Digital Radios

---

[2] A.V. Oppenheim and R.W. Schafer, Discrete Time Signal Processing, (3rd Ed), Prentice Hall, 2009.
[3] J.G. Proakis and D.K. Manolakis, Digital Signal Processing, (4th Ed), Prentice Hall, 2006.

Slide 7.3

To motivate filter design, we first talk about typical radio architectures and derive specifications for the hardware implementation of the filters. As a starting example, we will discuss the raised-cosine filter and its implementation. Basic architectural concepts are introduced, and their use will be illustrated on a digital baseband for ultra-wideband radio.

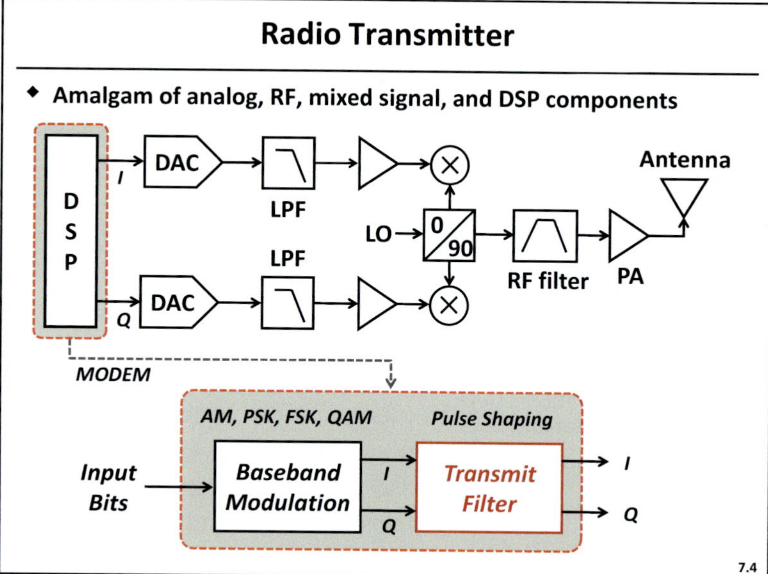

Slide 7.4

The slide shows a typical radio transmitter chain. The baseband signal from the MODEM is converted into an analog signal and then low-pass filtered before mixing. The MODEM is a DSP block responsible for modulating the raw input bit stream into AM, PSK, FSK, or QAM modulated symbols. Baseband modulation can be viewed as a mapping process where one or more bits of the input stream are mapped to a symbol from a given constellation. The constellation itself is decided by the choice of the modulation scheme. Data after modulation is still composed of bits (*square-wave*), which if directly converted to the analog domain will occupy a large bandwidth in the transmit spectrum. The digital data must therefore be filtered to avoid this spectral leakage. The transmit filter executes this filtering operation, also known as pulse shaping, where the abrupt square wave transitions in the modulated data stream are smoothened out by low-pass filtering. The band limited digital signal is now converted to the analog domain using a digital-to-analog (D/A) converter. Further filtering in the analog domain, after D/A conversion, is necessary to suppress spectral images of the baseband signal at multiples of the sampling frequency. In Chap. 13, we will see how it becomes possible to implement this analog filter digitally. The transmit filter in the MODEM will be treated as a reference in the subsequent slides.

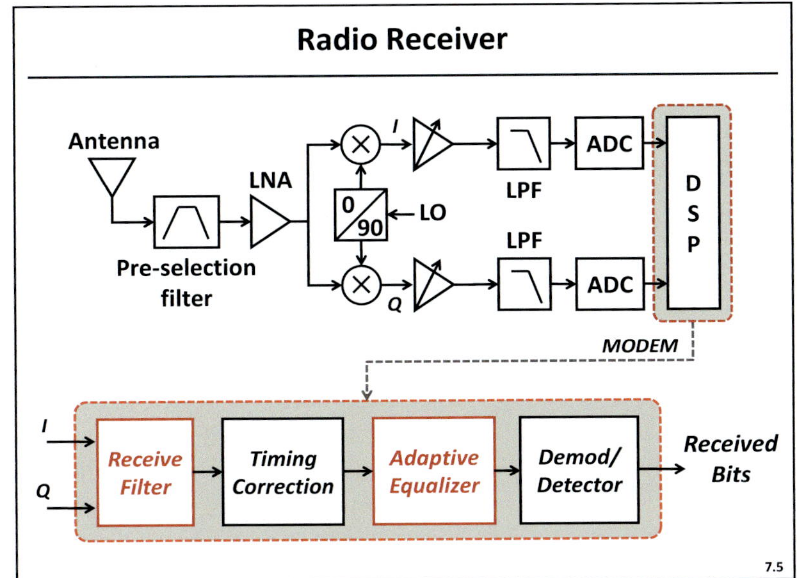

Slide 7.5

In the previous slide, we looked at filtering operations in radio transmission. Filters have their use in the radio receiver chain as well. Digital filtering is once again a part of the baseband MODEM block, just after the analog signal is digitized using the ADC. The receive filter in the MODEM attenuates any out-of-band spectral components (interference, channel noise etc.) that are not in the specified receive bandwidth. This receive filter is matched in its specifications to the transmit filter shown in the previous slide. After filtering, the baseband data goes through a timing correction block, which is followed by equalization. Equalizers are responsible for reversing the effects of the frequency-selective analog transmit channel. Hence, equalizers are also frequency selective in nature and are implemented using filters. They are an important part of any communication system, and their implementation directly affects the quality of service (QoS) of the system. We will look at various algorithms and architectures for equalizers in a later section of the chapter.

Design Procedure

◆ <u>**Assume:**</u> **Modulation, T$_s$ (symbol period), bandwidth specified**

  – Algorithm design

    • Transmit / receive filters

    • Modulator

    • Demodulator

    • Detector

    • Timing correction

    • Adaptive equalizer

  – Implementation architecture

  – Wordlength optimization

  – Hardware mapping

7.6

Slide 7.6

Radio design is a multi-stage process. After specifying the modulation scheme, sample period, and signal bandwidth, algorithm designers have to work on the different blocks in the radio chain. These include transmit and receive filters, modulators, demodulators, signal detectors, timing correction blocks, and equalizers. Significant effort is spent on the design of filter components: transmit and receive filters, and adaptive equalizers. After the algorithm design phase, the design needs to be described in a sample- and bit-accurate form for subsequent hardware implementation. This process includes choosing the optimal number of bits to minimize the impact of fixed-point accuracy (discussed in Chap. 10), and choosing an optimal architecture for implementation (topic of Chap. 11). These steps can be iterated to develop a hardware-friendly design. The final step is mapping the design onto hardware.

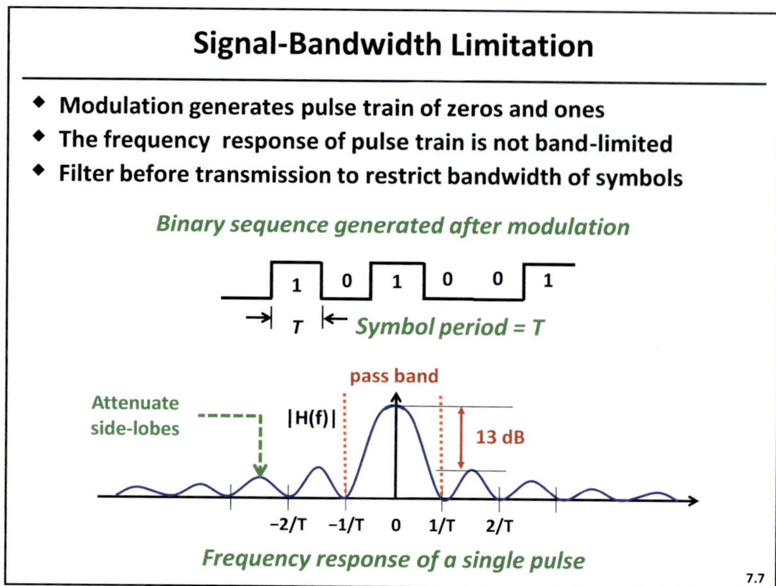

Slide 7.7

As noted earlier, the result of baseband modulation is a sequence of symbols, which are still in digital bit-stream form. The frequency domain representation of this bit-stream will be the well-known *sinc* function, which has about 13 dB of attenuation at the first side lobe. Typically, the allowable bandwidth for transmission is restricted to the main lobe, and any data transmission in the adjacent spectrum amounts to spectral leakage, which must be suppressed. A transmit filter is therefore required before transmission, to restrict the data to the available bandwidth. If the symbol period after modulation is T_s, then the allowable bandwidth (pass band) for transmission is $2/T_s$.

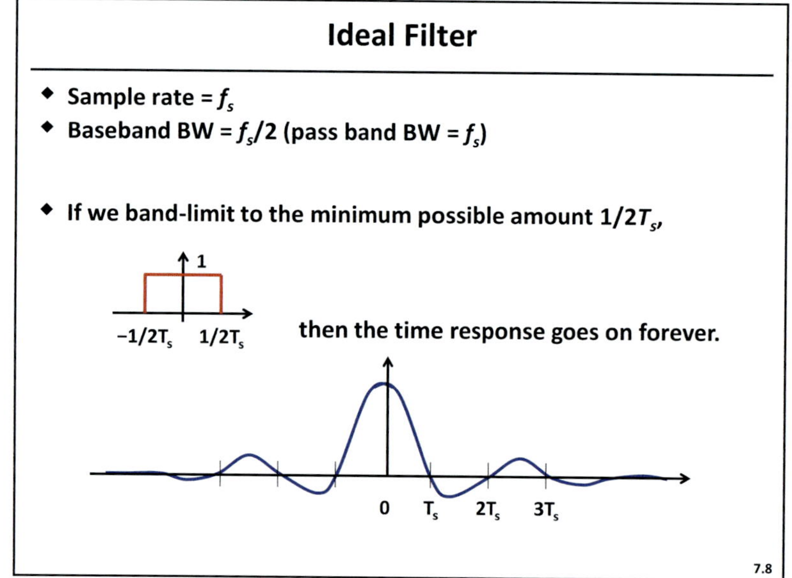

Slide 7.8

The diagram on the left of the slide illustrates the ideal frequency response of a transmit/receive filter. The rectangular response shown in red will only allow signals in the usable bandwidth to pass, and will completely attenuate any signal outside this bandwidth. We have seen in the previous slide that the frequency response of a square wave bit-stream is a *sinc* function. Similarly, the impulse response of the brick-wall filter shown in the figure is the infinitely long *sinc* function. Thus, a long impulse response is needed to realize the sharp roll-off characteristics (*edges*) of the brick-wall filter. A practical realization is infeasible in FIR form, when the impulse response is infinitely long.

Practical Transmit / Receive Filters

- **Transmit filters**
  - Restrict the transmitted signal bandwidth to a specified value
- **Receive filters**
  - Extract the signal from a specified bandwidth of interest

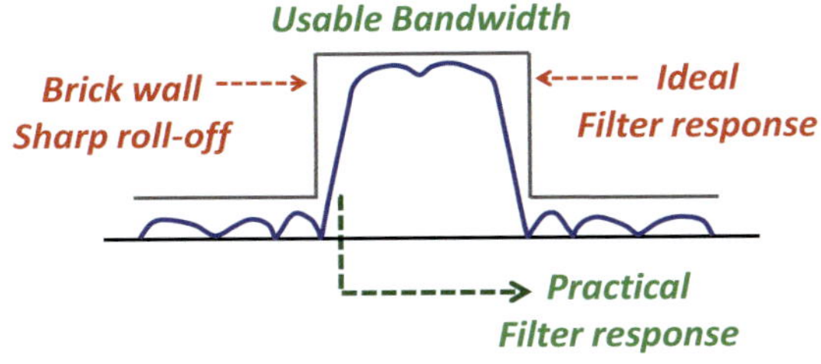

Ideal filter response has infinitely long impulse response.
Raised-cosine filter is a practical realization.

7.9

Slide 7.9

Practical FIR implementations of transmit/receive filters typically adopt a truncated version of the impulse response. Truncation of the impulse response results in a gentler roll-off in the frequency domain, in contrast to the sharp roll-off in the brick-wall filter. One such truncated realization is the raised-cosine filter, which will be discussed in the following slides.

Raised-Cosine Filter

- **Frequency response of raised-cosine filter**

$$H_{RC}(f) = \begin{cases} 1 \Leftrightarrow |f| \le \dfrac{1-\alpha}{2T_s} \\[2mm] \dfrac{T_s}{2}\left[1 + \cos\dfrac{\pi T_s}{\alpha}\left(|f| - \dfrac{1-\alpha}{2T_s}\right)\right] \Leftrightarrow \dfrac{1-\alpha}{2T_s} < |f_s| < \dfrac{1+\alpha}{2T_s} \\[2mm] 0 \Leftrightarrow |f| > \dfrac{1+\alpha}{2T_s} \end{cases}$$

7.10

Slide 7.10

The raised-cosine filter is a popular implementation of the transmit/receive (Tx/Rx) frequency response. The equations shown on the top of the slide describe the frequency response as a function of frequency and the parameter α. The roll-off slope can be controlled using the parameter α, as illustrated in the figure. The roll-off becomes smoother with increasing values of α. As expected, the filter complexity (the length of the impulse response) also decreases with higher α value. The corresponding impulse response of the raised cosine function decays with time, and can be truncated with negligible change in the shape of the response. The system designer can choose the extent of this truncation as well as the wordlength of the filter operations; in this process he/she sacrifices filter ideality for reduced complexity.

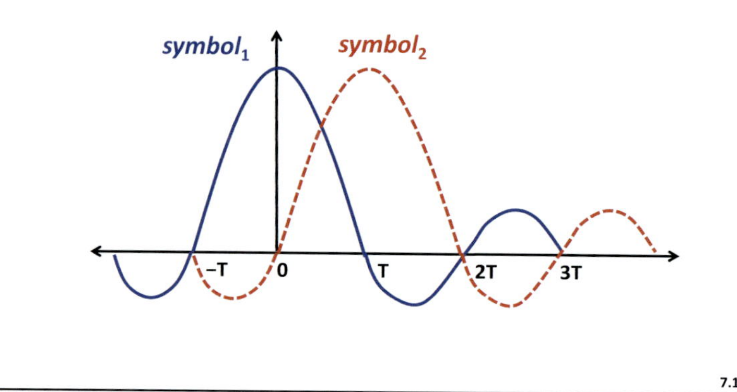

Slide 7.11

The diagram shows the time-domain pulse shaping of a digital bit stream by the raised-cosine filter. As expected, low-pass filters smoothen out the abrupt transitions of the square pulses. When square wave pulses are smoothened, there is a possibility of overlap between adjacent symbols in the time-domain. If the overlap occurs at the sampling instants, then this can lead to erroneous signal samples at the receiver end. An interesting feature of raised cosine pulse shaping is the contribution of the adjacent symbols at the sampling instants. The slide illustrates how *symbol₂* does not have any contribution at time $T = 0$ when *symbol₁* is sampled at the receiver. Similarly *symbol₁* has no contribution at time T when *symbol₂* is sampled. As such, no inter-symbol interference (ISI) results after pulse shaping using raised-cosine filters.

$$H_{RC}(f) = H_{Tx}(f) \cdot H_{Rx}(f)$$

$$H_{Tx}(f) = H_{Rx}(f) = \sqrt{H_{RC}(f)}$$

If $f_{D/A} = f_{A/D} = 4 \cdot (1/T_{symbol})$

$$h_{Tx}(n) = \sum \sqrt{H_{RC}(4m/NT_s)} \cdot e^{j(2\pi mn)/N} \text{ for } -(N-1)/2 \leq n \leq (N-1)/2$$

Slide 7.12

It is simpler to split the raised-cosine filter response symmetrically between the transmit and receive filters. The resulting filters are known as square-root raised-cosine filters. The square-root filter response is broader than the raised-cosine one, and there is often additional filtering needed to meet specifications of spectral leakage or interference attenuation. These additional filters, however, do not share properties of zero inter-symbol interference, as was true for the raised-cosine filter, and will require equalization at the receiver end.

Over-sampling of the baseband signals will result in more effective pulse shaping by the filter. This also renders the system less susceptible to timing errors during sampling at the receiver end. It is common practice to over-sample the signal by 4 before sending it to the raised-cosine filter. The sampling frequency of the filter equals that of the ADC on the receive side and that of the DAC on the transmit side. The impulse response of the filter can be obtained by taking an N-point inverse Fourier transform of the square-root raised-cosine frequency response (as shown in the slide). The impulse response is finite (= N samples), and the attenuation and roll-off characteristics of the filter will depend on the values of α and N. Next, we take a look at the structures used for implementing this FIR filter.

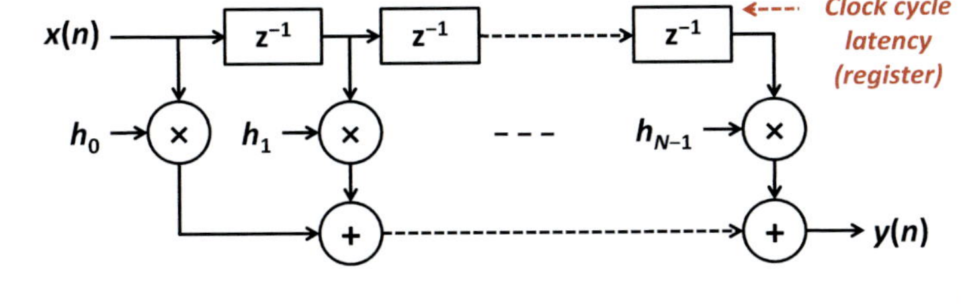

Slide 7.13

This slide illustrates a basic architecture used to implement an FIR filter. An N-tap finite impulse response function can be symbolically represented by a series of additions and multiplications as shown in the slide. The simplest way to implement such a filter is the brute-force approach of drawing a signal-flow graph of the filter equation and then using the same graph as a basis for hardware implementation. This structure, known as the direct-form implementation for FIR filters, is shown in the slide. The diagram shows N multiply operations for the N filter taps and $N-1$ addition operations to add the results of the multiplications. The clock delay is implemented using memory units (registers), where z^{-D} represents D clock cycles of delay.

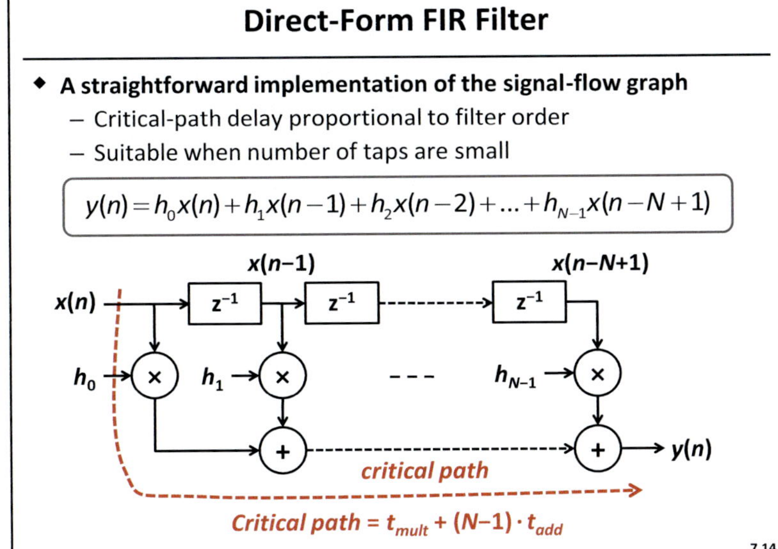

Slide 7.14

The direct-form architecture is simple, but it is by no means the optimal implementation of the FIR. The main drawback is the final series of $N-1$ additions that result in a long critical-path delay, when the number of filter taps N is large. This directly impacts the maximum achievable throughput of the direct-form filters. Several techniques can be used to optimize the signal-flow graph of the FIR so as to ensure high speed, without compromising on area or power dissipation.

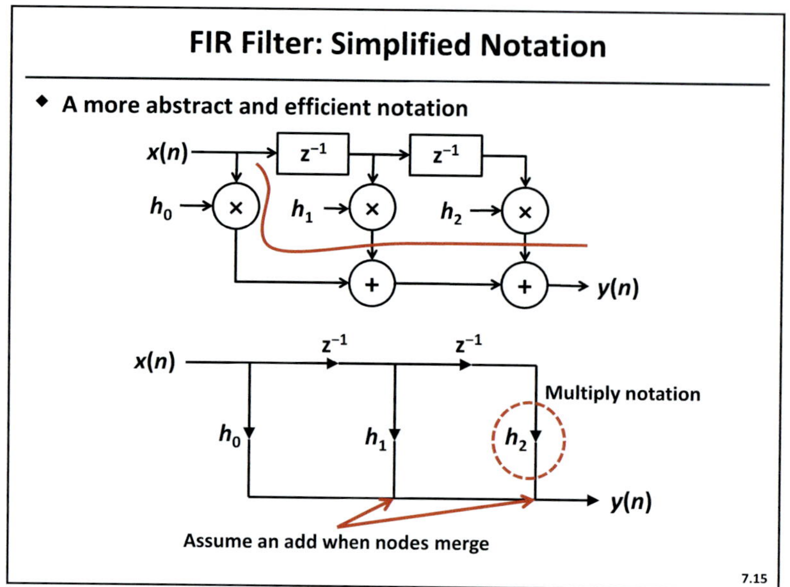

Before we look at optimization techniques for filter design, it is useful to become familiar with the compact signal-flow graph notation commonly used to represent DSP algorithms. The signal-flow graph can be compressed to the line-diagram notation shown at the bottom of the slide. Arrows with notations of z^{-D} over them are clocked register elements, while arrows with letters h_i next to them are notations for multiplication by h_i. When nodes come together, they represent an add operation, and nodes branching out from a point indicate fan-out in multiple directions.

Signal-flow graph manipulations can result in multiple architectures with the same algorithm functionality. For example, we can fold the signal-flow diagram shown in the slide, since the coefficients are symmetrical around the center. We can also move register delays around, resulting in a retimed graph, which vastly reduces the critical path. Constant coefficient multiplications can be implemented using hardwired shifts rather than multipliers, which results in significant area and power reduction.

Pipelining

- **Direct-form architecture is throughput-limited**
- **Pipelining: can be used to increase throughput**
- **Pipelining:** adding same number of delay elements in each forward cut-set (in the data-flow graph) from the input to the output
  - **Cut-set:** set of edges in a graph that if removed, graph becomes disjoint [4]
  - **Forward cut-set:** all edges in the cut-set are in the same direction
- **Increases latency**
- **Register overhead (power, area)**

[4] K.K. Parhi, VLSI Digital Signal Processing Systems: Design and Implementation, John Wiley & Sons Inc., 1999.

7.16

The first optimization technique we look at is pipelining. It is one of the simplest and most effective ways to speed up any digital circuit. Pipelining adds extra registers to the feed-forward portions of the signal-flow graph (SFG). These extra registers can be placed judiciously in the flow-graph, to reduce the critical path of the SFG. One of the overheads associated with pipelining is the additional I/O latency equal to the number of extra registers introduced in the SFG. Other overheads include increased area and power due to the higher register count in the circuit. Before going into pipelining examples, it is important to understand the definition of "cut-sets" [4] in the context of signal-flow graphs. A cutest can be defined as a set of edges in the SFG that hold together two disjoint parts G_1 and G_2 of the graph. In other words, if these edges are removed, the graph is no longer a connected whole, but is split into two disjoint parts. A forward cut-set is one where all the edges in the cut-set are oriented in the same direction (either from G_1 to G_2 or vice versa). This will become clear with the example in the next slide.

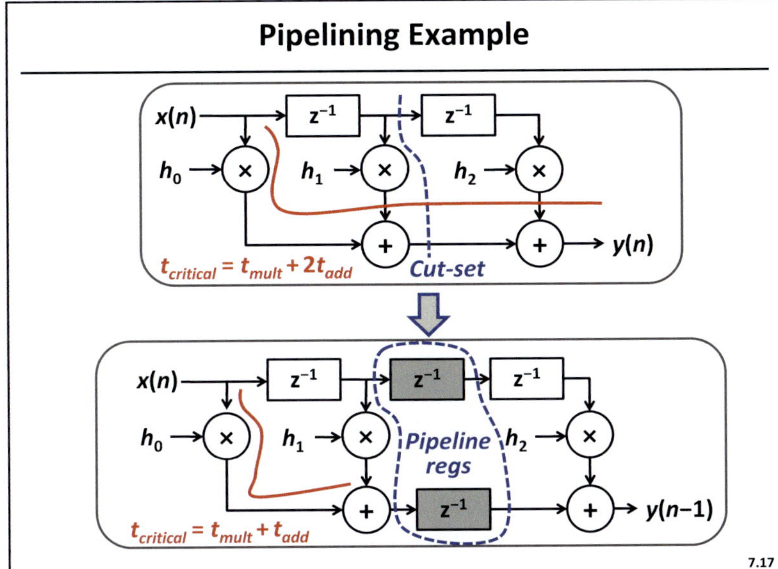

Slide 7.17

The figure shows an example of pipelining using cut-sets. In the top figure, the dashed blue line represents the cut, which splits the signal-flow graph into two distinct parts. The set of edges affected by the cut constitute the cut-set. Pipeline registers can be inserted on each of these edges without altering the functionality of the flow graph. The same number of pipeline registers must be inserted on each cut-set edge. In the bottom figure, we see that one register is inserted on the cut-set edges, which increases the I/O latency of the signal-flow graph. The output of the bottom graph is no longer the same as the top graph but is delayed by one clock cycle. The immediate advantage of this form of register insertion is a reduction in the length of the critical path. In the top figure, the critical path includes one multiply and two add operations, while in the second figure the critical path includes only one multiply and add operation after pipelining. In general, this form of pipelining can reduce the critical path to include only one add and one multiply operation for any number of filter taps N. The tradeoff is the increased I/O latency of up to N clock cycles for an N-tap direct-form filter. The second disadvantage is an excessive increase in area and power of the additional registers, when the number of taps N is large.

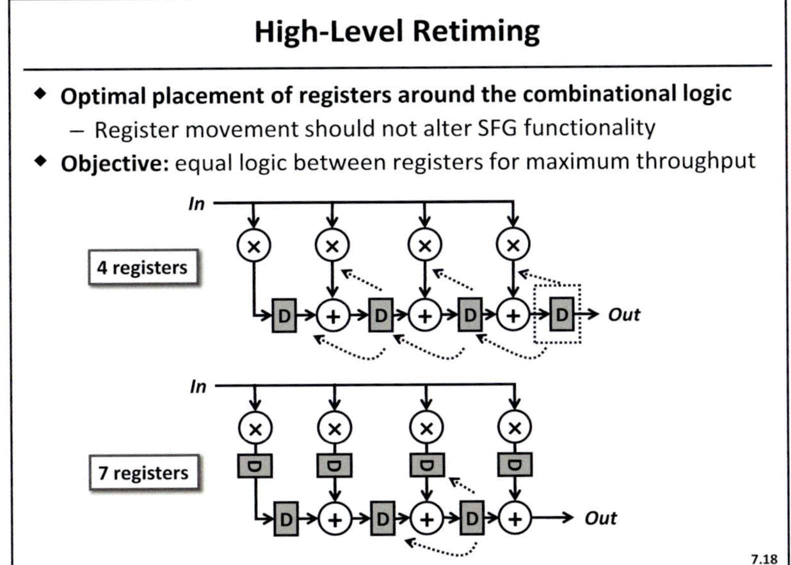

Slide 7.18

Pipelining results in the insertion of extra registers in the SFG, with the objective of increasing throughput. But it is also possible to move the existing registers in the SFG to reduce critical path, without altering functionality or adding extra registers. This form of register movement is called retiming, and it does not result in any additional I/O latency. The objective of retiming is to balance the combinational logic between delay elements to maximize the throughput. The figure shows two signal-flow graphs, where the top graph has a critical path of one adder and multiplier. The *dashed black lines* show register retiming moves for the *top* graph. After retiming, we find that the second graph has a reduced critical path of one multiply operation. No additional I/O latency was introduced in the process, although the number of registers increased from 4 to 7.

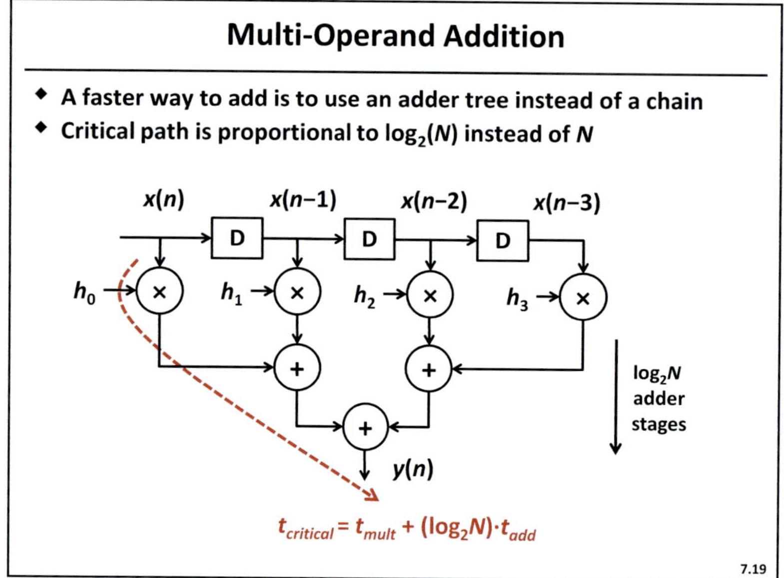

Slide 7.19

We saw earlier that the main bottleneck in the direct form implementation of the FIR was the final series of $N-1$ additions (N is the number of filter taps). A more efficient way to implement these additions is by using a logarithmic adder tree instead of the serial addition chain used earlier. An example of a logarithmic tree is shown in the figure, where each adder takes two inputs. The critical path is now a function of $t_{add} \cdot (\log_2(N))$, a logarithmic dependence on N.

Further optimization is possible through the use of compression trees to implement the final $N-1$ additions. The 3:2-compression adder, which was introduced in Chap. 5, takes 3 inputs and generates 2 outputs, while the 4:2 compression adder takes in 4 inputs and generates 2 outputs. The potential advantage is the reduction in the length of the adder tree from $\log_2(N)$ to $\log_3(N)$ or $\log_4(N)$. But the individual complexity of the adders also increase when compression adders are used, resulting in a higher t_{add} value. The optimized solution depends on where the product of $\log_k(N)$ and t_{add} is minimum (k refers to the number of adder inputs).

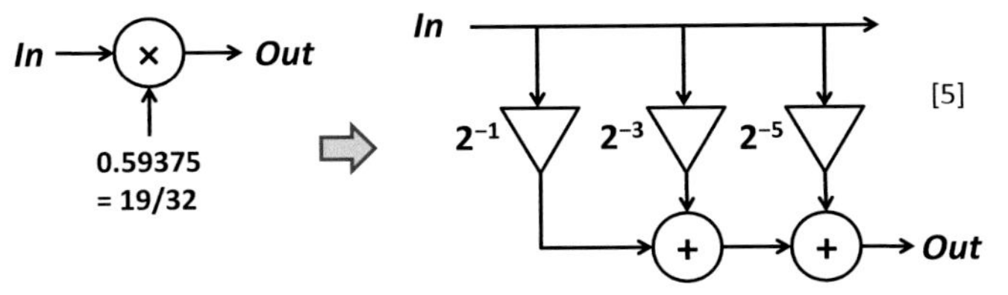

Slide 7.20

A major cost in implementing FIR filters comes from the bulky multiplier units required for coefficient multiplication. If the frequency response of the filter is stringent (sharp roll-off) then the number of taps can be quite large, exceeding 100 taps in some cases. Since the number of multiply operations is equal to the number of taps, such long FIR filters become area and power hungry even at slow operating frequencies. One way to reduce this cost is the use of power-of-two multiplication units. If the coefficients of the filter can be expressed as powers-of-two, then the multiplication is confined to shift and add operations. Shown in the figure is an example of one such coefficient 19/32. This coefficient can be written as $2^{-1} + 2^{-3} - 2^{-5}$. The resultant shift-and-add structure is shown in the figure on the right. It is to be noted that multiplications by powers of two are merely hardwired shift operations which come for free, thus making the effective cost of the multiply operation to be only 2 adder units.

Of course, it is not always possible to express all the coefficients in a power-of-two form. In such cases, a common optimization technique is to round off the coefficients to the nearest power of two, and analyze the performance degradation in the resulting frequency response. If the frequency response is still acceptable, the cost reduction for the FIR can be very attractive. If not, then some form of search algorithm [5] can be employed to round off as many coefficients as possible without degrading the response beyond acceptable limits. The filter may also be overdesigned in the first phase, so that the response is still acceptable after the coefficients are rounded off.

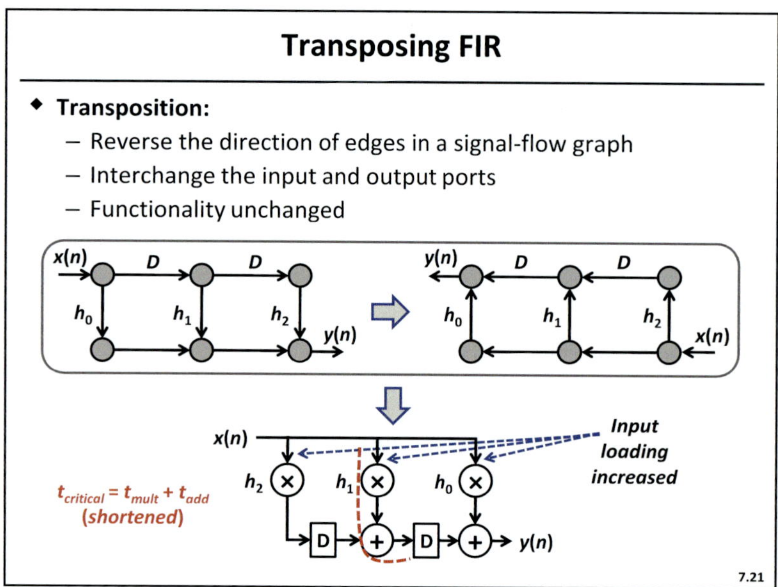

Slide 7.21

Probably the most effective technique for increasing the throughput of FIRs, without significant power or area overhead is the use of transposition. This concept comes from the observation that reversing the direction of arrows in a signal-flow graph does not alter the functionality of the system. The diagram in this slide shows how transposition works. The direction of arrows in the direct-form structure (*left-hand side*) is reversed, resulting in the signal-flow graph shown on the *right-hand side*. In the transposed architecture, the signal branching points of the original SFG are replaced with adders and vice versa, also the positions of input and output nodes are reversed, leading to the final transposed structure shown at the *bottom* of the slide.

The transposed form has a much shorter critical path. The critical path for this architecture has reduced to $t_{add} + t_{mult}$. Note that up to $2N-2$ pipeline register insertions were required to achieve the same critical path for the direct-form architecture, while the transposed structure achieves the same critical path with no apparent increase in area or power. One of the fallouts of using a transposed architecture is the increased loading on the input $x(n)$, since the input now branches out to every multiplier unit. If the number of taps in the filter is large, then input loading can slow down $x(n)$ considerably and lower the throughput. To avoid this slow-down, the input can be buffered to support the heavy loading, which results in some area and power overhead after all. Despite this issue, the transposed architecture is one of the most popular implementations of the FIR filters.

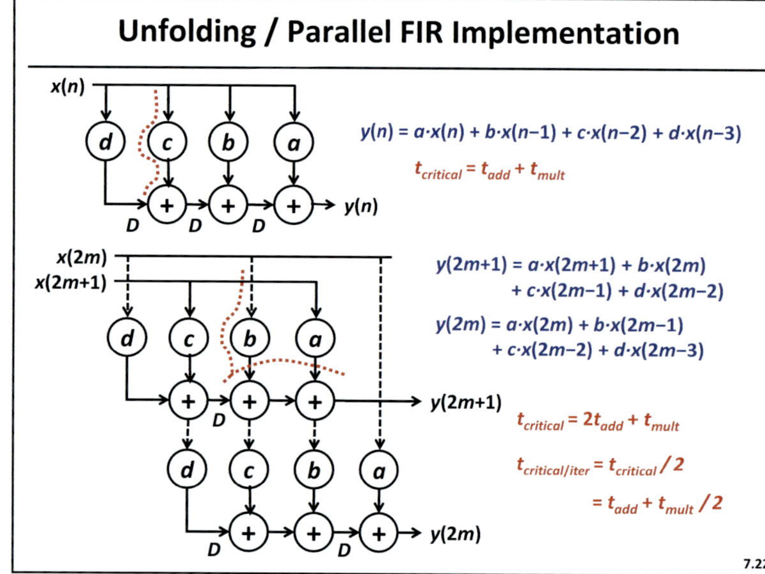

Slide 7.22

Like pipelining, parallelism is another architecture optimization technique targeted for higher throughput or lower power applications. Low power is achieved by speeding up the circuit through the use of concurrency and then scaling the supply voltage to maintain the original throughput, with a quadratic reduction in power. For most feed-forward DSP applications, parallelism can result in an almost linear increase in throughput.

Parallelism can be employed for

both direct- and transposed-forms of the FIR. This slide shows parallelization of the transposed-form filter. Parallel representation of the original flow graph is created by a formal approach called "unfolding". More details on unfolding will be discussed in Chap. 11. The figure shows a transposed FIR filter at the top of the slide, which has a critical path of $t_{add} + t_{mult}$. The figure on the bottom is a twice-unfolded (parallel $P=2$) version of this filter. The parallel FIR will accept two new inputs every clock cycle while also generating two outputs every cycle. Functional expressions for the two outputs, y($2m$) and $y(2m+1)$, are shown on the slide. The parallel filter takes in the inputs $x(2m)$ and $x(2m+1)$ every cycle. The multiplier inputs are delayed versions of the primary inputs and can be generated using registers. The critical path of the unfolded filter is $2 \cdot t_{add} + t_{mult}$. Since this filter produces two outputs per clock cycle while also taking in two inputs, $x(2m)$ and $x(2m+1)$, every cycle, the effective throughput is determined by $T_{clk}/2 = t_{add} + t_{mult}/2$, where T_{clk} is the clock period (throughput).

The same concept can be extended to multiple parallel paths $(P>2)$ for a greater increase in throughput. A linear tradeoff exists between the area increase and the throughput increase for parallel FIR filters. Alternatively, for constant throughput, we can use excess timing slack to lower the supply voltage and reduce power quadratically.

Recursive Filters

Slide 7.23

After a detailed look at FIR implementations, we will now discuss the use of infinite impulse response (IIR) filters. This class of filters provides excellent frequency selectivity, although the occurrence of feedback loops constrains their implementation.

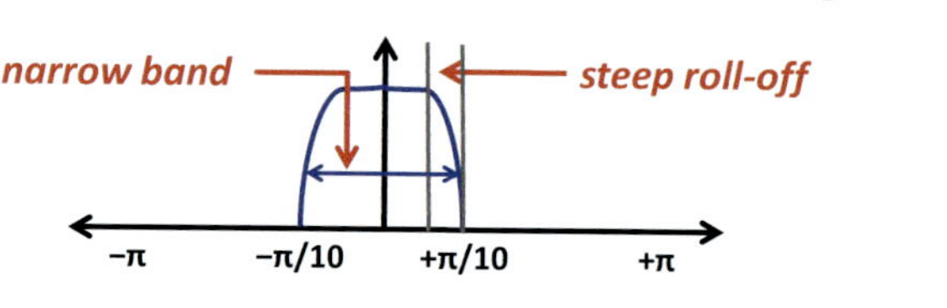

Slide 7.24

FIR filters have distinct advantages in terms of ease of design and guaranteed stability. The stability arises from the fact that all the poles of the system are inside the unit circle in the z-domain. This is because the poles of the z-domain transfer function are all located at the origin. Also, the phase response of FIRs can be made linear by ensuring that the tap coefficients are symmetric about the central tap. FIR filters are attractive when the desired frequency response characteristics are less stringent, implying low attenuation and gentle roll-off characteristics. In the extreme case of realizing steep roll-off filter responses, the FIR filter will end up having a large number of taps and will prove to be very computationally intensive. In these cases, a more suitable approach is the use of infinite impulse response (IIR) filters.

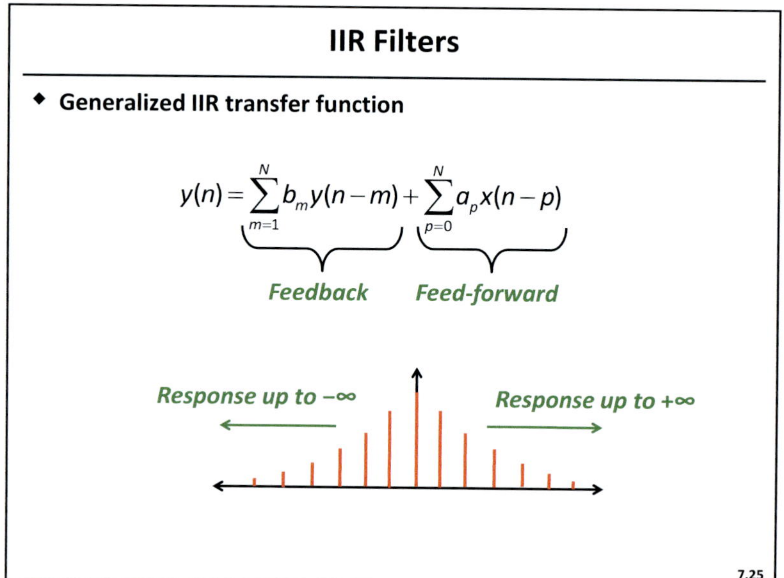

Slide 7.25

As the name suggests, IIR filters have an infinite impulse response ranging from $-\infty$ to $+\infty$ as shown in the figure. The FIR filter was completely feed-forward, in other words, the output of the filter was dependent only on the incoming input and its delayed versions. In the IIR filter, on the other hand, the expression for the filter output has a feed-forward as well as a recursive (feedback) part. This implies that the filter output is dependent on both the incoming input samples and the previous outputs $y(n-i)$, $i>0$. It is the feedback portion of the filter that is responsible for the infinitely long impulse response. The extra degree of freedom coming from the recursive part allows the filter to realize high-attenuation, sharp roll-off frequency responses with relatively low complexity.

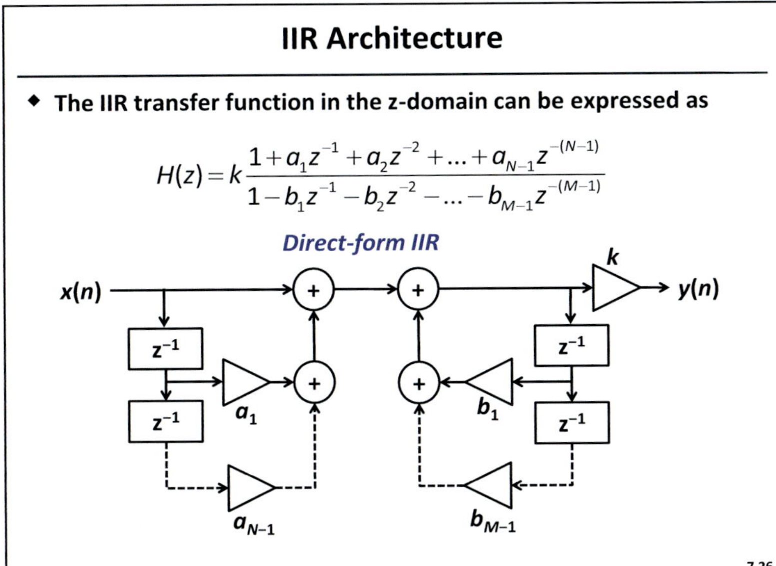

$$H(z) = k\,\frac{1 + a_1 z^{-1} + a_2 z^{-2} + \ldots + a_{N-1} z^{-(N-1)}}{1 - b_1 z^{-1} - b_2 z^{-2} - \ldots - b_{M-1} z^{-(M-1)}}$$

Slide 7.26

The z-domain transfer function $H(z)$ of a generic IIR filter is shown in this slide. The numerator of $H(z)$ represents the feed-forward section of the filter, while the feedback part is given by its denominator. A direct-form implementation of this expression is shown in the figure. The derivation of the output expression for $y(n)$ from the filter structure should convince the reader that the filter does indeed implement the transfer function $H(z)$. As in the FIR case, the direct-form structure is intuitive to understand, but is far from being the optimal implementation of the filter expression. In the next slide, we will look at optimization techniques to improve the performance of this filter structure.

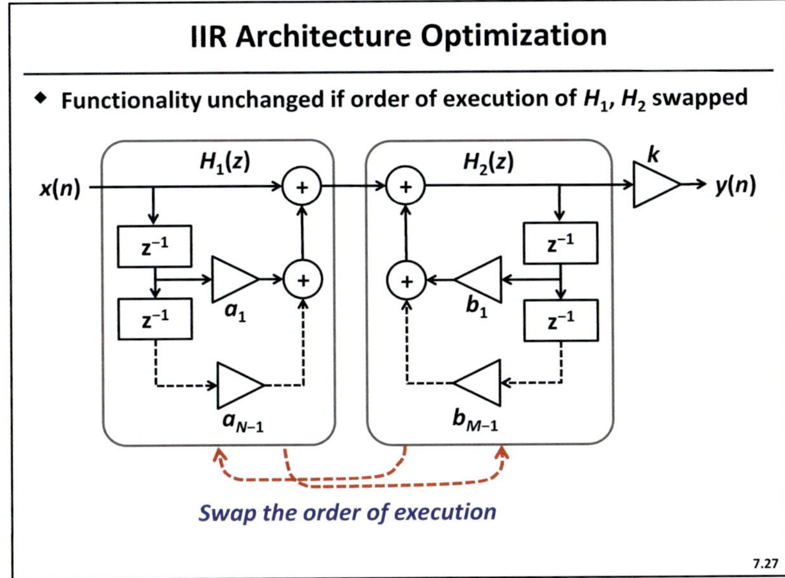

Slide 7.27

An interesting property of cascaded, linear, time-invariant systems lies in the fact that the functionality of the system remains unchanged even after the order of execution of the blocks is changed. In other words, if a transfer function $H(z) = H_1(z)\cdot H_2(z)\cdot H_3(z)$, then the output is not affected by the position of the three units H_1, H_2 or H_3 in the cascaded chain. We use this fact to our advantage in transforming the signal-flow graph for the IIR filter. The direct-form structure is split into two cascaded units H_1 and H_2, and their order of execution is swapped.

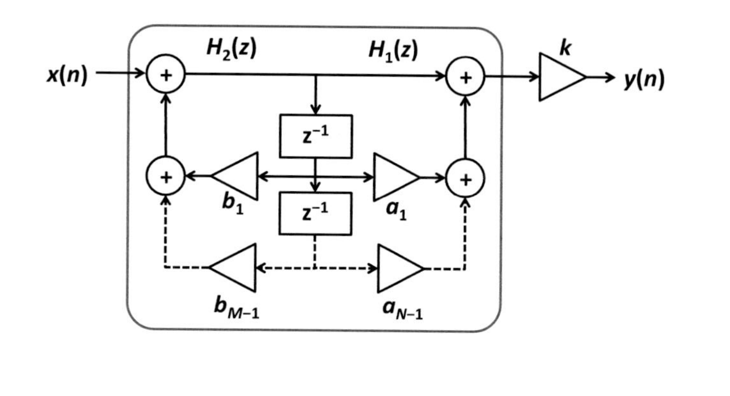

Slide 7.28

After swapping the positions of units H_1 and H_2, we can see that the bank of registers in both the units can be shared to form a central register chain. It is, of course, not mandatory that both units have the same number of registers in them, in which case the register chain will have the maximum number required by either unit. The resultant architecture is shown in the slide. Compared to the direct-form implementation, we can see that a substantial reduction in area can be achieved by reducing the register count.

Cascaded IIR

- Implement IIR transfer function as cascade of 2nd order sections
  - Shorter wordlengths in the feedback loop adders
  - Less sensitive towards finite precision arithmetic effects
  - More area and power efficient architecture

$$H(z) = k_1 \frac{1 + a_{11}z^{-1} + a_{21}z^{-2}}{1 - b_{11}z^{-1} - b_{21}z^{-2}} \cdots k_p \frac{1 + a_{1p}z^{-1} + a_{2p}z^{-2}}{1 - b_{1p}z^{-1} - b_{2p}z^{-2}}$$

Slide 7.29

IIR filters typically suffer from various problems associated with their recursive nature. The adder in the feedback accumulator ends up requiring long wordlengths, especially if the order of recursion (value of M) is high. The frequency response of the IIR is very sensitive to quantization effects. This would mean that arbitrarily reducing the wordlength of the adder or multiplier units, or truncating the coefficients can have a drastic effect on the resultant frequency response. These problems can be mitigated to some extent by breaking the transfer function $H(z)$ into a cascade of second- or first-order sections, as shown in the slide. The main advantage of using second-order sections is the reduced wordlength of the adder units in the feedback accumulator. Also, the cascaded realization is less sensitive to coefficient quantization effects, making the filter more robust.

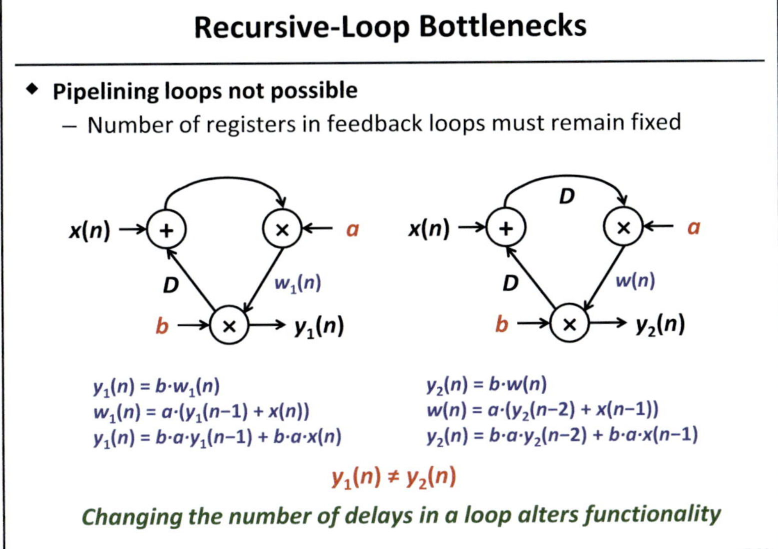

Slide 7.30

Another bottleneck associated with recursive filters is their inability to support pipelining. Inserting registers in any feedback loop changes the functionality of the filter disallowing the insertion of pipeline registers. For example, the first-order IIR filter graph shown in the left of the slide has one register in the feedback loop. The input-output relation is given by $y_1(n) = b \cdot a \cdot y_1(n-1) + b \cdot a \cdot x(n)$. An extra register is inserted in the feedback loop for the signal-flow graph shown on the right. This graph now has an input-output relation given by $y_2(n) = b \cdot a \cdot y_2(n-2) + b \cdot a \cdot x(n-1)$, which is different from the filter on the left. Restriction on delay insertion, therefore, makes feedback loops a throughput bottleneck in IIR filters.

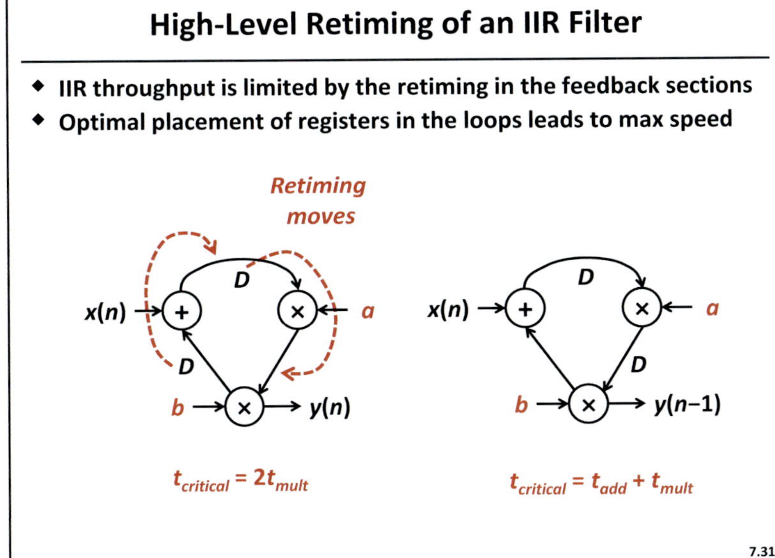

Slide 7.31

Maximizing the throughput in IIR systems entails the optimal placement of registers in the feedback loops. For example, the two signal-flow graphs shown in the slide are retimed versions of each other. The graph shown on the left has a critical path of $2t_{mult}$. Retiming moves shown with red dashed lines are applied to obtain the flow graph on the right. The critical path for this graph is $t_{add} + t_{mult}$. This form of register movement can ensure that the filter functions at the maximum possible throughput. Details on algorithms to automate retiming will be discussed in Chapter 11.

Unfolding: Constant Throughput

♦ **Unfolding recursive flow graphs**
  – Maximum attainable throughput limited by iteration bound
  – Unfolding does not help if iteration bound already achieved

$$y(n) = x(n) + ay(n-1)$$

$$t_{critical} = t_{add} + t_{mult}$$

$$y(2m) = x(2m) + ay(2m-1)$$

$$y(2m+1) = x(2m+1) + ay(2m)$$

$$t_{critical} = 2t_{add} + 2t_{mult}$$

$$t_{critical/iter} = t_{critical} / 2$$

7.32

Slide 7.32

The feedback portions of the signal-flow graph also restrict the use of parallelism in IIR filters. For example, the first-order IIR filter shown in the slide is parallelized with $P=2$. After unfolding, the critical path for the parallel filter doubles, while the number of registers in the feedback loop remains the same. Although we generate two outputs every clock cycle in the parallel filter, it takes a total of $2 \cdot (t_{add} + t_{mult})$ time units to do so. This will result in a throughput of $1/(t_{add} + t_{mult})$ per output sample, which is the same as that of the original filter. Hence, the maximum achievable throughput for the IIR is still restricted by the optimal placement of registers in the feedback loops. More details on unfolding IIR systems will be covered in Chap. 11.

IIR Summary

♦ **Pros**
  – Suitable when filter response has sharp roll-off, has narrow-band, or large attenuation in the stop-band
  – More area- and power-efficient compared to FIR realizations

♦ **Cons**
  – Difficult to ensure filter stability
  – Sensitive to finite-precision arithmetic effects (limit cycles)
  – Does not have linear phase response unlike FIR filters
  – All-pass filters required if linear phase response desired
  – Difficult to increase throughput
    • Pipelining not possible
    • Retiming and parallelism has limited benefits

7.33

Slide 7.33

In summary, IIR filters are a computationally efficient way to realize attenuation characteristics that require sharp roll-off and high attenuation. From an area and power perspective, they are superior to the corresponding FIR realization. But the recursive nature of the filter tends to make it unstable, since it is difficult to ensure that all poles of the system lie inside the unit circle in the z-domain. The filter phase response is non-linear and subsequent all-pass filtering may be required to linearize the phase response. All-pass filters have constant amplitude response in the frequency domain, and can compensate for the non-linear phase response of IIR filters. Their use, however, reduces the area efficiency of IIR filters. The IIR filter structure is difficult to optimize for very high-speed applications owing to the presence of feedback loops. Hence, a choice between FIR and IIR realization should be made depending on system constraints and design objectives.

<table>
<tr><td>

Multi-Rate Filtering

</td><td>

Slide 7.34

We now take a look at a new class of filters used for sample-rate conversion in DSP systems. Any digital signal processing always takes place at a predefined sample rate f_s. It is not mandatory, however, for all parts of a system to function at the same sample rate. Often for the sake of lower throughput or power consumption, it becomes advantageous to operate different blocks in a system at different sampling frequencies. In this scenario, sample-rate conversion has to be done without

</td></tr>
</table>

loss in signal integrity. Multi-rate filters are designed to enable such data transfers.

<table>
<tr><td>

Multi-Rate Filters

- **Data transfer between systems at different sampling rate**
  - Decimation
    - Higher sampling rate f_{s1} to lower rate f_{s2}
  - Interpolation
    - Lower sampling rate f_{s2} to higher rate f_{s1}

- **For integer f_{s1}/f_{s2}**
  - Drop samples when decimating
    - Leads to aliasing in the original spectrum
  - Stuff zeros when interpolating
    - Leads to images at multiples of original sampling frequency

7.35

</td><td>

Slide 7.35

Multi-rate filters can be subdivided into two distinct classes. If the data transfer takes place from a higher sample rate f_{s1} to a lower sample rate f_{s2}, then the process is called "decimation." On the other hand, if the transfer is from a lower rate f_{s1} to a higher rate f_{s2}, then the process is called "interpolation." For integer-rate conversion (i.e., $f_{s1}/f_{s2} \in I^+$), the rate conversion process can be interpreted easily.

Decimation implies a reduced number of samples. For example, if we have 500 samples at the rate of

</td></tr>
</table>

500 MHz (f_{s1}), then we can obtain 250 samples at the rate of 250 MHz (f_{s2}) by dropping alternate samples. The ratio $f_{s1}/f_{s2} = 2$, is called the "decimation factor." But only skipping alternate samples from the original signal sequence does not guarantee reliable data transfer. Aliasing occurs if the original frequency spectrum has data content beyond 250 MHz. We will talk about solutions to this problem in the next slide.

Similarly, interpolation can be interpreted as increasing the number of samples. Stuffing zeros between the samples of the original signal sequence can obtain this increase. For example, a data sequence at 500 MHz can be obtained from a sequence at 250 MHz by stuffing zeros between adjacent samples to double the number of samples in the data stream. However, this process will also result in images of the original spectrum at multiples of 250 MHz. We will take a look at these issues in the next few slides.

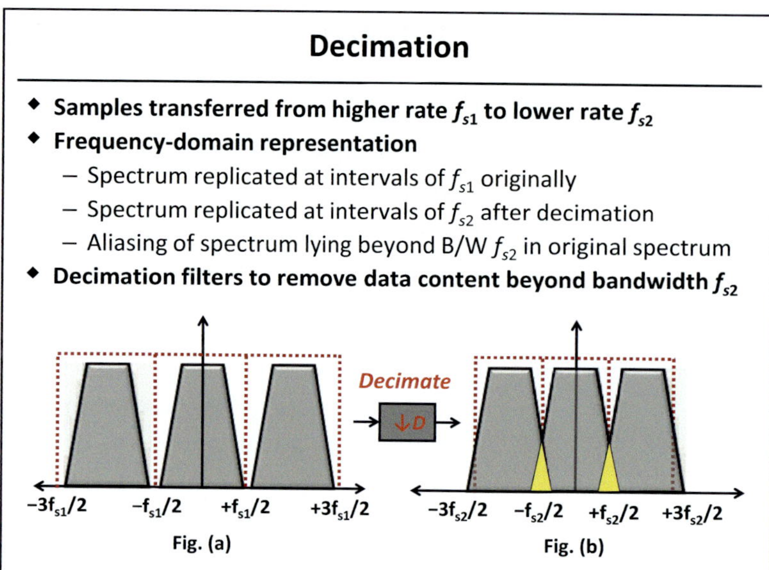

Slide 7.36

In digital systems, the frequency spectrum of a signal is contained within the band $-f_s/2$ to $f_s/2$, where f_s is the sampling frequency. The same spectrum also repeats at intervals of f_s, as shown in Fig (a). This occurs because the Fourier transform used to compute the spectrum of a periodic signal is also periodic with f_s. The reader is encouraged to verify this property from the Fourier equations discussed in the next chapter. After decimation to a lower frequency f_{s2}, the same spectrum will repeat at intervals of f_{s2} as shown in Fig. (b). A problem arises if the bandwidth of the original data is larger than f_{s2}. This will lead to an overlap between adjacent spectral images, shown by the yellow regions in Fig. (b). This overlap or aliasing can corrupt the data sequence, so care should be taken to avoid such an overlap. The only alternative is to remove any spectral content beyond bandwidth f_{s2} from the original spectrum, to avoid aliasing. Decimation filters, discussed in the next slide, are used for this purpose.

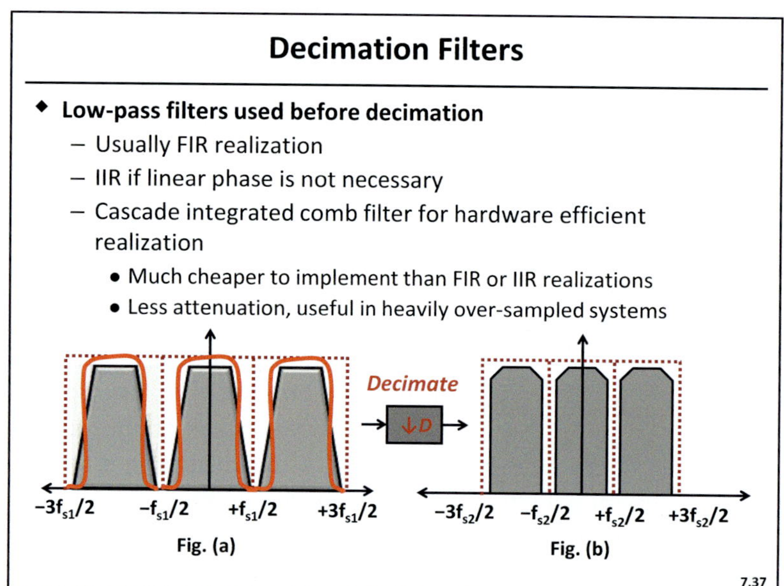

Slide 7.37

The decimation filter must ensure that no aliasing corrupts the signal after rate conversion. This would require the removal of any signal content beyond the bandwidth of f_{s2}, making the decimation filter a low-pass filter. Figure (a) shows the original spectrum of the data at sampling frequency f_{s1}. The multiples of the original spectrum are formed at frequency f_{s1}. Before decimating to sampling frequency f_{s2}, the signal is low-pass filtered to restrict the bandwidth to f_{s2}. This will ensure that no aliasing occurs after down sampling by a factor D, as shown in Fig. (b). The low-pass filter is usually an FIR, if the decimation factor is small (2–4). For large decimation factors, the filter response has a sharp roll-off. Implementing FIR filters for such a response can get computationally expensive. For large decimation factors, a viable alternative is the use of IIR filters to realize the sharp roll-off frequency response. However, the non-linear phase in IIR filters may require the use of additional all-pass filters for phase compensation. An alternative implementation is the use of cascade integrated comb

(CIC) filters.   CIC filters are commonly used for decimation due to their hardware-efficient realization. Although their stop-band attenuation is small, they are especially useful in decimating over-sampled systems. More details on CIC filtering will be discussed in Part IV of the book.

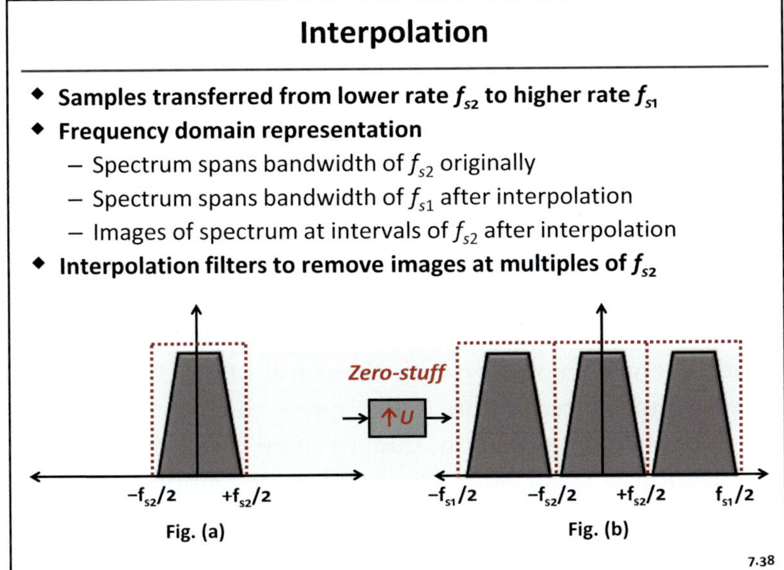

Slide 7.38

Interpolation requires the introduction of additional data samples since we move from a lower sampling frequency to a higher one. The easiest way to do this is the zero-stuffing approach. For example, when moving from sampling frequency f_{s2} to a higher rate $f_{s1}=3\cdot f_{s2}$, we need three samples in the interpolated data for every one sample in the original data sequence. This can be done by introducing two zeros between adjacent samples in the data sequence. The main problem with this approach is the creation of spectral images at multiples of f_{s2}. Figure (a) shows the spectrum of the original data sampled at f_{s2}. In Fig. (b) the spectrum of the new data sequence after zero stuffing is shown. However, we find that images of the original spectrum are formed at $-f_{s2}$ and $+f_{s2}$ in Fig. (b). This phenomenon occurs because zero-stuffing merely increases the number of samples without additional information about the values of the added samples. If we obtain a Fourier transform of the zero-stuffed data we will find that the spectrum is periodic with sample rate f_{s2}. The periodic images have to be removed to ensure that only the central replica of the spectrum remains after interpolation. The solution is to low-pass filter the new data sequence and remove the images. The use of interpolation filters achieves this image rejection. Low-pass filtering will ensure that the zeros between samples are transformed into a weighted average of adjacent samples, which gives an approximation of the actual value of the input signal at the interpolated instances.

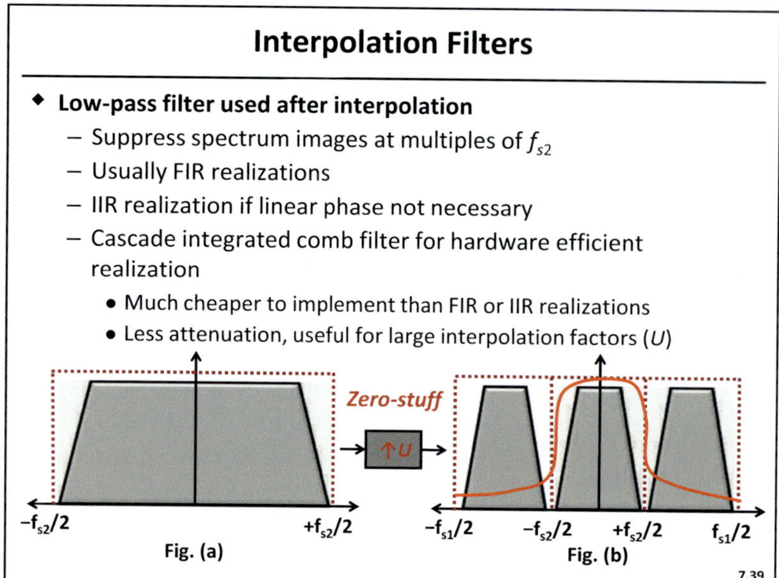

Slide 7.39

The interpolation filter is essentially a low-pass filter that must suppress the spectral images in the zero-stuffed data sequence. A typical frequency response of this filter is shown in *red* in Fig. (b). For small interpolation factors (2–4), the frequency response has a relaxed characteristic and is usually realized using FIR filters. IIR realizations are possible for large up-sampling ratios (U), but to achieve linear phase additional all-pass filters may be necessary. However, this method would be preferable to the corresponding FIR filter, which tends to become very expensive with sharp roll-off in the response. Another alternative is the use of the cascade integrated comb (CIC) filters that were discussed previously. CIC filters are commonly used for interpolation due to their hardware-efficient realization. Although their stop-band attenuation is small, they are useful in up-sampling by large factors (e.g. 10–20 times). More details on CIC filters, particularly their implementation, will be discussed in Chap. 14.

Application Example:

Adaptive Equalization

Slide 7.40

Let us now go back to the original problem of designing radio systems and look at the application of digital filters in the realization of equalizers. The need for equalization stems from the frequency selective nature of the communication channel. The equalizer must compensate for the non-idealities of the channel response. The equalizer examples we looked at earlier assumed a time-invariant channel, which makes the tap-coefficient values of the equalizer fixed. However, the channel need not be strictly time-invariant. In fact, wireless channels are always time-variant in nature. The channel frequency response then changes with time, and the equalizer must also adjust its tap coefficients to accurately cancel the channel response. In such situations the system needs adaptive equalization, which is the topic of discussion in the next few slides. The adaptive equalization problem has been studied in detail, and several classes of these types of equalizers exist in literature. The most prominent ones are zero-forcing and least-mean-square both of which are feed-forward structures. The feedback category includes decision feedback equalizers. We will also take a brief look at the more advanced fractionally spaced equalizers.

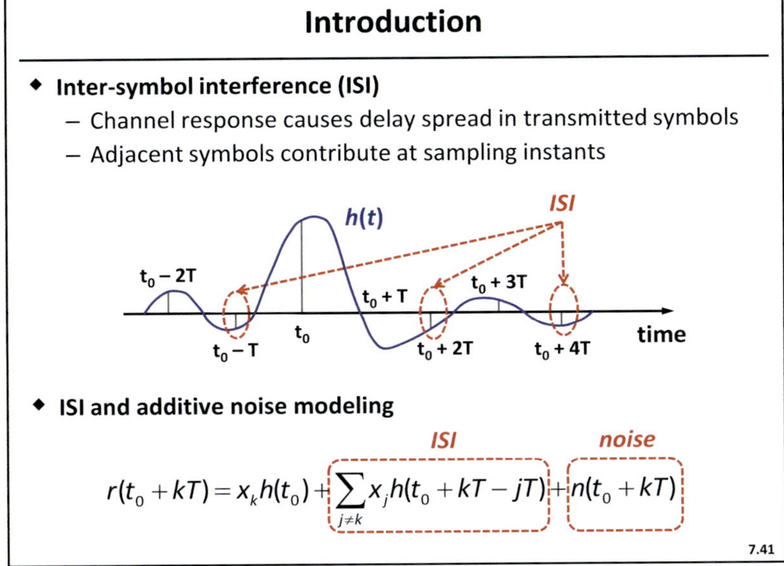

Slide 7.41

The figure in this slide shows a channel impulse response with non-zero values at multiples of the symbol period T. These non-zero values, marked by dashed ovals, will cause the adjacent symbols to interfere with the symbol being sampled at the instant t_0. The expression for the received signal r is given at the bottom of the slide. The received signal includes the contribution from inter-symbol interference (ISI) as well as additive white noise from the channel.

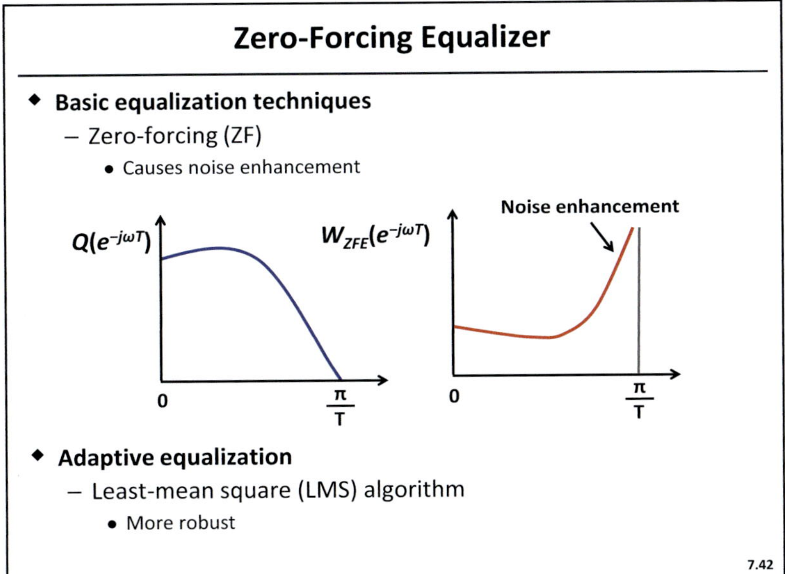

Slide 7.42

The most basic technique for equalization is creating an equalizer transfer function, which nullifies the channel frequency response. This method is called the zero-forcing approach. Although this will ensure that the equalized data has no inter symbol interference, it can lead to noise enhancement for a certain range of frequencies. Noise enhancement can degrade the bit error rate performance of the decoder, which follows the equalizer. For example, the left figure in the slide shows a typical low-pass channel response. The figure on the right shows the zero-forcing equalizer response. The high-pass nature of the equalizer response will enhance the noise at higher frequencies. For this reason, zero forcing is not popular for practical equalizer implementations. The second algorithm we will look at is the least-mean-square, or LMS, algorithm.

Adaptive Equalization

- **Achieve minimum mean-square error (MMSE)**
  - Equalized signal: z_k, transmitted signal: x_k
  - Error: $e_k = z_k - x_k$
  - Objective:
    - Minimize expected error, **min $E[e_k^2]$**
  - Equalizer coefficients at time instant k: $c_n(k)$, $n \in \{0,1,...,N\}$
  - For real optimality: set

$$\frac{\partial E[e_k^2]}{\partial c_n(k)} = 0$$

- **Equalized signal z_k given by:**

$$z_k = c_0 r_k + c_1 r_{k-1} + c_2 r_{k-2} + ... + c_{n-1} r_{k-n+1}$$

- **z_k is convolution of received signal r_k with tap coefficients**

7.43

Slide 7.43

The least-mean-square algorithm adjusts the tap coefficients of the equalizer so as to minimize the error between the received symbol and the transmitted symbol. However, in a communication system, the transmitted symbol is not known at the receiver. To solve this problem a predetermined training sequence known at the receiver can be transmitted periodically to track the changes in the channel response and re-calculate the tap coefficients. The expression for the error value e_k and the objective function are listed in the slide. The equalizer is feed-forward with N taps denoted by c_n, $n \in \{0,1,2,...,N\}$, and the set of N tap coefficients at time k is denoted by $c_n(k)$. Minimizing the expected error requires solving N equations where the partial derivative of $E[e_k^2]$ with respect to each of the N tap coefficients is set to zero. Solving these equations give us the value of the tap coefficient at time k.

To find the partial derivative of the objective function, we have to express the error e_k as a function of the tap coefficients. The error e_k is the difference between the equalized signal z_k and the transmitted signal x_k, which is a constant. The equalized signal z_k, on the other hand, is the convolution of the received signal r_k and the equalizer tap coefficients c_n. This convolution is shown at the end of the slide. Now we can take the derivative of the expectation of error with respect to the tap coefficients c_n. This computation will be illustrated in the next slides.

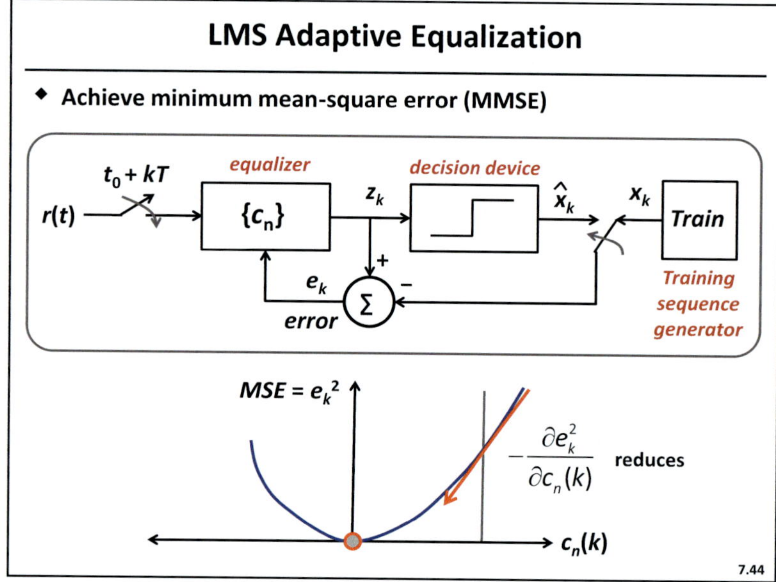

Slide 7.44

The block diagram of the LMS equalizer is shown in the slide. The equalizer operates in two modes. In the training mode, a known sequence of transmitted data is used to calculate the tap coefficients. In the active operational mode, normal data transmission takes place and equalization is performed with the pre-computed tap coefficients. Shown at the *bottom* of the slide is the curve for the mean square error (MSE), which is convex in nature. The partial derivative of the MSE curve with respect to any tap coefficient is a tangent to the curve. When we set the partial derivative equal to zero, we reach the bottom flat portion of the convex curve, which represents the position of the optimal tap coefficient (red circle in the figure). The use of steepest-descent method in finding this optimal coefficient is discussed next.

<table>
<tr><td>

LMS Adaptive Equalization (Cont.)

♦ **Alternative approach to minimize MSE**
  – For computational optimality
    • Set: $\partial E[e_k^2]/\partial c_n(k) = 2e_k r_{k-n} = 0$
  – Tap update equation: $e_k = z_k - x_k$
  – Step size: Δ
  – Good approximation if using:

  $c_n(k+1) = c_n(k) - \Delta e_k r(k-n), n = 0, 1, ..., N-1$

    • Small step size
    • Large number of iterations

  – Error-signal power comparison: $\sigma_{LME}^2 \leq \sigma_{ZF}^2$

7.45
</td></tr>
</table>

Slide 7.45

The gradient function is a vector of dimension N with the jth element being the partial derivative of the function argument with respect to c_j. The (+) symbol represents the convolution operator and r is the vector of received signals from time instant $k-N+1$ to k. From the mathematical arguments discussed, we can see that the gradient function for the error e_k is given by the received vector r where $r=[r_k, r_{k-1},...,r_{k-N+1}]$. The expression for the partial derivative of the objective function E with respect to c_n is given by $2e_k r_{k-n}$, as shown in the slide. Now, $\partial E[e_k^2]/\partial c_n(k)$ is a vector, which points towards the steepest ascent of the cost function. To find the minimum of the cost function, we need to take a step in the opposite direction of $\partial E[e_k^2]/\partial c_n(k)$, or, in other words, to make the steepest descent. In mathematical terms, we get the coefficient update equation given by:

$$c_n(k+1)=c_n(k)-\Delta e_k r(k-n), n=0,1,...,N-1.$$

Here Δ is the step size by which the gradient is scaled and used in the update equation. If the step size Δ is small then a large number of iterations are required to converge to the optimal coefficients. If Δ is large, then the number of iterations is small. However, the tap coefficients will not converge to the optimal points but to somewhere close to it. A good approximation of the tap coefficients is found when the error signal power for the LMS coefficients becomes less than or equal to the zero-forcing error-signal power.

Decision-Feedback Equalizers

- ◆ **To cancel the interference from previously detected symbols**
  - Pre-cursor channel taps and post-cursor channel taps
  - Feed-forward equalizer
    - Remove the pre-cursor ISI
    - FIR (linear)
  - Feedback equalizer
    - Remove the post-cursor ISI
    - Like a ZF equalizer
      - If previous symbol detection is correct
  - Feedback equalizer coefficient update equation:

$$b_{m+1}(k+1) = b_m(k) - \Delta e_k d_{k-m}$$

7.46

Slide 7.46

Up until now we have looked at the FIR realization of the equalizer. The FIR filter computes the weighted sum of a sequence of incoming data. However, a symbol can be affected by inter-symbol interference from symbols transmitted before as well as after it. The interference caused by the symbols coming before in time is referred to as "pre-cursor ISI," while the interference caused by symbols after is termed "post-cursor ISI." An FIR realization can remove the pre-cursor ISI, since the preceding symbols are available and can be stored in registers. However, removal of post-cursor ISI will require some form of feedback or recursion. The post-cursor ISI translates to a non-causal impulse response for the equalizer, which requires recursive cancellation. Such equalizers are known as "decision-feedback equalizers." The name stems from the fact that the output of the decoder/decision device is fed back into the equalizer to cancel post-cursor ISI. The accuracy of equalization is largely dependent upon the accuracy of the decoder in detecting the correct symbols. The update equation for the feedback coefficients is once again derived from the steepest-descent method and is shown on the slide. Instead of using the received signal r_{k-n}, the update equation uses previously detected symbols d_{k-m}.

Decision-Feedback Equalizers (Cont.)

- ◆ **Less noise enhancement compared with ZF or LMS**
- ◆ **More freedom in selecting coefficients of feed-forward equalizer**
  - Feed-forward equalizer need not fully invert channel response
- ◆ **Symbol decision may be incorrect**
  - Error propagation (slight)

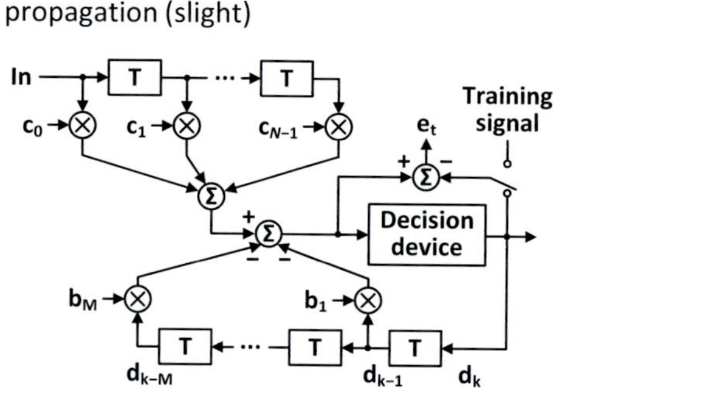

7.47

Slide 7.47

The decision-feedback equalizer (DFE) is shown on this slide. The architecture has two distinct parts; the feed-forward filter with coefficients c_n to cancel the pre-cursor ISI, and the feedback filter with coefficients b_m to cancel the post-cursor ISI. The DFE has less noise enhancement as compared to the zero-forcing and LMS equalizers. Also, the feed-forward equalizer need not remove the ISI fully now that the feedback portion is also present. This offers more freedom in the selection of the feed-forward coefficients c_n. Error propagation is possible if previously detected symbols are incorrect and the feedback filter is unable to cancel the post-cursor ISI correctly. This class of equalization is known to be robust and is very commonly used in communication systems.

<table>
<tr><td>

Fractionally-Spaced Equalizers

- **Sampling at symbol period**
  - Equalizes the aliased response
  - Sensitive to sampling phase

- **Can over-sample (e.g. 2x higher rate)**
  - Avoid spectral aliasing at the equalizer input
  - Sample the input signal of Rx at higher rate (e.g. 2x faster)
  - Produce equalizer output signal at symbol rate
  - Can update coefficients at symbol rate
  - Less sensitive to sampling phase

$$c_n(k+1) = c_n(k) - \Delta e_k \cdot r(t_0 + kT - NT/2)$$

7.48

</td></tr>
</table>

Slide 7.48

Until now we have looked at equalizers that sample the received signal at the symbol period T. However, the received signal is not exactly limited to a bandwidth of $1/T$. In fact, it usually spans a 10–40% larger bandwidth after passing through the frequency-selective channel. Hence, sampling at symbol period T can lead to signal cancellation and self-interference effects that arise when the signal components alias into a band of $1/T$. This sampling is also extremely sensitive to the phase of the sampling clock. The sampling clock should always be positioned at the point where the eye pattern of the received signal is the widest.

To avoid these issues, a common solution is to over-sample the incoming signal at rates higher than $1/T$, for example at twice the rate of $2/T$. This would mean that the equalizer impulse response is spaced at a time period of $T/2$ and is called a "fractionally-spaced equalizer." Over-sampling will lead to more computational requirements in the equalizer, since we now have two samples for a symbol instead of one. However it simplifies the receiver design since we now avoid the problems associated with spectral aliasing, which is removed with the higher sampling frequency. The sensitivity to the phase of the sampling clock is also mitigated since we now sample the same symbol twice. The attractiveness of this scheme has resulted in most equalizer implementations being at least 2x faster than the symbol rate. The equalizer output is still T-spaced, which makes these equalizers similar to decimating or re-sampling filters.

<table>
<tr><td>

Summary

- **Use equalizers to reduce ISI and achieve high data rate**
- **Use adaptive equalizers to track time-varying channel response**
- **LMS-based equalization prevails in MODEM design**
- **Decision feedback equalizer makes use of previous decisions to estimate current symbol**
- **Fractionally spaced equalizers resilient to sampling phase variation**
- **Properly select step size for convergence**

---

[6] S. Qureshi, "Adaptive Equalization," *Proceedings of the IEEE*, vol. 73, no. 9, pp. 1349-1387, Sep. 1985.

7.49

</td></tr>
</table>

Slide 7.49

To summarize, we have looked at the application of FIR and feedback filters as equalizers to mitigate the effect of inter-symbol interference in communication systems. The equalizer coefficients can be constant if the channel is time-invariant. However, for most practical cases an adaptive equalizer [6] is required to track the changes in a time-variant channel. Most equalizers use the least-mean-square (LMS) criterion to converge to the optimum tap coefficients. The LMS equalizer has an FIR structure and is capable of cancelling pre-cursor ISI generated from preceding symbols. The decision-feedback equalizer (DFE) is used to cancel post-cursor ISI using feedback from previously detected symbols. Error propagation is possible in the DFE if previously detected symbols are incorrect. Fractionally-spaced equalizers which sample the incoming symbol at rates in excess of the symbol period T, are used to reduce the problems associated with aliasing and sample phase variation in the received signal.

Implementation Technique:

Distributed Arithmetic

Slide 7.50

FIR filters may consume a large amount of area when a large number of taps is required. For applications where the sampling speed is low, or where the speed of the underlying technology is significantly greater than the sampling speed, we can greatly reduce silicon area by resource sharing. Next, we will discuss an area-efficient filter implementation technique based on distributed arithmetic.

Distributed Arithmetic: Concept

- ◆ FIR filter response

$$y_k = \sum_{n=0}^{N-1} h_n \cdot x_{k-n}$$

- ◆ Filter parameters
  - N: number of taps
  - W: wordlength of x
  - $|x_{k-n}| \leq 1$

- ◆ Equivalent representation
  - Bit-level decomposition

$$y_k = \sum_{n=0}^{N-1} h_n \cdot \left(-x_{k-n,W-1} + \sum_{i=1}^{W-1} x_{k-n,W-1-i} \cdot 2^{-i} \right)$$

MSB *Remaining bits*

- ◆ <u>Next step:</u> interchange summations, unroll taps

7.51

Slide 7.51

In cases when the filter performance requirement is below the computational speed of technology, the area of the filter can be greatly reduced by bit-level processing. Let's analyze an FIR filter with N taps and an input wordlength of W (index $W-1$ indicates MSB, index 0 indicates LSB). Each of the x_{k-n} terms from the FIR filter response formula can be expressed as a summation of individual bits. Assuming $|x_{k-n}| \leq 1$, the bit-level representation of x_{k-n} is shown on the slide. Next,

we can interchange the summations and unroll filter taps.

Distributed Arithmetic: Concept (Cont.)

- **FIR filter response: bit-level decomposition**

$$y_k = \sum_{n=0}^{N-1} h_n \cdot \left(-x_{k-n,W-1} + \sum_{i=1}^{W-1} x_{k-n,W-1-i} \cdot 2^{-i} \right)$$

$\underbrace{}$ **MSB** $\underbrace{}$ **Remaining bits**

- **Interchange summations, unroll taps into bits**

$$y_k = -\sum_{n=0}^{N-1} h_n \cdot x_{k-n,W-1} + \sum_{i=1}^{W-1} \left(\sum_{n=0}^{N-1} h_n \cdot x_{k-n,W-1-i} \right) \cdot 2^{-i}$$

MSB
$$= -H_{W-1} \left(x_{-k,W-1}; \; x_{k-1,W-1}; \; ...; \; x_{k-N-1,W-1} \right)$$
tap 1   **tap 2**   **tap N**

Other bits
$$+ \sum_{i=1}^{W-1} H_{W-1-i} \left(x_{k,W-1-i}; x_{k-1,W-1-i}; ...; x_{k-N-1,W-1-i} \right) \cdot 2^{-i}$$
tap 1   **tap 2**   **tap N**   **Bit: MSB – i**

7.52

Slide 7.52

After the interchange of summations and unrolling of the filter taps, we obtain an equivalent representation expressed as a sum over all filter coefficients ($n=0, \ldots, N-1$) bit-by-bit, where x_{k-Nj}, represents j^{th} bit of coefficient x_{k-N}. In other words, we create a bit-wise expression for the filter response where each term H_{W-1-i} in the bit-sum has contributions from all N filter taps. In this notation, i is the offset from the MSB ($i=0, \ldots, W-1$), and H is the weighted bit-level sum of filter coefficients.

Example: 3-tap FIR

x_k	x_{k-1}	x_{k-2}	H_i (bit slice i)
0	0	0	0
0	0	1	h_2
0	1	0	h_1
0	1	1	$h_1 + h_2$
1	0	0	h_0
1	0	1	$h_0 + h_2$
1	1	0	$h_0 + h_1$
1	1	1	$h_0 + h_1 + h_2$

H_i **is the weighted bit-level sum of filter coefficients**

MSB i LSB

x_k	0	...	0	...	0
x_{k-1}	0	...	0	...	1
x_{k-2}	0	...	1	...	1

3 bits → address

LUT

0	0
h_2	1
...	
$h_0 + h_1 + h_2$	7

$$y_k = \left((...(0 + H_0) \cdot 2^{-1} + H_1) \cdot 2^{-1} + ... + H_{W-2} \right) \cdot 2^{-1} - H_{W-1}$$

7.53

Slide 7.53

Here's an example of a 3-tap filter where input wordlength is also $W=3$. Consider bit slice i. As described in the previous slide, H_i is the bit-level sum of filter coefficients h_k, as shown in the truth table on this slide. The last column of the table can be treated as memory, whose address space is defined by the bit values of filter input. For example, at bit position i, where $x=[0\ 0\ 1]$, the value of bit-wise sum of coefficients is equal to h_2. The filter can, therefore, be simply viewed as a look-up table containing pre-computed coefficients.

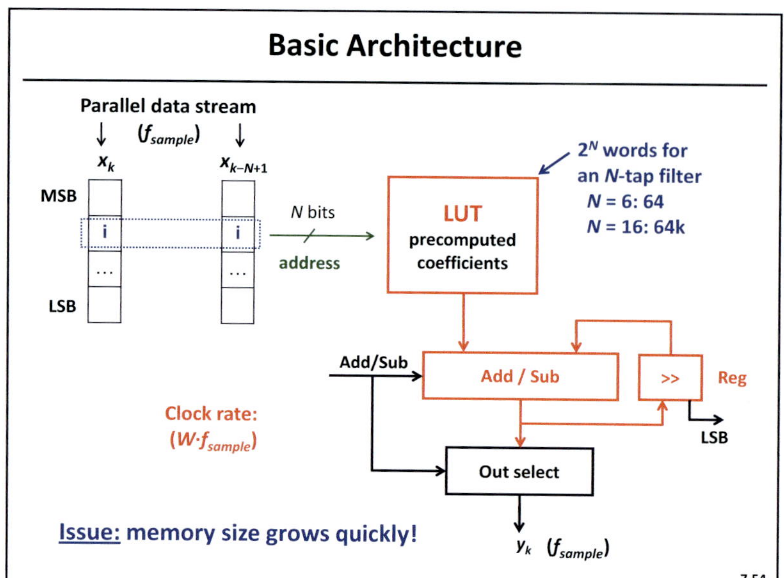

Slide 7.54

The basic filter architecture is shown on this slide. A W-bit parallel data stream x sampled at rate f_{sample} forms an N-bit address (x_k, ..., x_{k-N+1}) for the LUT memory. In each clock cycle, we shift the address by one bit and add (subtract if the bit is the MSB) the result. Since we sample the input at f_{sample} and process one bit at a time, we need to up-sample the computation W times. All blocks in red boxes operate at rate $W \cdot f_{sample}$. Due to the bit-serial nature of the computation, it takes W clock cycles to do one filtering operation. While this architecture provides a compact realization of the filter, the area savings from the bit-serial approach may be outweighed by the large area of the LUT memory. For N taps we need 2^N words. If $N=16$, for example, LUT memory with 64k words is required! Memory size grows exponentially with the number of taps. There are several ways to address this problem.

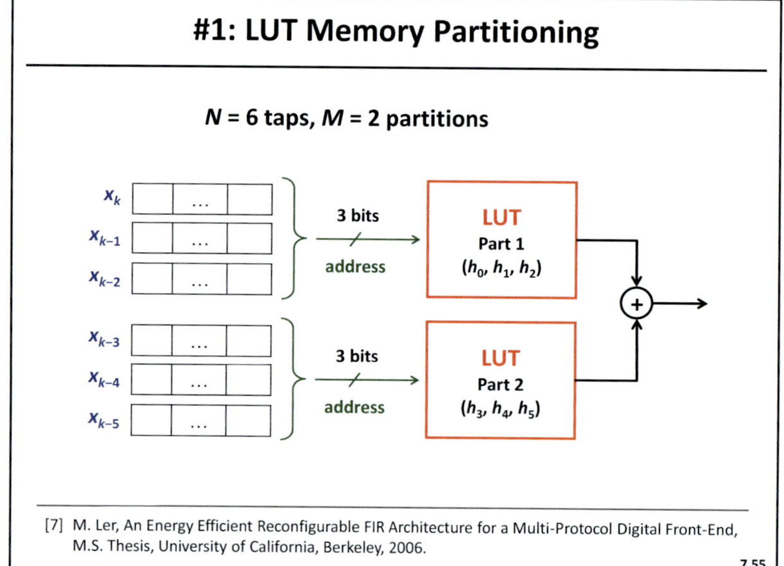

Slide 7.55

Memory partitioning can be used to address the issue of the LUT area. The concept of memory partitioning is illustrated in this slide. Assume $N=6$ taps and $M=2$ partitions. A design with $N=6$ taps would require $2^6=64$ LUT words without partitioning. Assume that we now partition the LUT memory into $M=2$ segments. Each LUT segment uses 3 bits for the address space, thus requiring $2^3=8$ words. The total number of words for the segmented memory architecture is $2 \cdot 2^3=16$ words. This is a 4x (75%) reduction in memory size as compared to the direct memory realization. Segmented memory requires an adder at the output to sum the partial results from the segments.

#2: Memory Code Compression

- **Idea:** $x = \dfrac{1}{2}[x - \underbrace{(-x)}_{\overline{x}}]$

- **Signed-digit offset binary coding:** {1, −1} instead of {1, 0}

- **Bit-level expression**

$$x = \frac{1}{2}[-\underbrace{(x_{W-1} - \overline{x}_{W-1})}_{c_{W-1}}]$$

$$+\frac{1}{2}[\sum_{i=1}^{W-1} \underbrace{(x_{W-1-i} - \overline{x}_{W-1-i})}_{c_{W-1-i}} \cdot 2^{-i} - 2^{-(W-1)}]$$

- <u>**Next step:**</u> plug this inside expression for y_k

7.56

Slide 7.56

Another idea to consider is memory code compression. The idea is to use signed-digit offset binary coding that maps the {1, 0} base into {1, −1} base for computations. Using the newly formed base, we can manipulate bit-level expressions using the $x = 0.5 \cdot (x - (-x))$ identity.

#2: Memory Code Compression (Cont.)

- **Use:** $x = \dfrac{1}{2}\left[\displaystyle\sum_{i=0}^{W-1} c_{W-1-i} \cdot 2^{-i} - 2^{-(W-1)}\right]$

- **Another representation of y_k**

$$y_k = \frac{1}{2}\sum_{n=0}^{N-1} h_n \left[\sum_{i=0}^{W-1} c_{k-n,W-1-i} \cdot 2^{-i} - 2^{-(W-1)}\right]$$

$$= \sum_{i=0}^{W-1} H_{W-1-i}(c_{k,W-1-i}; c_{k-1,W-1-i}; \ldots; c_{k-N-1,W-1-i}) \cdot 2^{-i}$$

$$+ H(0;0;\ldots;0) \cdot 2^{-(W-1)}$$

- **Term H_{W-1-i} has only 2^{N-1} values**
  - Memory requirement reduced from 2^N to 2^{N-1}

7.57

Slide 7.57

Substituting the new representation into the y_k formula yields the resulting bit-level coefficients. The bit-level coefficients could take 2^{N-1} values. Compared to the original formulation that maps the coefficients into a 2^N-dimensional space, code compression yields a 2x reduction in the required memory. Code compression can be combined with LUT segmentation to minimize LUT area.

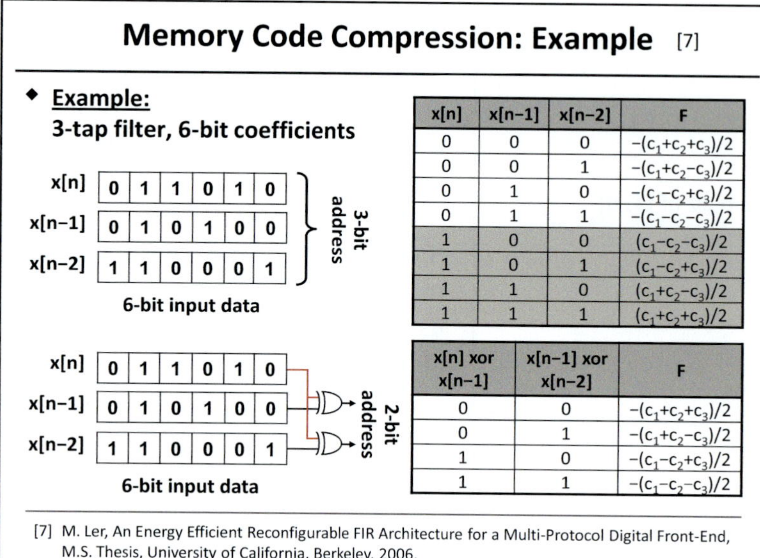

$x[n]$	$x[n-1]$	$x[n-2]$	F
0	0	0	$-(c_1+c_2+c_3)/2$
0	0	1	$-(c_1+c_2-c_3)/2$
0	1	0	$-(c_1-c_2+c_3)/2$
0	1	1	$-(c_1-c_2-c_3)/2$
1	0	0	$(c_1-c_2-c_3)/2$
1	0	1	$(c_1-c_2+c_3)/2$
1	1	0	$(c_1+c_2-c_3)/2$
1	1	1	$(c_1+c_2+c_3)/2$

$x[n]$ xor $x[n-1]$	$x[n-1]$ xor $x[n-2]$	F
0	0	$-(c_1+c_2+c_3)/2$
0	1	$-(c_1+c_2-c_3)/2$
1	0	$-(c_1-c_2+c_3)/2$
1	1	$-(c_1-c_2-c_3)/2$

[7] M. Ler, An Energy Efficient Reconfigurable FIR Architecture for a Multi-Protocol Digital Front-End, M.S. Thesis, University of California, Berkeley, 2006.

7.58

Slide 7.58

An example of memory code compression is illustrated in this slide [3]. For a 3-tap filter (3-bit address), we can observe similar entries in memory locations corresponding to MSB = 0 and MSB = 1. The only difference is the sign change. In terms of gate-level implementation, the sign bit can be arbitered by the MSB bit (shown in *red*) and two XOR gates that control reduced address space.

Summary

- **Digital filters are key building elements in DSP systems**
- **FIR filters can be realized in direct or transposed form**
  - Direct form has long critical-path delay
  - Transposed form has large input loading
  - Multiplications can be simplified by using coefficients that can be derived as sum of power-of-two numbers
- **Performance of recursive IIR filters is limited by the longest loop delay (iteration bound)**
  - IIR filters are suitable for sharp roll-off characteristics
  - More power and area efficient than FIR
- **Multi-rate filters are used for decimation and interpolation**
- **Distributed arithmetic can effectively reduce the size of coefficient memory in FIR filters by using bit-serial arithmetic**

7.59

Slide 7.59

Digital filters are important building blocks in DSP systems. FIR filters can be realized in direct or transposed form. The direct form realization has long critical-path delay (if no pipelining is used) while transposed form has large input loading. Multiplier coefficients can be approximated with a sum of power-of-two numbers to simplify the implementation. IIR filters are used where sharp roll-off and compact realization are required. Multi-rate decimation and interpolation filters, which are standard in wireless transceivers, are briefly introduced. The chapter finally discussed FIR filter implementation based on distributed arithmetic. The idea is to compress coefficient memory by performing bit-serial arithmetic.

References

- J. Proakis, Digital Communications, (3rd Ed), McGraw Hill, 2000.

- A.V. Oppenheim and R.W. Schafer, Discrete Time Signal Processing, (3rd Ed), Prentice Hall, 2009.

- J.G. Proakis and D.K. Manolakis, Digital Signal Processing, 4th Ed, Prentice Hall, 2006.

- K.K. Parhi, VLSI Digital Signal Processing Systems: Design and Implementation, John Wiley & Sons Inc., 1999.

- H. Samueli, "An Improved Search Algorithm for the Design of Multiplierless FIR Filters with Powers-of-Two Coefficients," *IEEE Trans. Circuits and Syst.*, vol. 36 , no. 7 , pp. 1044-1047, July 1989.

- S. Qureshi, "Adaptive Equalization," *Proceedings of the IEEE,* vol. 73, no. 9, pp. 1349-1387, Sep. 1985.

- M. Ler, An Energy Efficient Reconfigurable FIR Architecture for a Multi-Protocol Digital Front-End, M.S. Thesis, University of California, Berkeley, 2006.

Additional References

- R. Jain, P.T. Yang, and T. Yoshino, "FIRGEN: A Computer-aided Design System for High Performance FIR Filter Integrated Circuits," *IEEE Trans. Signal Processing,* vol. 39, no. 7, pp. 1655-1668, July 1991.

- R.A. Hawley *et al.*, "Design Techniques for Silicon Compiler Implementations of High-speed FIR Digital Filters," *IEEE J. Solid-State Circuits,* vol. 31, no. 5, pp. 656-667, May 1996.

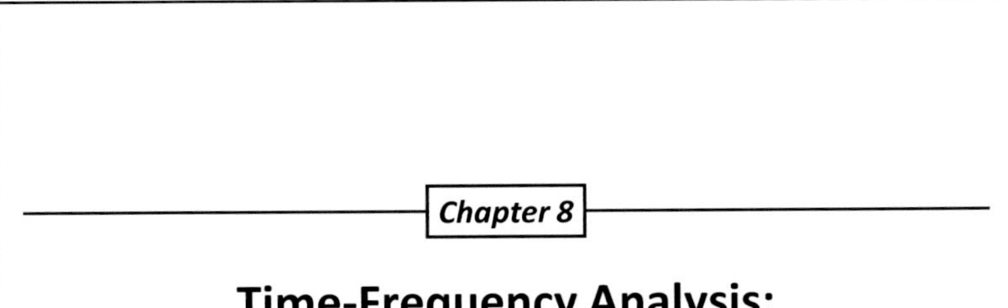

Slide 8.1

In this chapter we will discuss the methods for time frequency analysis and the DSP architectures for implementing these methods. In particular, we will use the FFT and the wavelet transform as our examples for this chapter. The well-known Fast Fourier Transform (FFT) is applicable to the frequency analysis of stationary signals. Wavelets provide a flexible time-frequency grid to analyze signals whose spectral content changes over time. An analysis of algorithm complexity and implementation is presented.

Slide 8.2

The Discrete Fourier Transform (DFT) was discovered in the early nineteenth century by the German mathematician Carl Friedrich Gauss. More than 150 years later, the algorithm was rediscovered by the American mathematician James Cooley, who came up with a recursive approach for calculating the DFT in order to reduce the computation time and make the algorithm more practical.

The FFT is a fast way to compute DFT. It transforms a time-domain data sequence into frequency components and vice versa. The FFT is one of the key building blocks in digital signal processors for wireless communications, media/image processing, etc. It is also used for the analysis of spectral content in Electroencephalography (EEG) for brain-wave studies and many other applications.

Many FFT implementation techniques exist to tradeoff the power and area of the hardware. We will first look into the basic architectural tradeoffs for a fixed FFT size (i.e., fixed number of discrete points) and then extend the analysis to programmable FFT kernels, for applications such as software-defined and cognitive radios.

D. Marković and R.W. Brodersen, *DSP Architecture Design Essentials*, Electrical Engineering Essentials,
DOI 10.1007/978-1-4419-9660-2_8, © Springer Science+Business Media New York 2012

Fourier Transform: Concept

- A complex function can be approximated with a weighted sum of basis functions

- Fourier used sinusoids with varying frequencies as the basis functions

$$X(\omega) = \frac{1}{\sqrt{2\pi}} \int_{-\infty}^{\infty} x(t) e^{-j\omega t} dt$$

$$x(t) = \frac{1}{\sqrt{2\pi}} \int_{-\infty}^{\infty} X(\omega) e^{j\omega t} d\omega$$

- This representation provides the frequency content of the original function

- Fourier transform assumes that all spectral components are present at all times

8.3

Slide 8.3

The Fourier transform approximates a function with a weighted sum of basis functions. Fourier transform uses sinusoids as the basis functions. The Fourier formulas for the time-domain $x(t)$ and frequency-domain $X(\omega)$ representations of a signal x are given by the equations on this slide. The equations assume that the spectral or frequency content of the signal is time-invariant, i.e. that x is stationary.

The Fast Fourier Transform (FFT)

- Efficient method for calculating discrete Fourier transform (DFT)

- N = length of transform, must be composite
  - $N = N_1 \cdot N_2 \cdot \ldots \cdot N_m$

Transform length	DFT ops	FFT ops	DFT ops / FFT ops
64	4,096	384	11
256	65,536	2,048	32
1,024	1,048,576	10,240	102
65,536	4,294,967,296	1,048,576	4,096

8.4

Slide 8.4

The FFT is used to calculate the DFT with reduced numerical complexity. The key idea of the FFT is to recursively compute the DFT by breaking a sequence of discrete samples of length N into sub-sequences N_1, N_2, ..., N_m such that $N = N_1 \cdot N_2 \cdot \ldots \cdot N_m$. This recursive procedure results in a greatly reduced computation time. The time complexity is $O(N \cdot \log N)$ as compared to $O(N^2)$ for the direct computation.

Consider the various transform lengths shown in the table. The number of operations for the DFT and the FFT is compared for $N = 64$ to $65,536$. For a 1,024-point FFT, which is typical in advanced communication systems and EEG analysis, the FFT achieves a 100x reduction in the number of arithmetic operations as compared to DFT. The savings are even higher for larger N, which can vary from a few hundred in communication applications, to a million in radio astronomy applications.

The DFT Algorithm

- **Converts time-domain samples into frequency-domain samples**

$$X_k = \sum_{n=0}^{N-1} x_n e^{-j\frac{2\pi}{N}nk} \qquad k = 0, \ldots, N-1$$

complex number

- **Implementation options**
  - Direct-sum evaluation: $O(N^2)$ operations
  - FFT algorithm: $O(N \cdot \log N)$ operations

- **Inverse DFT: frequency-to-time conversion**
  - DFT with opposite sign in the exponent
  - A 1/N factor

$$x_k = \frac{1}{N} \sum_{n=0}^{N-1} X_n e^{j\frac{2\pi}{N}nk} \qquad k = 0, \ldots, N-1$$

8.5

Slide 8.5

The DFT algorithm converts a complex discrete-time sequence $\{x_n\}$ into the discrete samples $\{X_k\}$ which represent its spectral content, as given by the formula near the top of the slide. As discussed before, the FFT is an efficient way to compute the DFT. A direct-sum DFT evaluation would take $O(N^2)$ operations, while the FFT takes $O(N \cdot \log N)$ operations. The same algorithm can also be used to compute inverse-FFT (IFFT), i.e. to convert frequency-domain symbol $\{X_k\}$ back into the time-domain sample $\{x_n\}$, by changing the sign of the exponent and adding a normalization factor of $1/N$ before the sum, as given by the formula at the bottom of the slide. The similarity between the two formulas also means that the same hardware can be programmed to perform either FFT or IFFT.

The Divide-and-Conquer Approach

- **Map the original problem into several sub-problems in such a way that the following inequality holds:**

$\sum$Cost(sub-problems) + Cost(mapping) < Cost(original problem)

- **DFT:**

$$X_k = \sum_{n=0}^{N-1} x_n W_N^{nk} \qquad k = 0, \ldots, N-1 \qquad\qquad X(z) = \sum_{n=0}^{N-1} x_n z^{-n}$$

  - X_k = evaluation of $X(z)$ at $z = W_N^{-k}$
  - $\{x_n\}$ and $\{X_k\}$ are periodic sequences

- **So, how shall we divide the problem?**

8.6

Slide 8.6

The FFT is based on a divide-and-conquer approach, where the original problem (a sequence of length N) is decomposed into several sub-problems (i.e. sub-sequences of shorter durations), such that the total cost of the sub-problems plus the cost of mapping is less than the cost of the original problem. Let's begin with the original DFT, as described by the formula for X_k on this slide. Here X_k is a frequency sample of $X(z)$ at $z = W_N^{-k}$, where $\{x_n\}$ and $\{X_k\}$ are N-periodic sequences.

The Divide and Conquer Approach (Cont.)

- ◆ **Procedure:**
  - Consider sub-sets of the initial sequence
  - Take the DFT of these sub-sequences
  - Reconstruct the DFT from the intermediate results
- ◆ **#1: Define:** I_t, $t = 0, \dots, r - 1$ partition of $\{0, \dots, N - 1\}$ that defines G different subsets of the input sequence

$$X(z) = \sum_{i=0}^{N-1} x_i z^{-i} = \sum_{t=0}^{r-1} \sum_{i \in I_t} x_i z^{-i}$$

- ◆ **#2: Normalize the powers of z w.r.t. x_{0t} in each subset I_t**

$$X(z) = \sum_{t=0}^{r-1} z_{i_{0t}} \sum_{t} x_i z^{-i + i_{0t}}$$

  - Replace z by w_N^{-k} in the inner sum

8.7

Slide 8.7

The DFT computation is organized by firstly partitioning the samples into sub-sets, taking the DFT of these subsets and finally reconstructing the DFT from the intermediate results. The first step is to define the subsets. Suppose that we have a partitioning I_t ($t = 0, 1, \dots, r-1$) which defines G subsets with I_t elements in their respective subsets. The original sequence can then be written as given by the first formula on this slide. In the second step, we normalize the powers of z in each subset I_t, as shown by the second formula. Finally, we replace z by w_N^{-k} for compact notation.

Cooley-Tukey Mapping

- ◆ **Consider decimated versions of the initial sequence, $N = N_1 \cdot N_2$**

$$I_{n_1} = \{n_2 \cdot N_1 + n_1\} \qquad n_1 = 0, \dots, N_1 - 1; \quad n_2 = 0, \dots, N_2 - 1$$

- ◆ **Equivalent description:**

$$X_k = \sum_{n_1=0}^{N_1-1} W_N^{n_1 k} \sum_{N_2=0}^{N_2-1} x_{n_2 N_1 + n_1} \cdot W_N^{n_2 N_1 k} \qquad \textbf{(8.1)}$$

$$W_N^{iN_1} = e^{-j2\pi \frac{iN_1}{N}} = e^{-j2\pi \frac{i}{N_2}} = W_{N_2}^{i} \qquad \textbf{(8.2)}$$

- ◆ **Substitute (2) into (1)**

$$X_k = \sum_{n_1=0}^{N_1-1} W_N^{n_1 k} \underbrace{\sum_{N_2=0}^{N_2-1} x_{n_2 N_1 + n_1} \cdot W_{N_2}^{n_2 k}}_{\text{DFT of length } N_2}$$

DFT of length N_2

8.8

Slide 8.8

Using the notation from the previous slide, we can express a sequence of N samples as a composition of two sub-sequences N_1 and N_2 such that $N = N_1 \cdot N_2$. The original sequence is divided into N_2 sequences $I_{n1} = \{n_2 \cdot N_1 + n_1\}$ with $n_2 = 0, 1, \dots, N_2$, as given by the formula for X_k, (8.1). Note that $W_N^{iN_1} = W_{N_2}^i$, as in (8.2). Substituting (8.2) into (8.1), we obtain an expression for X_k that consists of N_1 DFTs of length N_2. Coefficients W_N^k are called "twiddle factors."

Cooley-Tukey Mapping (Cont.)

Define: $Y_{n_1,k} = k^{th}$ output of n_1^{th} length-N_2 DFT

$$X_k = \sum_{n_1=0}^{N_1-1} Y_{n_1,k} W_N^{n_1 k} \qquad k = k_1 N_2 + k_2 \qquad \begin{array}{l} k_1 = 0, ..., N_1 - 1 \\ k_2 = 0, ..., N_2 - 1 \end{array}$$

- $Y_{n_1,k}$ can be taken modulo N_2

$$W_{N_2}^k = W_{N_2}^{N_2+k'} = W_{N_2}^{N_2} \cdot W_{N_2}^{k'} = W_{N_2}^{k'}$$

> All the X_k for k being congruent modulo N_2 are from the same group of N_1 outputs of $Y_{n_1,k}$

$Y_{n_1,k} = Y_{n_1,k_2}$ since k can be taken modulo N_2

- **Equivalent description:**

$$X_{k_1 N_2 + k_2} = \sum_{n_1=0}^{N_1-1} Y_{n_1,k} W_N^{n_1(k_1 N_2 + k_2)} = \sum_{n_1=0}^{N_1-1} \underbrace{Y_{n_1,k_2} W_N^{n_1 k_2}}_{Y'_{n_1,k_2}} W_{N_1}^{n_1 k_1}$$

From N_2 DFTs of length N_1 applied to Y'_{n_1,k_2}

8.9

Slide 8.9

The result from the previous slide can be re-written by introducing $Y_{n_1,k}$ as a short notation for the kth output of the n_1^{th} DFT. Frequency samples X_k can be expressed as shown on the slide. Given that the partitioning of the sequence $N = N_1 \cdot N_2$, $Y_{n_1,k}$ can be taken modulo N_2. This implies that all of the X_k for k being congruent modulo N_2 are from the same group of N_1 outputs of $Y_{n_1,k}$. This representation is also known as Cooley-Tukey mapping. Equivalently, X_k can be described by the formula on the bottom of the slide, by taking N_2 DFTs of length N_1 and applying them to $Y_{n_1,k} W_N^{n_1 k_2}$. The Cooley-Tukey mapping allows for practical implementation of the DFT.

Cooley-Tukey Mapping, Example: $N_1 = 5$, $N_2 = 3$

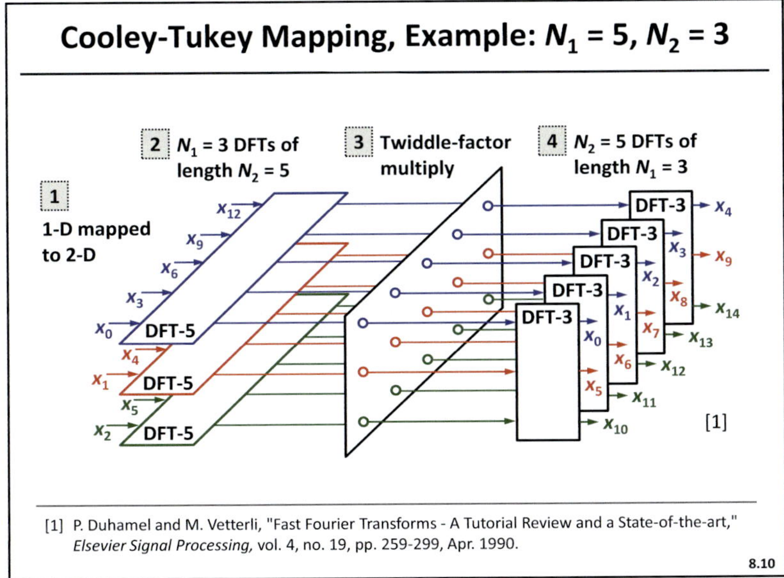

[1] P. Duhamel and M. Vetterli, "Fast Fourier Transforms - A Tutorial Review and a State-of-the-art," *Elsevier Signal Processing*, vol. 4, no. 19, pp. 259-299, Apr. 1990.

8.10

Slide 8.10

The Cooley-Tukey FFT can be graphically illustrated as shown on this slide for $N = 15$ samples [1]. The computation is divided into length-3 ($N_1 = 3$) and length-5 ($N_2 = 5$) DFTs. The first stage (inner sum) computes three 5-point DFTs, the second stage performs multiplication by the twiddle factors, and the third stage (outer sum) computes five 3-point DFTs to combine the partial results. Note that the ordering of the output samples (indices) is different from the ordering of the input samples, so re-ordering is needed to match the corresponding samples.

$N_1 = 3$, $N_2 = 5$ versus $N_1 = 5$, $N_2 = 3$

- **Original 1-D sequence**

$$x_0 \ x_1 \ x_2 \ x_3 \ x_4 \ x_5 \ x_6 \ x_7 \ x_8 \ x_9 \ x_{10} \ x_{11} \ x_{12} \ x_{13} \ x_{14}$$

1-D to 2-D mapping

- $N_1 = 3$, $N_2 = 5$

x_0	x_3	x_6	x_9	x_{12}
x_1	x_4	x_7	x_{10}	x_{13}
x_2	x_5	x_8	x_{11}	x_{14}

- $N_1 = 5$, $N_2 = 3$

x_0	x_5	x_{10}
x_1	x_6	x_{11}
x_2	x_7	x_{12}
x_3	x_8	x_{13}
x_4	x_9	x_{14}

- **1D-2D mapping can't be obtained by simple matrix transposition**

8.11

Slide 8.11

This slide shows the 1D to 2D mapping procedure for the cases of $N_1=3$, $N_2=5$ and $N_1=5$, $N_2=3$. The samples are organized sequentially into N_1 columns and N_2 rows, column by column. As shown on the slide, $N_1 \cdot N_2 = 3 \cdot 5$ and $N_1 \cdot N_2 = 5 \cdot 3$ are not the same, and one cannot be obtained from the other by simple matrix transposition.

Radix-2 Decimation in Time (DIT)

- $N_1 = 2$, $N_2 = 2^{N-1}$ **divides the input sequence into the sequence of even- and odd-numbered samples ("decimation in time" (DIT))**

$$X_{k_2} = \sum_{n_2=0}^{N/2-1} x_{2n_2} \cdot W_{N/2}^{n_2 k_2} + W_N^{k_2} \sum_{n_2=0}^{N/2-1} x_{2n_2+1} \cdot W_{N/2}^{n_2 k_2}$$

$$X_{N/2+k_2} = \sum_{n_2=0}^{N/2-1} x_{2n_2} \cdot W_{N/2}^{n_2 k_2} - W_N \sum_{n_2=0}^{N/2-1} x_{2n_2+1} \cdot W_{N/2}^{n_2 k_2}$$

- X_{k2} **and** $X_{k2+N/2}$ **obtained by 2-pt DFTs on the outputs of length N/2 DFTs of the even- and odd-numbered sequences, one of which is weighted by twiddle factors**

8.12

Slide 8.12

For the special case of $N_1=2$, $N_2 =2^{N-1}$ divides the input sequence into the sequences of even and odd samples. This partitioning is called "decimation in time" (DIT). The two sequences are given by the equations on this slide. Samples X_{k2} and $X_{k2+N/2}$ are obtained by radix-2 DFTs on the outputs of $N/2$-long DFTs. Weighting by twiddle factors (highlighted by the dashed boxes) is needed for the X_{k2} sequence.

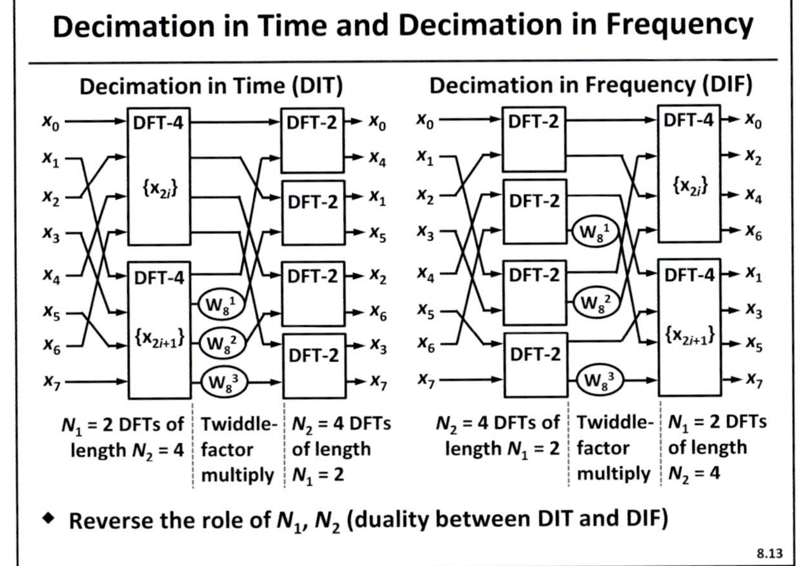

Slide 8.13

The previously described decimation-in-time approach is graphically illustrated on the left for $N=8$. It consists of 2 DFT-4 blocks, followed by twiddle-factor multiplies and then $N/2=$ four DFT-2 blocks at the output. The top DFT-4 block takes only even samples $\{x_{2i}\}$ while the bottom block takes only odd samples $\{x_{2i+1}\}$. By reversing the order of the DFTs (i.e., moving the 2-point DFTs to the first stage and the $N/2$-point DFTs to the third stage), we get decimation in frequency (DIF), as shown on the right. The output samples of the top DFT-4 block are even-ordered frequency samples, while the outputs of the bottom DFT-4 blocks are odd-ordered frequency samples, hence the name "decimation in frequency." Samples must be arranged, as shown on the slide, to ensure proper decimation in time/frequency.

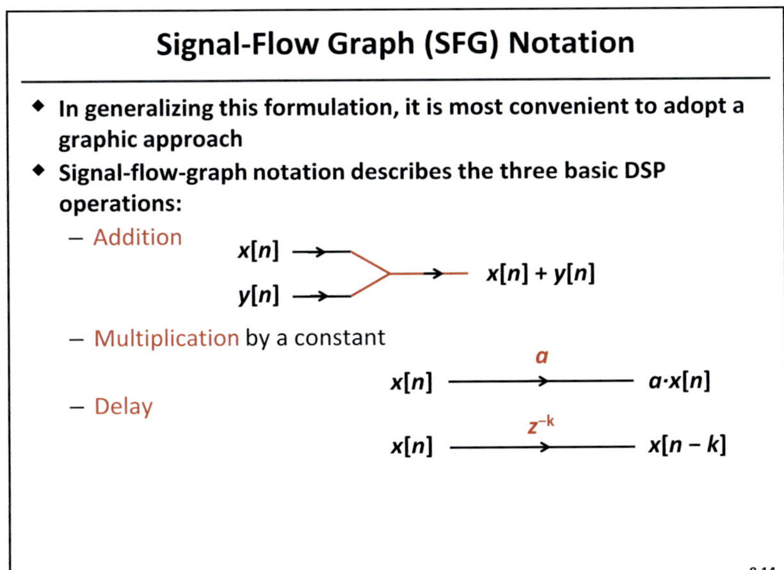

Slide 8.14

For compact representation, signal-flow-graph (SFG) notation is adopted. Weighed edges represent multiplication while vertices represent addition. A delay by k samples is annotated by a z^{-k} along the edge. This simple notation allows for quick modeling and comparison of various FFT architectures.

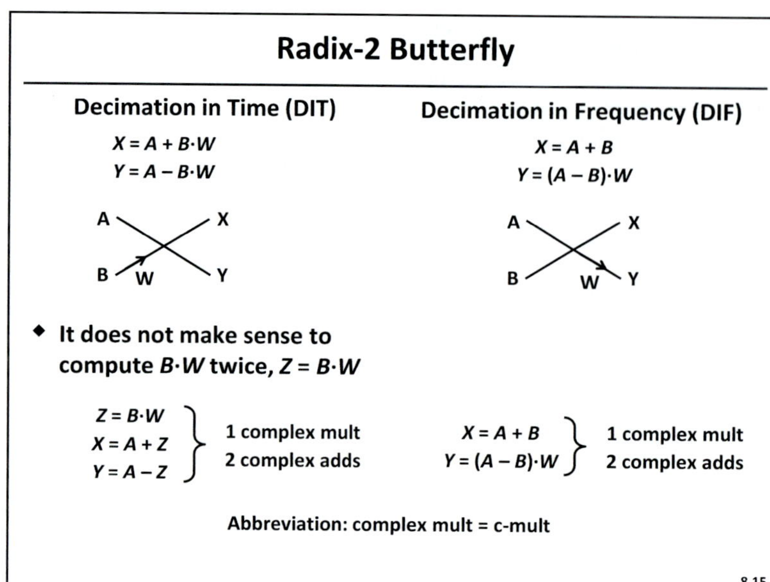

Slide 8.15

Using the SFG notation, a radix-2 butterfly can be represented as shown in this slide. The expressions for decimation in time and frequency can be further simplified by taking the common expression $B \cdot W$ and by rewriting $A \cdot W - B \cdot W$ as $(A - B) \cdot W$. This reduces the number of multipliers from 2 to 1. As a result of expression sharing, both the DIT and the DIF operations can be computed with just 1 complex multiplier and 2 complex adders.

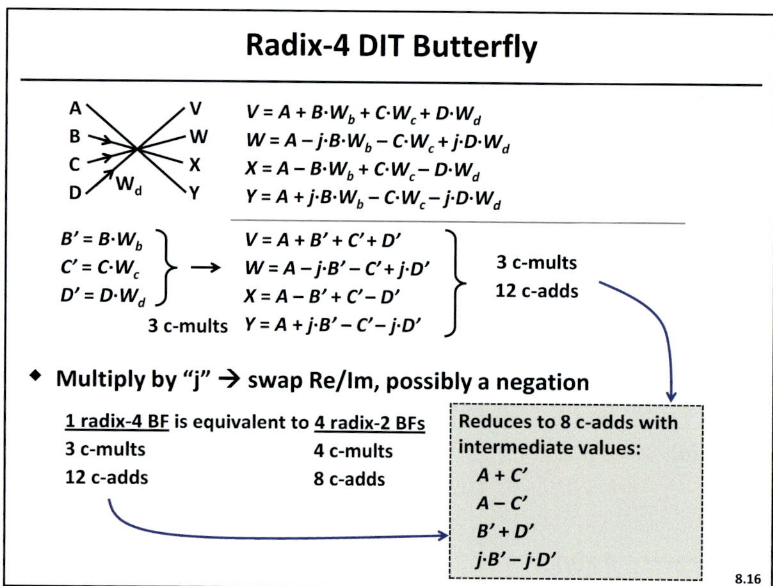

Slide 8.16

Similarly, the radix-4 butterfly can be represented using a SFG, as shown in this slide. The radix-4 butterfly requires 3 complex-multiply and 12 complex-add operations. Taking into account the intermediate values ($A + C$, $A - C$, $B' + D$, and $j \cdot B' - j \cdot D'$), the number of add operations can be reduced from 12 to 8. In terms of numerical complexity, one radix-4 operation is roughly equivalent to 4 radix-2 operations. The numbers of atomic add and multiply operations can be used for quick comparison of different FFT architectures, as shown on the slide.

Comparison of Radix-2 and Radix-4

- **Radix-4 has about the same number of adds and 75% the number of multiplies compared to radix-2**

- **Higher radices**
  - Possible, but rarely used (complicated control)
  - Additional frequency gains diminish for $r > 4$

- **Computational complexity:**
  - Number of mults = reasonable 1st estimate of algorithmic complexity
    - $M = \log_r(N) \rightarrow N_{mults} = O(M \cdot N)$
  - Add N_{adds} for more accurate estimate

- **1 complex mult = 4 real mults + 2 real adds**

8.17

Slide 8.17

Radix-4 is numerically simpler than radix-2 since it requires the same number of additions but only needs 75% the number of multipliers. Higher radices (>4) are also possible, but the use of higher radices alone or mixed with lower radices has been largely unexplored. Radix 2 and 4 are most commonly in FFT implementations. The computational complexity of an FFT can be quickly estimated from the number of multipliers, since multipliers are much larger than adders, which is $O(N \cdot \log_2 N)$. These estimates can be further refined by also accounting for the number of adders.

FFT Complexity

N	FFT Radix	# Re mults	# Re adds
256	2	4096	6,144
256	4	3072	5,632
256	16	2,560	5,696
512	2	9,216	13,824
512	8	6,144	13,824
4096	2	98,304	147,456
4096	4	73,728	135,168
4096	8	65,536	135,168
4096	16	61,440	136,704

For $M = \log_r(N)$
$N_{mult} = O(M \cdot N)$

Decreases monotonically with radix increase

Decreases, reaches min, increases

8.18

Slide 8.18

Considering the number of multiply and add operations, we can quickly estimate the numerical complexity of an FFT block for varying numbers of points. The table compares FFTs with 256, 512, and 4,096 points for radices of 2, 4, 8, and 16. We can see that the number of real multipliers monotonically decreases with increasing radix while the number of adders has a minimum as a function of radix. The number of multipliers and adders are of the same order of magnitude, so we can quickly approximate the hardware cost by the number of multipliers. Also of note is that the number of points N dictates the possible radix factorizations. For example, $N = 512$ can be realized with radix 2 and radix 8. Mixed-radix implementations are also possible and offer more degrees of freedom in the implementation as compared to single-radix designs. Mixed-radix realizations will be discussed in Part IV of the book.

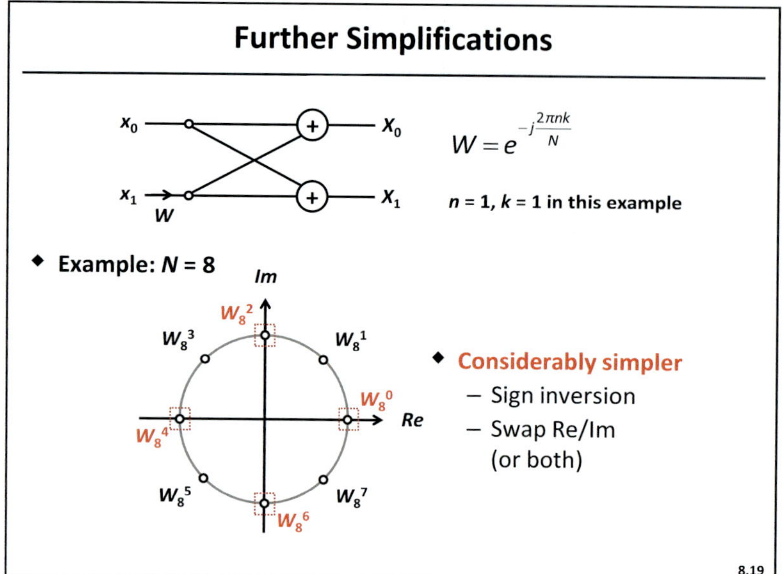

Slide 8.19

Further simplifications are possible by observing regularity in the values of the twiddle factors. This slide illustrates twiddle factors for $N=8$. We can see that w_8^0, w_8^2, w_8^4, and w_8^6 reduce to trivial multiplications by ±1 and $\pm j$, which can be implemented with simple sign inversion and/or swapping of real and imaginary components. Using these observations leads to a simplified implementation and reduced hardware area.

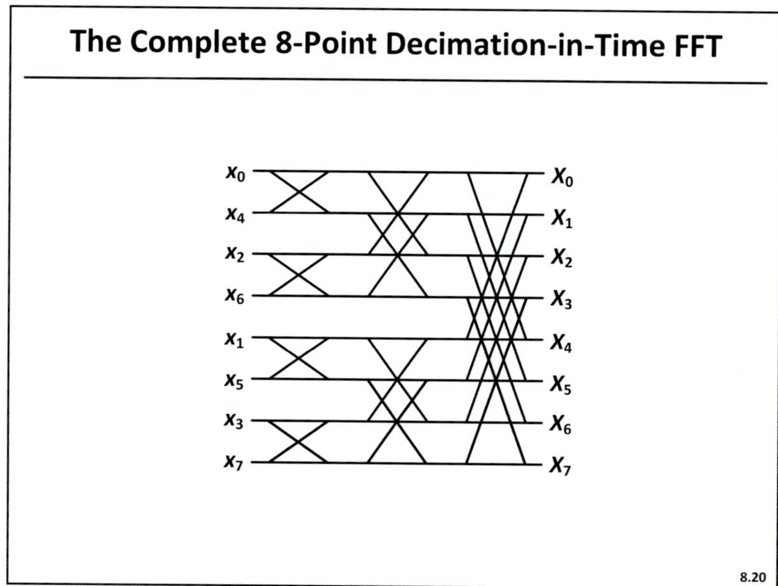

Slide 8.20

The SFG diagram of an 8-point FFT is illustrated here. With a radix-2 implementation, an N-point FFT requires $\log_2(N)$ stages. Each stage has $N/2$ butterflies. The overall complexity of the FFT is $N/2 \cdot \log_2(N)$ butterfly operators. The input sequence has to be ordered as shown to produce the ordered $(0, 1, \ldots, N-1)$ output sequence.

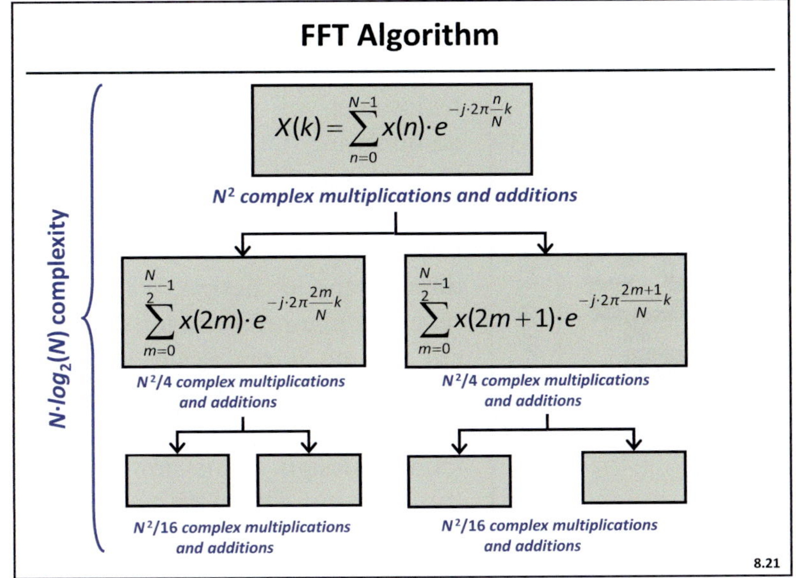

Slide 8.21

This slide shows the radix-2 organization of the FFT algorithm. Each branch is divided into two equal radix-2 partitions. Due to the power-of-2 factorization, an N-point FFT requires $\log_2(N)$ butterfly stages. The total numerical complexity for an N-point FFT is $N \cdot \log_2(N)$. As discussed previously, radix-2 is a simple implementation approach, but may not be the least complex. Multiple realizations have to be considered for large N to minimize hardware area.

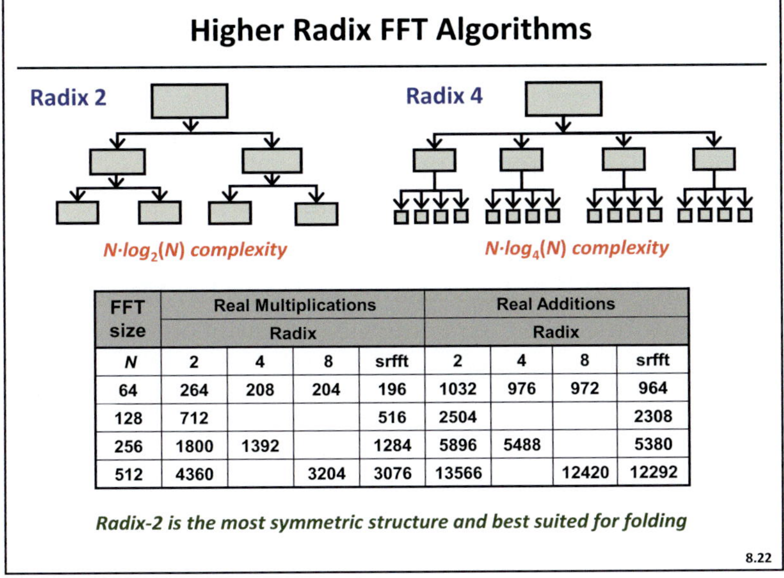

| FFT size | Real Multiplications | | | | Real Additions | | | |
| | Radix | | | | Radix | | | |
N	2	4	8	srfft	2	4	8	srfft
64	264	208	204	196	1032	976	972	964
128	712			516	2504			2308
256	1800	1392		1284	5896	5488		5380
512	4360		3204	3076	13566		12420	12292

Slide 8.22

Higher radices reduce the number of operations as shown in the table on this slide. The radix-r design has a complexity of $N \cdot \log_r(N)$. Higher radices also come with a larger granularity of the butterfly operation, making them suitable only for $N = 2^r$. To overcome the granularity issue, implementations with mixed radices are possible. This is also known as the split-radix FFT (SRFFT). The SRFFT has the lowest numerical complexity (see table), but it is the most complex to design due to a variety of building blocks. While radix-2 may not give a lower operation count, its structure is very simple. The modularity of the radix-2 architecture makes it attractive for hardware realization. For this reason, radix-2 and radix-4 realizations are the most common in practice. In Part IV, we will discuss an optimized SRFFT implementation with radices ranging from 2 to 16 to be used in a multi-band radio DSP front end.

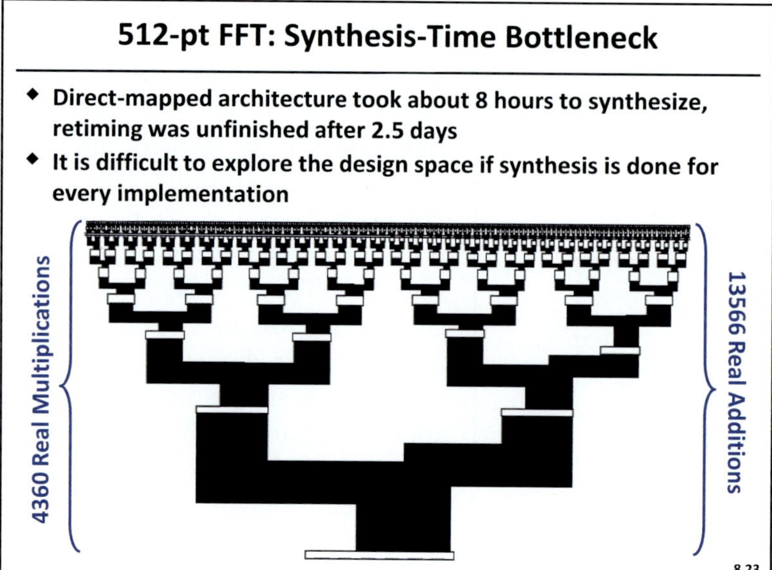

Slide 8.23

This slide shows the direct realization of a 512-point FFT architecture targeting a 2.5 GS/s throughput. The design consists of 4360 multipliers and 13566 adders that are organized in nine stages. This direct-mapped design will occupy several square millemeter of silicon are making it practically infeasible. The logic synthesis process itself will be extremely time intensive for this kind of design. This simple design took about 8hr to synthesize using standard chip design tools, because of difficulty in meeting the timing. The top-level retiming operation on the design did not converge after 2.5 days!

The FFT is a good example of an algorithm where high sample rate can be traded off for area reduction. The FFT takes in N samples of discrete data and produces N output samples in each iteration. Suppose, the FFT hardware functions at clock rate of f_{clk}, then the effective throughput becomes $N \cdot f_{clk}$ samples/second. For $N=512$ and $f_{clk}=100$MHz, we would have an output sample rate of 51.2 Gs/s which is larger than required for most applications. A common approach is to exploit the modular nature of the architecture and re-use hardware units like the radix-2 butterfly. A single iteration of the algorithm is folded and executed in T clock cycles instead of a single cycle. The number of butterfly units would reduce by a factor of T, and will be re-used to perform distinct operations in the SFG in each of these T cycles. The effective sample rate now becomes $N \cdot f_{clk}/T$. More details on folding will be discussed in Part III of the book.

To estimate power and performance for an FFT it is best to explore hierarchy. We will next discuss hierarchical estimation of power and performance of the underlying butterfly units.

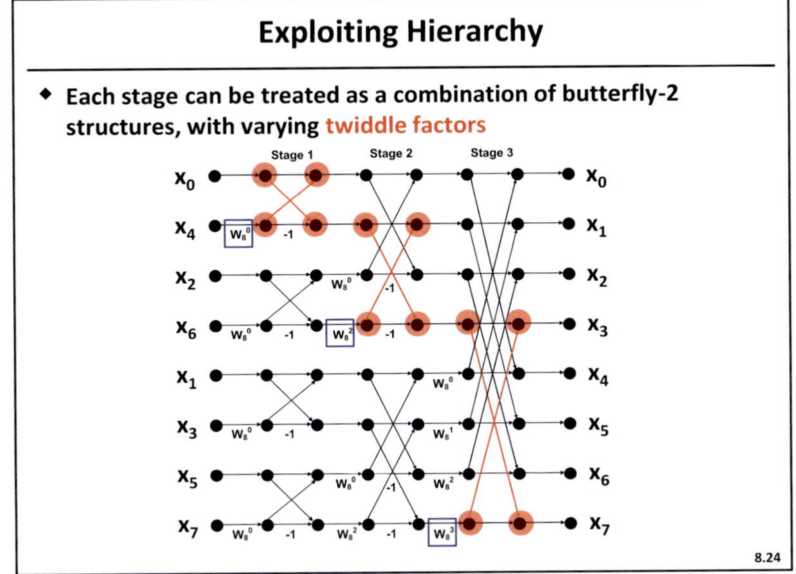

Slide 8.24

Exploiting design hierarchy is essential for efficient hardware mapping and architectural optimization. Each stage of the FFT can be treated as a combination of radix-2 modules. An 8-point FFT module is shown in the slide to illustrate this concept. The radix-2 blocks are highlighted in each of the three stages. The butterfly stages have different twiddle factors and wiring complexity. Modeling area, power, and performance of these butterfly blocks aids in estimation of the FFT hardware.

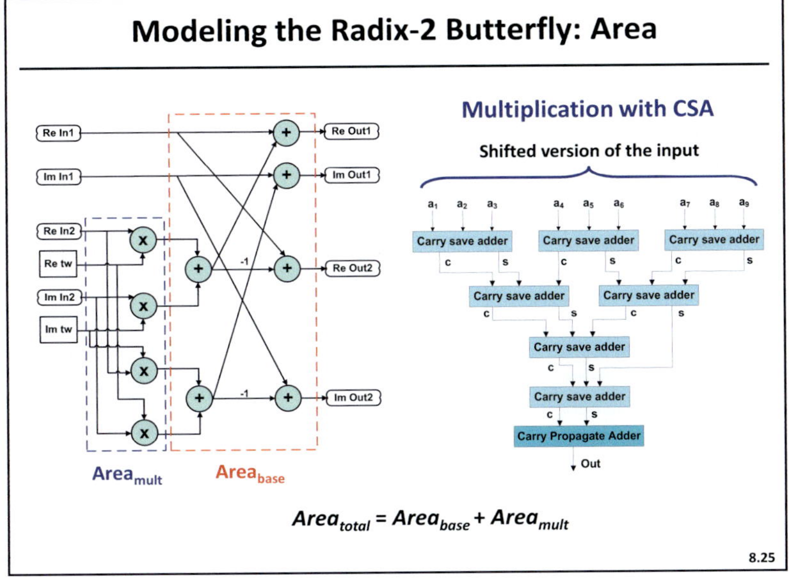

$$Area_{total} = Area_{base} + Area_{mult}$$

Slide 8.25

Each radix-2 butterfly unit consists of 4 real multipliers and 6 real adders. The area of the butterfly block is modeled by adding up the areas of the individual components. A better/smaller design can be made if multipliers are implemented with carry-save adders (CSAs). These implementation parameters are included in the top-level model of the radix-2 block in order to evaluate multiple architectural realizations. The interconnect area can be also included in these block-level estimates. This concept can be hierarchically extended.

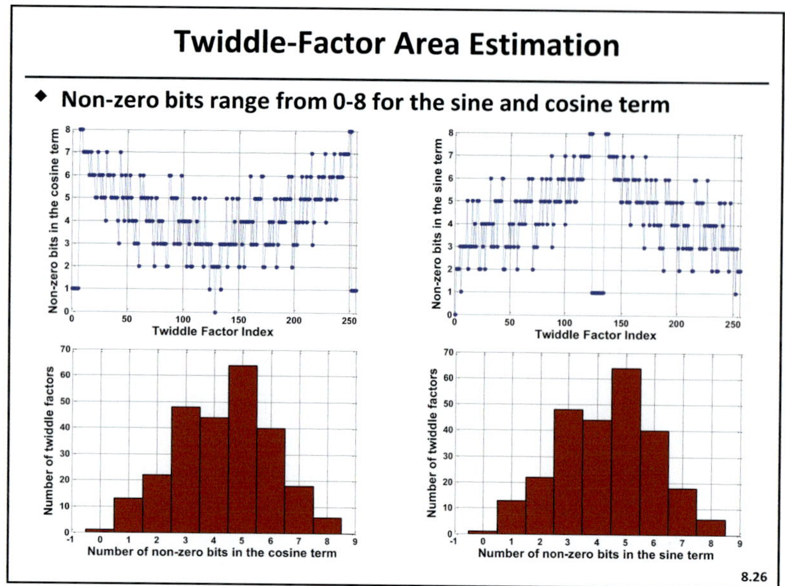

Slide 8.26

In addition to the baseline radix-2-butterfly block, we need to estimate hardware parameters for twiddle factors. The twiddle factors are typically stored in memory and read out when required as inputs for the butterfly units. Their storage contributes substantially to the area of the whole design and should be optimized as far as possible. In a 512-point FFT, there are 256 twiddle factors, but there are regularities that can be exploited. First, the twiddle factors are symmetric around index 128, so we need to consider only 128 factors. Second, among these 128 factors, there is significant regularity in the number of non-zero bits in the sine and cosine terms, as shown on the plots. The histograms indicate that 5 non-zero bits occur with the highest frequency. These observations can be used to develop analytical models for the twiddle-factor area estimation.

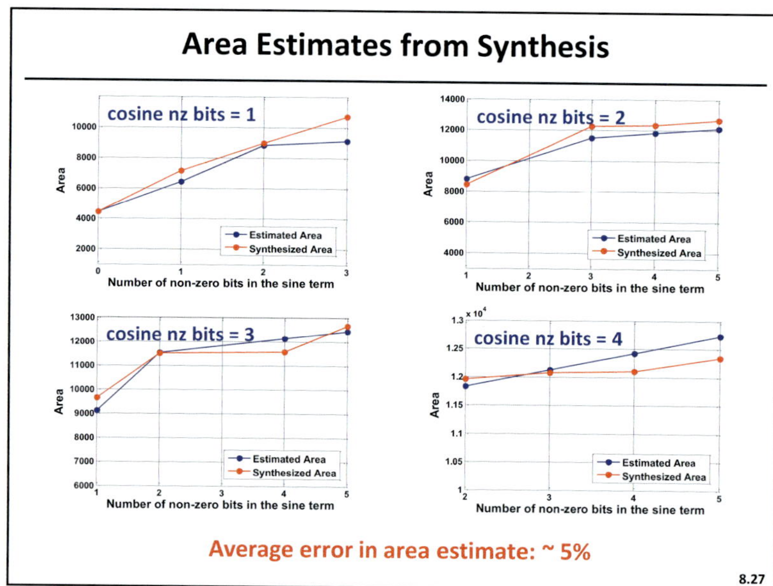

Slide 8.27

Twiddle factor area estimation can be performed by calculating the cost of each non-zero bit. Logic synthesis is performed and the area estimates are used to develop analytical models for all possible coefficient values. Results from this modeling approach are shown on this slide for the sine term with non-zero sine bits ranging from 1 to 4 and non-zero cosine bits varying from 1 to 4 (the sub-plots). The results show a 5% error between the synthesis and analytical models.

FFT Timing Estimation

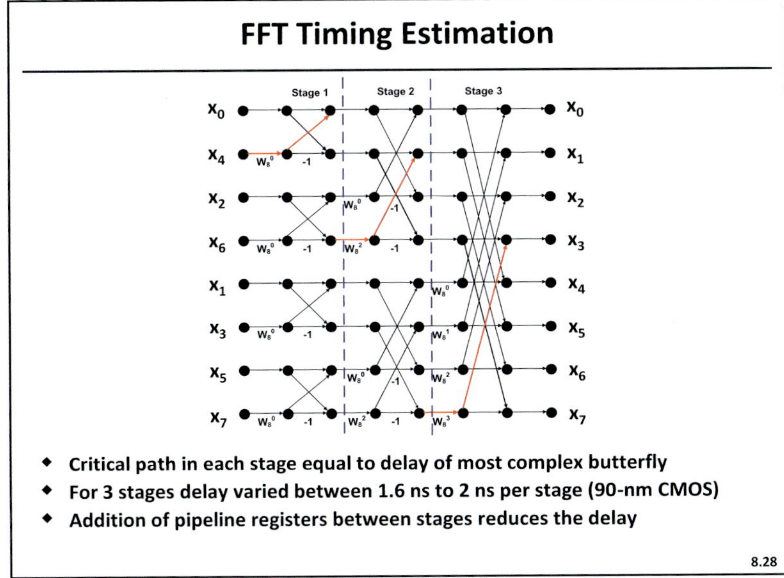

- **Critical path in each stage equal to delay of most complex butterfly**
- **For 3 stages delay varied between 1.6 ns to 2 ns per stage (90-nm CMOS)**
- **Addition of pipeline registers between stages reduces the delay**

8.28

Slide 8.28

The butterfly structure is also analyzed for timing since the butterfly performance defines the performance of the FFT. The critical path in each stage is equal to the delay of the most complex butterfly computation in that stage (twiddle factors and wiring included). There could be variations in delay across stages, which would necessitate retiming. For an 8-point design in a 90 nm CMOS technology, the critical-path delay varies between 1.6 ns and 2 ns per stage. Retiming or pipelining can then be performed to balance the path delay to improve performance and/or to reduce power through voltage scaling.

FFT Energy Estimation

- **Challenge:** Difficult to estimate hierarchically. Carrying out switch level simulations in MATLAB or gate level simulation in RC is time consuming.
- **Solution:** Derive analytical expressions for energy in each butterfly as a function of input switching activity and nonzero bits in twiddle factor.

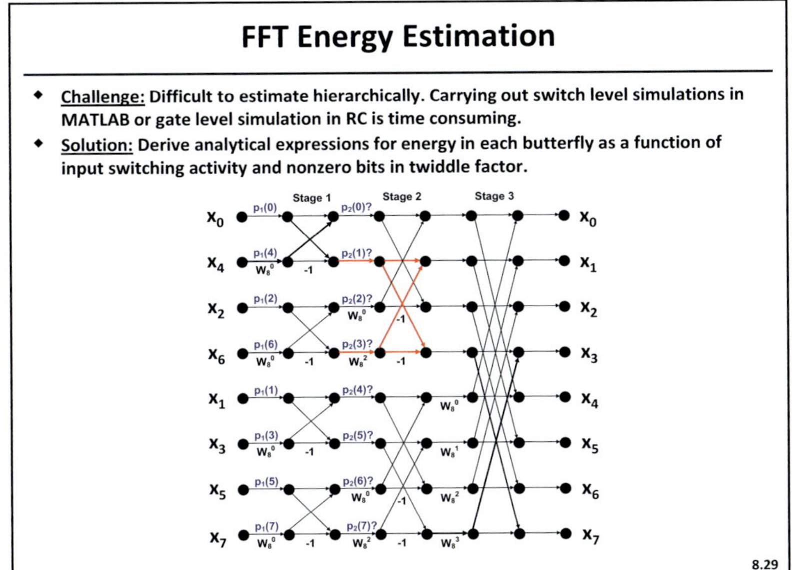

8.29

Slide 8.29

Finally, we can create a compact analytical model for energy consumption. Energy estimation is limited by accurate estimation of activity factors on the switching nodes in the design. The activity factor can be estimated by switch-level simulations in MATLAB or by gate-level simulations in the synthesis environment, both of which are very time consuming. Instead, we can derive analytical expressions for energy in each butterfly as a function of input switching activity and the number of non-zero bits in the twiddle factor. One such computation is highlighted in the slide for the second-stage butterfly, which takes $p_2(1)$ and $p_2(3)$ as inputs. This approach is propagated through the entire FFT to compute the transition probabilities of the intermediate nodes. With node switching probabilities, area (capacitance), and timing information, we can estimate the top-level power/energy of the datapath elements.

Shortcomings of the Fourier Transform (FT)

- **FT gives information about the spectral content of the signal but loses all time information**
  - FT assumes all spectral components are present at all times
  - Time-domain representation does not provide information about spectral content of the signal
- **Non-stationary signals whose spectral content changes in time cannot be supported by the Fourier-domain representation**
  - Non-stationary signals are abundant in nature
  - Examples: Neural action potentials, seismic signals, etc.
- **In order to have an accurate representation of these signals, a time-frequency representation is required**
- **In this section, we review the wavelet transform and analyze VLSI implementations of the discrete wavelet transform**

8.30

Slide 8.30

The Fourier transform is a tool that is widely used and well known to electrical engineers. However, it is often overlooked that the Fourier transform only gives us information about the spectral content of the signal while sacrificing time-domain information. The Fourier transform of a signal provides information about which frequencies are present in the signal, but it can never tell us which frequencies are present in the signal at which times. Thus, the Fourier transform cannot give an accurate representation of a signal whose spectral content changes with time. These signals are also known as non-stationary signals. There are many examples of non-stationary signals, such as neural action potentials and seismic signals. These signals have time-varying spectra. In order to accurately represent these signals, a method to analyze time-varying spectral content is needed. The wavelet transform that we will discuss in the remaining portion of this chapter is one such tool for time-frequency analysis.

Ambiguous Signal Representation with FT

$$x_1(t) = \begin{cases} 2 \cdot sin(2\pi \cdot 50t) & 0 < t \le 6s \\ 2 \cdot sin(2\pi \cdot 120t) & 6s < t \le 12s \end{cases}$$

Not all spectral components exist at all times

$$x_2(t) = sin(2\pi \cdot 50t) + sin(2\pi \cdot 120t)$$

- **The power spectral density (PSD) is similar, illustrating the inability of the FT to handle non-stationary spectral content**

8.31

Slide 8.31

Let us look at an example that illustrates how the Fourier transform cannot accurately represent time-varying spectral content. We consider two signals $x_1(t)$ and $x_2(t)$ described as follows:

$$x_1(t) = 2 \cdot sin(2\pi 50t) \text{ for } 0 < t \le 6$$
and $x_1(t) = 2 \cdot sin(2\pi 120t)$ for $6 < t \le 12$

Thus, in each of these individual periods the signal is a monotone.

$$x_2(t) = sin(2\pi 50t) + sin(2\pi 120t)$$

In this signal both tones of 50 Hz and 120 Hz are present at all times.

The plots in this slide show the power spectral density (PSD) of x_1 and x_2. As seen from the plots, both signals have identical PSDs although they are completely different signals. Thus, the Fourier transform is not a good way to represent signal $x_1(t)$, since we cannot express the change of frequency with time seen in $x_1(t)$.

<table>
<tr><td>

Support for Non-Stationary Signals

- ◆ **<u>A work-around:</u> modify FT to allow analysis of non-stationary signals by slicing in time – Short-time Fourier transform (STFT)**

  - Segment in time by applying windowing functions and analyze each segment separately

  - Many approaches (between late 1940s and early 1970s) differing in the choice of windowing functions

8.32

</td></tr>
</table>

Slide 8.32

A possible solution to the previous limitation is to apply windowing functions to divide up the time scale into multiple segments and perform FFT analysis on each of the segments. This concept is known as the short-time Fourier transform (STFT). It was the subject of considerable research for about 30 years, from the 1940s to the 1970s, where researchers focused on finding suitable windowing functions for a variety of applications. Let's illustrate this with an example.

<table>
<tr><td>

Short-Time Fourier Transform (STFT)

- ◆ **Time-frequency representation**

$$S(\tau, f) = \int x(t) w^*(t - \tau) e^{-j2\pi ft} dt$$

  $w(t)$: windowing function

  τ: translation parameter

$$x(t) = \int_\tau \int_f S(\tau, f) w^*(t - \tau) e^{j2\pi ft} d\tau df$$

- ◆ **Multiply signal by a window and then take a FT of the result (segment into stationary short-enough pieces)**
  - $S(\tau, f)$ is STFT of $x(t)$ at frequency f and translation τ

- ◆ **Translate the window to get spectral content of signal at different times**
  - $w(t - \tau)$ does time-segmentation of $x(t)$

8.33

</td></tr>
</table>

Slide 8.33

The STFT executes FFTs on time-partitions of a signal $x(t)$. The time partitioning is implemented by multiplying the signal with different time-shifted versions of a window function $\omega(t-\tau)$, where τ is the time shift. In the example from Slide 8.31, we can use a window that is 6 seconds wide. When the window is at $\tau=0$, the STFT would be a monotone at 50 Hz. When the window is at $\tau=6$, the STFT would be a monotone at 120 Hz.

Heisenberg's Uncertainty Principle

- **The problem with previous approach is uniform resolution for all frequencies (same window for *x*(*t*))**

- **There is a tradeoff between the resolution in time and frequency**
  - High-frequency components for a short time span require a narrow window for time resolution, but this results in wider frequency bands (cost: poor frequency resolution)
  - Low-frequency components of longer time span require a wider window for frequency resolution, but this results in wider time bands (cost: poor time resolution)

- **This is an example of Heisenberg's uncertainty principle**
  - FT is an extreme case where all time domain information is lost to get precise frequency information
  - STFT offers fixed time/frequency resolution, which needs to be chosen keeping the above tradeoff in mind

8.34

Slide 8.34

The problem with the approach described on the previous slide is that, since it uses the same windowing function, it assumes uniform resolution for all frequencies. There exists a fundamental tradeoff between the frequency and the time resolution. When a signal has high-frequency content for a short time span, a narrow window is needed for time resolution, but this results in wider frequency bands and, hence, poor frequency resolution. On the other hand, if the signal has low-frequency content of longer time span, a wider window is needed, but this results in narrow frequency bands and, hence, poor time resolution. This time-frequency resolution tradeoff is a demonstration of Heisenberg's uncertainty principle. The continuous wavelet transform, which we shall now describe, provides a way to avoid the problem of fixed resolution posed by the STFT.

Continuous Wavelet Transform

- **Addresses the resolution problem of STFT by evaluating the transform for scaled versions of the window function**
  - Varying time and frequency resolutions are varied by using windows of different lengths
  - The transform is defined by the following equation

$$W(a,b) = \frac{1}{\sqrt{a}} \int x(t)\Psi^*\left(\frac{t-b}{a}\right) dt$$

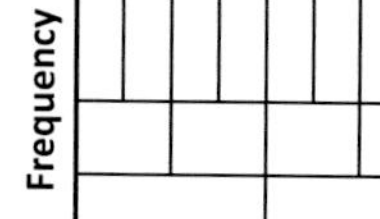

  - *a* > 0, *b*: scale and translation parameters

- **Design problem: find the range of *a* and *b***
  - Ideally, infinitely many values of *a* and *b* would be required to fully characterize the signal
  - Can limit the range based on a priori knowledge of the signal

8.35

Slide 8.35

This slide mathematically illustrates the continuous wavelet transform (CWT). The wavelet function Ψ^* plays a dual role of the window and basis functions. Parameter b is the translation parameter (providing windowing) and parameter a is the scaling parameter (providing multi-resolution). The wavelet coefficients are evaluated as the inner products of the signal with the scaled and translated versions of Ψ^*, which is referred to as the mother wavelet. Shifted and translated versions of the mother wavelet form the basis functions for the wavelet transform. For a complete theoretical representation of a signal, we need infinitely many values of a and b, which is not practically feasible. For practical realization, a and b are varied in finite steps based on a priori knowledge of the signal in a way that provides adequate resolution. The basis functions implement multi-resolution partitioning of the time and frequency scales as shown on the plot on the right. The practical case of simple binary partitioning is shown in the plot.

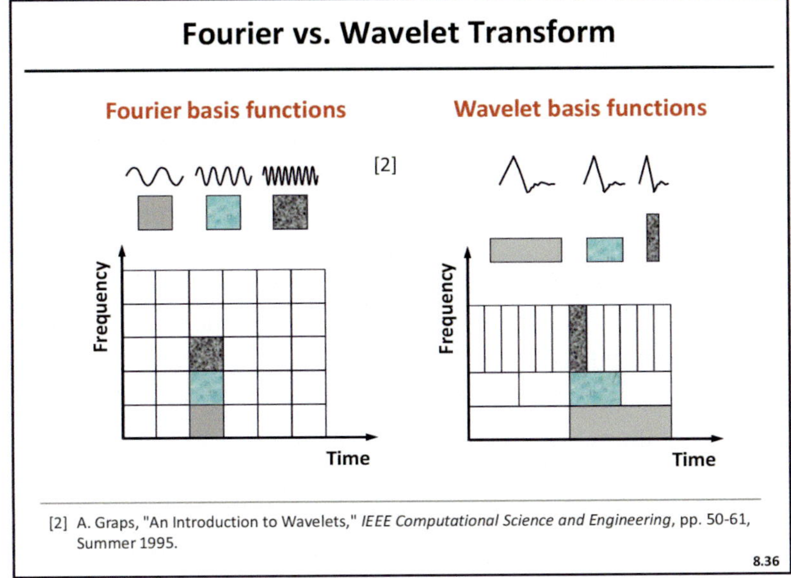

Slide 8.36

This slide gives a simplified illustration of the time-frequency scales used in Fourier and wavelet analyses [2]. The Fourier transform assumes the presence of all frequency components (basis functions) at all times, as shown on the left. The wavelet transform removes this restriction to allow for the presence of only a subset of the frequency components at any given time. Wavelet partitioning thus allows for better time-frequency representation of non-stationary signals.

The plot on this slide illustrates the difference between the wavelet transform and the STFT. The STFT has a fixed resolution in the time and the frequency domain as shown in the plot on the left. The wavelet transforms overcomes this limitation by assigning higher time resolution for higher frequency signals. This concept is demonstrated in the plot on the right.

Orthonormal Wavelet Basis

- Wavelet representation, in general, has redundant data representation

- We would like to find a mother wavelet that when translated and scaled leads to orthonormal basis functions

- Several orthonormal wavelets have been developed
  – Morlet, Meyer, Haar, etc.

- The Morlet wavelet is an example of an orthonormal wavelet

Slide 8.37

While the CWT provides multiple time-frequency resolutions, it does not generally provide concise signal representation. This is because the inner product with the wavelet at different scales carries some redundancy. In other words, basis functions are non-orthogonal. In order to remove this redundancy we need to construct a set of orthogonal basis functions. Several wavelets have been derived that meet this requirement. The Morlet wavelet shown on this slide is an example of such a wavelet. It is a constant subtracted from a plane-wave and multiplied by a Gaussian window. The wavelet provides an orthonormal basis, thereby allowing for efficient signal representations. In some sense, application-specific wavelets serve as matched filters that seek significant features of the signal. The wavelet functions need to be discretized for implementation in a digital chip.

Discrete Wavelet Series

- **We need discrete-domain transforms for digital implementation**
- **Discretize the translation and scale parameters (a, b)**
  - Example: Daubechies ($a = 2^j$, $b = 2^j k$)

- **Can this be implemented on digital a circuit?**
  - Input signal is not yet discretized
  - Number of scaling and translation parameters are still infinite

- **Key to efficient digital implementations**
  - Need to limit the number of scales used in the transform
  - Allow support for discrete-time signals

8.38

Slide 8.38

A discrete-wavelet series can be obtained by discretizing parameters a and b in the mother wavelet. The scaling parameter is quantized as $a = 2^j$ and the translation parameter is quantized as $b = k \cdot 2^j$. Since the parameters are varied in a dyadic (powers-of-two) series, the discrete wavelet series is also referred to as the dyadic discrete wavelet series.

Discretization of the mother wavelet is necessary, but not sufficient for digital implementation. The input is still a continuous-time signal and it also needs to be discretized. Further, the quantization of a and b reduces the infinite set of continuous values of a and b to an infinite set of discrete values. This means that we still have to evaluate the inner products for an infinite set of values. In order to arrive at a digital implementation of the wavelet transform, we seek an implementation that efficiently limits the numbers of scales required for the wavelet transform. As such, the transform needs to be modified to support discrete-time implementation.

Towards Practical Implementations

- **How to limit the number of scales for analysis?**

$$F\left(f(at)\right) = \frac{1}{|a|} F\left(\frac{\omega}{a}\right)$$

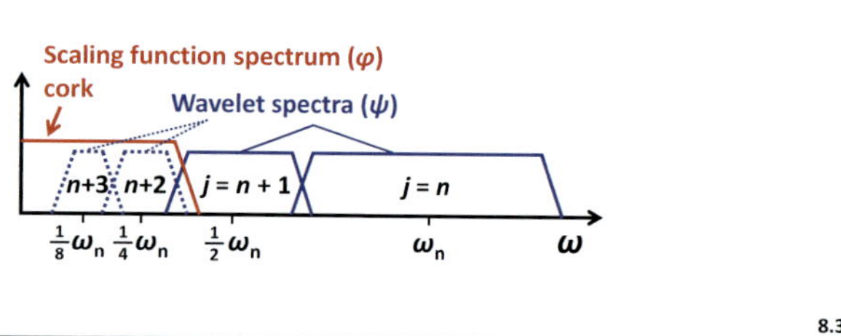

- **Each wavelet is a like a constant – Q filter**

- **Scaling function**

8.39

Slide 8.39

In practice, each wavelet function can be seen as a band-pass filter of progressively narrower bandwidth and lower center frequencies. Thus the wavelet transform can be implemented with constant Q filter banks. In the example shown on the slide, bandwidth of Ψ_1 is twice the bandwidth of Ψ_2, etc. The standard wavelet series (Ψ_1, Ψ_2, Ψ_3, Ψ_4, etc.) nicely covers all but very low frequencies; taken to the limit, an infinitely large n is needed to represent DC. To overcome this issue, a cork function φ is used to augment wavelet spectra at low frequencies.

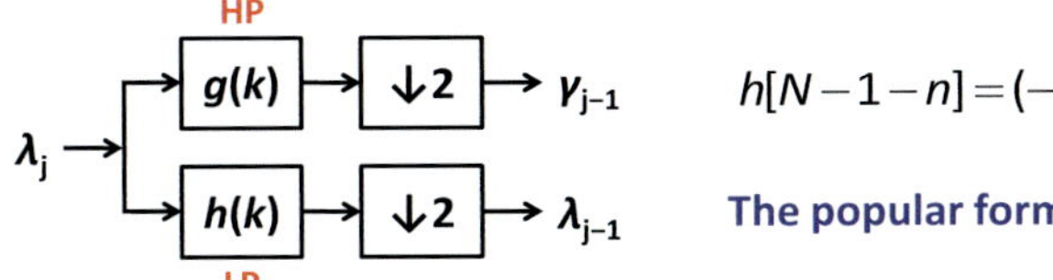

Slide 8.40

Mallat's wavelet is one of the most attractive wavelets for digital implementation, because it describes the wavelet transform in terms of digital filtering and sampling. The filtering is done iteratively using low-pass (h) and high-pass (g) components, whose sampling ratios are powers of 2. The equations define high-pass and low-pass filtering operations. A cascade of HP/LP filters leads to simple and modular digital implementations. Due to the ease of implementation, this is one of the most popular forms of the discrete wavelet transform.

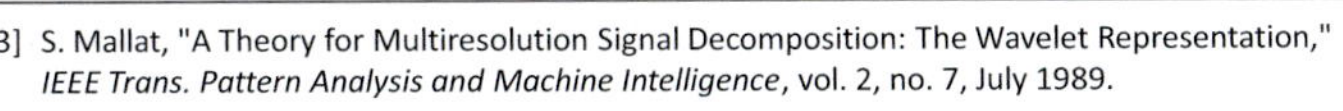

Slide 8.41

The concept described on the previous slide is illustrated here in more detail [3]. The filter bank for the wavelet transform is shown on the right. The original signal, which has a bandwidth of $(0, \pi)$ (digital frequency at a given f_s) is passed through two half-band filters. The high-pass filter has a transfer function denoted by $G(z)$ and impulse response $g(n)$. The low-pass filter has a transfer function denoted by $H(z)$ and an impulse response denoted by $h(n)$. The output of $G(z)$ has a bandwidth from $(\pi/2, \pi)$, while the output from $H(z)$ occupies $(0, \pi/2)$. The down-sampled output of the first-stage HPF is the first level of wavelet coefficients. The down-sampled LPF output is applied to the next stage, which splits the signal content into frequency bands $(0, \pi/4)$ and $(\pi/4, \pi/2)$. At each stage, the HPF output calculates the DWT coefficients for that level/stage.

This architecture is derived from the basics of quadrature-mirror filter theory. In his seminal paper on the implementation of wavelets, Mallat showed that the wavelet transform of a band-limited signal can be implemented using a bank of quadrature-mirror filters.

Haar Wavelet

- **One of the simplest and most popular wavelets**

- **It can be implemented as a filter bank shown on the previous slide**
- **$g(n)$ and $h(n)$ are 2-tap FIR filters with**
  $g(n)$ = [0.70701 –0.7071]
  $h(n)$ = [0.70701 0.7071]

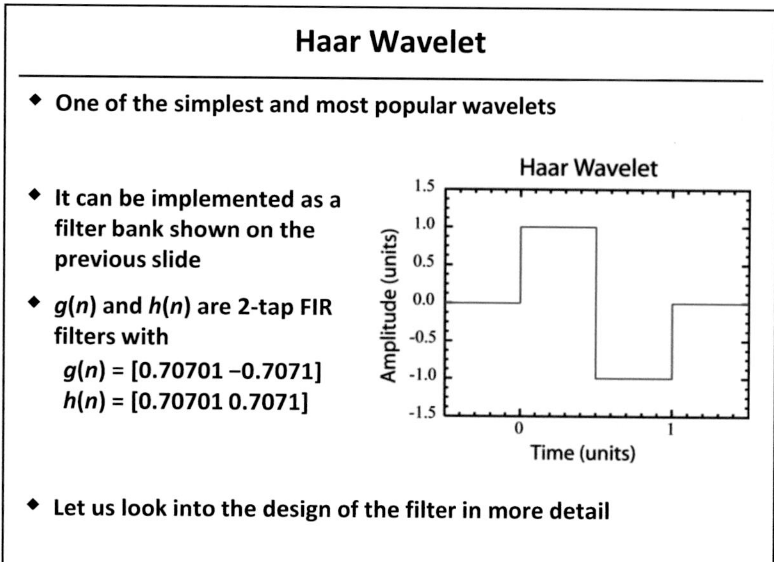

- **Let us look into the design of the filter in more detail**

8.42

Slide 8.42

Let us illustrate the design of the filter bank with the Haar wavelet filter. It can be implemented using the filter bank discussed in the previous slide. The filters for the Haar wavelet are simple 2-tap FIR filters given by $g(n) = 1/\mathrm{sqrt}(2) * [1 \ -1]$ and $h(n) = 1/\mathrm{sqrt}(2) * [1 \ 1]$. The Haar wavelet is sometimes used for spike detection of neural signals.

Haar Wavelet: Direct-Mapped Implementation

- **Part (a) shows direct-mapped implementation of a single stage of the Haar wavelet filter**

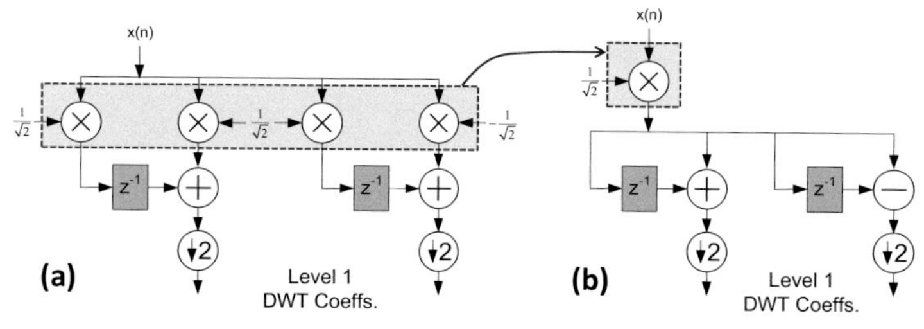

- **We can exploit the symmetry in the coefficients to share a single multiplier for a stage**

8.43

Slide 8.43

This figure shows one stage of the Haar wavelet filter. Several stages are cascaded to form a Haar wavelet filter. The depth is decided based on the application. Since the stage shown above is replicated many times, any savings in power/area of a single stage will linearly affect the total filter power/area. Figure (a) shows the direct-mapped implementation of the wavelet filter, which consists of 4 multipliers, 2 adders, 2 delays and 2 down-samplers. We can exploit the symmetric nature of the coefficients in the Haar wavelet filter to convert the implementation in (a) to the one in (b), which uses only 1 multiplier. The number of adders, delays and down-samplers remains the same. This simplification in multiplication has a considerable effect on the area since the area of a multiply is much greater than that of an adder or register. We can also look into further possibilities for area reduction in adders and down-samplers.

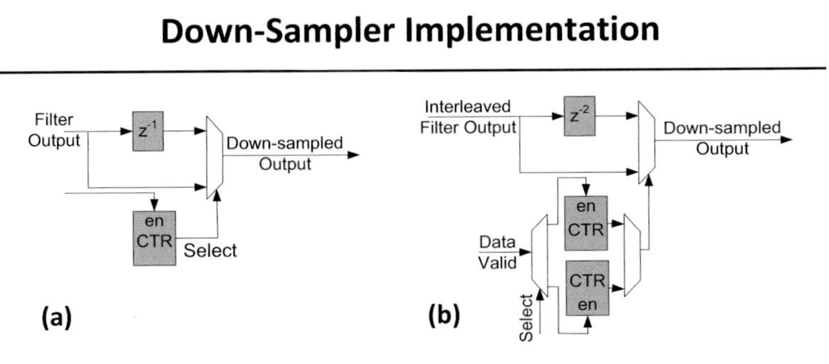

- **Down-sampler implementation**
  - Allows for selection of odd/even samples by controlling enable (en) signal of the 1-bit counter
- **Interleaved down-sampler implementation**
  - Allows odd / even sample selection with different data arrival time for each channel

8.44

Slide 8.44

Let's now consider the down-sampler implementation for single and multiple data streams. Figure (a) shows the single-stream architecture. The multiplexer selects between the current (select = 0) and the previous (select = 1) sample of the filter output. A one-bit counter toggles the select line each clock cycle, thus giving a down-sampled version of the input stream at the output. The down-sampler thus requires a register (N flip-flops for an N-bit word), a 1-bit counter (implemented as a flip-flop) and a multiplexer ($\sim 3N$ AND gates). Thus, the down-sampler has the approximate complexity of a delay element (N flip-flops). This implementation of the down-sampler allows us to control the sequence of bits that are sent to the output, even when the data is discontinuous. A "data valid" signal can be used to trigger the enable of the counter with a delay to change between odd/even bit streams.

It is often desirable to interleave the filter architecture to support multiple streams. The down-sampler in (a) can be modified as shown in (b) to support multiple data streams. The counter is replicated in order to allow for the selection of odd/even samples on each stream and to allow for non-continuous data for both streams.

Polyphase Filters for DWT

- **The wavelet filter computes outputs for each sample, half of which are discarded by the down-sampler**

- **Polyphase implementation**
  - Splits input to odd and even streams
  - Output combined so as to generate only the output which would be needed after down-sampling

- **Split low-pass and high-pass functions as**

  $$H(z) = H_e(z^2) + z^{-1}H_o(z^2)$$
  $$G(z) = G_e(z^2) + z^{-1}G_o(z^2)$$

- **Efficient computation strategy: reduces switching power by 50%**

8.45

Slide 8.45

The down-sampling operation in the DWT implies that half of the outputs are discarded. Since half of the outputs are discarded eventually we can avoid half the computations altogether. The "polyphase" implementation of the DWT filter bank indeed allows us to do this. In the polyphase implementation, the input data stream is split into odd and even streams upfront. The filter is modified such that only the outputs that need to be retained are computed. The mathematical formulation for a polyphase implementation is described as:

$$H(z) = H_e(z^2) + z^{-1}H_o(z^2),$$
$$G(z) = G_e(z^2) + z^{-1}G_o(z^2).$$

The filters $H(z)$ and $G(z)$ are decomposed into filters for the odd and even streams. The polyphase implementation decreases the switching power by 50%.

Polyphase Implementation of Haar Wavelet

♦ **Consider the highlighted section of the Haar DWT filter**

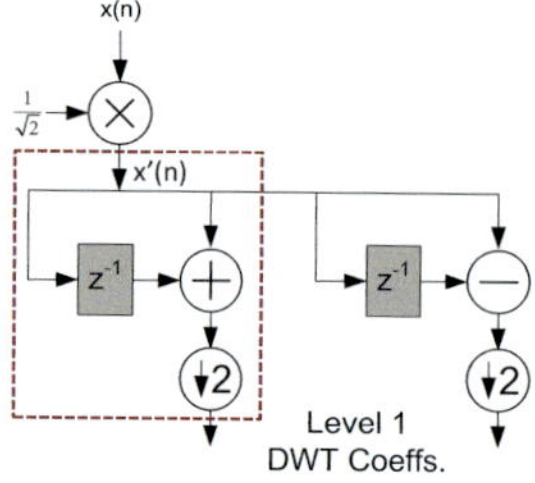

♦ **Due to down-sampling half of the computations are unnecessary**

♦ **Need for a realization that computes only the outputs that would be needed after down-sampling**

8.46

Slide 8.46

Let us now examine the polyphase implementation of the Haar wavelet. Consider the section of the Haar wavelet filter highlighted by the red box in this figure. While the adder and the delay toggle every clock cycle, the down-sampler at the output of the adder renders half of the switching energy wasted, as it discards half of the outputs. The polyphase implementation allows us to work around the issue of surplus energy consumption.

Haar Wavelet: Sequence of Operations for $h(n)$

♦ **Consider the computation for $h(n)$ in the Haar wavelet filter**

$x'(n)$	$x'(1)$	$x'(2)$	$x'(3)$	$x'(4)$	$x'(5)$
$x'(n-1)$	$x'(0)$	$x'(1)$	$x'(2)$	$x'(3)$	$x'(4)$
$x'(n) + x'(n-1)$	$x'(1) + x'(0)$	$x'(2) + x'(1)$	$x'(1) + x'(0)$	$x'(2) + x'(1)$	$x'(5) + x'(4)$
↓2	$x'(1) + x'(0)$		$x'(3) + x'(2)$		$x'(5) + x'(4)$

♦ **Computations in red are redundant**
♦ **Polyphase implementations**
 – Split input into odd / even streams
 – No redundant computations performed

$x'(2n+1)$	$x'(1)$	$x'(3)$	$x'(5)$
$x'(2n)$	$x'(0)$	$x'(2)$	$x'(4)$
$x'(2n) + x'(2n+1)$	$x'(1) + x'(0)$	$x'(3) + x'(2)$	$x'(5) + x'(4)$

8.47

Slide 8.47

This slide shows the value of $x'(n) = x(n)/\mathrm{sqrt}(2)$ at each clock cycle. The output of the adder is $x'(n) + x'(n-1)$ while the down-sampler only retains $x'(2n+1) + x'(2n)$ for $n = 0,\ 1,\ 2,\ \ldots$ Thus half the outputs (marked in red) in the table are discarded. In the polyphase implementation, the input is split into odd and even streams as shown in the second table. The adder then operates on a divide-by-2 clock and only computes the outputs that would eventually be retained in the first table. Thus there are no redundant computations, giving us an efficient implementation.

Applications of DWT

- **Spike Sorting**
  - Useful in detection and feature extraction

- **FBI fingerprints**
  - Efficient compression of finger prints without loosing out on information needed to distinguish between finger prints

- **Image compression**
  - 2D wavelet transform provide efficient representation of images

- **Generality of the WT lets us take a pick for the wavelet used**
  - Since a large number of wavelets exist, we can pick the right wavelet useful for the application

8.48

Slide 8.48

The discrete wavelet transform (DWT) has several applications. It is useful in neural spike sorting, in detection and feature-extraction algorithms. DWT also provides a very efficient way of image compression. The DWT coefficients need far fewer bits than the original image while retaining most of the information necessary for good reconstruction of the image. The 2-D DWT is the backbone of the JPEG 2000 image compression standard. It is also widely used by the FBI in the compression of fingerprint images. As opposed to the Fourier transform, the DWT allows us to choose a basis best suitable for a given application. For instance while Haar wavelet is best suitable for neural-spike feature extraction, the cubic-spline wavelet is most suitable for neural-spike detection.

Summary

- **FFT is a technique for frequency analysis of stationary signals**
  - It is a key building block in radio systems and many other apps
  - FFT is an economical implementation of DFT that leverages shorter sub-sequences to implement DFT on N discrete samples
  - Key building block of an FFT is butterfly operator, which can be realized with 2^N radix (2 and 4 being the most common)
  - FFT does not work well with non-stationary signals
- **Wavelet is a technique for time-frequency analysis of non-stationary signals**
  - Multi-resolution in time and frequency is used based on orthonormal wavelet basis functions
  - Wavelets can be implemented as a series of decimation filters
  - Used in applications such as fingerprint recognition, image compression, neural spike sorting, etc.

8.49

Slide 8.49

Fast Fourier transform (FFT) is a well-known technique for frequency analysis of stationary signals. FFT is a standard building block in radio receivers and many other applications. FFT is an economical implementation of discrete Fourier transform (DFT). FFT based on the use of shorter sub-sequences to realize DFT. FFT has a compact hardware realization, but does not work for non-stationary signals. Wavelet is a technique used for analysis of stationary signals. Wavelets are based on orhonormal basis functions that provide varying time and frequency resolution. They can be implemented as a series of decimation filters and are widely used in fingerprint recognition, image compression, neural spike sorting and other applications. Having analyzed basic DSP algorithms, next four chapters will focus on modeling and optimization of DSP architectures.

References

- P. Duhamel and M. Vetterli, "Fast Fourier Transforms - A Tutorial Review and a State-of-the-art," *Elsevier Signal Processing,* vol. 4, no. 19, pp. 259-299, Apr. 1990.

- A. Graps, "An Introduction to Wavelets," *IEEE Computational Science and Engineering,* pp. 50-61, Summer 1995.

- S. Mallat, "A Theory for Multiresolution Signal Decomposition: The Wavelet Representation," *IEEE Trans. Pattern Analysis and Machine Intelligence,* vol. 2, no. 7, July 1989.

Additional References

- R. Polikar, "The Story of Wavelets," *in Proc. IMACS/IEEE CSCC,* 1999, pp. 5481-5486.

- T.K. Sarkar *et al.,* "A Tutorial on Wavelets from an Electrical Engineering Perspective, Part 1: Discrete Wavelet Techniques," *IEEE Antennas and Propagation Magazine,* vol. 40, no. 5, pp. 49-70, Oct. 1998.

- V. Samar *et al.,* "Wavelet Analysis of Neuroelectric Waveforms: A Conceptual Tutorial," *Brain and Language 66,* pp. 7–60, 1999.

Part III

Architecture Modeling and Optimized Implementation

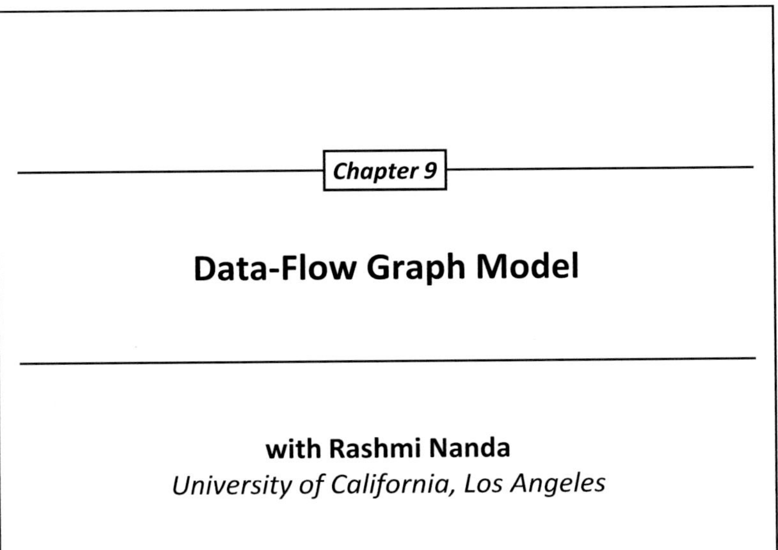

This chapter discusses various representations of DSP algorithms. We study how equations describing the algorithm functionality are translated into compact graphical models. These models enable efficient implementation of the algorithm in hardware while also enabling architectural transformations through matrix manipulations. Common examples of graphical representations are flow graphs and block diagrams. An example of block diagrams using Simulink models will be shown at the end of the chapter.

Iteration

- **Iterative nature of DSP algorithms**
  - Executes a set of operations in a defined sequence
  - One round of these operations constitutes an iteration
  - Algorithm output computed from result of these operations

- **Graphical representations of iterations** [1]
  - Block diagram (BD)
  - Signal-flow graph (SFG)
  - Data-flow graph (DFG)
  - Dependence graph (DG)

  [1] K.K. Parhi, VLSI Digital Signal Processing Systems: Design and Implementation, John Wiley & Sons Inc., 1999.

- **Example: 3-tap filter iteration**
  - $y(n) = a \cdot x(n) + b \cdot x(n-1) + c \cdot x(n-2)$, $n = \{0, 1, ..., \infty\}$
  - Iteration: 3 multipliers, 2 adders, 1 output $y(n)$

9.2

DSP algorithms are defined by iterations of a set of operations, which repeat in real time to continuously generate the algorithm output [1]. More complex algorithms contain multiple independent inputs and multiple outputs. An iteration can be graphically represented using block diagram (BD), signal-flow graph (SFG), data-flow graph (DFG) or dependence graph (DG). These representations capture the signal-flow properties and operations of the algorithm. The slide shows an example of an iteration for a 3-tap FIR filter. The inputs are $x(n)$ and delayed versions of $x(n)$. Each iteration executes 2 additions and 3 multiplications (a, b, c are operands) to compute a single output $y(n)$. In the rest of the chapter we look at some of these graphical representations and their construction in Simulink.

D. Marković and R.W. Brodersen, *DSP Architecture Design Essentials*, Electrical Engineering Essentials,
DOI 10.1007/978-1-4419-9660-2_9, © Springer Science+Business Media New York 2012

Block Diagram Representation

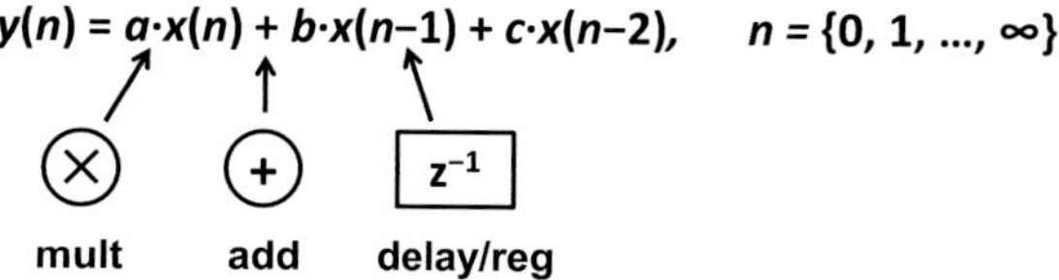

- **Block diagram of 3-tap FIR filter**

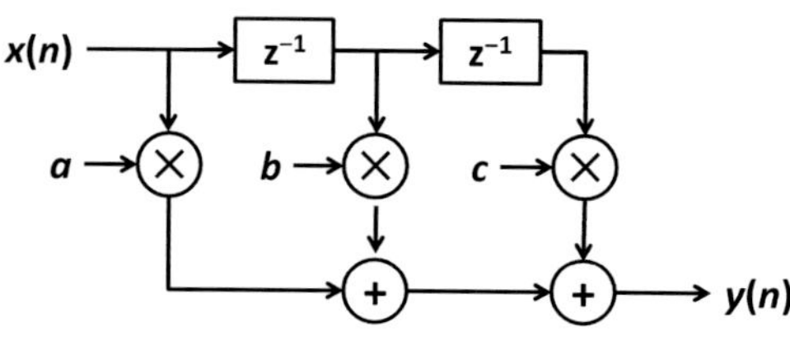

9.3

Slide 9.3

The slide shows the block diagram for the 3-tap filter discussed previously. The block diagram explicitly shows all computations and delay elements in the algorithm. This diagram is similar to an actual hardware implementation of the filter. Such block diagram representations are also used in Simulink to model DSP algorithms as will be shown later in the chapter. We next look at the more symbolic signal-flow graph.

Signal-Flow Graph Representation

- **Network of nodes and edges**
  - Edges are signal flows or paths with non-negative # of regs
    - Linear transforms, multiplications or registers shown on edges
  - Nodes represent computations, sources, sinks
    - Adds (> 1 input), sources (no input), sinks (no output)

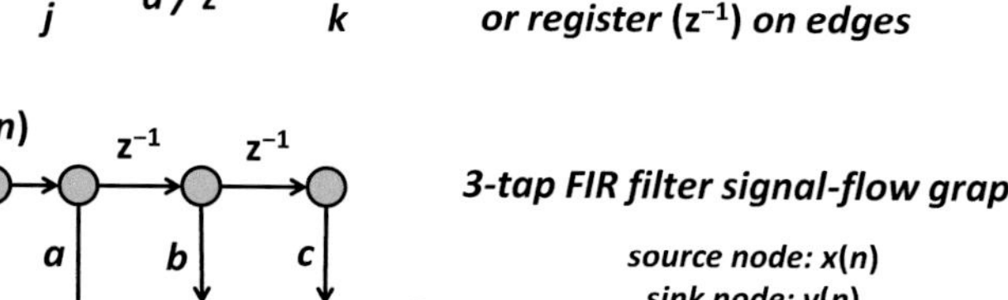

9.4

Slide 9.4

Signal-flow graphs primarily indicate the direction of signal flows in the algorithm. Signal-flow graphs are useful in describing filtering operations, which are mainly composed of additions and constant multiplications. Other graphical methods such as block diagrams and data-flow graphs better describe DSP algorithms with non-linear operations. The nodes in the graph represent signal sources, sinks and computations. When several inputs merge at a node, they indicate an add operation. A node without an input is a signal source, while one without an output branch is a sink. Nodes with a single input and multiple outputs are branch nodes, which distribute the incoming signal to multiple nodes. Constant multiplication and delay elements are treated as linear transforms, which are shown directly on the edges. An example of an SFG for a 3-tap FIR is shown.

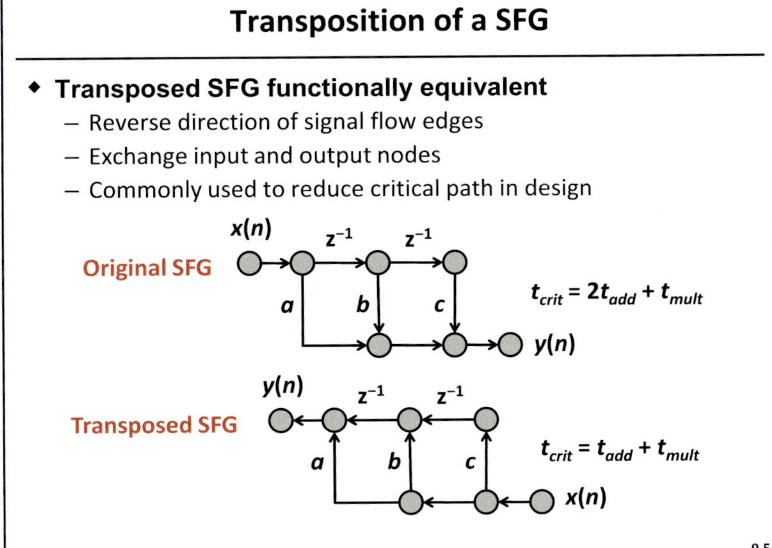

Slide 9.5

SFGs are amenable to transposition simply by reversing the direction of the signal flow on the branches. An example of transposition is shown in the slide for the 3-tap FIR filter. The transposed SFG replaces the branch nodes with additions and vice versa while maintaining the algorithm functionality. The input source nodes are exchanged with the output sink nodes, while the direction of signal flow on all the branches is reversed. The transposed SFG in the slide is equivalent in functionality to the original SFG. The main advantage of such a transposition in FIRs is a near constant critical path independent of the number of filter taps (ignoring the input loading), while the original design has a critical path that grows linearly with the number of taps.

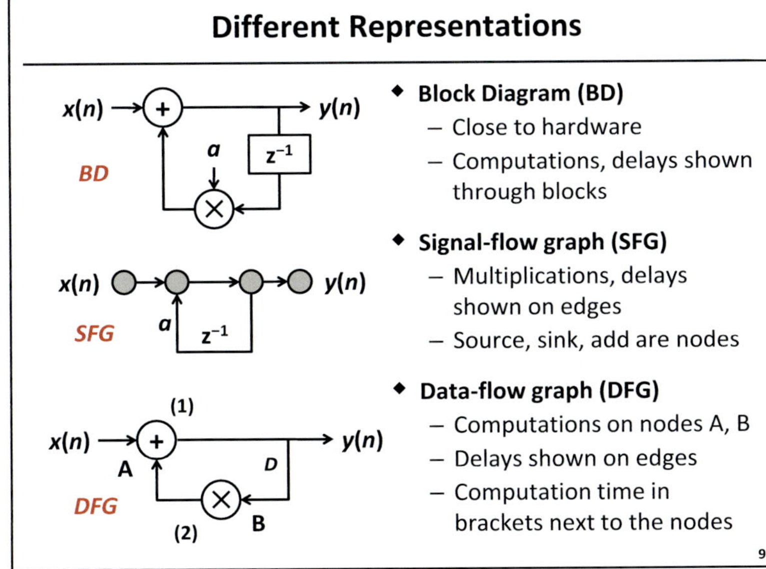

Slide 9.6

This slide compares three representations of a first-order IIR filter. The main difference between block diagrams and flow graphs lies in the compact nature of the latter owing to the use of symbolic notation for gain and delay elements. The DFG differs from the SFG. The DFG does not include explicit source, sink or branch nodes, and depicts multiply operations through dedicated nodes. The DFG can optionally show the normalized computation time of each node in brackets. We use the DFG notation extensively in Chap. 11 to illustrate the automation of architectural transformations. A key reason for this choice is that the DFG connectivity information can be abstracted away in matrices (Slides 9 and 10), which are amenable to transformations. The DFG will be discussed at length in the rest of the chapter.

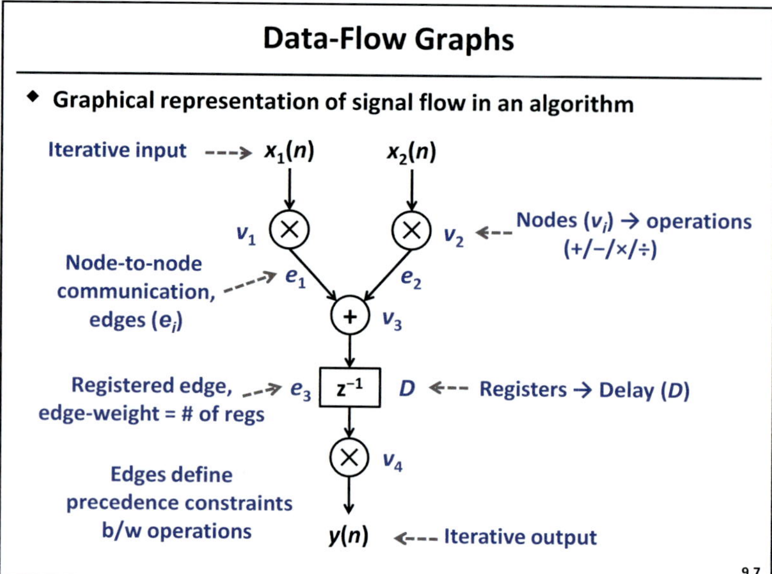

Slide 9.7

To begin, let's look again at data-flow-graph models, and explain the terminology that will be used throughout this chapter. In a DFG, the operations will be referred to as *nodes* and denoted as v_i, $i \in \{1, 2, 3, ...\}$; the edges e_i, $i \in \{1, 2, 3, ...\}$, indicate the signal flow in the graph. The signal flow can be between the nodes or flow from/to the input/output signals. For iterative DSP functions, certain edges on the flow graph can have registers on them. These registers are referred to as *delays* and are denoted by D. The number of registers on any edge is referred to as the weight of the edge $w(e_i)$.

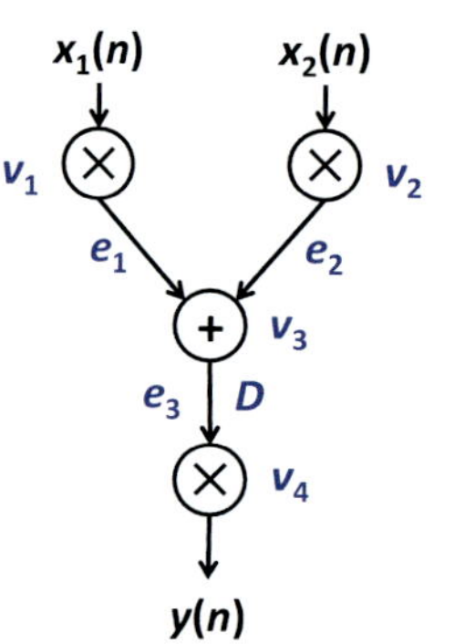

Formal Definition of DFGs

A directed DFG is denoted as G = <V,E,d,w>

• *V: Set of vertices (nodes) of G. The vertices represent operations.*

• *d: Vector of logic delay of vertices. d(v) is the logic delay of vertex v.*

• *E: Set of directed edges of G. A directed edge e from vertex u to vertex v is denoted as e:u → v.*

• *w(e) : Number of sequential delays (registers) on the edge e, also referred to as the weight of the edge.*

• *p:u→ v: Path starting from vertex u, ending in vertex v.*

• *D: Symbol for registers on an edge.*

Slide 9.8

This slide summarizes the formal model of the DFGs. Vectors V, E and w are one-dimensional lists of the vertices, edges and edge-weights respectively. The logic delay (not to be confused with register delay) is stored in the vector d where $d(v_i)$ is the logic delay of node v_i. The operations in the graph are associated with timing delays, which have to be taken in as input during architecture optimization in order to guarantee that the final design meets timing constraints. The logic delay can either be in absolute units (nanoseconds, picoseconds, etc.), or normalized to some reference like a clock period. The latter normalization is useful during time multiplexing, when operations are executed on pipelined hardware units. It is important to note that the edges express the precedence relation of the DSP function. They define the sequence in which the operations must be executed to keep the functionality unchanged. This edge constraint can either be an inter-iteration constraint, where $w(e_i) = 0$, or an intra-iteration constraint, where $w(e_i) > 0$. For example, the edge e_3 on the graph has a delay on it indicating inter-iteration constraint. This means that the output of node v_3 is taken as input to node v_4 after one iteration of the algorithm.

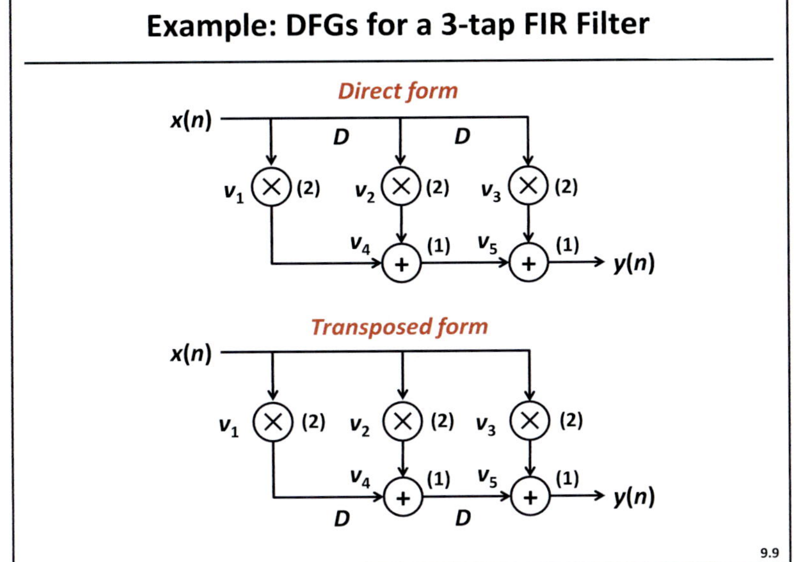

Slide 9.9

The figures in the slide show example DFGs for the direct and transposed form of the 3-tap FIR filter discussed earlier in the chapter. Nodes associated with names and computation times represent the operations in the filter equation. The registers are represented by delay elements D annotated next to their respective edges. We will see more examples of DSP data-flow graphs and their manipulations in Chap. 11, when we discuss architectural transformations.

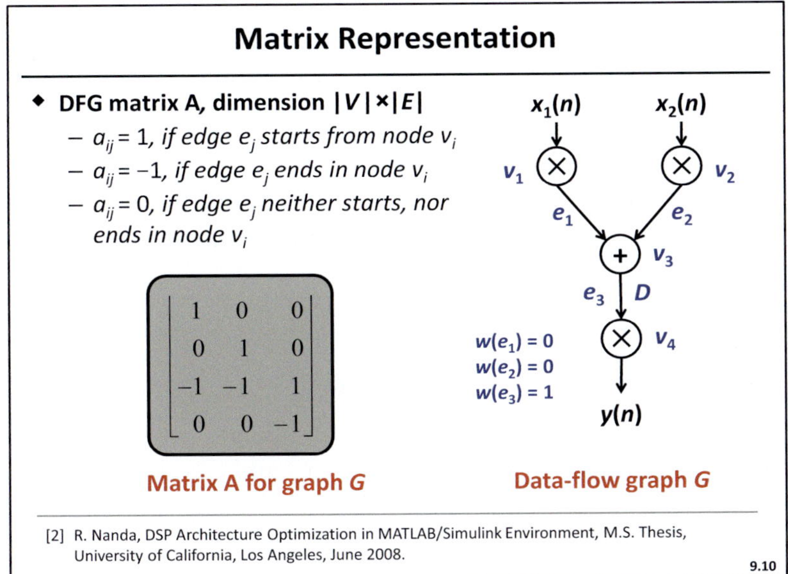

Slide 9.10

There are several ways to implement the data-flow-graph model using data structures. Structures like arrays or linked lists make it easy to implement and automate the execution of the architectural transformation algorithms. We discuss a simple array/matrix based representation [2]. The DFG matrix A is of dimension $|V| \times |E|$, where the operator $|\cdot|$ is the number of elements in the vector. The column a_i of the matrix defines the edge e_i of the graph. For edge e_i: $v_k \rightarrow v_j$, the element $a_{ki} = 1$ (source node) and element $a_{ji} = -1$ (destination node). All other entries in the column a_i are set to 0. For example, the edge e_3: $v_3 \rightarrow v_4$ is represented by column $a_3 = [0\ 0\ 1\ -1]^{\text{T}}$ in the matrix A.

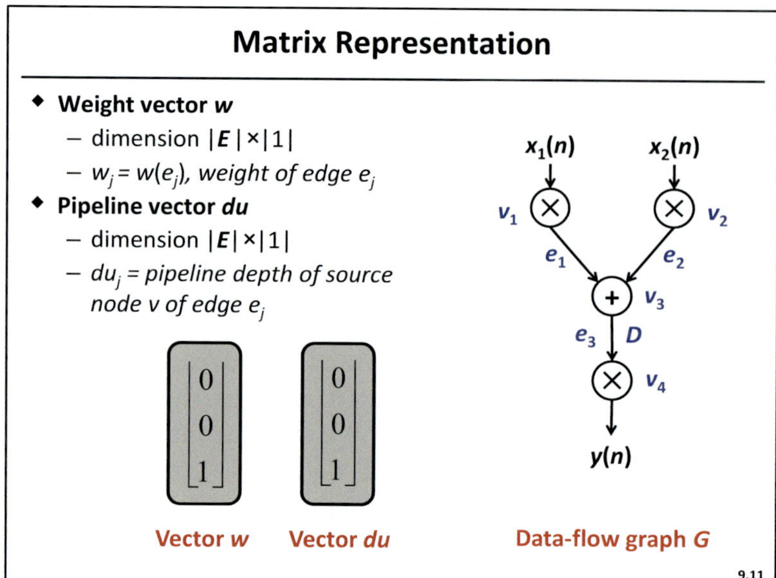

Slide 9.11

The other three vectors of interest are the weight vector $\mathbf{w}$, the logic delay vector $\mathbf{d}$ and the pipeline vector $\mathbf{du}$. The weight and logic delay vectors remain the same as described in the DFG model. Though closely related, the pipeline vector $\mathbf{du}$ is not the same as the logic-delay vector $\mathbf{d}$. This vector can be assigned values only after the operations have been mapped to hardware units. For example, if a multiply operation v_i is to be executed on a hardware unit which has two pipeline stages, then all edges e_i with source node v_i will have $du(e_i) = 2$. In other words, we characterize the delay of the operations not in terms of their logic delay, but in terms of their pipeline stages or register delay. The value of $\mathbf{du}$ changes depending on the clock period, since for a shorter clock period the same operation will have to be mapped onto a hardware unit with an increased number of pipeline stages. For example, in the graph $\mathbf{G}$ the add and multiply operations have logic delays of 200 and 100 units, respectively, which makes the vector $\mathbf{d} = [200\ 200\ 100\ 200]^{\mathrm{T}}$. The add and multiply operations are mapped onto hardware units with one and two pipeline stages, making the value of $du = 2$ for edges e_1 and e_2 and $du = 1$ for e_3.

To provide a user-friendly approach, an automated flow can be used to extract these matrices and vectors from a Simulink block diagram. We look at construction of Simulink block diagrams in the next slides.

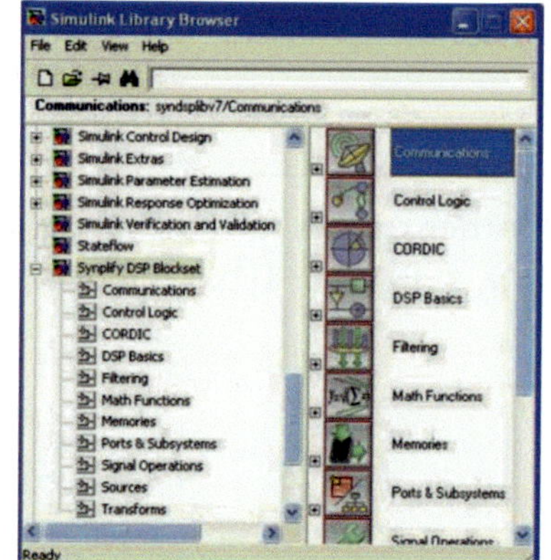

Slide 9.12

Simulink allows for easy construction and verification of block diagrams using a drag-and-drop push-button flow. The Simulink libraries have several predefined macros like FFT, signal sources, and math operations etc., commonly used in creating DSP systems. Commercially available add-ons like Xilinx System Generator (XSG) and Synplify DSP (now Synphony HLS) can also be used to setup a cycle-accurate, finite-wordlength DSP system. A snapshot of the Synplify DSP library is shown on the right side of the slide. A big advantage of having predefined macros available is the ease with which complex systems can be modeled and verified. An example of this will be shown in the next slide. On-the-fly RTL generation for the block diagrams is also made simple with the push-button XSG or Synplify DSP tools.

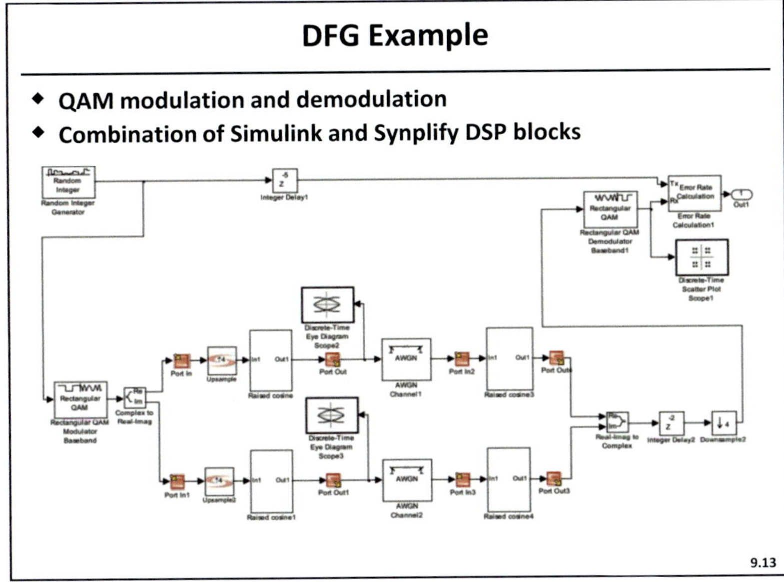

noise sources, random number generators, etc., as well as discrete eye-diagram plotters required for simulating the model. The low-pass filter, which limits the transmitted signal bandwidth, is implemented using a raised-cosine FIR block in Synplify DSP.

Slide 9.13

This figure shows an example of a quadrature amplitude modulation (QAM) design. The model uses full-precision Simulink blocks as sources and sinks. The input and output ports in the model define the boundaries between the full-precision Simulink blocks and the finite-wordlength Synplify DSP blocks. The input ports quantize the full-precision input data while the output port converts the finite-wordlength data back to integer or double format. The Simulink blockset has provisions for AWGN

Summary

- **Graphical representations of DSP algorithms**
  - Block diagrams
  - Signal-flow graphs
  - Data-flow graphs

- **Matrix abstraction of data-flow graph properties**
  - Useful for modeling architectural transformations

- **Simulink DSP modeling**
  - Construction of block diagrams in Simulink
  - Functional simulation, RTL generation
  - Data-flow property extraction

9.14

Slide 9.14

To summarize, this chapter describes the graphical representations for iterative DSP algorithms. Examples of block diagrams and signal-flow graphs were shown. The data-flow graphs are discussed in some length, since they are the preferred representation used in Chap. 11. The matrix abstraction of DFGs was briefly mentioned and will be discussed again later. An example of system modeling and simulation in Simulink was introduced near the end of the chapter for completeness. More complex system modeling using Simulink will also be addressed later.

References

- K.K. Parhi, VLSI Digital Signal Processing Systems: Design and Implementation, John Wiley & Sons Inc., 1999.

- R. Nanda, DSP Architecture Optimization in Matlab/Simulink Environment, M.S. Thesis, University of California, Los Angeles, 2008.

Slide 10.1

Chapter 10

Wordlength Optimization

with Cheng C. Wang
University of California, Los Angeles
and Changchun Shi
Independent Researcher, Incline Village, NV

This chapter discusses wordlength optimization. Emphasis is placed on automated floating-to-fixed point conversion. Reduction in the number of bits without significant degradation in algorithm performance is an important step in hardware implementation of DSP algorithms. Manual tuning of bits is often performed by designers. Such approach is time-consuming and results in sub-optimal results. This chapter discusses an automated optimization approach.

Slide 10.2

Number Systems: Algebraic

Algebraic Number

e.g. $a = \pi + b$

[1]

- **High level abstraction**
- **Infinite precision**
- **Often easier to understand**
- **Good for theory/algorithm development**
- **Hard to implement**

[1] C. Shi, Floating-point to Fixed-point Conversion, Ph.D. Thesis, University of California, Berkeley, 2004.

10.2

Mathematical computations are generally computed by humans in decimal radix due to its simplicity. For the example shown here, if $b = 5.6$ and $\pi = 3.1416$, we can compute a with relative ease. Nevertheless, most of us would prefer not to compute using binary numbers, where $b = 1\text{'b}101.1001100110$ and $\pi = 1\text{'b}11.0010010001$ (a prefix of 1'b is used here to distinguish binary from decimal numbers). With this abstraction, many algorithms are developed without too much consideration for the binary representation in actual hardware, where something as simple as $0.3 + 0.6$ can never be computed with full accuracy. As a result, the designer may often find the actual hardware performance to be different from expected, or that implementations with sufficient precision incur high hardware costs [1]. The hardware cost of interest depends on the application, but it is generally a combination of performance, energy, or area for most VLSI and DSP designs. In this chapter, we discuss some methods of optimizing the number of bits (wordlength) used in every logic block to avoid excessive hardware cost while meeting the precision requirement. This is the basis of wordlength optimization.

D. Marković and R.W. Brodersen, *DSP Architecture Design Essentials*, Electrical Engineering Essentials,
DOI 10.1007/978-1-4419-9660-2_10, © Springer Science+Business Media New York 2012

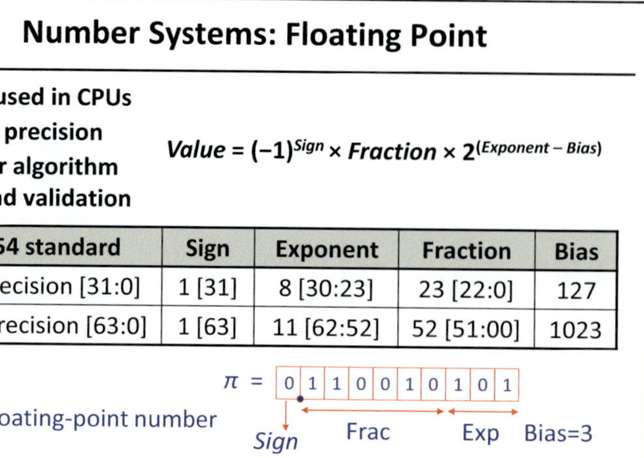

IEEE 754 standard	Sign	Exponent	Fraction	Bias
Single precision [31:0]	1 [31]	8 [30:23]	23 [22:0]	127
Double precision [63:0]	1 [63]	11 [62:52]	52 [51:00]	1023

Slide 10.3

Let us first look into how numbers are represented by computers. A common binary representation is floating point, where a binary number is divided into a sign bit, fractional bits, and exponent bits. The sign bit of 0 or 1 represents a positive or negative number, respectively. The exponents have an associated pre-determined bias for offset adjustment. In the example here, a bias of 3 is chosen, which means the exponent bits of 1′b000 to 1′b111 (0 to 7) represent actual exponent of −3 to 4. The fractional bits always start with an MSB of 2^{-1}, and the fraction is multiplied by $2^{Exponent-Bias}$ for the actual magnitude of the number.

It is apparent here that a floating-point representation is very versatile, and that the full precision of the fractional bits can often be utilized by adjusting the exponent. However, in the case of additions and subtractions, the exponents of the operands must be the same. All the operands are shifted so that their exponents match the largest exponent. This causes precision loss in the fractional bits (e.g. $1'b100101 \times 2^{-2}$ needs to shift to $1'b001001 \times 2^{0}$ if it is being added to a number with an exponent of 0).

This representation is commonly used in CPUs due to its wide input range, as general-purpose computers cannot make any assumptions about the dynamic range of its input data. While versatile, floating point is very expensive to implement in hardware and most modern CPU designs require dedicated floating-point processors.

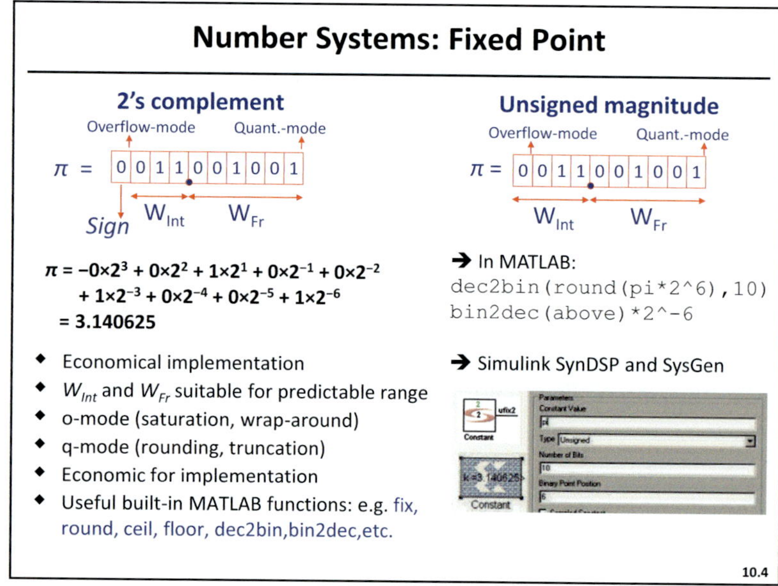

Slide 10.4

In most hardware designs, a high-precision floating-point implementation is often a luxury. Hardware requirements such as area and power consumption, and operating frequency demand more economical representation of the signal. A lower-cost (and higher hardware performance) alternative to floating-point arithmetic is fixed point, where the binary point is "fixed" for each datapath. The bit-field is divided into a sign bit, W_{Int} integer bits, and W_{Fr} fractional bits.

The maximum precision is $2^{-W_{Fr}}$, and the dynamic range is limited to $2^{W_{Int}}$. While these limitations may seem unreasonable for a general-purpose computer (unless very large wordlengths are used), it is acceptable for many dedicated-application designs where the input-range and precision requirements are well-defined.

Information regarding the binary point of each fixed-point number is stored separately. For manual conversions between decimal and binary, make sure the decimal point is taken into account: a scaling factor of 2^6 is needed in this example, since the binary point (W_{Fr}) is 6 bits. Simulink blocks such as Synplify DSP and Xilinx System Generator perform the binary-point alignment automatically for a given binary point.

In the Simulink environment, the designer can also specify overflow and quantization schemes when a chosen numerical representation has insufficient precision to express a value. It should be noted, however, that selecting saturation and rounding modes increase hardware usage.

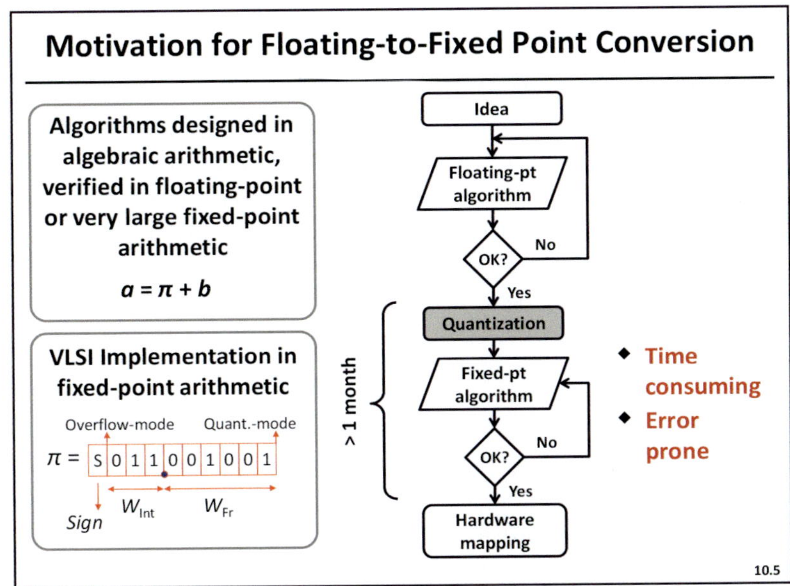

Slide 10.5

Most algorithms are built assuming infinite (or sufficient) precision, either from a floating-point or a long-wordlength fixed-point system. From a design perspective, to efficiently convert a floating-point design to a fixed-point design requires careful allocation of wordlengths. Excessive wordlength leads to slower performance, larger area and higher power, while insufficient wordlength introduces large quantization errors, and can heavily degrade the precision of the system or even cause system failure (such as divide-by-0 due to insufficient W_{Fr} in the denominator).

Floating-point to fixed-point conversion is an important task, and a series of questions rises from this process: How many fractional bits are required to meet my precision requirement? Is it cost-effective to perform rounding to gain the extra LSB of precision, or is it better to add fractional bits in the datapath and use truncation? How to determine the dynamic range throughout the system to allocate sufficient integer bits? Is it cost-effective to perform saturation and potentially use fewer integer bits, or should I determine the maximum-possible integer bits to avoid adding the saturation logic? Answering these questions for each design requires numerous iterations, and meeting the quantization requirements while minimizing wordlengths throughout the system becomes a tedious task, imposing a large penalty on both man-hour and time-to-market, as each iteration is a change in the system-level, and system-level specifications ought to be frozen months before chip fabrication. A systematic tool has therefore been developed to automate this conversion process, and is discussed in this chapter.

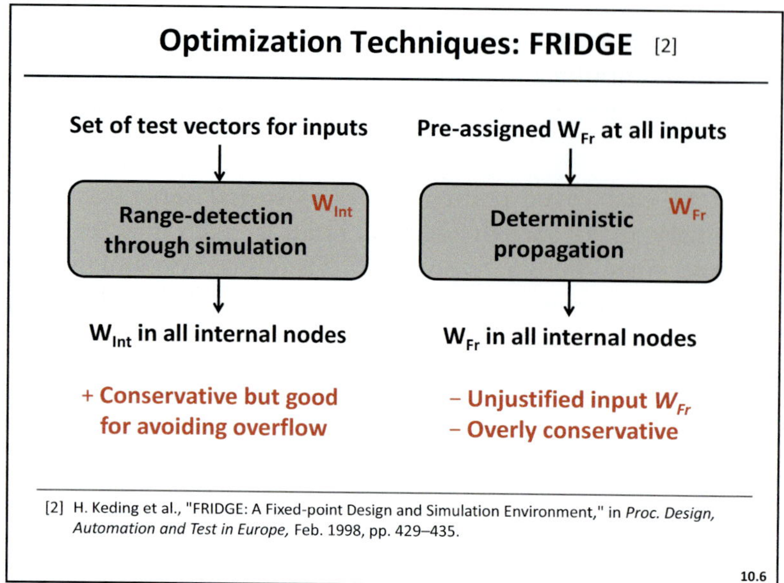

Slide 10.6

In the recent 15 years or so, much attention was given to addressing the wordlength optimization problem. Before the investigation of analytical approaches, we will review representative approaches from earlier efforts.

One past technique for determining both W_{Int} and W_{Fr} is Fixed-point pRogrammIng DesiGn Environment (FRIDGE) [2]. To find W_{Int}, the user is required to provide a set of test vectors that resemble worst-case operating conditions (i.e. cases that cause maximum wordlengths), or provide enough practical test vectors so the maximum wordlengths can be estimated with sufficient confidence. For each logic block, W_{Int} is determined by placing a range-detector at its output to record its maximum value. Since each internal node is only driven by one logic block, W_{Int} of every internal node can be determined by only one simulation of the test vector(s). One drawback of this conservative approach is that occasional overflows are not allowed. However, since overflows tend to have a much greater effect on quantization error than truncation, attempting to save 1 MSB by allowing overflow may not be a less conservative method. The user can still manually reduce the optimized wordlength by 1 to observe the effect of overflow through simulations.

FRIDGE optimizes W_{Fr} using deterministic propagation. The user specifies W_{Fr} at every input node, and then every internal node is assigned a W_{Fr} large enough to avoid any further quantization error. As examples, W_{Fr} for an adder is the maximum of its inputs' W_{Fr}'s and W_{Fr} of a multiplier is the sum of its inputs' W_{Fr}'s. This wordlength propagation approach is also overly conservative and has numerous drawbacks: First, the input W_{Fr} is chosen by the user. Because this is unverified by the optimization tool, choosing different W_{Fr} at the input can lead to sub-optimal results. In addition, not all W_{Fr} can be determined through propagation, so some logic blocks (e.g., a feedback multiplier) require user interaction. Due to the limitations of the FRIDGE technique, it is only recommended for W_{Int} optimization. Methods of W_{Fr} optimization will be discussed next.

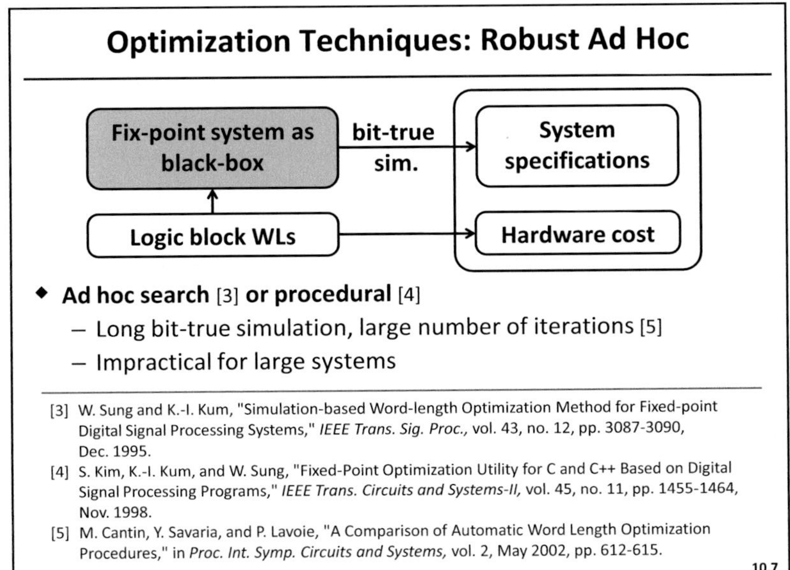

Slide 10.7

Another approach for W_{Fr} optimization is through iterative bit-true simulations by Sung et al. [3, 4]. The fixed-point system can be modeled using software (e.g., C or SystemC) or Simulink (SynDSP, XSG) with the W_{Fr} for every node described as a variable. With each simulation, the quantization error of the system (e.g., bit-error rate, signal-to-noise ratio, mean-squared error) is evaluated along with the hardware costs (e.g., area, power, delay), which is computed as a function of wordlengths. Since the relationship between wordlength and quantization is not characterized for the target system, wordlengths for each iteration are determined in an ad-hoc fashion, and numerous iterations are often needed to determine the wordlength-critical blocks [5]. Even after these blocks are located, more iterations are required to determine their suitable wordlength vaiables.

It is apparent that this kind of iterative search is impractical for large systems. However, two important concepts are introduced here. First is the concept of bit-true simulations. Although algorithms are generally developed in software, it is not difficult to construct the same functions in fixed-point, such MATLAB/Simulink blocks. The benefit of Simulink is the allowance of bit-true and cycle-true simulations to model actual hardware behavior using functional blocks, and third-party blocksets such as Xilinx System Generator and Synopsys Synphony HLS. These tools allow direct mapping into hardware-description language (HDL), which eliminates the error-prone process of manually converting software language into HDL. Another important optimization concept is "cost efficiency", where the goal of each iteration is to minimize the hardware cost (as a function of wordlengths) while meeting the quantization error requirements. To achieve a wordlength-optimal design, it is necessary to locate the logic blocks that provide the largest hardware-cost reduction with the least increase in quantization error. The formulation for wordlength optimization is founded on this concept, but a non-iterative approach is required to achieve acceptable results within a reasonable timeframe.

Problem Formulation: Optimization

Minimize hardware cost:
$$f(W_{Int,1}, W_{Fr,1}; W_{Int,2}, W_{Fr,2}; \ldots; o\text{-}q\text{-}modes)$$

Subject to quantization-error specifications:
$$S_j(W_{Int,1}, W_{Fr,1}; W_{Int,2}, W_{Fr,2}; \ldots; o\text{-}q\text{-}modes) < spec, \ \forall j$$

Feasibility:
$$\exists \ N \in Z^+, \text{ s.t. } S_j(N, N; \ldots; any\ mode) < spec, \ \forall j$$

Stopping criteria:
$$f < (1+a)f_{opt} \text{ where } a > 0.$$

From now on, concentrate on W_{Fr} [1]

[1] C. Shi, Floating-point to Fixed-point Conversion, Ph.D. Thesis, University of California, Berkeley, 2004.

10.8

Slide 10.8

The details of the theory behind the optimization approach are given extensively in [1]. Here we summarize key results used in practice. The framework of the wordlength optimization problem is formulated as follows: a hardware cost function is created as a function of every wordlength (actually every *group* of wordlengths; more on this later). This function is to be minimized subject to all quantization error specifications. These specifications may be defined for more than one output, in which case all j requirements need to be met. Since the optimization focuses on wordlength reduction, a design that meets the quantization-error requirements is required to start the optimization. Since a spec-meeting design is not guaranteed from users, a large number is chosen to initialize W_{Fr} of every block in order to make the system practically full-precision. This leads to a feasibility requirement where a design with wordlength N must meet the quantization error specification, else a more relaxed specification or a larger N is required. As with most optimization programs, tolerance a is required for the stopping criteria. A larger a decreases optimization time, but in wordlength optimization, simulation time far outweighs the actual optimization time. Since W_{Int} and the use of overflow mode are chosen in one simulation, this optimization is only required to determine quantization modes and W_{Fr}.

Perturbation Theory On MSE   [6]

Output MSE Specs:

$$MSE = E[(\text{Infinite-precision-output} - \text{Fixed-point-output})^2]$$

$$= \bar{\mu}^T B \bar{\mu} + \sum_{i=1}^{p} c_i 2^{-2W_{Fr,i}} \text{ for a datapath of } p, \text{ WL } \quad B \in \mathbb{S}_+^p, C \in \mathbb{R}_+^p$$

$$\mu_i = \begin{cases} -\dfrac{1}{2}q_i w^{-W_{Fr,i}}, & \text{datapath} \\ \text{fix-pt}(c_i, 2^{-W_{Fr,i}}) - c_i, & \text{const } c_i \end{cases} \qquad q_i = \begin{cases} 0, & \text{round-off} \\ 1, & \text{truncation} \end{cases}$$

[6] C. Shi and R.W. Brodersen, "A Perturbation Theory on Statistical Quantization Effects in Fixed-point DSP with Non-stationary Input," in *Proc. IEEE Int. Symp. Circuits and Systems,* vol. 3, May 2004, pp. 373-376.

10.9

Slide 10.9

To avoid iterative simulations, it is necessary to model the quantization noise of interest (generally at the output) as a function of wordlengths. Based on the original perturbation theory [6], we observe that such MSE follows an elegant formula for the fixed-point datatypes. In essence, the theory linearizes a smooth non-linear time-varying system. The result states that the MSE error, as defined here, can be modeled as a function of all the fractional wordlengths, all the quantization modes, and all the constant coefficients to be quantized in the system. The non-negative nature of MSE implies that B is positive semi-definite and C is non-negative. Both B and C depend on the system architecture and input statistics, but they can be estimated numerically.

While a large design could originally contain hundreds or thousands independent wordlengths to be optimized at the beginning, we will see that the design complexity can be drastically reduced by employing grouping of related blocks to have the same wordlength. In practice, after reducing the number of independent wordlengths, a complex system may only have a few or few tens of independent wordlengths. The new matrix B and vector C are directly related to the original B and C by combining the corresponding terms. In the FFC problem, we often are only interested in estimating the new B and C that has considerably less number of entries to estimate, which reduces the number of simulations required.

Many details, including the reason for adopting a MSE-based specification with justifications for assuming non-correlation based on the perturbation theory, are included in [1].

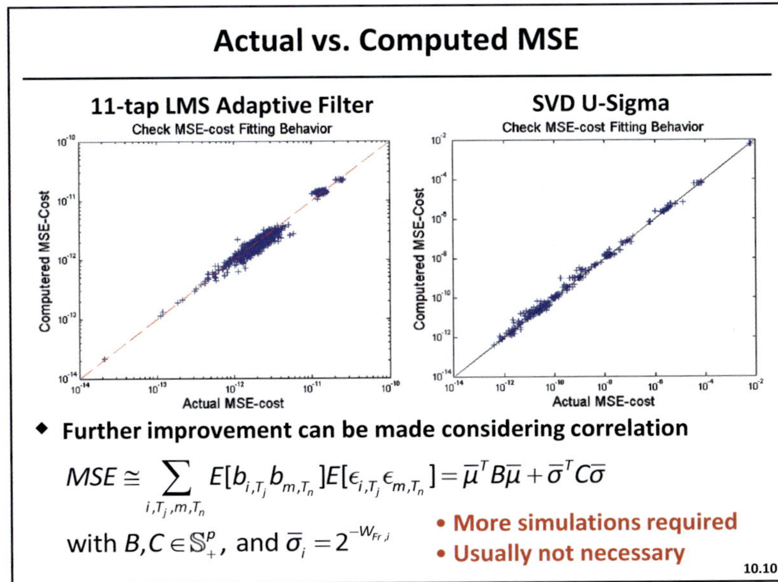

$$MSE \cong \sum_{i,T_j,m,T_n} E[b_{i,T_j} b_{m,T_n}] E[\epsilon_{i,T_j} \epsilon_{m,T_n}] = \bar{\mu}^T B \bar{\mu} + \bar{\sigma}^T C \bar{\sigma}$$

with $B,C \in \mathbb{S}^p_+$, and $\bar{\sigma}_i = 2^{-W_{Fr,i}}$

Slide 10.10

Once the B and C are estimated, the MSE can be predicted at different combinations of practical wordlengths and quantization modes. This predicted MSE should match closely to the actual MSE as long as the underlying assumptions used in perturbation theory still apply reasonably. The actual MSE is estimated by simulating the system with the corresponding fixed-point datatypes. This slide demonstrates the validity of the non-correlation assumption. Shown here is an adaptive filter design and an SVD design where simulations are used to fit the coefficients B and C, which in turn are used to directly obtain the "computed" MSE. The actual MSE in the x-axis is from simulation of the corresponding fixed-point data-types. By varying the fixed-point data-types we see that the computed MSE from the once estimated B and C fits well with the actual MSE across the broad range of MSEs. A more accurate calculations can be achieved by including correlation, which requires B and C to both be positive-semi-definite matrices. This increases both simulation requirements and computation complexity, and it is usually unnecessary.

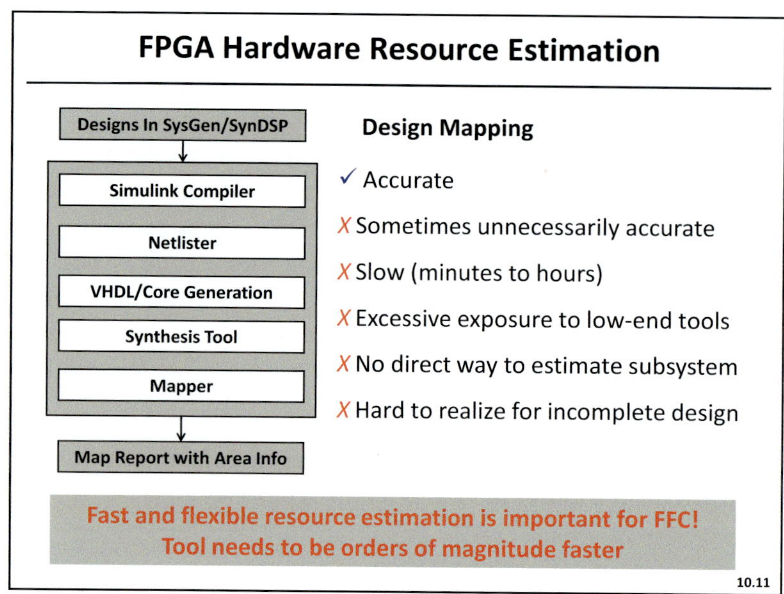

Slide 10.11

Having an accurate MSE model is not sufficient for wordlength optimization. Recapping from Slide 10.8, the optimization goal is to minimize the hardware cost (as a function of wordlength) while meeting the criteria for MSE. Therefore hardware cost is evaluated just as frequently as MSE cost, and needs to be modeled accurately. When the design target is an FPGA, hardware cost generally refers to area, but for ASIC designs it is generally power or performance that defines the hardware cost.

Traditionally, the only method for area estimation is design mapping or synthesis, but such method is very time consuming. The Simulink design needs to be first compiled and converted to a Verilog or VHDL netlist, then the logic is synthesized as look-up-tables (LUTs) and cores for FPGA, or standard-cells for ASIC. The design is then mapped and checked for routability within the area constraint. Area information and hardware usage can then be extracted from the mapped design. This approach is very accurate, for it only estimates the area after the design is routed, but the entire process can take minutes to hours, and needs to be re-executed with even the slightest change in the design. These drawbacks limit the utility of design mapping in our optimization, where fast and flexible estimation is the key — each resource estimation step cannot consume more than a fraction of a second.

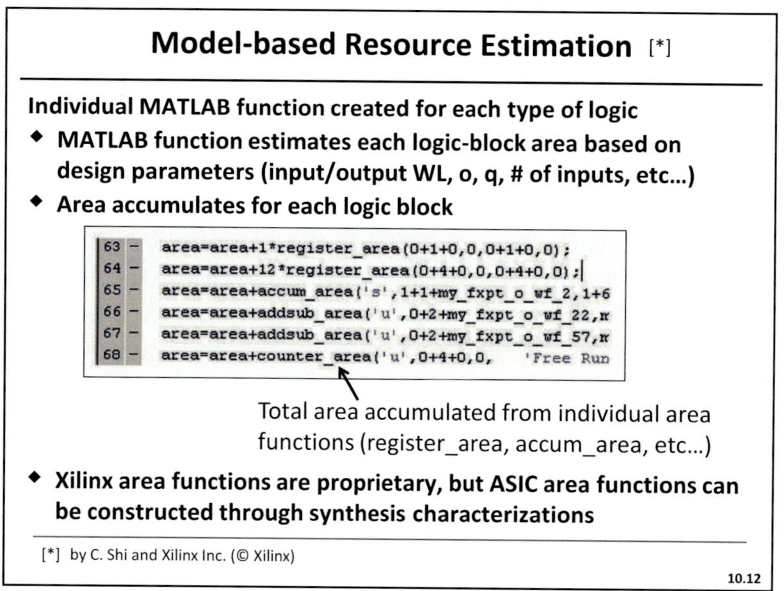

Slide 10.12

To avoid repetitive design mapping to estimate area, a model-based resource estimation is developed to provide area estimations based on cost functions. Each cost function returns an estimated area based on its functionality and design parameters such as input and output wordlengths, overflow and quantization modes, and the number of inputs. These design parameters are automatically extracted from Simulink. The total area is determined by iteratively accumulating the individual area functions. This dramatically speeds up the area-estimation process, as only simple calculations are required per logic block.

Although the exact area function of each XSG logic block is proprietary to Xilinx, the end-user may create similar area functions for ASIC designs by characterizing synthesis results, as shown on the next slide.

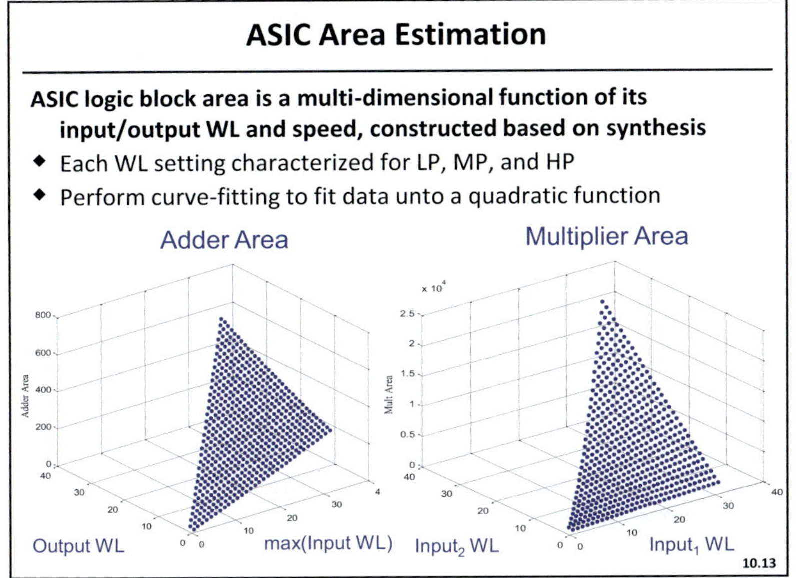

Slide 10.13

For FPGA applications, area is the primary concern, but for ASIC applications, the cost function can also be changed to model energy or circuit performance. Due to the core usage and LUT structures on the FPGA, the logic area on FPGA may differ significantly from ASIC, where no cores are used and all logic blocks are synthesized in standard-cell gates. This means an area-optimal design for the FGPA flow is not necessarily area-optimal for the ASIC flow.

ASIC area functions also depend on design parameters such as wordlengths and quantization modes, but they are dependent on circuit performance as well. For example, the area of a carry-ripple adder is roughly linear to its wordlength, but the area of a carry-look-ahead adder tends to be on the order of $O(N{\cdot}logN)$. Therefore, it is recommended that three data points be gathered for each design parameter: a high-performance (HP) mode that requires maximum performance, a low-power (LP) mode that requires minimum area, and a medium performance (MP) mode where the performance criteria is roughly 1/3 to 1/2 between HP and LP. The area functions for each performance mode can be fitted into a multidimensional function of its design parameters by using a least-squares curve-fit in MATLAB. Some parameters impact the area more than others. For the adder example, the longer input wordlength and output wordlength have the largest impact on area. They are roughly linearly proportional to the area. For the multiplier, the two input wordlengths have the largest impact on area, and the relationship is roughly quadratic. Rounding can increase area by as much as 30%. For many users, ASIC power is more important than area, therefore the area function can be used to model energy-per-operation instead of area.

Alternatively, the user may create area, energy, or delay cost-functions based on standard-cell documentations to avoid extracting synthesis data for each process technology. From the area or power information of each standard-cell, some estimation can be modeled. For example, an N-bit accumulator can be modeled as the sum of N full-adder cells and N registers, a N-bit, M-input mux can be modeled as $N{\cdot}M$ 2-input muxes, and a N-bit by M-bit multiplier can be modeled as $N{\cdot}M$ full-adder cells. The gate sizes can be chosen based on performance requirements. Low-power designs are generally synthesized with gate sizes of 2× (relative to a unit-size gate) or smaller, while high-performance designs typically require gate sizes of 4× or higher. Using these approximations,

ASIC area can be modeled very efficiently. Although the estimation accuracy is not as good as the fitting from the synthesis data, it is often sufficient for wordlength optimization purposes.

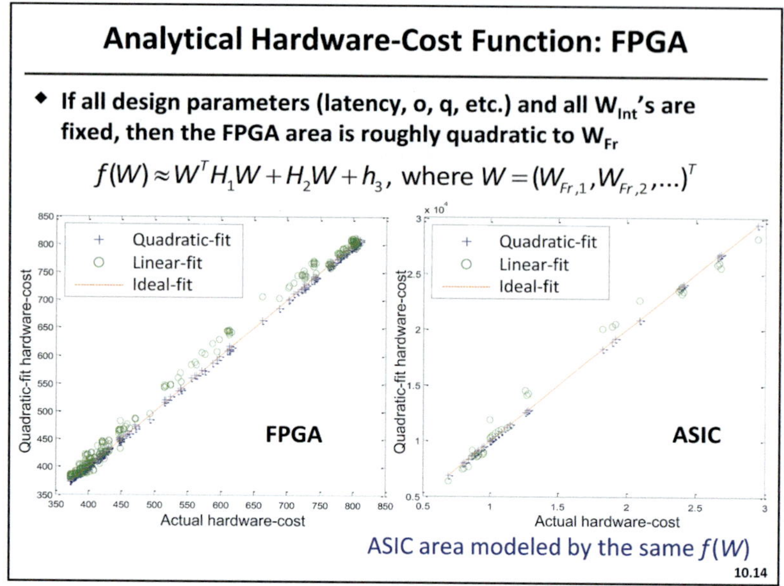

Slide 10.14

The hardware-cost function can be modeled as a quadratic function of W_{Fr}, assuming all other design parameters are fixed. Such assumption is reasonable given that W_{Int} and overflow-modes are determined prior to W_{Fr} optimization. Coefficient matrices H_1 and H_2 and vector h_3 are fitted from the area estimations. From the plot on the left, it is apparent that a quadratic-fit provides sufficient accuracy, while a linear-fit is subpar. Linear fitting is only recommended when the quadratic estimation takes too long to complete. Since the area and energy for most ASIC blocks can also be modeled as quadratic functions of wordlengths, the quadratic hardware-cost function $f(W)$ also fits nicely for ASIC designs, as shown in the plot on the right.

The equation shown in the slide is:

$$f(W) \approx W^T H_1 W + H_2 W + h_3, \text{ where } W = (W_{Fr,1}, W_{Fr,2}, \ldots)^T$$

With adequate models for both MSE cost and hardware cost, we can now proceed with automated wordlength optimization. The remainder of the chapter will cover both the optimization flow and usage details of the wordlength optimization tool.

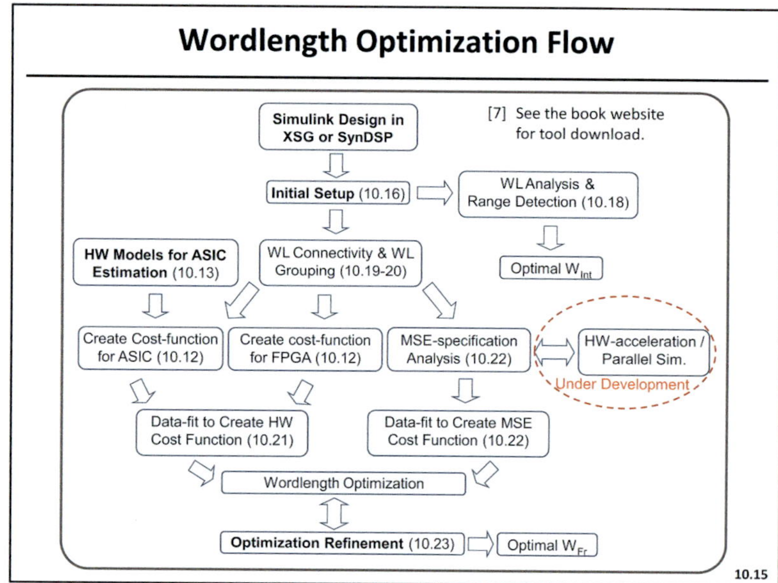

Slide 10.15

Each design imported for wordlength optimization is processed with a fixed methodology. An outline of the wordlength optimization flow is shown here. Some key steps such as integer wordlength analysis, hardware cost analysis, and MSE analysis were already introduced in previous slides. Each step will be discussed individually in the following slides. The tool is publicly available for download [7].

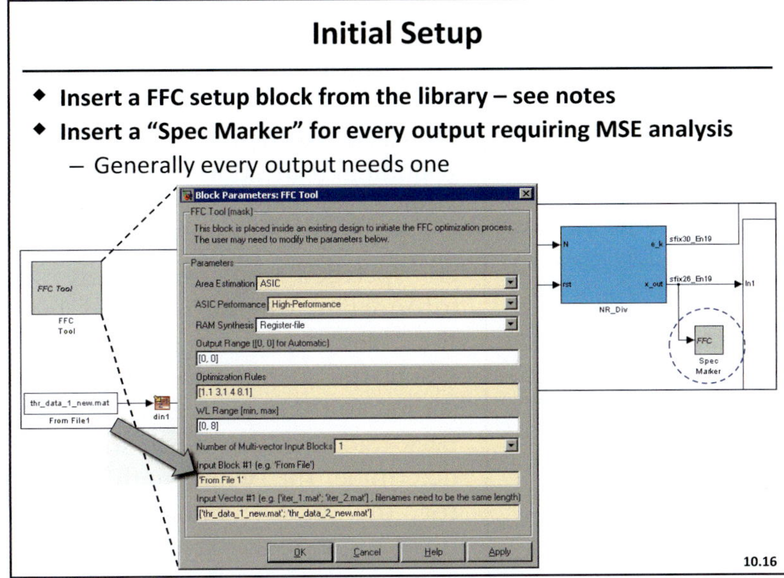

Slide 10.16

Before proceeding to the optimization, an initial setup is required. A setup block needs to be added from the optimization library, and the user should open the setup block to specify parameters. The area target of the design (FPGA, or ASIC of HP, MP, or LP) should be defined. Some designs have an initialization phase during start-up that should not be used for MSE characterization, so the user may then specify the sample range of outputs to consider: if the design has 1,000 samples per second, and the simulation runs for 10 seconds, then the range [2,000, 1,0000] specifies the timeframe between 2 and 10 seconds for MSE characterization. If left at [0, 0], the characterization starts at 25% of the simulation time. The optimization rules apply to wordlength grouping and are introduced in Slide 10.20. Default rules of [1.1 3.1 4 8.1] is a good set.

The user needs to specify the wordlength range to use for MSE characterization. For example, [8, 40] specifies a W_{Fr} of 40 to be "full-precision", and each MSE iteration will "minimize" one W_{Fr} to 8 to determine its impact on total MSE. Depending on the application, a "full-precision" W_{Fr} of 40 is generally sufficient, though smaller values improve simulation time. A "minimum" W_{Fr} of 4 to 8 is generally sufficient, but designs without high sensitivity to noise can even use minimum W_{Fr} of 0. If multiple simulations are required to fully characterize the design, the user needs to specify the input vector for each simulation in the parameter box.

The final important step is the placement of specification markers. The tool characterizes MSE only at the location where Spec Marker is placed, therefore it is generally useful to place markers at all outputs, and at some important intermediate signals as well. The user should open each Spec Marker to ensure that a unique number is given for each marker, else an optimization error may occur.

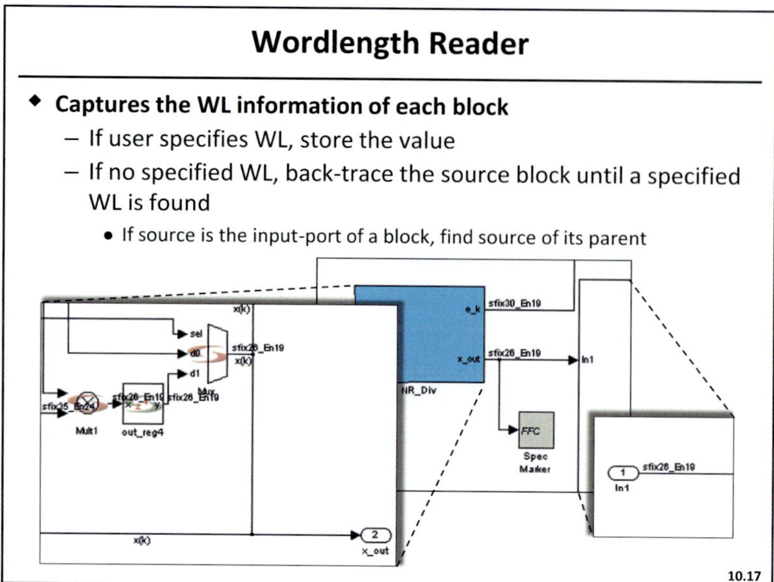

Slide 10.17

The wordlength reader gathers W_{Int} and W_{Fr}, along with overflow/saturation and rounding/truncation information from every block. The simplest case is when the wordlength of the block is specified by the user. Its information can be obtained from the block parameters. In cases where the wordlength parameter is set to "Automatic", it is necessary to back-trace to the block-source to determine the wordlength needed for the block. For example, a register should have the same wordlength as its input, while a 2-input adder would require a wordlength of $\max(W_{Int,input})+1$. Sometimes it may be necessary to trace back several blocks to find a block with a specified wordlength. An illustration of the back-trace process is shown on the left inset: the wordlength of the mux is set to "Automatic", so the tool back-traces to the register, then to the multiplier to gather the wordlength. In cases where back-tracing reaches the input of a submodule, a hierarchical back-trace is required. In the right inset, the wordlength from input port "In1" is determined by the mux from the previous logic block.

In the current tool, overflow/saturation and rounding/truncation options are chosen by the user, and are not yet a part of the optimization flow, and the tool chooses overflow and truncation by default.

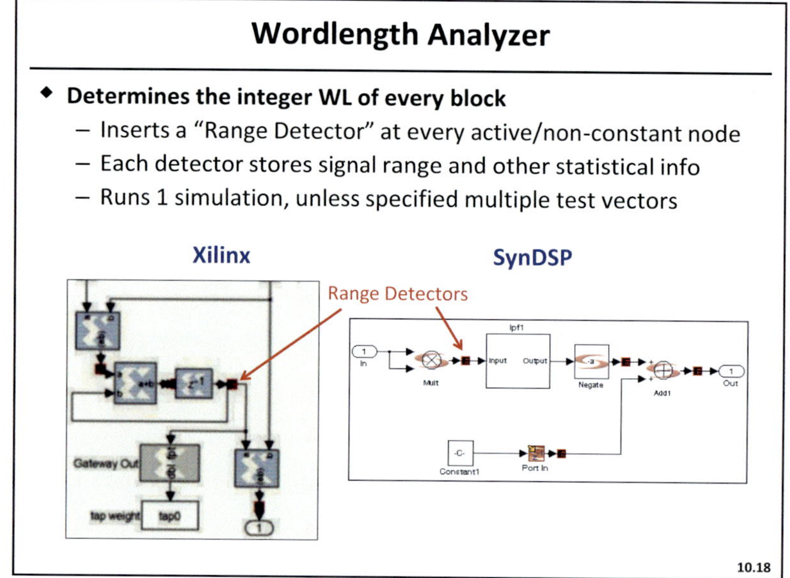

Slide 10.18

During wordlength analysis, a "Range Detector" is inserted at each active node automatically. Passive nodes such as subsystem input and output ports, along with constant numbers and non-datapath signals (e.g. mux selectors, enable/valid signals) are not assigned a range-detector.

Based on the FRIDGE algorithm in Slide 10.6, the wordlength analyzer determines W_{Int} based on a single iteration of the provided test-vector(s). Therefore, it is important that the user provides input test vectors that cover the entire input range. Failure to provide adequate information can result in the removal of bits.

The range-detector block gathers information such as the mean, variance, and the maximum value at each node. The number of integer bits is determined by the 4th standard-deviation of the value, or its maximum value, whichever is greater. If the calculated W_{Int} is greater than the original W_{Int}, then the original W_{Int} is used. Signed outputs have W_{Int} of at least 1, and W_{Int} of constants are determined based on their values.

Due to the optimization scheme, the wordlength analyzer does not correct overflow errors that occur in the original design. As a result, the user must ensure that the design behaves as expected before using this tool.

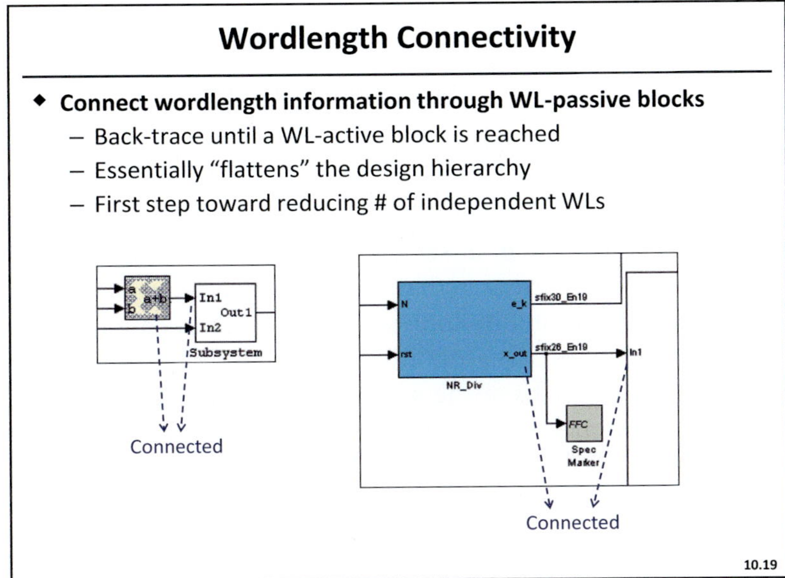

Slide 10.19

With W_{Int} determined by the wordlength analyzer, the remaining effort aims to optimize W_{Fr} in the shortest time possible. Since the number of iterations for optimizing W_{Fr} is proportional to the number of wordlengths, reducing the number of wordlengths is attractive for speeding up the optimization. The first step is to determine the wordlength-passive blocks, which are blocks that do not have physical area, such as input and output ports of submodules in the design, and can be viewed as wordlength feed-throughs.

These connectivity optimizations essentially flatten the design so every wordlength only defines an actual hardware block. As a result, the number of wordlengths is reduced. Further reduction is obtained by wordlength grouping.

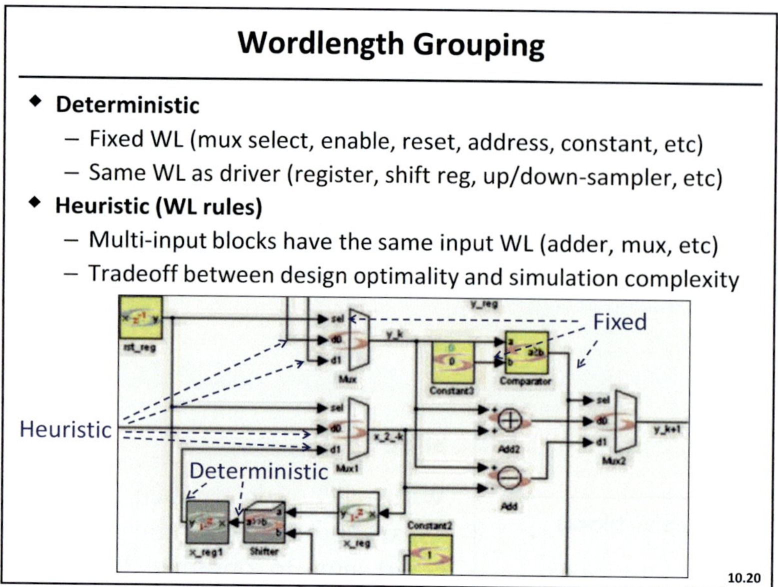

Slide 10.20

The purpose of the wordlength grouping function is to locate wordlength dependencies between blocks and group the related blocks together under one wordlength variable. Deterministic wordlength grouping includes blocks whose wordlength is fixed, such as mux-select, enable, reset, address, comparator and constant signals. These wordlengths are marked as fixed, as shaded in yellow. Some blocks such as shift registers, up- and down-samplers do not result in a change in wordlength. The wordlengths of these blocks can be grouped with their drivers, as shaded in gray.

Another method of grouping wordlengths is by using a set of heuristic wordlength rules. Grouping these inputs into the same wordlength can further reduce simulation time, though at a small cost of design optimality. For example, in the case of a multiplexer, allowing each data input of a mux to have its own wordlength group may result in a slightly more optimal design, but it can generally be assumed that all data inputs to a multiplexer have the same wordlength. The same case applies to adders and subtractors. Grouping these inputs into the same wordlength can further reduce simulation complexity, albeit at a small cost to design optimality. These heuristic wordlength groupings are defined as eight general types of "rules" for the optimization tool, with each rule type subdivided to more specific rules. Currently there are rules 1 through 8.1. These rules are defined in the tool's documentation [9], and can be selectively enabled in the initialization block.

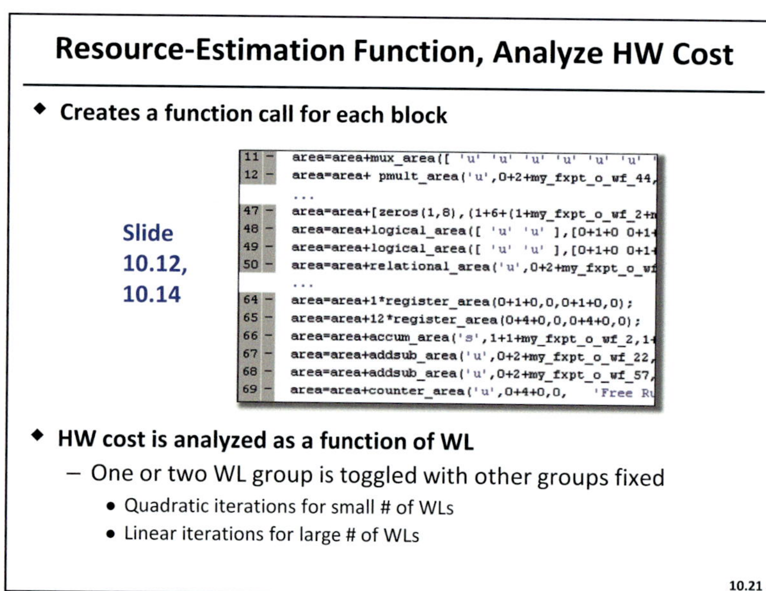

Slide 10.21

To accurately estimate the hardware cost of the current design, it is necessary to first build a cost function based on all existing blocks in the design. This function-builder tool will build a cost function that returns the total area as the sum of the areas from each block. The area-estimation function discussed on Slide 10.12 is used here. The resource-estimation tool examines every block in the design and first obtains its input and output wordlength information, which could be fixed numbers or wordlength variables. Each block is then assigned a resource-estimation function to estimate its area.

As shown in the slide, each type of block has its own resource-estimation function, which returns the hardware cost of that block based on its input and output wordlength information, along with properties such as number of inputs, latency, and others.

The constructed resource estimation function is then evaluated iteratively by the hardware-cost analyzer. Since each wordlength group defines different logic blocks, they each contribute differently towards the total area. It is therefore necessary to iterate through different wordlength combinations to determine the sensitivity of total hardware cost to each wordlength group. Quadratic number of iterations is usually recommended for a more accurate curve-fitting of the cost function (Slide 10.14). In each iteration only two wordlength variables are changed while the other variables remain constant (usually fixed at minimum or maximum wordlength). However, if there are too many wordlength groups (e.g. more than 100), a less accurate linear fit will be used to save time. In this instance only one variable will be changed per iteration. There are continuous research interests to extend the hardware cost function to include power estimation and speed requirement. Currently these are not fully supported in our FFC tool, but can be implemented without structural change to the optimization flow.

<table>
<tr><td>

Analyze Specifications, Analyze Optimization

♦ **Computes MSE's sensitivity to each WL group**
 – First simulate with all WL at maximum precision
 – WL of each group is reduced individually   **Slide 10.9, 10.10**

♦ **Once MSE function and HW cost function are computed, user may enter the MSE requirement**
 – Specify 1 MSE for each Spec Marker

♦ **Optimization algorithm summary**
 1) Find the minimum W_{Fr} for a given group (others high)
 2) Based on all the minimum W_{Fr}'s, increase all WL to meet spec
 3) Temporarily decrease each W_{Fr} separately by one bit, only keep the one with greatest HW reduction and still meet spec
 4) Repeat 3) until W_{Fr} cannot be reduced anymore

10.22

</td></tr>
</table>

Slide 10.22

The MSE-specification analysis is described in Slides 10.9 and 10.10. While the full B matrix and C vector are needed to be estimated to fully solve the FFC problem, this would imply an order of $O(N^2)$ number of simulations for each test vector, which sometimes could still be too slow to do. However, it is often possible to drastically reduce the number of simulations needed by exploring design-specific simplifications. One such example is if we are only interested in rounding mode along the datapath.

Ignoring the quantization of constant coefficients for now, the resulting problem is only related to the C vector, thus only $O(N)$ simulations are needed for each test vector. For smaller designs and short test vectors, the analysis is completed within minutes, but larger designs may take hours or even days to complete this process, though no intermediate user interaction is required. Fortunately, all simulations are independent of each other, thus many runs can be performed in parallel. Parallel-simulation support is currently being implemented. FPGA-based acceleration is a much faster approach, but requires mapping the full-precision design to an FPGA first, and masking off some of the fractional bits to 0 to imitate a shorter-wordlength design. The masking process must be performed by programming registers to avoid reforming synthesis with each change in wordlength.

After MSE-analysis, both MSE and hardware cost functions are available. The user is then prompted to enter an MSE requirement for every Spec Marker in the design. It is advised to have a more stringent MSE for control signals and important datapaths. The details of choosing MSE requirements are in [2]. A good general starting point is 10^{-6} for datapath and 10^{-10} for control signals.

The MSE requirements are first examined for feasibility in the "floating point" system, where every wordlength variable is set to its maximum value. Once the requirements are considered feasible, the wordlength tool employs the following algorithm for wordlength reduction:

While keeping all other wordlengths at the maximum value, each wordlength group is reduced individually to find the minimum-possible wordlength while still meeting the MSE requirements. Each wordlength group is then assigned its minimum-possible wordlength which is likely to be an infeasible solution that does not meet the MSE requirements. In order to find a feasible solution all wordlengths are increased uniformly. Finally, the wordlength for each group is reduced temporarily, and the group that results in the largest hardware reduction while meeting the MSE requirements is chosen. This step is then iterated until no further hardware reduction is feasible, and the wordlength-optimal solution is created. There are likely other more efficient algorithms to explore the simple objective function and constraint function, but since we now have the analytical format of the optimization problem, any reasonable optimization procedure will yield the near-optimal point.

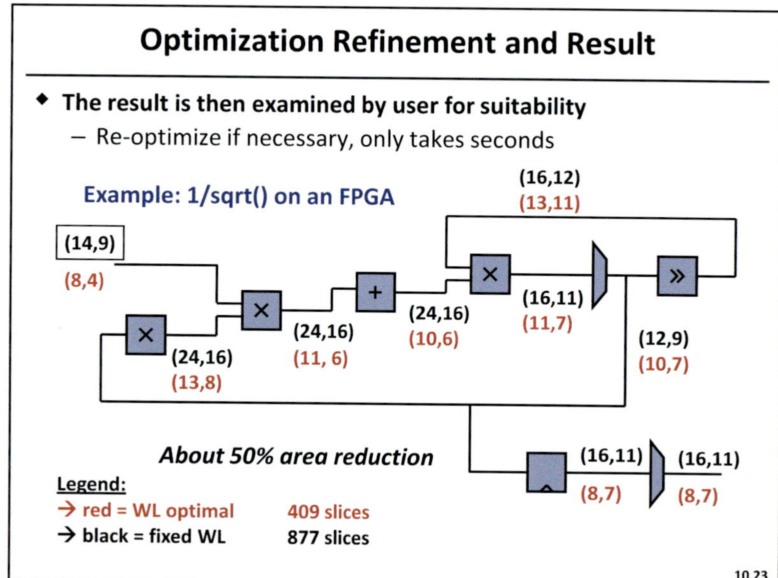

Slide 10.23

The MSE requirement may require a few refinements before arriving at a satisfactory design, but one key advantage of this wordlength optimization tool is its ability to rapidly refine designs without restarting the characterization and simulation process, because both the hardware and MSE cost are modeled as simple functions. In fact, it is now practical to easily explore the tradeoff between hardware cost and MSE performance.

Furthermore, given an optimized design for the specified MSE requirement, the user is then given the opportunity to simulate and examine the design for suitability. If unsatisfied with the result, a new MSE requirement can be entered, and a design optimized for the new MSE is created immediately. This step is still important as the final verification stage of the design to ensure full compliance with all original system specifications.

A simple 1/sqrt() design is shown as an example. Note that both W_{Int} and W_{Fr} are reduced, and a large saving of FPGA slices is achieved through wordlength reduction.

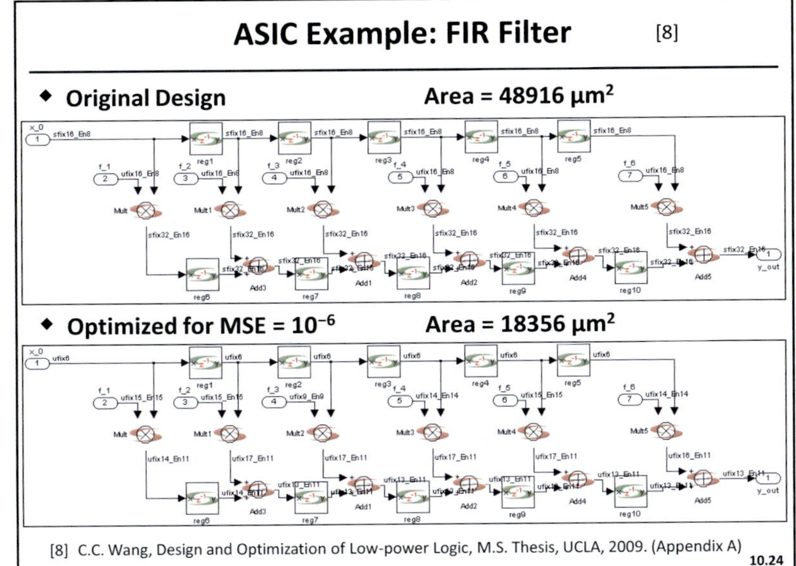

Slide 10.24

Since the ASIC area estimation is characterized for adders, multipliers, and registers, a pipelined 6-tap FIR filter is used as a design example [8]. The design is optimized for an MSE of 10^{-6}, and area savings from the optimized design is more than 60% compared to all-16-bit design. The entire optimization flow for this FIR design is under 1 minute. For more complex non-linear systems characterization may take overnight, but no intermediate user interaction is required.

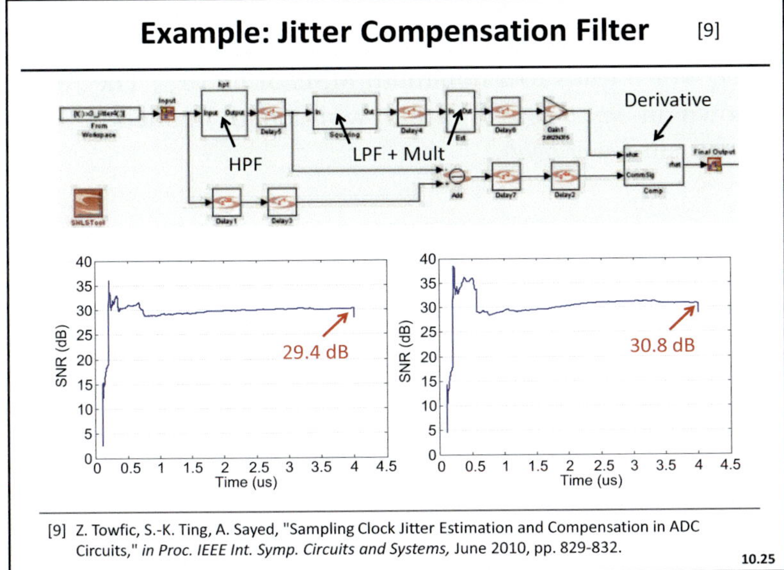

Slide 10.25

The design of a state-of-the art jitter compensation unit using high-frequency training signal injection [9] is shown. Its main blocks include high-pass and low-pass filters, multipliers, and derivative computations. The designer spent many iterations in finding a suitable wordlength, but is still unable to reach a final SNR of 30 dB, as shown in lower left. This design consumes ~14k LUTs on a Virtex-5 FPGA. Using the wordlength optimization tool, we finalized on a MSE of 4.5×10^{-9} after a few simple refinements. Shown in lower right, the optimized design is able to achieve a final SNR greater than 30 dB while consuming only 9.6k LUTs, resulting in a 32% savings in area *and* superior performance.

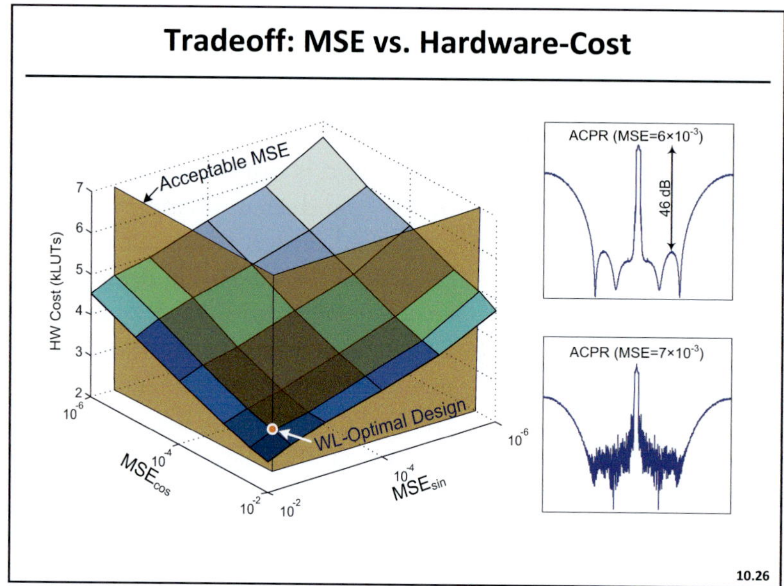

Slide 10.26

The final detailed example is a high-performance reconfigurable digital front end for cellular phones. Due to the GHz-range operational frequency required by the transceiver, a high-precision design simply cannot meet the performance requirement. The authors had to explore the possible architectural transformations and wordlength optimization to make the performance feasible. Since high-performance designs often synthesize to parallel logic architectures (e.g. carry look-ahead adder), the wordlength-optimized design results in 40% savings in area.

We now explore the tradeoff between MSE and hardware cost, which in this design directly translates to power, area, and timing feasibility. Since this design has two outputs (sine and cosine channels), the MSE at each output can be adjusted independently. The adjacent-channel-power-ratio (ACPR) requirement of 46 dB must be met, which leads to a minimum MSE of 6×10^{-3}. The ACPR of the wordlength-optimal design is shown in *upper-right*. Further wordlength reduction from higher MSE (7×10^{-3}) violates ACPR requirement (*lower-right*).

Summary

- **Wordlength minimization is important in the implementation of fixed-point systems in order to reduce area and power**
  - Integer wordlength can be simply found by using range detection, based on input data
  - Fractional wordlengths require more elaborate perturbation theory to minimize hardware cost subject to MSE error due to quantization
- **Design-specific information can be used**
  - Wordlength grouping (e.g. in multiplexers)
  - Hierarchical optimization (with fixed input/output WLs)
  - WL optimizer for recursive systems takes longer due to the time require for algorithm convergence
- **FPGA/ASIC hardware resource estimation results are used to minimize WLs for FPGA/ASIC implementations**

10.27

Slide 10.27

Wordlength optimization is an important step in algorithm verification. Reduction in the number of bits can have considerable effect on the chip power and area and it is the first step in algorithm implementation. Manual simulation-based tuning is time consuming and infeasible for large systems, which necessitates automated approach. Wordlength reduction can be formulated as optimization problem where hardware cost (typically area or power) is minimized subject to MSE

error due to quantization. Integer bits are determined based on the dynamic range of input data by doing node profiling to determine the signal range at each node. Fractional bits can be automatically determined by using perturbation-based approach. The approach is based on comparison of outputs from "floating-point-like" (the number of bits is sufficiently large that the model can be treated as floating point) and fixed-point designs. The difference due to quantization is subject to user-defined MSE specification. The perturbation-based approach makes use of wordlength groupings and design hierarchy to simplify optimization process. Cost functions for FPGA and ASIC hardware targets are evaluated to support both hardware flows. After determining wordlengths, the next step is architectural optimization.

We encourage readers to download the tool from the book website and try it on their designs. Due to constant updates in Xilinx and Synopsys blockset, some version-compatibility issues may occur, though we aim to provide updates with every major blockset release (support for Synopsys Synphony blockset is recently added). It is open-source, so feel free to modify it and make suggestions, but please do not use it for commercial purposes without our permission.

References

- C. Shi, Floating-point to Fixed-point Conversion, Ph.D. Thesis, University of California, Berkeley, 2004.

- H. Keding *et al.*, "FRIDGE: A Fixed-point Design and Simulation Environment," in *Proc. Design, Automation and Test in Europe,* Feb. 1998, pp. 429–435.

- W. Sung and K.-I. Kum, "Simulation-based Word-length Optimization Method for Fixed-point Digital Signal Processing Systems," *IEEE Trans. Sig. Proc.,* vol. 43, no. 12, pp. 3087-3090, Dec. 1995.

- S. Kim, K.-I. Kum, and W. Sung, "Fixed-Point Optimization Utility for C and C++ Based on Digital Signal Processing Programs," *IEEE Trans. Circuits and Systems-II,* vol. 45, no. 11, pp. 1455-1464, Nov. 1998.

- M. Cantin, Y. Savaria, and P. Lavoie, "A Comparison of Automatic Word Length Optimization Procedures," in *Proc. Int. Symp. Circuits and Systems,* vol. 2, May 2002, pp. 612-615.

- C. Shi and R.W. Brodersen, "A Perturbation Theory on Statistical Quantization Effects in Fixed-point DSP with Non-stationary Input," in *Proc. IEEE Int. Symp. Circuits and Systems,* vol. 3, May 2004, pp. 373-376.

- See the book supplement website for tool download.

  Also see:
  http://bwrc.eecs.berkeley.edu/people/grad_students/ccshi/research/FFC/documentation.htm

- C.C. Wang, Design and Optimization of Low-power Logic, M.S. Thesis, UCLA, 2009. (Appendix A)

- Z. Towfic, S.-K. Ting, A.H. Sayed, "Sampling Clock Jitter Estimation and Compensation in ADC Circuits," in *Proc. IEEE Int. Symp. Circuits and Systems,* June 2010, pp. 829-832.

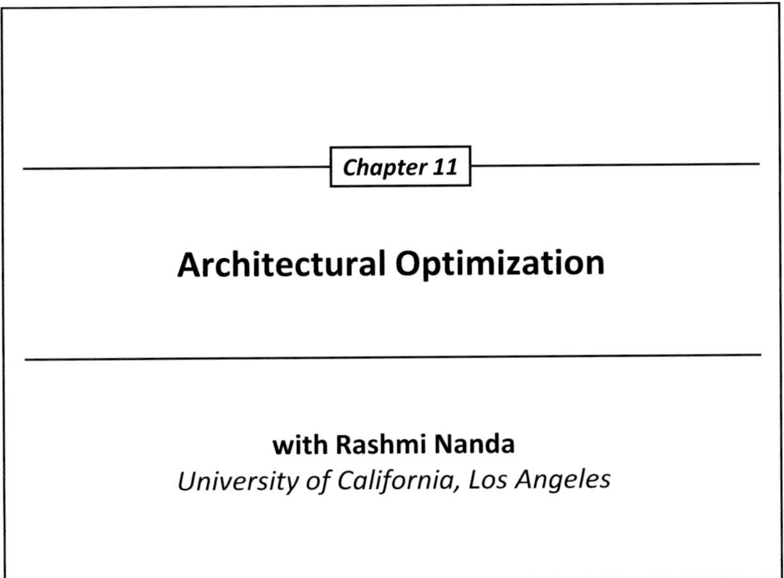

Slide 11.1

Automation of the architectural transformations introduced in Chap. 3 is discussed here. The reader will gain insight into how data-flow graphs are mathematically modeled as matrices and how transformations such as retiming, pipelining, parallelism and time-multiplexing are implemented at this level of abstraction. Reasonable understanding of algorithms used for automation can prove to be very useful, especially for designers working with large designs where manual optimization can become tedious. Although a detailed discussion on algorithm complexity is beyond the scope of this book, some of the main metrics are optimality (in a sense of algorithmic accuracy) and time-complexity of the algorithms. In general, a tradeoff exists between the two and this will be discussed in the following slides.

DFG Realizations

- **Data-flow graphs can be realized with several architectures**
  - Flow-graph transformations
  - Change structure of graph without changing functionality
  - Observe transformations in energy-area-delay space

- **DFG Transformations**
  - Retiming
  - Pipelining
  - Time-multiplexing/folding
  - Parallelism

- **Choice of the architecture**
  - Dictated by system specifications

11.2

Slide 11.2

We will use DFG model to implement architectural techniques. Architecture transformations are used when the original DFG fails to meet the system specifications or optimization target. For example, a recursive DFG may not be able to meet the target timing constraints of the design. In such a scenario, the objective is to structurally alter the DFG to bring it closer to the desired specifications without altering the functionality. The techniques discussed in this chapter will include retiming, pipelining, parallelism and time multiplexing. Benefits of these transformations in the energy-area-delay space have been discussed in Chap. 3. Certain transformations like pipelining and parallelism may alter the datapath by inserting additional latency, in which case designer should set upper bounds on the additional latency.

D. Marković and R.W. Brodersen, *DSP Architecture Design Essentials*, Electrical Engineering Essentials,
DOI 10.1007/978-1-4419-9660-2_11, © Springer Science+Business Media New York 2012

Retiming [1]

♦ **Registers in a flow graph can be moved across edges**

♦ **Movement should not alter DFG functionality**

♦ **Benefits**
  – Higher speed
  – Lower power through V_{DD} scaling
  – Not very significant area increase
  – Efficient automation using polynomial-time CAD algorithms [2]

[1] C. Leiserson and J. Saxe, "Optimizing synchronous circuitry using retiming," *Algorithmica,* vol. 2, no. 3, pp. 211–216, 1991.
[2] R. Nanda, DSP Architecture Optimization in MATLAB/Simulink Environment, M.S. Thesis, University of California, Los Angeles, 2008.

11.3

Slide 11.3

Retiming, introduced by Leiserson and Saxe in [1], has become one of the most powerful techniques to obtain higher speed or lower power architectures without trading off significant area. They showed that it was possible to move registers across a circuit without changing the input-output relationship of a DFG. They developed algorithms, which could guide the movement of registers in a manner as to shorten the critical path of the graph. One key benefit of the technique is in its ability to obtain an optimal solution with algorithms that have polynomial-time complexity. As a result, most logic synthesis tools and a few high-level synthesis tools have adopted this technique [2].

Retiming

♦ **Register movement in the flow graph without functional change**

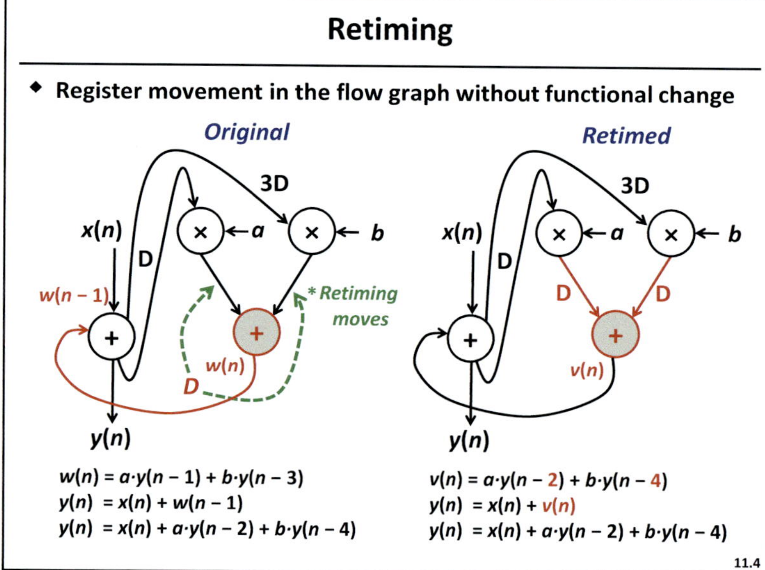

11.4

Slide 11.4

Retiming moves come from two basic results for iterative DSP algorithms:

If output $y(n)$ of a node is delayed by k units to obtain $y(n-k)$, then $y(n-k)$ can also be obtained by delaying the inputs by k units.

$$y(n - k) = x_1(n) + x_2(n) = z(n) = x_1(n - k) + x_2(n - k) \quad (11.1)$$

If all inputs to a node are delayed by at least k units, then the node output $y(n)$ can be obtained by removing k delays from all the inputs and delaying the output by k units.

$$y(n) = x_1(n - k) + x_2(n - k - 2) = g(n - k) \text{ where } g(n) = x_1(n) + x_2(n - 2) \quad (11.2)$$

This concept is illustrated in the figures where the green arrows represent the retiming moves. Instead of delaying $w(n)$ to obtain $w(n-1)$, the retimed flow-graph delays the inputs $a·y(n-1)$ and $b·y(n-3)$ by one unit to compute $v(n) = w(n-1)$.

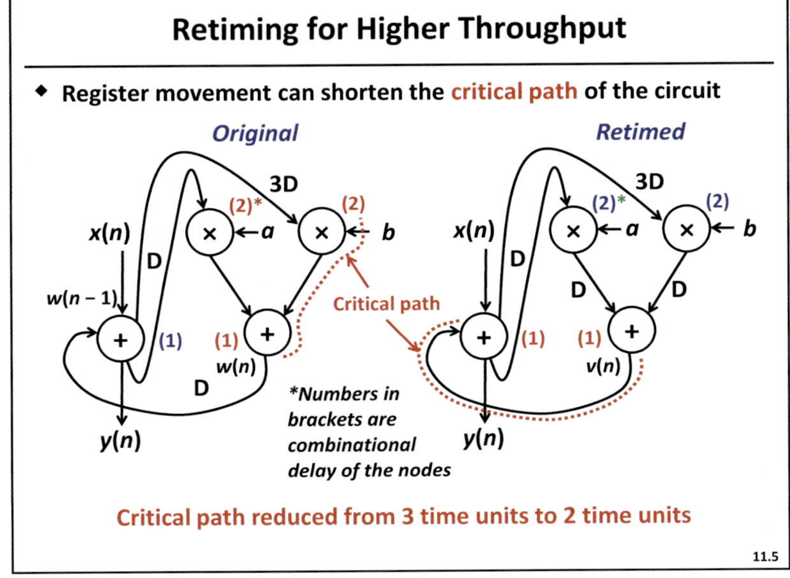

Slide 11.5

The biggest benefit of register movement in a flow graph is the possible reduction in the critical path. The maximum operating frequency is constrained by the worst-case delay between two registers. However, the original flow graph may have an asymmetrical distribution of logic between registers. By optimally placing registers in the DFG, it is possible to balance the logic depth between the registers to shorten the critical path, thus maximizing the throughput of the design. Of course, a balanced logic depth may not be possible to obtain in all cases due to the structure of the DFG itself and the restriction on the valid retiming moves. An example of a reduction in the critical-path delay is shown on the slide, where the critical path is highlighted in red. The numbers in brackets next to the nodes are the logic delays of the operations. In the original graph the critical path spans two operations (an add and a multiply). After retiming, shown in the retimed graph, the critical path is restricted to only one multiply operation.

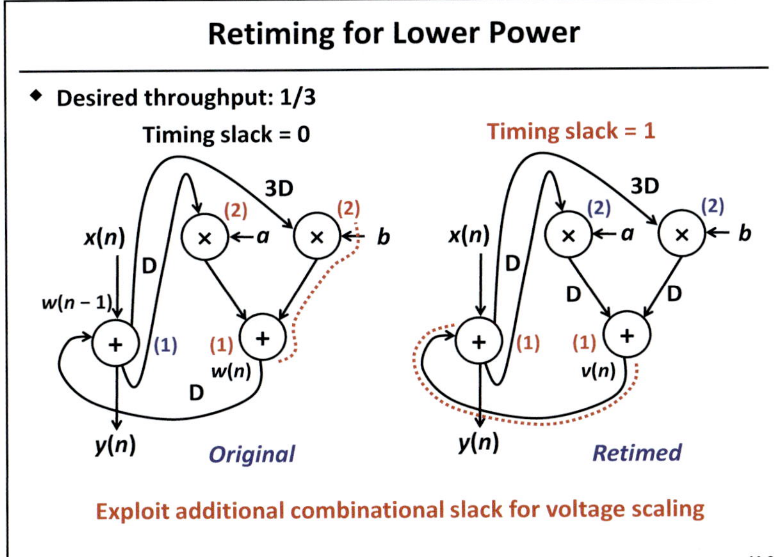

Slide 11.6

A natural extension to maximizing throughput lies in minimizing power by exploiting the additional combinational slack in the design. This slack can be leveraged for power reduction through voltage scaling. The reduction in power is quadratic with a decrease in voltage; even a small combinational slack can result in significant power reduction. Retiming at a higher level of abstraction (DFG level) can also lead to less aggressive gate sizing at the logic synthesis level. In other words, the slack achieved through retiming makes it easier to achieve the timing constraints using smaller gates. Having smaller gates results in less switching capacitance, which is beneficial from both a power and area perspective.

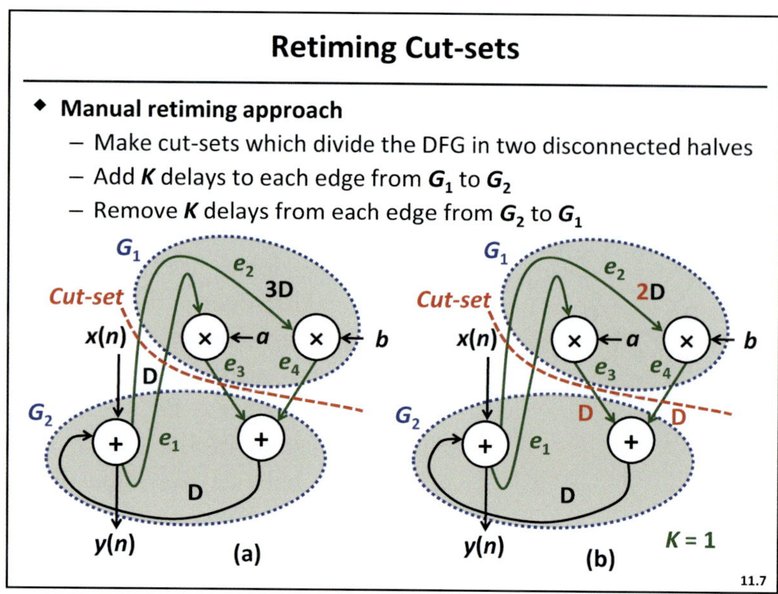

Slide 11.7

Cut-sets can be used to retime a graph manually. The graph must be divided into two completely disconnected halves through a cut. To check whether a cut is valid, remove all of the edges lying on the cut from the graph, and check if the resulting two graphs G_1 and G_2 are disjointed. A valid retiming step is to remove K ($K > 0$) delays from each edge that connects G_1 to G_2, then add K delays to each edge that connects from G_2 to G_1 or vice versa. The retiming move is valid only if none of the edges have negative delays after the move is complete. Hence, there is an upper limit on the value of K which is set by the minimum weight of the edges connecting the two sets.

In this example, a cut through the graph splits the DFG into disjointed halves G_1 and G_2. Since the two edges e_3 and e_4 from G_1 to G_2 have 0 delays on them, retiming can only remove delays from edges going from G_2 to G_1 and add delays on the edges in the opposite direction. The maximum value of K is restricted to 1 because the edge e_1 has only 1 delay on it. The retimed DFG is shown in figure (b).

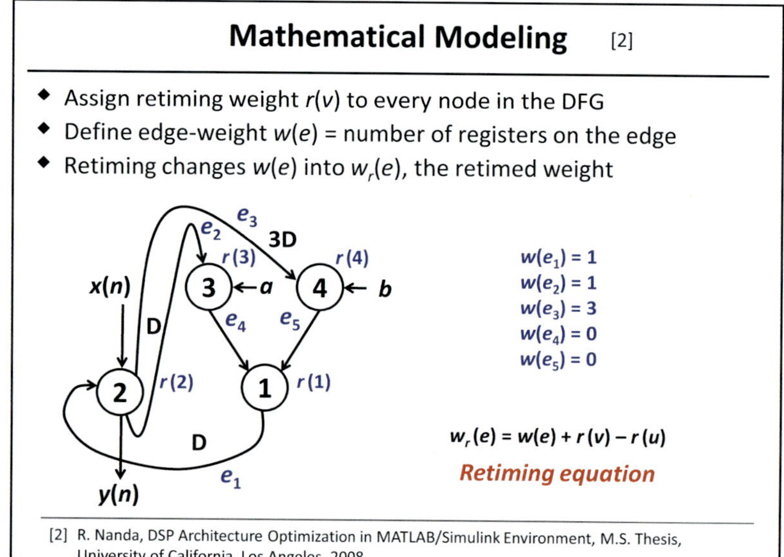

Slide 11.8

Now that we have looked at how to manually retime a graph, we discuss how retiming is automated using the model proposed in [2]. The model assigns a retiming weight $r(v_i)$ to every node v_i in the DFG. After any retiming move, an edge e: $v_1 \rightarrow v_2$ may either gain or lose delays. This gain or loss is computed by taking the difference between the retiming weights of its destination and source nodes, in this case $r(v_2) - r(v_1)$. If this difference is positive, then retiming adds delays to the edge and if negative then delays are removed from the edge. The number of delays added or subtracted is equal to $abs(r(v_2) - r(v_1))$. The new edge-weight $w_r(e)$ is given by $w(e) + r(v_2) - r(v_1)$, where $w(e)$ is the edge-weight in the original graph.

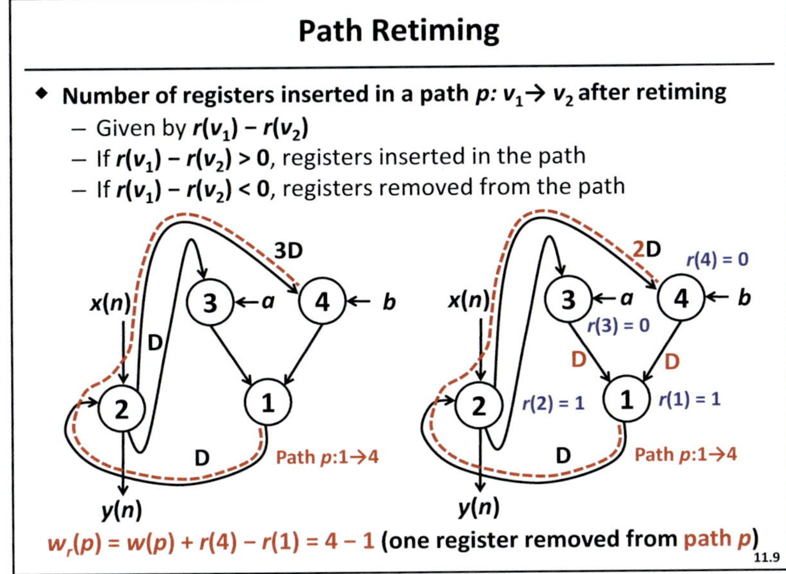

Slide 11.9

It can be shown that the properties of the retiming equation hold for paths in the same way as for the edges. For example, let $p: v_1 \rightarrow v_4$ be a path through edges $e_1(v_1 \rightarrow v_2)$, $e_2(v_2 \rightarrow v_3)$ and $e_3(v_3 \rightarrow v_4)$. The total number of delays on the path p after retiming will be given by the sum of the retimed edge weights e_1, e_2 and e_3.

$$w_r(p) = w_r(e_1) + w_r(e_2) + w_r(e_3)$$

$$w_r(p) = w(e_1) + r(v_2) - r(v_1) + w(e_2) + r(v_3) - r(v_2) + w(e_3) + r(v_4) - r(v_3)$$

$$w_r(p) = w(p) + r(v_4) - r(v_1)$$

Effectively, the number of delays on the path after retiming is given by the original path weight plus the difference between the retiming weights of the destination and source nodes of the path. Hence, we see that the retiming equation holds for path weights as well.

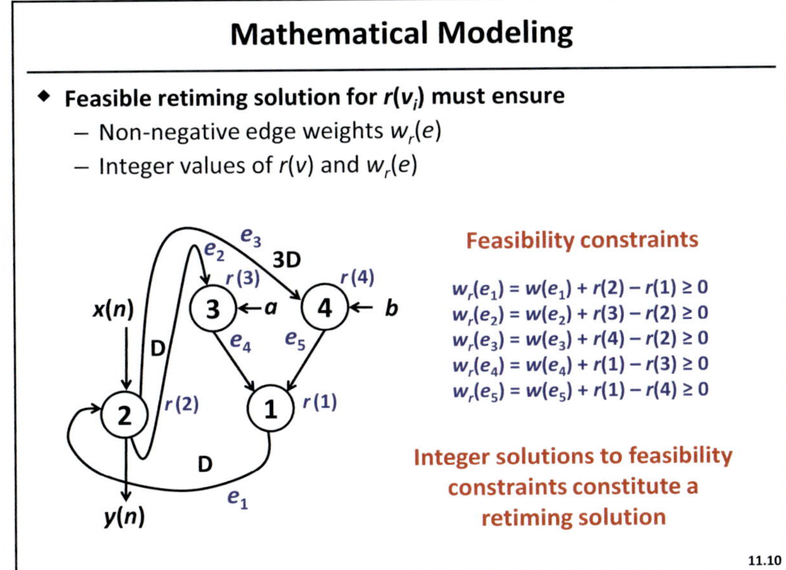

Slide 11.10

The retiming equation relates the retiming weights of the nodes to the number of delays lost or gained by a path or edge. Therefore, the retiming weights can capture register movements across the flow-graph. A more important question is ascertaining whether the register movement is valid. In other words, given a set of values for the retiming weights, how do we know that the retiming moves they represent are valid and do not violate the DFG functionality? The main constraint is imposed by the non-negativity requirement of the retimed edge-weights. This would imply a feasibility constraint of the form $w(e) + r(v_2) - r(v_1) \geq 0$ for every edge. Also, the retimed edge weights would have to be integers, which impose integer constraints on the edge weights.

Retiming with Timing Constraints

- **Find retiming solution which guarantees critical path in DFG ≤ _T_**
  - Paths with logic delay > **_T_** must have at least one register

- **Define**
  - **_W(u,v)_**: minimum number of registers over all paths b/w nodes **_u_** and **_v_**, _min {w(p) | p : u → v}_
  - If no path exists between the vertices, then **_W(u,v) = 0_**
  - **_Ld(u,v)_**: maximum logic delay over all paths b/w nodes **_u_** and **_v_**
  - If no path exists between vertices **_u_** and **_v_** then **_Ld(u,v) = −1_**

- **Constraints**
  - Non-negative weights for all edges, $W_r(v_i, v_j) \geq 0$, $\forall$ i,j
  - Look for nodes **_(u,v)_** with **_Ld(u,v) > T_**
  - Define in-equality constraint $W_r(u,v) \geq 1$ for such nodes

11.11

Slide 11.11

We now describe the problem formulation for retiming a design so that the critical path is less than or equal to a user-specified period T. The formulation should be such that it is possible to identify all paths with logic delay $> T$. These paths are critical and must have at least one register added to them. The maximum logic delay between all pairs of nodes (u, v) is called $Ld(u, v)$. If no path exists between two nodes, then the value of $Ld(u, v)$ is set to -1. In the same way, $W(u, v)$ captures the minimum number of registers over all paths between the nodes (u, v), $W(u, v)$ is set to zero, if no path exists between two nodes. Since we are dealing with directed DFGs, the value of $Ld(u, v)$ is not necessarily equal to the value of $Ld(v, u)$, so both these values must be computed separately. The same holds for $W(u, v)$ and $W(v, u)$.

Leiserson-Saxe Algorithm [1]

- **Algorithm for feasible retiming solution with timing constraints**

```
Algorithm {r(vᵢ), flag} ← Retime(G,d,T)
  k ← 1
  for u = 1 to |V|
    for v = 1 to |V| do
      if Ld(u,v) > T then
        Define inequality Iₖ : W(u,v) + r(v) − r(u) ≥ 1
      else if Ld(u,v) > −1 then
        Define inequality Iₖ : W(u,v) + r(v) − r(u) ≥ 0
      endif
      k ← k+1
    endfor
  endfor
```

[1] C. Leiserson and J. Saxe, "Optimizing synchronous circuitry using retiming," Algorithmica, vol. 2, no. 3, pp. 211-216, 1991.
[2] R. Nanda, DSP Architecture Optimization in MATLAB/Simulink Environment, M.S. Thesis, University of California, Los Angeles, 2008.

Use Bellman-Ford algorithm to solve the inequalities I_k [2]

11.12

Slide 11.12

This slide shows the algorithm proposed by Leiserson and Saxe in [1]. The algorithm scans the value of $Ld(u, v)$ between all pairs of nodes in the graph. For nodes with $Ld(u, v) > T$, an inequality of the form in (11.3) is defined,

$$W(u, v) + r(v) - r(u) \geq 1. \quad (11.3)$$

For nodes with $-1 < Ld(u, v) \leq T$, an equality of the form shown below is defined,

$$W(u, v) + r(v) - r(u) \geq 0. \quad (11.4)$$

A solution to the two sets of inequality constraints will generate a retiming solution for $r(v_i)$ which ensures the non-negativity of the edge weights and that the critical paths in the DFG have at least one register. The inequalities can be solved using polynomial-time algorithms like Bellman-Ford and Floyd-Warshall described in [2]. The overall time-complexity of this approach is bounded by $O(|V|^3)$ in the worst case. Iteratively repeating this algorithm and setting smaller values of T each time can solve the critical-path minimization problem. The algorithm fails to find a solution if T is set to a value smaller than the minimum possible critical path for the graph.

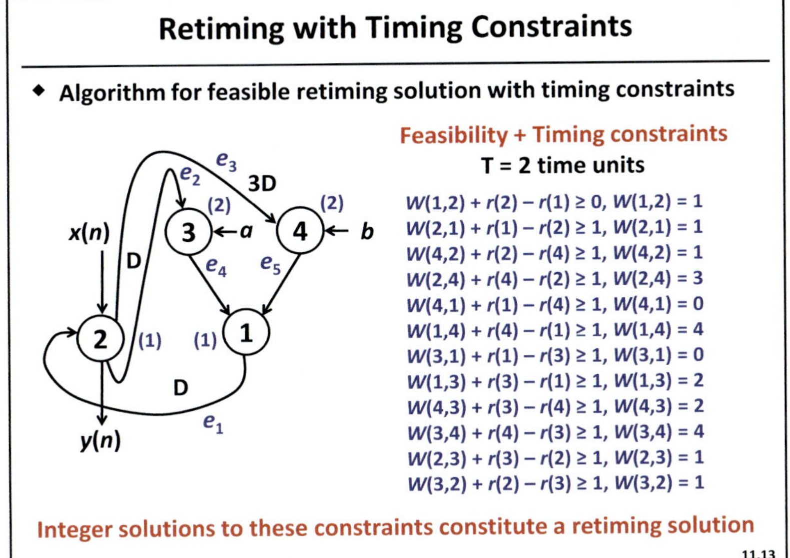

Slide 11.13

This slide shows an example run of the retiming algorithm. The timing constraint dictates that the maximum logic depth between registers should be less than or equal to 2 time units. The logic delay of nodes 1 and 2 are 1, while that of nodes 3 and 4 are 2. The first step is to compute the values of $Ld(u, v)$ and $W(u, v)$ for all pairs of nodes in the graph. From the graph, it is clear that the only non-critical path is the one with edge e_1. All other paths and edges have logic delay greater than 2. Hence the value of $Ld(u, v)$ is greater than 2 for all pairs of nodes except node-pair (1, 2). Accordingly, the inequalities are formulated to ensure the feasibility and timing constraints.

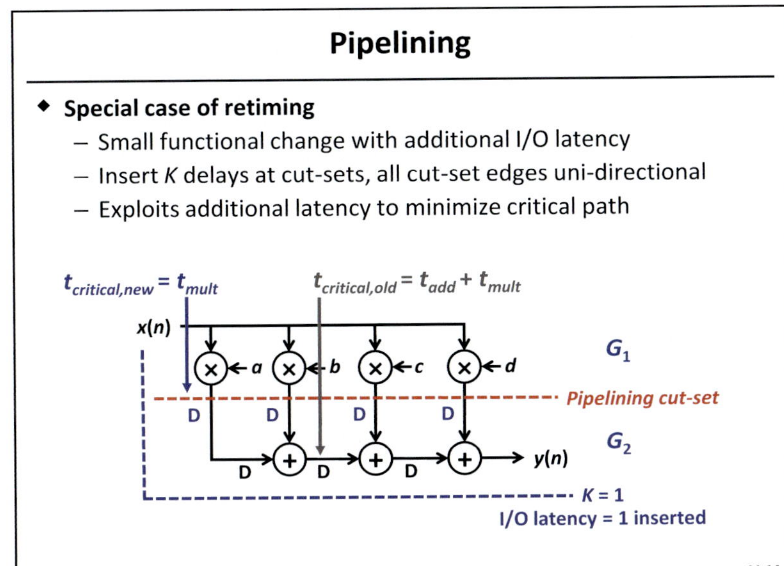

Slide 11.14

We now take a look at pipelining which can be treated as a special case of retiming. Retiming re-positions the registers in a flow-graph in order to reduce the critical path. However, to strictly maintain functional equivalence with the original graph, no additional latency can be inserted in the paths from the input to the output. Our previous discussion did not look at such a constraint. Additional inequalities will need to be generated to constrain the retiming weights so that no additional latency is inserted. In certain situations the system can tolerate additional latency. For such cases, we can add extra registers in paths from the input to the output and use them to further reduce the critical path. This extension to retiming is called pipelining.

A simple example of a pipelining cut-set is shown for a transposed FIR filter. The cut divides the graph into two groups G_1 and G_2. If K additional delays are added on the edges from G_1 to G_2, then K delays will have to be removed from edges going from G_2 to G_1. The interesting thing to note here is that no edges exist in the direction from G_2 to G_1. This graph is completely feed-forward in the direction from G_1 to G_2. Hence, there is no upper bound on the value of K. This special unconstrained cut-set is called a pipelining cut-set. With each added delay on the edges from G_1 to G_2 we add extra latency to output $y(n)$. The upper bound on the value of K comes from the maximum input-to-output (I/O) latency that can be tolerated by the FIR filter. The value of K is set to 1 in the slide. The additional registers inserted reduce the critical path from $t_{add} + t_{mult}$ to t_{mult}.

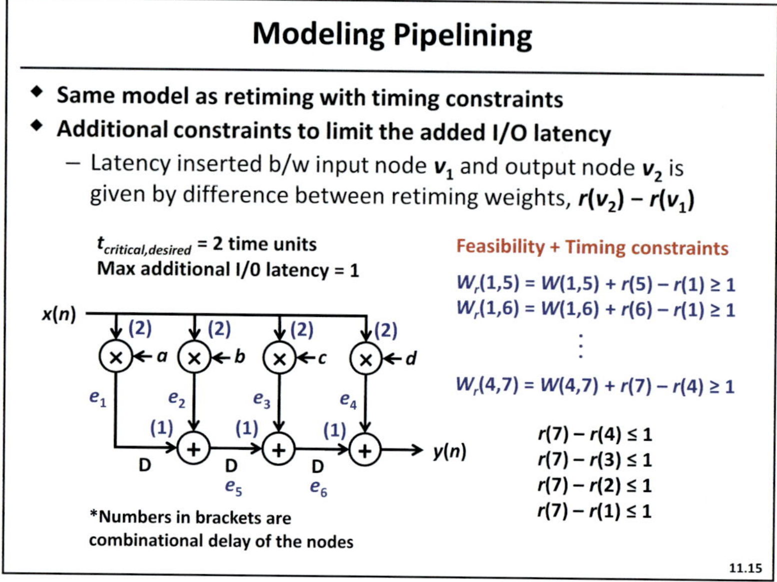

To model the pipelining algorithm, we first take a look at how the additional I/O latency insertion can be captured. We showed earlier that the retiming equation holds for paths in the DFG. That is, any additional registers inserted in a path p can be computed by the difference between $r(v)$ and $r(u)$, where v and u are the destination and source nodes of the path. So the additional I/O latency inserted in a DFG can be computed by identifying all the paths from the input to the output, then computing the difference in the retiming weights of the corresponding output and input nodes. For example, the DFG in the slide has 4 input nodes 1, 2, 3 and 4. All four nodes have a path to the single output node 7. The I/O latency inserted in each of these I/O paths can be restricted to 0 if a new set of retiming constraints is formulated:

$$r(7) - r(1) \leq 0 \qquad (11.5)$$

$$r(7) - r(2) \leq 0 \qquad (11.6)$$

$$r(7) - r(3) \leq 0 \qquad (11.7)$$

$$r(7) - r(4) \leq 0 \qquad (11.8)$$

On the other hand, if the system allows up to K units of additional latency, then the RHS of (11.5), (11.6), (11.7), and (11.8) can be replaced with K. The feasibility and timing constraints for pipelining remain the same as that of retiming. The only difference is that $K > 0$ for the latency constraints ($K = 1$ in the slide).

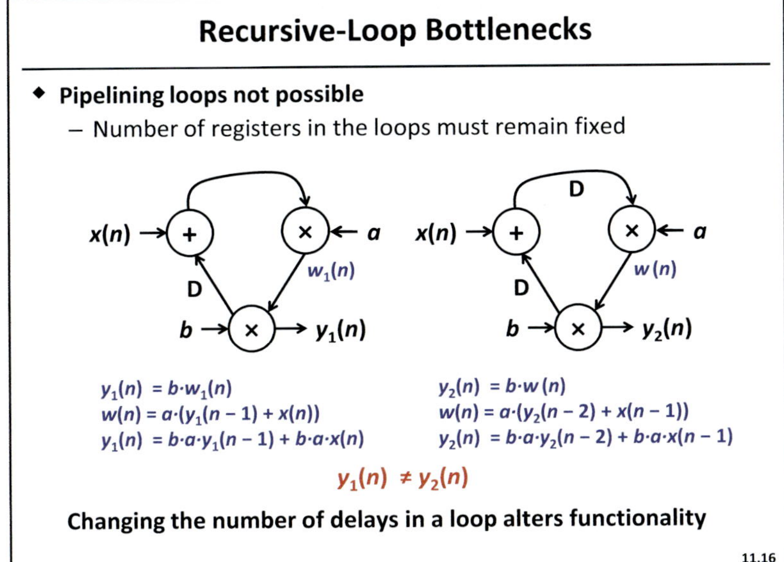

Slide 11.16

In the previous discussion it was shown that purely feed-forward sections of the flow graph make pipelining cut-sets possible. This raises an interesting question about recursive flow graphs when such cuts are not possible. Recursive flow graphs contain loops, which limit the maximum achievable critical path in the graph. It can be shown that no additional registers can be inserted in loops in a graph. Consider a loop L of a DFG to be a path starting and ending at the same node. For example, a path p going through nodes $v_1 \rightarrow v_2 \rightarrow v_3 \rightarrow v_1$ constitutes a loop. The number of registers inserted on this path after retiming is given by $r(v_1) - r(v_1) = 0$. This concept is further illustrated in the slide, where it is shown that if extra registers are inserted in a loop, the functionality would be altered. Therefore, in a recursive loop, the best retiming can do is position the existing registers to balance the logic depth between the registers. Fundamentally, the loops represent throughput bottlenecks in the design.

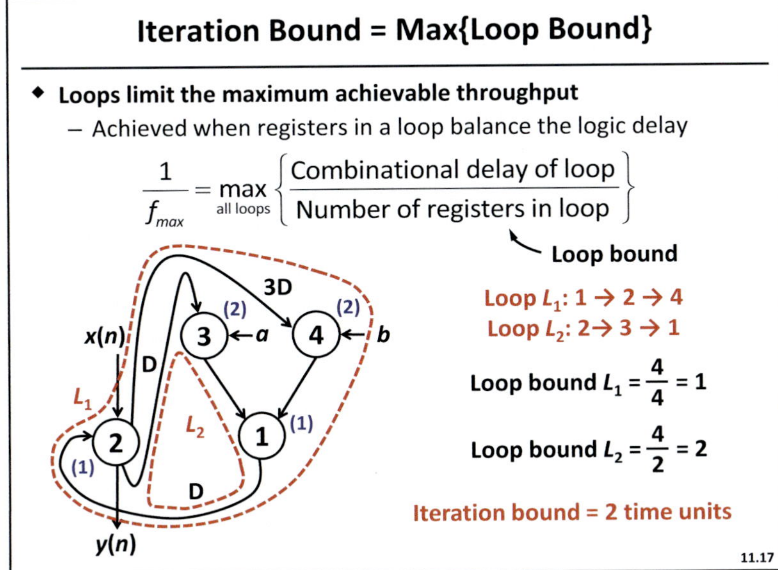

Slide 11.17

Now that we have seen how loops affect the minimum critical path achievable in a design, it is useful to know in advance what this minimum value can be. This aids in setting of the parameter T in the retiming algorithm. The term *iteration bound* was defined to compute the maximum throughput achievable in a DFG. This is a theoretical value achieved only if all of the registers in a loop can be positioned to balance the logic depth between them. The iteration bound will be determined by the slowest loop in the DFG, given by (11.9).

$$\text{Iteration bound} = \max_{\text{all loops}} \{\text{combinational delay of loop/number of register in loop}\} \qquad (11.9)$$

The example in the slide shows a DFG with two loops, L_1 and L_2. L_1 has 4 units of logic delay (logic delay of nodes 1, 2 and 4) and 4 registers in it. This makes the maximum speed of the loop equal to 1 unit time if all logic can be balanced between the available registers. However, L_2 is slower owing to the smaller number of registers in it. This sets the iteration bound to 2 time units.

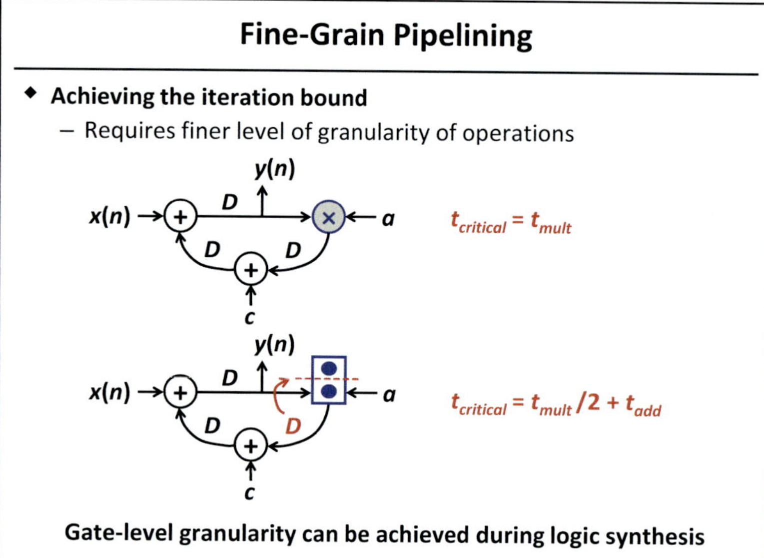

Slide 11.18

The maximum throughput set by the iteration bound is difficult to obtain unless the granularity of the DFG is very fine. At a granularity level of operations (like adders and multipliers), registers can be placed only on edges between the operations. Hence, a second limitation on the critical path will be imposed by the maximum logic-block delay among all operations. A more optimized approach is to allow registers to cut through the logic within the operations. In effect, we are reducing the granularity level of the graph from logic blocks to logic gates such as AND, OR, NOT, etc. In this scenario the registers have much more freedom to balance the logic between them. This approach is known as fine-graine pipelining, and is commonly used in logic synthesis tools where retiming is done after the gate-level netlist has been synthesized. The slide shows an example to illustrate the advantage of fine-grain pipelining. This fine level of granularity comes at the expense of tool complexity, since the number of logic cells in a design far outnumbers the number of logic blocks.

Parallelism [2]

♦ **Unfolding of the operations in a flow-graph**
  – Parallelizes the flow-graph
  – Higher speed, lower power via V_{DD} scaling
  – Larger area

♦ **Describes multiple iterations of the DFG signal flow**
  – Symbolize the multiple number of iterations by **P**
  – Unfolded DFG constructed from the following **P** equations

$$y_i = y(Pm + i), i \in \{0, 1, ..., P - 1\}$$

  – DFG takes the inputs **x(Pm), x(Pm + 1), ..., x(Pm + P – 1)**
  – Outputs are **y(Pm), y(Pm + 1), ..., y(Pm + P – 1)**

[2] K.K. Parhi, VLSI Digital Signal Processing Systems: Design and Implementation, John Wiley & Sons Inc., 1999.

11.19

Slide 11.19

We now take a look at the next architectural technique, which targets either higher throughput or lower power architectures. Like retiming, parallelism is aimed at speeding up a design to gain combinational slack and then using this slack to either increase the throughput or lower the supply voltage. Parallelism, however, is much more effective than retiming. It creates several instances of the original graph to take multiple inputs and deliver multiple outputs in parallel. The tradeoff is an increase in area in order to accommodate multiple instances of the design. For iterative DSP algorithms, parallelism can be implemented through the unfolding transformation. Given a DFG with input $x(n)$ and output $y(n)$, the parallel DFG computes the values of $y(Pm), y(Pm+1), y(Pm+2)$, ..., $y(Pm+P-1)$ given the inputs $x(Pm), x(Pm+1), x(Pm+2), ..., x(Pm+P-1)$, where P is the required degree of parallelism. The parallel outputs are recombined using a selection multiplexer.

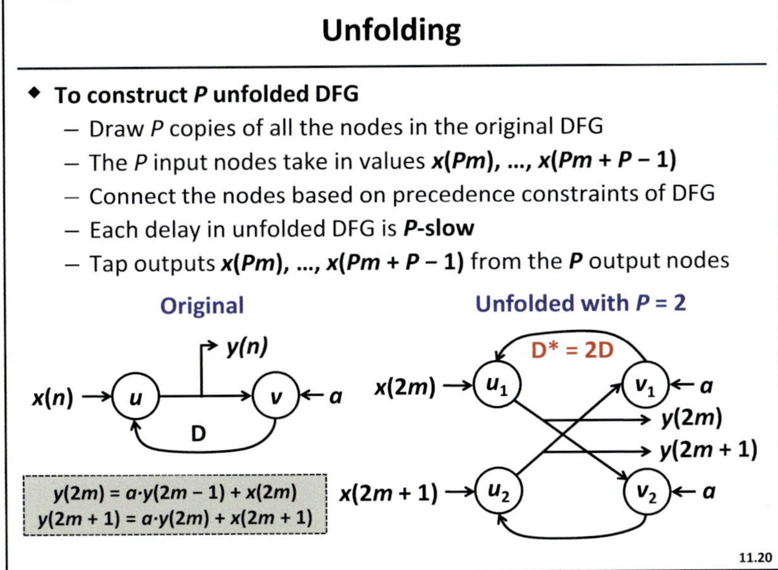

Slide 11.20

The unfolded DFG has P copies of the nodes and edges in the original DFG. Every node u in the DFG will have P copies $u_0, u_1, \ldots, u_{P-1}$ in the unfolded DFG. If input $x(n)$ is utilized by node u in the original DFG, then the node u_j takes in the input $x(Pm+j)$ in the unfolded graph. Similarly, if the output $y(n)$ is tapped from the node u in the original DFG, then outputs $y(Pm+j)$ are tapped from nodes u_j, where j takes values from $\{0, 1, 2, \ldots, P-1\}$.

Interconnecting edges without registers (intra-iteration edges) between nodes u and j will map to P edges between u_j and v_j in the unfolded graph. For interconnecting edges between u and v with D registers on it, there will be P corresponding edges between u_j and v_k where $k = (i + D)$ modulo P. Each of these edges has $(i + D)$ modulo-P registers on them. An example of a 2-unfolded DFG is shown in the figure for a second-order IIR filter.

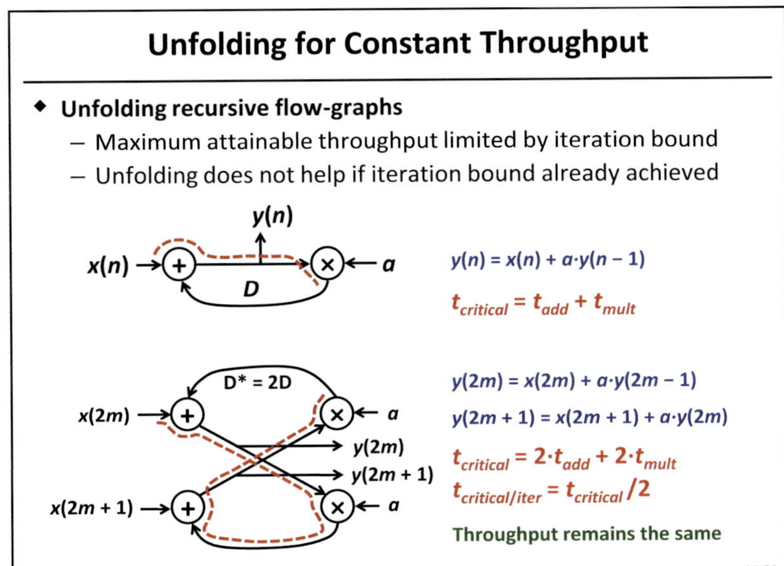

Slide 11.21

Now that we have looked at the mechanics of unfolding, a discussion on its limitations is also of significance. As discussed earlier, the maximum throughput attained by a flow graph is limited by its iteration bound. It is interesting to note that the exact same limit holds true for unfolded graphs as well. In fact, it can be shown that unfolding cannot change the iteration bound of a DFG at all. This stems from the fact that even after unfolding the graph, the logic depth in the loops increases by P times while the number of registers in them remains constant. Therefore, the throughput per iteration is still limited by the value of the iteration bound. This concept is illustrated in the slide for the second-order IIR filter example discussed previously.

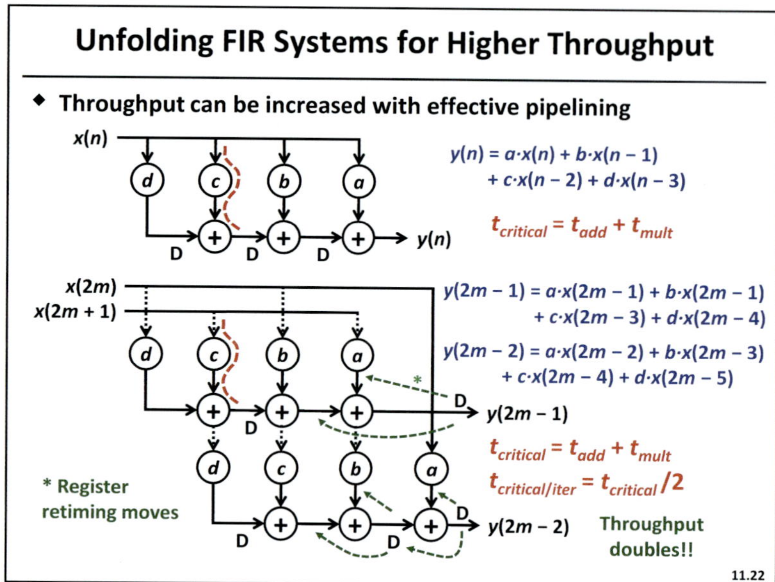

Slide 11.22

For feed-forward structures where pipeline registers can be inserted, there is no iteration bound as a throughput bottleneck. In this case, unfolding combined with pipelining can be used to increase the throughput significantly (P times in some cases). This is illustrated for the 2-unfolded FIR filters in the slide. The additional pipeline registers in the unfolded DFG are retimed (dashed green lines indicate retiming moves) to reduce the critical path. The tradeoff for this transformation lies in the additional area and I/O latency. The throughput increase by P can be used to improve the energy efficiency by an order of magnitude, as will be shown later in the chapter.

<hr>

Introduction to Scheduling

- **Dictionary definition**
  - The coordination of multiple related tasks into a time sequence
  - To solve the problem of satisfying time and resource constraints between a number of tasks

- **Data-flow-graph scheduling**
  - Data-flow-graph iteration
    - Execute all operations in a sequence
    - Sequence defined by the signal flow in the graph
  - One iteration has a finite time of execution T_{iter}
  - Constraints on T_{iter} given by throughput requirement
  - If required T_{iter} is long
    - T_{iter} can be split into several smaller clock cycles
    - Operations can be executed in these cycles
    - Operations executing in different cycles can share hardware

11.23

Slide 11.23

The final transformation discussed in this chapter will be scheduling. The data-flow graph can be treated as a sequence of operations, which have to be executed in a specified order to maintain correct functionality. In previous transformations, a single iteration of the DFG was executed per clock cycle. In the case of unfolding, P iterations are executed every clock cycle; the clock period can be slowed by a factor of P to maintain original throughput. We have seen how area can be traded-off for higher speed with unfolding. The opposite approach is to exploit the slack available in designs, which can operate faster than required. In this case, the iteration period T_{iter} of the DFG (1/throughput) is divided into several smaller clock cycles. The operations of the DFG are spread across these clock cycles while adhering to their original sequence of execution. The benefit is that the same level of operations executing in mutually exclusive clock cycles can share a single hardware resource. This leads to an area reduction since a single unit can now support multiple operations.

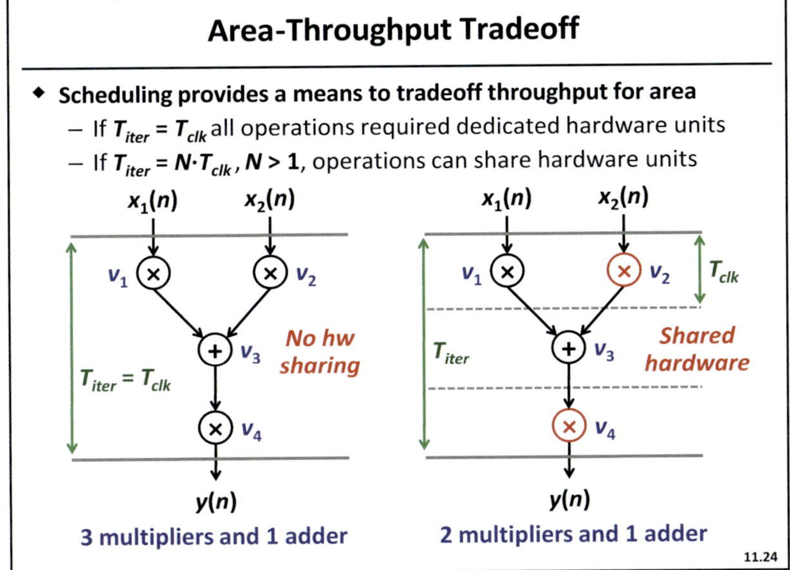

Slide 11.24

An example of scheduling is illustrated in this slide. The same DFG is implemented in two ways. On the left, each operation has a dedicated hardware unit for execution, and all operations execute simultaneously in a single cycle. A total of 3 multipliers and 1 adder will be required for this implementation. On the right, the iteration period T_{iter} is split into three clock cycles, with v_1 and v_2 executing in the first cycle, while v_3 and v_4 execute in the second and third cycle, respectively. Note that both cases maintain the same sequence of operations. Due to the mutually exclusive time of execution, v_2 and v_4 can share a multiplier unit. This brings the hardware count to 2 multipliers and 1 adder, which is one multiplier less compared to the DFG on the left. The number of clock cycles per iteration period will be referred to as N throughout the chapter. It is easy to see that larger value of N allows more operations to execute in different cycles, which leads to more area reduction. The logic depth of the hardware resources will determine minimum clock period and consequently the limit on the value of N.

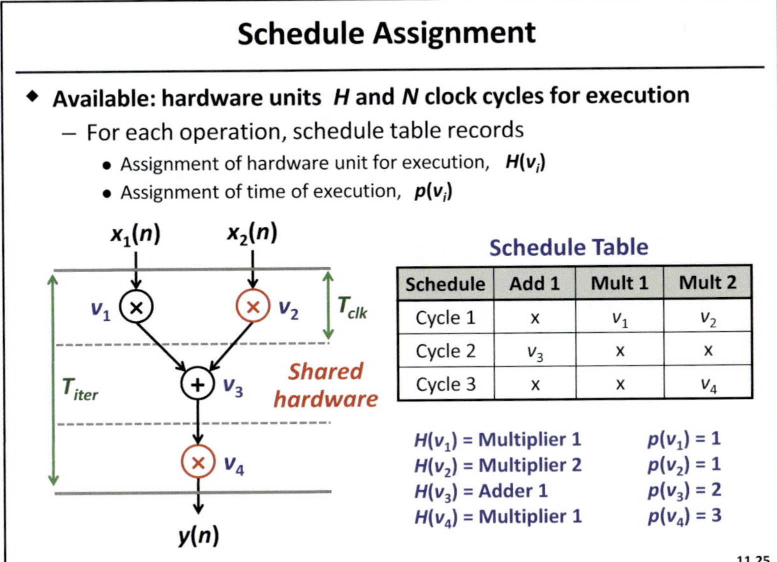

Schedule	Add 1	Mult 1	Mult 2
Cycle 1	x	v_1	v_2
Cycle 2	v_3	x	x
Cycle 3	x	x	v_4

Slide 11.25

We can formally define the scheduling problem as follows:

• For a given iteration period, set the value of clock period and $N = T_{\text{iter}}/$ clock period.

• Assign a cycle for execution $p(v_i)$, to each operation in the flow graph. The sequence of execution should maintain the DFG functionality.

• Assign a hardware unit H_i for executing the operation in its respective clock cycle. One hardware unit cannot execute more than one operation in the same cycle.

• The assignment should ensure that the operations are executed in N cycles, and the objective must be to reduce the area of the hardware units required.

The entire process can be visualized as a table with N rows representing N clock cycles. Each column of the table represents a hardware unit. If operation v executes in cycle j using hardware H_i, then the jth element of the ith column equals v. An example of such an assignment is shown in the slide for $N = 3$. An "x" entry in a column represents that the resource is free in that cycle.

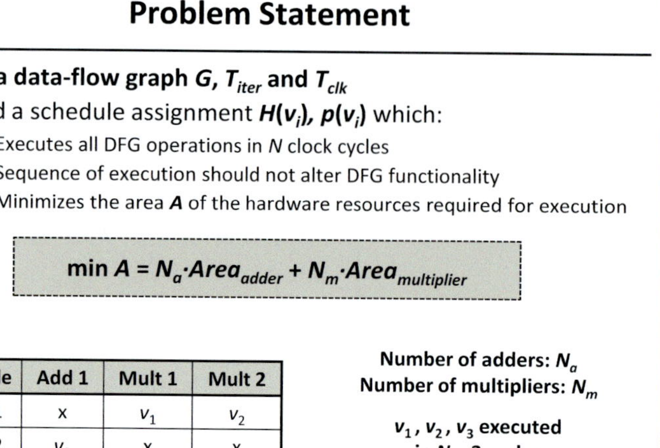

Schedule	Add 1	Mult 1	Mult 2
Cycle 1	x	v_1	v_2
Cycle 2	v_3	x	x
Cycle 3	x	x	v_4

Slide 11.26

This slide re-iterates the problem statement and defines an objective function of the scheduling phase. The area function is a weighted sum of the number of available hardware units. The weights depend on the relative area of the resource units. This scheduling problem is NP-hard in general. This means that finding the optimum schedule assignment, which minimizes area while satisfying the timing and precedence constraints, could take an infinitely long time in the worst case. Procedures like integer linear programming (ILP) have demonstrated that such an optimum can be reached regardless, with no bounds on CPU runtime. We will take a brief look at ILP modeling later. For now, we focus on simple heuristics for generating schedules, which although not fully optimal, are commonly used.

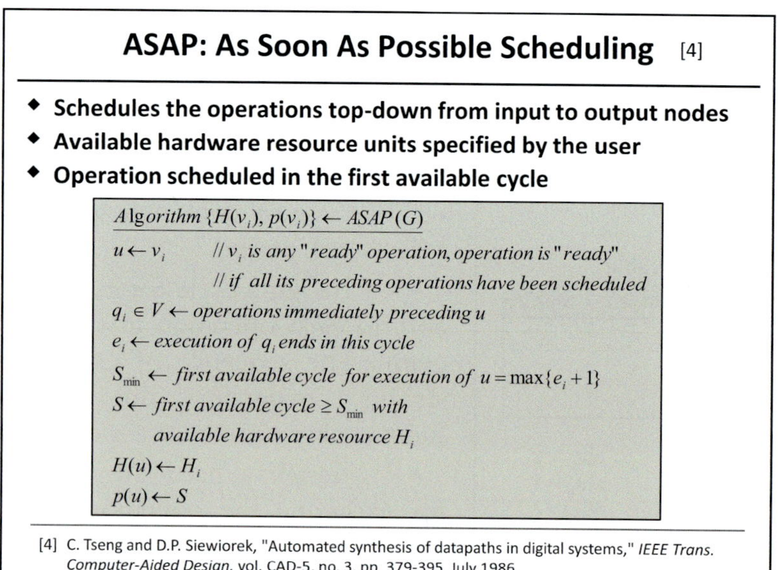

Slide 11.27

One such heuristic is known as "As Soon As Possible scheduling", also referred to as ASAP scheduling [4]. As its name suggests, this algorithm executes the operations of the DFG as soon as its operands are available and a free execution resource can be found. The user must initially specify the available hardware units. The practice is to start with a very small number of resources in the first pass of the algorithm and then increase this number if a schedule cannot be found. The algorithm looks for a "ready" node u in the list of unscheduled nodes. A "ready" operation is defined as a node whose operands are immediately available for use. Input nodes are always "ready" since incoming data is assumed to be available at all times. All other operations are ready only when their preceding operations have been completed. Next, the scheduler looks for the first possible clock cycle S_{min} where the node u can execute. If a hardware unit H_i is available in any cycle $S \geq S_{min}$, then the operation is scheduled in unit H_i at cycle

S. If no such cycle can be found, then scheduling is infeasible and the hardware resource count should be increased for the next pass of ASAP execution. This continues until all operations have been assigned an execution cycle S and a hardware unit H.

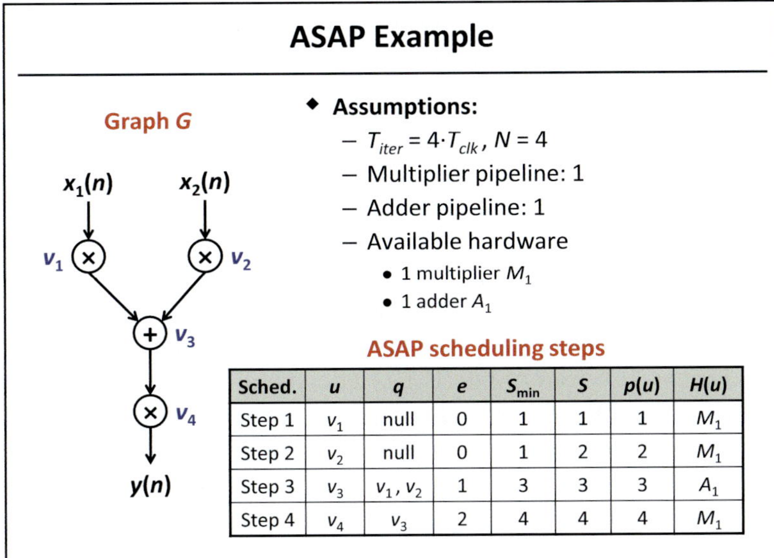

Sched.	u	q	e	S_{min}	S	$p(u)$	$H(u)$
Step 1	v_1	null	0	1	1	1	M_1
Step 2	v_2	null	0	1	2	2	M_1
Step 3	v_3	v_1, v_2	1	3	3	3	A_1
Step 4	v_4	v_3	2	4	4	4	M_1

Slide 11.28

An example of ASAP scheduling is shown for the graph G in the slide. The first "ready" operation is the input node v_1, which is followed by the node v_2. The node v_1 uses multiplier M_1 in cycle 1, which forces v_2 to execute in cycle 2. Once v_1 and v_2 are executed, node v_3 is "ready" and is assigned to cycle 3 and adder unit A_1 for execution. Following this, the operation v_4 becomes ready and is scheduled in cycle 4 and assigned to unit M_1. If $N<4$, scheduling would become infeasible with a single multiplier.

The scheduler would have to increase the hardware count for a second pass to ensure feasibility.

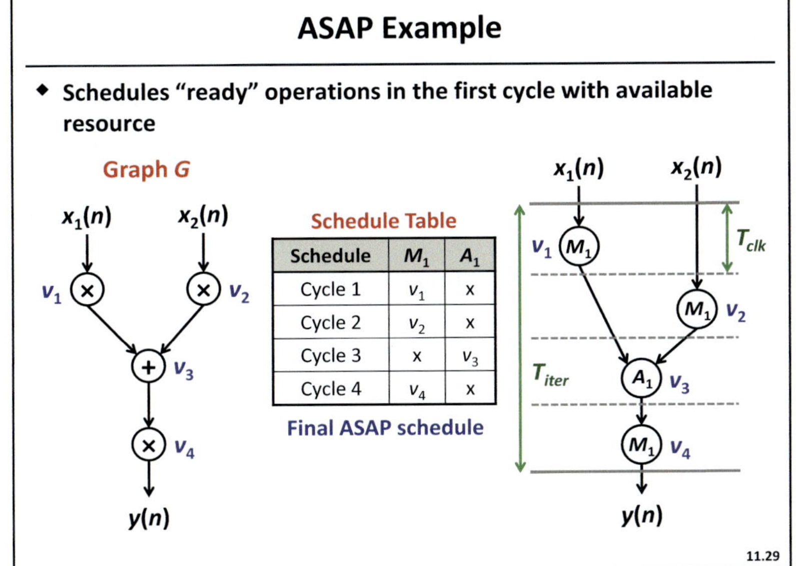

Schedule	M_1	A_1
Cycle 1	v_1	x
Cycle 2	v_2	x
Cycle 3	x	v_3
Cycle 4	v_4	x

Slide 11.29

This slide shows the scheduling table and the resultant sequence of operation execution for the ASAP scheduling steps shown in Slide 11.28.

Scheduling Algorithms

- **More heuristics**
  - Heuristics vary in their selection of next operation to scheduled
  - This selection strongly determines the quality of the schedule

  - **ALAP:** As Late As Possible scheduling
    - Similar to ASAP except operations scheduled from output to input
    - Operation "ready" if all its succeeding operations scheduled

  - **ASAP, ALAP** do not give preference to timing-critical operations
    - Can result in timing violations for fixed set of resources
    - More resource/area required to meet the T_{iter} timing constraint

  - **List scheduling**
    - Selects the next operation to be scheduled from a list
    - The list orders the operations according to timing criticality

11.30

Slide 11.30

"As Late As Possible" or ALAP scheduling is another popular heuristic similar to ASAP, the only difference is its definition of a "ready" operation. ALAP starts scheduling from the output nodes and moves along the graph towards the input nodes. Operations are considered "ready" if all its succeeding operations have been executed. Although heuristics like ASAP and ALAP are simple to implement, the quality of the generated schedules are very poor compared to optimal solutions from other formal methods like Integer Linear Programming (ILP). The main reason is that the selection of "ready" operations does not take the DFG's critical path into account. Timing-critical operations, if scheduled later, tend to delay the execution of a number of other operations. When the execution delays become infeasible, then the scheduler has to add extra resources to be able to meet the timing constraints. To overcome this drawback, an algorithm called "list scheduling" was proposed in [5]. This scheduler selects the next operation to be scheduled from a list, in which operations are ranked in descending order of timing criticality.

List Scheduling   [5]

- **Assign precedence height $P_H(v_i)$ to each operation**
  - $P_H(v_i)$ = length of longest combinational path rooted by v_i
  - Schedule operations in descending order of precedence height

$x_1(n)$ $x_2(n)$ $x_3(n)$ $t_{add} = 1, t_{mult} = 2$

$P_H(v_1) = T(v_4) = 1$
$P_H(v_2) = T(v_5) + T(v_6) = 3$
$P_H(v_3) = T(v_5) + T(v_6) = 3$
$P_H(v_5) = T(v_6) = 2$
$P_H(v_4) = 0, P_H(v_6) = 0$

Possible scheduling sequence	
ASAP	$v_1 \rightarrow v_2 \rightarrow v_3 \rightarrow v_4 \rightarrow v_5 \rightarrow v_6$
LIST	$v_3 \rightarrow v_2 \rightarrow v_5 \rightarrow v_1 \rightarrow v_4 \rightarrow v_6$

[5] S. Davidson *et. al.*, "Some experiments in local microcode compaction for horizontal machines," *IEEE Trans. Computers*, vol. C-30, no. 7, pp. 460-477, July 1981.

11.31

Slide 11.31

The LIST scheduling process is illustrated in this slide. The first step is to sort the operations based on descending their precedence height P_H. The precedence height of a node v is defined as the longest combinational path, which starts at v. In the figure, the longest path from node v_3 goes through $v_3 \rightarrow v_5 \rightarrow v_6$. The length of the path is defined by the sum of the logic delays, which is $2t_{mult} + t_{add} = 3$ in this case, making $P_H(v_3) = 3$. Once all operations have been sorted, the procedure is similar to ASAP scheduling. The second step is to schedule operations in descending order of P_H. The scheduling is done in the listed order in the first available cycle with a free hardware resource. The table in the slide shows different scheduling orders for ASAP and LIST. Note how LIST scheduling prefers to schedule node v_3 first, since it is the most timing critical ($P_H(v_3) = 3$). We will see the effect of this ordering on the final result in the next two slides.

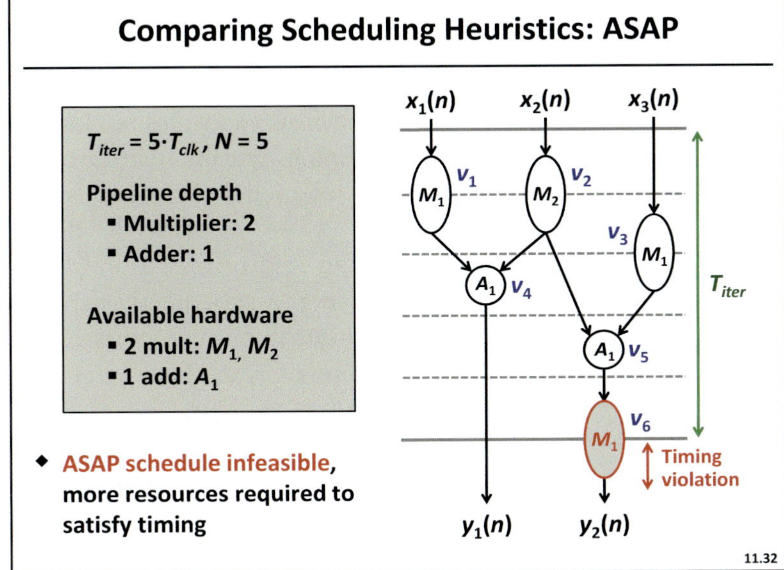

Slide 11.32

This slide shows ASAP scheduling example considering pipeline depth of the operators for $N = 5$. The adder and multiplier units have pipeline depth of 1 and 2, respectively. The clock period is set to the delay of a single adder unit. Hardware resources initially available are 2 multipliers and 1 adder. It is clear that scheduling v_3 later forces it to execute in cycle 2, which delays the execution of v_5 and v_6. Consequently, we face timing violations at the end when trying to schedule v_6. In such a scenario the scheduler will add an extra multiplier to the available resource set and re-start the process.

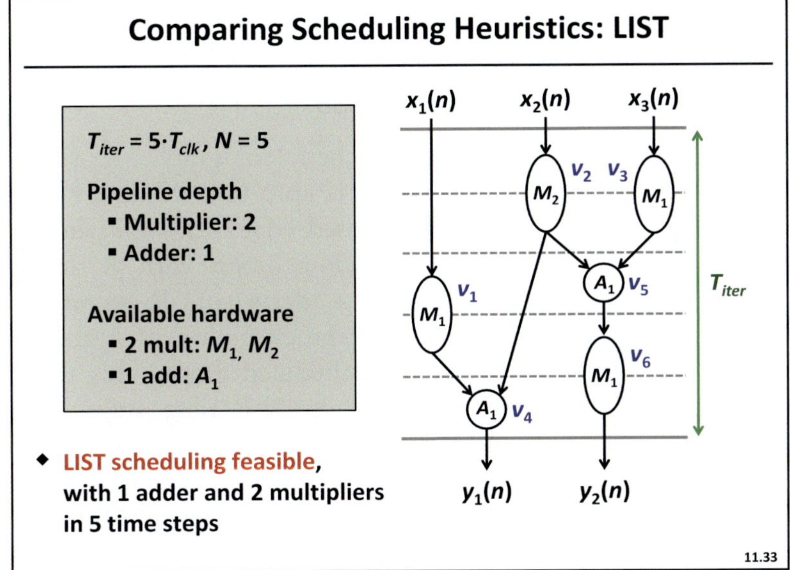

Slide 11.33

For the same example, the LIST scheduler performs better. Node v_3 is scheduled at the start of the scheduling process and assigned to cycle 1 for execution. Node v_1 is delayed to execute in cycle 3, but this does not cause any timing violations. On the other hand, scheduling v_3 first allows v_5 and v_6 to execute in the third and fourth cycle, respectively, making the schedule feasible. Notice that M_1 is active for both operations v_1 and v_6 in cycle 4. While v_6 uses the first pipeline stage of M_1, v_1 uses the second. Other scheduling heuristics include force-directed scheduling, details of which are not discussed here.

Inter-Iteration Edges: Timing Constraints

- Edge $e : v_1 \rightarrow v_2$ with zero delay forces precedence constraints
  - Result of operation v_1 is input to operation v_2 in an iteration
  - Execution of v_1 must precede the execution of v_2

- Edge $e : v_1 \rightarrow v_2$ with delays represent relaxed timing constraints
  - If R delays present on edge e
  - Output of v_1 in I^{th} iteration is input to v_2 in $(I + R)^{th}$ iteration
  - v_1 not constrained to execute before v_2 in the Ith iteration

- Delay insertion after scheduling
  - Use folding equations to compute the number of delays/registers to be inserted on the edges after scheduling

11.34

Slide 11.34

In the previous examples we looked at scheduling graphs without any memory elements on the edges. Edge e: $v_1 \rightarrow v_2$ with R delays requires an appropriate number of registers in the scheduled graph to satisfy inter-iteration constraints. The output of v_1 in the Ith iteration is the input to v_2 in the $(I + R)^{th}$ iteration when there are R registers on the edge e. This relaxes the timing constraints on the execution of v_2 once operation v_1 is completed. An example of this scenario is shown in the next slide. The number of registers to be inserted on the edge in the scheduled graph is given by the folding equations to be discussed later in the chapter.

Inter-Iteration Edges

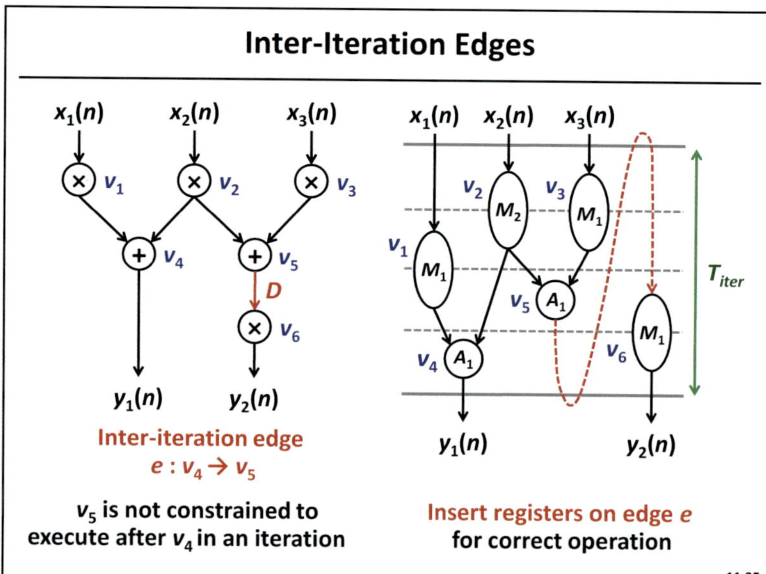

11.35

Slide 11.35

In a scheduled graph, a single iteration is split into N time steps. A registered edge, e: $v_1 \rightarrow v_2$, in the original flow graph with R delays implies a delay of R iterations between the execution of v_1 and v_2. In the scheduled graph, this delay translates to $N \cdot R$ time steps. The slide shows an example of scheduling with inter-iteration edge constraints. The original DFG on the left has one edge (e: $v_5 \rightarrow v_6$) with a single delay element D on it. In the scheduled graph (for $N=4$) on the right, we see that operations v_5 and v_6 start execution in the third time step. However, the output of v_5 must go to the input of v_6 only after $N=4$ time steps. This is ensured by inserting 4 register delays on this edge in the scheduled graph. The exact number of delays to be inserted on the scheduled edges can change; this occurs if the nodes were scheduled in different time steps or if the source node is pipelined and takes more than one time step to complete execution. The exact equations for the number of delay elements are derived in the next section.

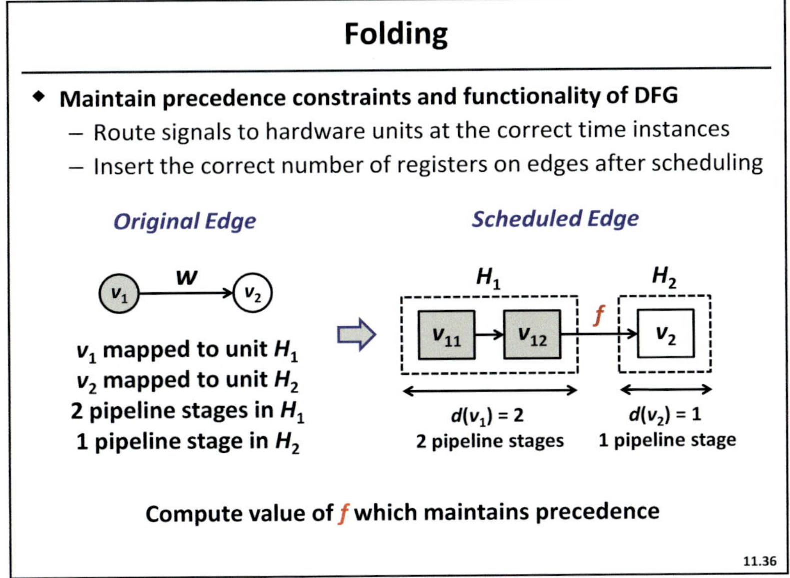

Slide 11.36

Folding equations ensure that precedence constraints and functionality of the original DSP algorithm are maintained in the scheduled architecture. The objective is to ensure that operations get executed on the hardware units at the correct time instance. This mainly requires that operands are routed to their respective resource units and that a correct number of registers are inserted on the edges of the scheduled graph. An example of a folded edge is shown in the slide.

The original DFG edge shown on the left is mapped to the edge shown on the right in a scheduled implementation. Operation v_1 is executed on H_1, which has two pipeline stages. Operation v_2 is executed on hardware H_2. To ensure correct timing behavior, the register delay f on the scheduled graph must be computed correctly so the inter-iteration constraint on the original edge is satisfied. The number of registers f is computed using folding equations that are discussed next.

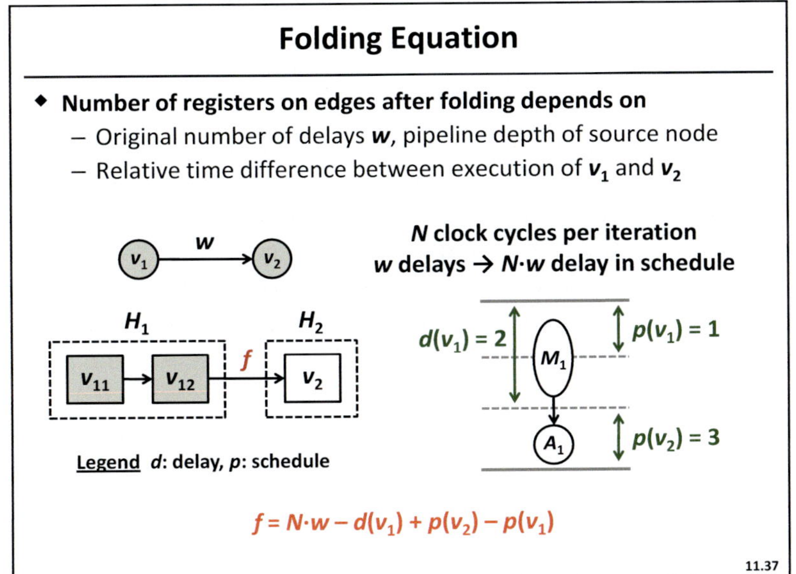

Slide 11.37

The number of registers f on a folded edge depends on three parameters: the delay w on the edge in the original DFG, the pipeline depth $d(v)$ of the source node and the relative difference between the times of execution of the source and destination nodes in the scheduled graph. We saw earlier that a delayed edge in the original flow-graph is equivalent to a delay of one iteration which translates to N time steps in the scheduled graph. Therefore a factor of $N \cdot w$ appears in the folding equation for f shown at the bottom of the slide. If the source node v_1 is scheduled to execute in time step $p(v_1)$, while v_2 executes in time step $p(v_2)$, then the difference between the two also contributes to additional delay, as shown in the scheduled edge on the right. Hence, we see the term $p(v_2) - p(v_1)$ in the equation. Finally, pipeline registers $d(v_1)$ in the hardware unit that executes the source node introduce extra registers on the scheduled edge. Hence $d(v_1)$ is subtracted from $N \cdot w + p(v_2) - p(v_1)$ to give the final folding equation. In the scheduled edge on the right, we see that the source node v_1 takes two time steps to complete execution and has two pipeline registers ($d(v_1) = 2$). Also, the difference between $p(v_1)$ and $p(v_2)$ is 2 for this edge, bringing the overall number of registers to $N\cdot w$ $+2-2=N\cdot w$. Each scheduled edge in the DFG must use this equation to compute the correct number of delays on it to ensure that precedence constraints are not violated.

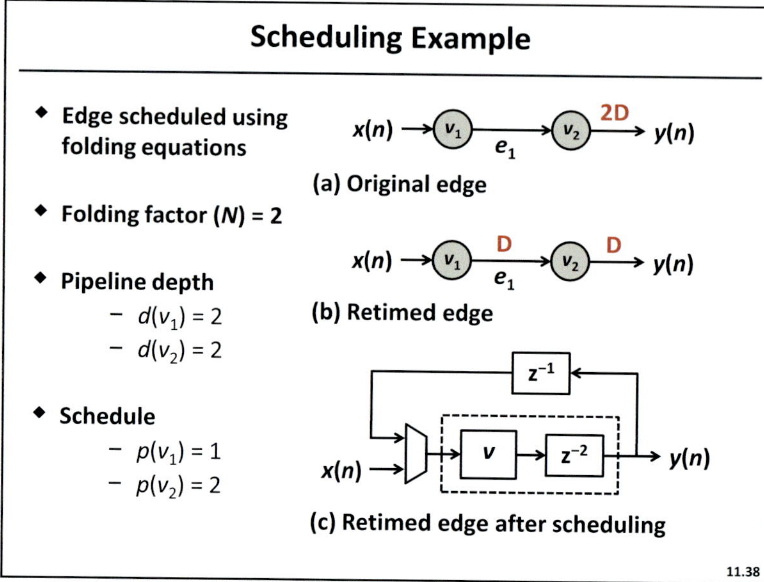

Slide 11.38

This slide shows an example of the architecture after an edge in a DFG has been scheduled. Nodes v_1 and v_2 represent operations of the same type which have to be scheduled onto the single available resource v in $N=2$ time steps. The edge is first retimed to insert a register on it (Figure (b)). The retimed edge is scheduled such that operation v_1 is scheduled in the first time step $(p(v_1)=1)$ while v_2 is scheduled in the second time step $(p(v_2)=2)$. The processing element v is pipelined by two stages $(d(v_1)=d(v_2)$ $=2)$ to satisfy throughput constraints. We get the following folding equation for the delays on edge e_1.

$$f = N\cdot w + p(v_2) - p(v_1) - d(v_1) = 2\cdot 1 + 2 - 1 - 2 = 1$$

The scheduled edge in Figure (c) shows the routing of the incoming signal $x(n)$ to resource v. The multiplexer chooses $x(n)$ as the output in the first cycle when operation v_1 is executed. The output of v_1 is fed back to v after a single register delay on the feedback path to account for the computed delay f. The multiplexer chooses the feedback signal as the output in the second cycle when the operation v_2 is executed. Scheduled architectures have a datapath, consisting of the resource units used to execute operations, and a control path, consisting of multiplexers and registers that maintain edge constraints. Sometimes the overhead of muxes and delay elements strongly negates the gains obtained from resource sharing in time-multiplexed architectures.

Efficient Retiming & Scheduling

- **Retiming with scheduling**
  - Additional degree of freedom associated with register movement results in less area or higher throughput schedules
- **Challenge: Retiming with scheduling**
  - Time complexity increases if retiming done with scheduling
- **Approach: Low-complexity retiming solution**
  - Pre-process data flow graph (DFG) prior to scheduling
  - Retiming algorithm converges quickly (polynomial time)
  - Time-multiplexed DSP designs can achieve faster throughput
  - Min-period retiming can result in reduced area as well
- **Result: Performance improvement**
  - An order of magnitude reduction in the worst-case time-complexity
  - Near-optimal solutions in most cases

11.39

Slide 11.39

The use of retiming during the scheduling process can improve the area, throughput, or energy efficiency of the resultant schedules. As we have seen earlier, retiming solutions are generated through the Leiserson-Saxe [1] algorithm, which in itself is an Integer Linear Programming (ILP) framework. To incorporate retiming within the scheduler, the brute-force approach would be to resort to an ILP framework for scheduling as well, and then introduce retiming variables in the folding equations. However, the ILP approach to scheduling is known to be NP-complete and can take exponential time to converge.

If retiming is done simultaneously with scheduling, the time complexity worsens further due to an increased number of variables in the ILP. To mitigate this issue, we can separate scheduling and retiming tasks. The idea is to first use the Bellman-Ford algorithm to solve the folding equations and generate the retiming solution. Following this, scheduling can be done using the ILP framework. This approach, called "Scheduling with BF retiming" in the next slide, has better convergence compared to an integrated ILP model, where scheduling and retiming are done simultaneously. Another approach is to pre-process the DFG prior to scheduling to ensure a near-optimal and time-efficient solution. Following this, scheduling can be performed using heuristics like ASAP, ALAP or list scheduling. The latter approach has polynomial complexity and is automated with the architecture optimization tool flow discussed in Chap. 12.

Results: Area and Runtime

DSP Design	N	Scheduling (current)		Scheduling (ILP) with BF retiming		Scheduling with pre-processed retiming	
		Area	CPU(s)	Area	CPU(s)	Area	CPU(s)
Wave filter	16	NA	NA	8	264	14	0.39
	17	13	0.20	7	777	8	0.73
Lattice filter	2	NA	NA	41	0.26	41	0.20
	4	NA	NA	23	0.30	23	0.28
8-point DCT	3	NA	NA	41	0.26	41	0.21
	4	NA	NA	28	0.40	28	0.39

NA – scheduling infeasible without retiming

Near-optimal solutions at significantly reduced worst-case runtime

11.40

Slide 11.40

The slide presents results of several scheduling and retiming approaches. We compare traditional scheduling, scheduling with Bellman-Ford retiming and scheduling with pre-processed retiming. ILP scheduling with BF retiming results in the most area-efficient schedules. This method, however, suffers from poor worst-case time complexity. The last method (scheduling with pre-processed retiming) is the most time-efficient and yields a very-close-to-optimal solution in all cases, as shown in the table. Up to a 100-times improvement in worst-case runtime is achieved over the Bellman-Ford method for an $N=17$ wave filter. The pre-processing method is completely decoupled from the scheduler, which means that we are no longer restricted to using the ILP models that have high time complexity. In the pre-processing phase the DFG is first retimed with the objective of reducing the hardware requirement in the scheduling phase.

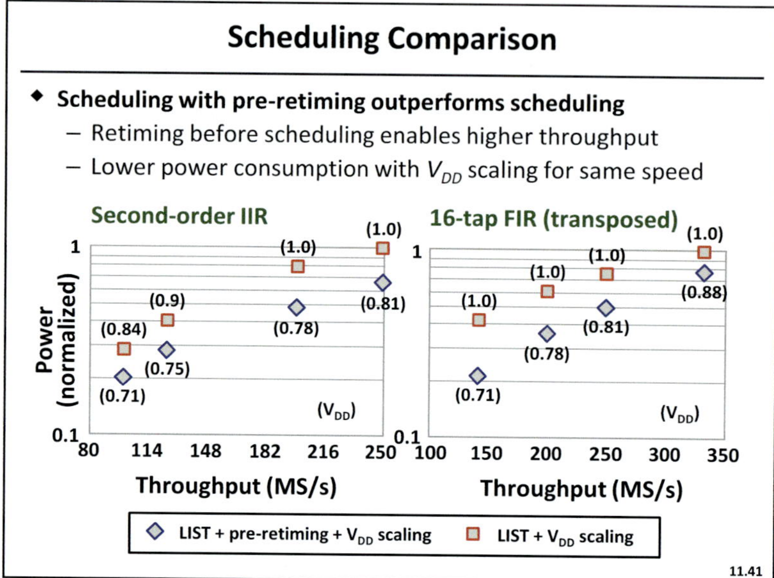

Slide 11.41

Exploring new methods to model architectural transformations, we show that a pre-processing phase with retiming can assist the scheduling process to improve throughput, area, and power. The proposed pre-processing is decoupled from the scheduler, which implies that we are no longer restricted to using the ILP models that have high time complexity. For standard benchmark examples (IIR and FIR filters), this pre-processing scheme can yield area improvements of more than 30%, over 2x throughput improvement, and power reduction beyond 50% using V_{DD} scaling. (Note: the scheduler attempts to minimize area, so power may increase for very low throughput designs, because the controller overhead becomes significant for small structures such as these filters.)

Slide 11.42

In summary, this chapter covered algorithms used to automate transformations such as pipelining, retiming, parallelism, and time multiplexing. The Leiserson-Saxe algorithm for retiming, unfolding for parallelism and various scheduling algorithms are described. In Chap. 12, we will discuss a MATLAB/Simulink based design flow for architecture optimization. This flow allows flexibility in the choice of architectural parameters and also outputs the optimized architectures as Simulink models.

References

- C. Leiserson and J. Saxe, "Optimizing Synchronous Circuitry using Retiming," *Algorithmica,* vol. 2, no. 3, pp. 211–216, 1991.

- R. Nanda, DSP Architecture Optimization in Matlab/Simulink Environment, M.S. Thesis, University of California, Los Angeles, 2008.

- K.K. Parhi, VLSI Digital Signal Processing Systems: Design and Implementation, John Wiley & Sons Inc., 1999.

- C. Tseng and D.P. Siewiorek, "Automated Synthesis of Datapaths in Digital Systems," *IEEE Trans. Computer-Aided Design,* vol. CAD-5, no. 3, pp. 379–395, July 1986.

- S. Davidson *et al.*, "Some Experiments in Local Microcode Compaction for Horizontal Machines," *IEEE Trans. Computers,* vol. C-30, no. 7, pp. 460–477, July 1981.

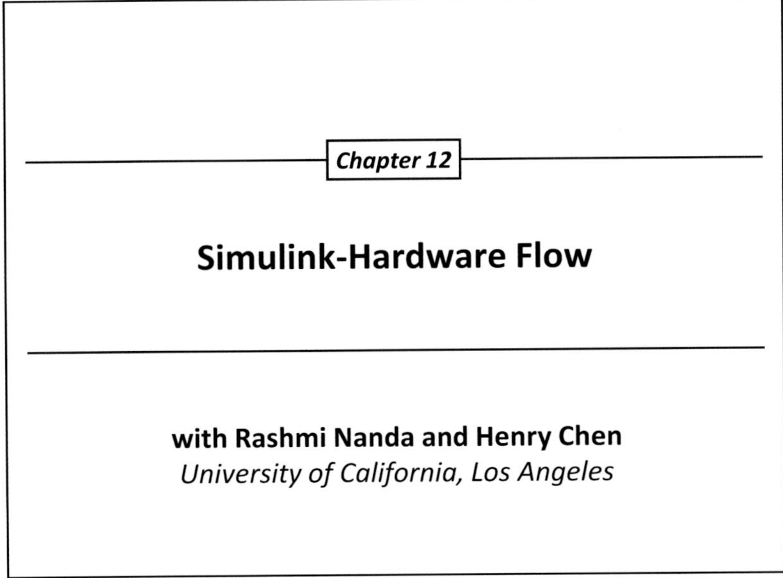

Slide 12.1

Having looked at algorithms for scheduling and retiming, we will now discuss an integrated design flow, which leverages these automation algorithms to create a user-friendly optimization framework. The flow is intended to address key challenges of ASIC design in scaled technologies: design complexity and design flexibility. Additionally, the design challenges are further underscored by complex and cumbersome verification and debugging processes. This chapter will present an FPGA-based methodology to manage the complexity of ASIC architecture design and verification.

Slide 12.2

Traditional ASIC development is partitioned among multiple engineering teams, which specialize in various aspects from algorithm development to circuit implementation. Propagating design changes across the abstraction layers is very challenging because the design has to be re-entered multiple times. For example, algorithm designers typically work with MATLAB or C. This description is then refined for fixed-point accuracy, mapped to an architecture in the RTL format for ASIC synthesis, and finally test vectors are adapted to the logic analysis hardware for final verification. The problem is that each translation requires an equivalence check between the descriptions, which is the key reason for the increasing cost of ASIC designs. One small logic error could cost months of production delay and a significant financial cost.

The MATLAB/Simulink environment can conveniently represent all design abstraction layers, which allows for algorithm, architecture, and circuit development within a unified description. System architects greatly benefit from improved visibility into the basic implementation tradeoffs early in the algorithm development. This way we can not only explore the different limitations of mapping algorithms to silicon, but also greatly reduce the cost of verification.

D. Marković and R.W. Brodersen, *DSP Architecture Design Essentials*, Electrical Engineering Essentials,
DOI 10.1007/978-1-4419-9660-2_12, © Springer Science+Business Media New York 2012

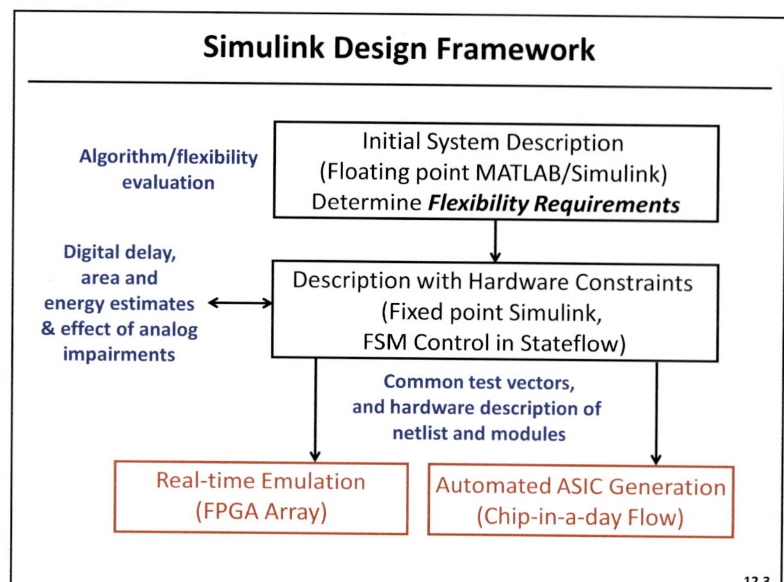

Slide 12.3

In Simulink, we can evaluate the impact of technology and architecture changes by observing simple implementation tradeoffs. The Simulink design editor is used for the entire design flow: (1) to define the system and determine the required level of flexibility, (2) to verify the system in a single environment that has a direct implementation path, (3) to obtain estimates of the implementation costs (area, delay, energy) and (4) to optimize the system architecture. This involves partitioning the system and modeling the implementation imperfections (e.g. analog distortions, A/D accuracy, finite wordlength). We can then perform real-time verification of the system description with prototyped analog hardware and, finally, automatically generate digital hardware from the system description.

Here is an example design flow. We describe the algorithm using floating-point for the initial description. At this point, we are not interested in the implementation; we are just exploring how well the algorithm works relative to real environments. Determining the flexibility requirements is another critical issue, because it will drive a large portion of the rest of the design effort. It could make few orders of magnitude difference in energy efficiency, area efficiency, and cost. Next, we take an algorithm and model it into an architecture, with constraints and fixed-point wordlengths. We decide how complex different pieces of hardware are, how to control them properly, and how to integrate different blocks. After the architectural model, we need an implementation path in order to know how well everything works, how much area, how much power is consumed, etc. There are two implementation paths we are going to discuss: one is through programming an FPGA; the other through building an ASIC.

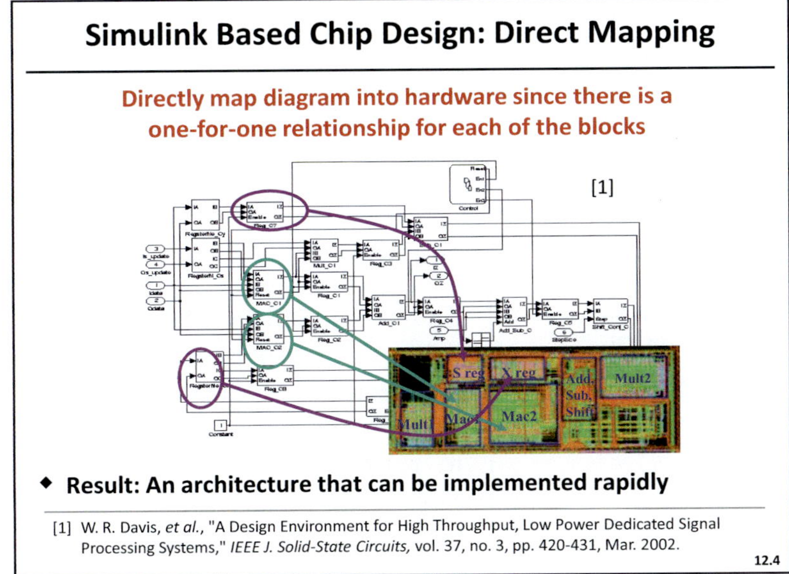

Slide 12.4

The Simulink chip-design approach was introduced in the early 2000s in the work by Rhett Davis et al. [1]. They proposed a 1-to-1 mapping between Simulink fixed-point blocks and hardware macros. Custom tools were developed to elaborate Simulink MDL model into an EDIF (electronic design interchange format) description. The ASIC backend was fully automated within the Simulink-to-silicon hierarchical tool flow. A key challenge in this work is to verify that the Simulink library blocks are hardware-equivalent: bit-true and cycle accurate.

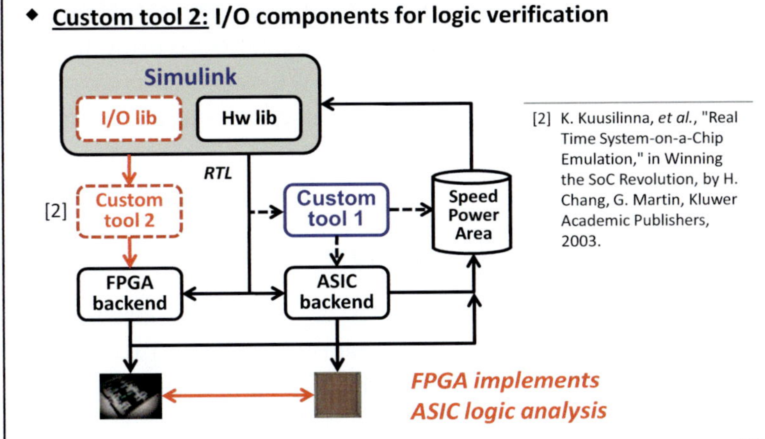

Slide 12.5

To address the issue of hardware accuracy, Kimmo Kuusilinna and others at the Berkeley Wireless Research Center (BWRC) extended the Simulink-to-silicon framework to include hardware emulation of a Simulink design on an FGPA [2]. The emulation is simply done by using the Xilinx hardware library and toolflow for FPGA mapping. A key component of this work was another custom tool that translates the RTL produced by Simulink into a language suitable for commercial ASIC backend tools. The tool also invokes post-synthesis logic-level simulation to confirm I/O equivalence between ASIC and FPGA descriptions. Architecture feedback about speed, power, and area is propagated to Simulink to further refine the architecture.

Using this methodology, we can close the loop by performing input/output verification of a fabricated ASIC, using an FPGA. Given functional equivalence between the two hardware platforms, we use FPGAs to perform real-time at-speed ASIC verification. Blocks from a custom I/O library are incorporated into an automated FPGA flow to enable a user-friendly test interface controlled entirely from MATLAB and Simulink. Effectively, we are building ASIC logic analysis functionality on an FPGA. This framework greatly simplifies the verification and debugging processes.

Custom tools can be developed for various optimization routines. These include automated wordlength optimization to minimize hardware utilization (custom tool 1), and a library of components to program electrical interfaces between the FPGA and custom chips (custom tool 2). Taken together, the out-of-box and custom tools provide a unified environment for design entry, optimized implementation, and functional verification.

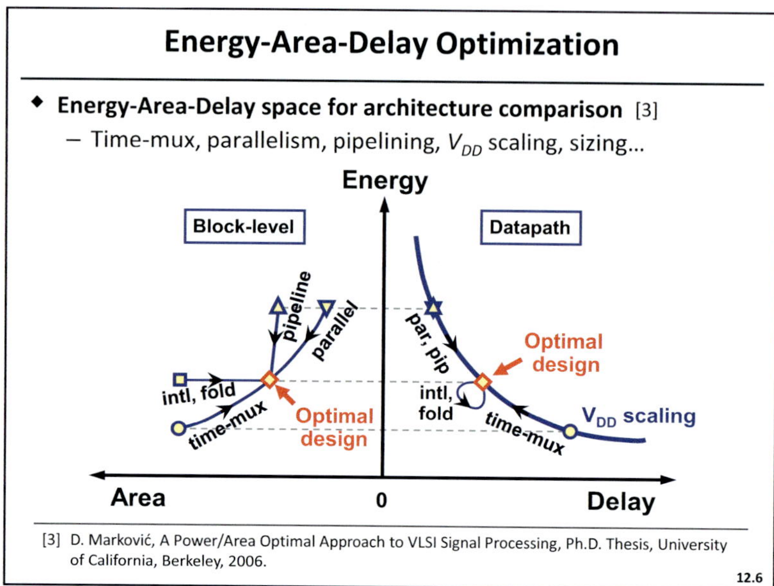

Slide 12.6

As discussed in Part II, the energy-delay of a datapath is used for architectural comparisons in the energy-area space [3]. The goal is to reach an optimal design point. For example, parallelism and pipelining relax the delay constraint to reduce energy at the expense of increased area. Time-multiplexing requires faster logic to tradeoff reduced area for an increased energy. Interleaving and folding introduce simultaneous pipelining and up-sampling to stay approximately at the same energy-delay point while reducing the area. Moving along the voltage scaling curve has to be done in such as way as to balance sensitivities to other variables, such as gate sizing. We can also incorporate wordlength optimization and register retiming. These techniques will be used to guide automated architecture optimization.

Slide 12.7

The idea behind architecture optimization is to provide an automated algorithm-to-hardware flow. Taking a systematic architecture evaluation approach illustrated in the previous slide, an algorithm block-based model can be transformed to RTL in a highly automated fashion. There could be many feasible architectural solutions that meet the performance requirements. The user can then determine the solution that best fits the target application. For example, estimates for the DSP sub-system can help system designers properly (re)allocate hardware resources to analog and mixed-signal components to minimize energy and area costs of the overall system. An automated approach helps explore many possible architectural realizations of the DSP and also provides faster turn-around time for the implementation.

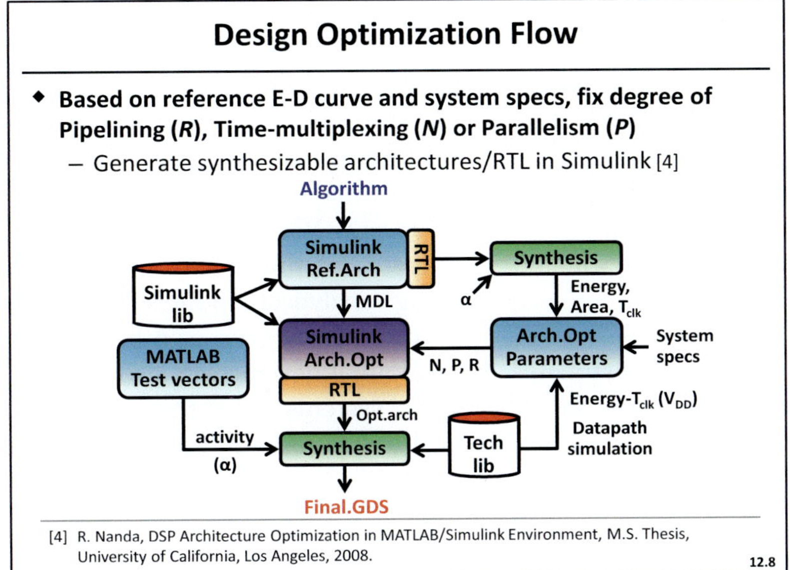

Slide 12.8

An optimization flow for automating architectural transformations is detailed in this slide [4]. The flow starts from the direct-mapped block diagram representation of the algorithm in Simulink. This corresponds to the Simulink reference architecture in the flow chart. RTL for the Simulink model (MDL) is generated using Synplify DSP or XSG and synthesized to obtain energy, area and performance estimates. If these estimates satisfy the desired specifications, no further optimization is needed. If the estimates fall short of the desired specs, then the architecture optimization parameters are set. These parameters are defined by the degree of time multiplexing (N), the degree of parallelism (P) and pipelining (R). The parameter values are set based on optimized targets of the design. For example, to trade-off throughput and minimize area, the time multiplexing parameter N must be increased. Alternately, if the speed of the design has to be increased or the power reduced through voltage scaling then pipelining or parallelism must be employed. Trends for scaling power and delay with reduced supply voltage and architectural parameters (N, P, R) are obtained from the pre-characterized energy-delay sensitivity curves in the technology library (Tech. Lib in the flow chart).

The optimization parameters, along with the Simulink reference model, are now input to the architecture optimization phase. This phase uses the DFG automation algorithms discussed earlier to optimize the reference models. This optimizer first extracts the DFG matrices from the reference model in MATLAB. The matrices are then transformed depending on the optimization parameters. The transformed matrices are converted into a new Simulink model which corresponds to the optimized architecture in the flow chart. The optimized model can be further synthesized to check whether it satisfies the desired specifications. The architectural parameters can be iterated until the specifications are met. Switching activities extracted from MATLAB test vectors serve as inputs to the synthesis tool for accurate power estimation. Optimization results using this flow for an FIR filter are presented in the next slide.

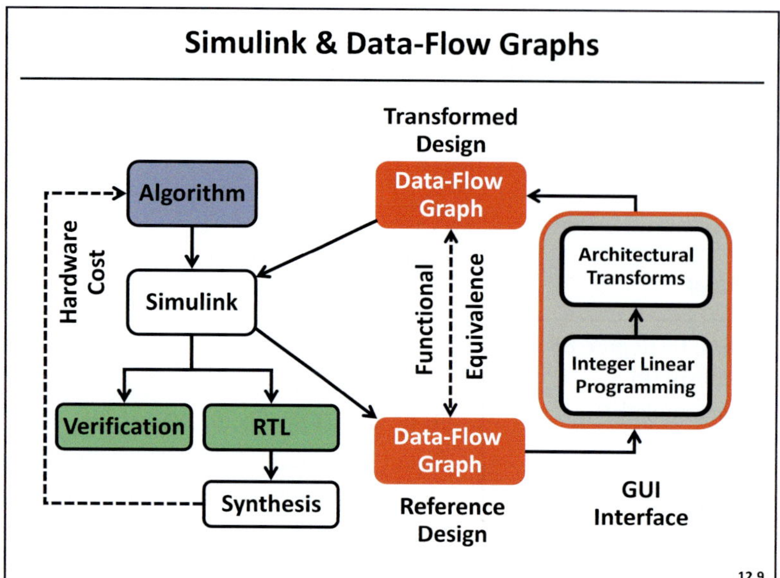

Slide 12.9

The architecture optimization approach is illustrated here. We first generate hardware estimates for a direct-mapped design to establish a reference point. The reference design is modeled by a data-flow graph (DFG) as described in Chap. 9. The DFG can be transformed to many possible realizations by using architectural transformations described in Chap. 11. We can estimate energy and area of the transformed design using analytical models for sub-blocks, or we can synthesize the design using backend tools to obtain more refined estimates. The resulting designs can also serve as hierarchical blocks for rapid design exploration of more complex systems. The transformation routines described in Chap. 11 can be presented within a graphical user interface (GUI) as will be shown later in this chapter.

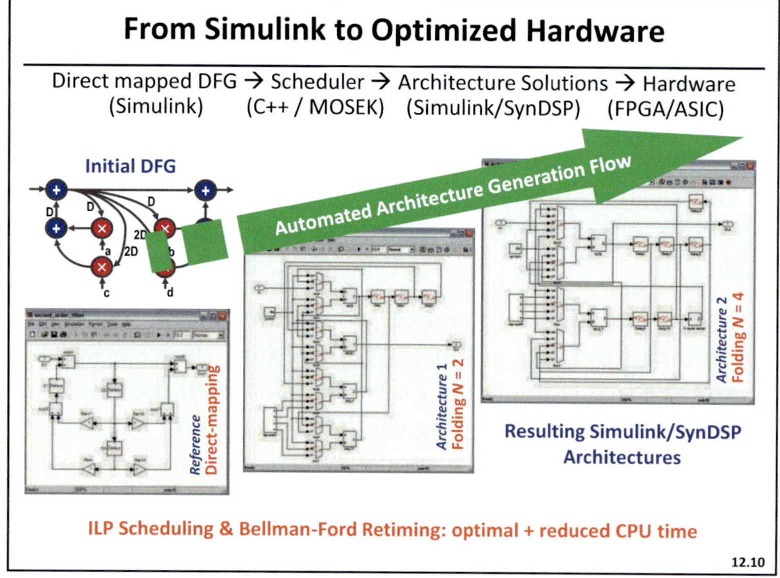

Slide 12.10

Architecture transformations will be demonstrated using the software that is publicly available on the book website. The software is based on LIST scheduling and has a GUI for ease of demonstration. It supports only data-flow graph structures with adders and multipliers. Users are encouraged to write further extensions for control-flow graphs. The tool has been tested on Matlab 2007b and SynDSP 3.6. The authors welcome comments, suggestions, and further tool extensions.

Tool Demo (Source Code Available) [5]

[5] See the book website for tool download.

- ◆ **GUI based demo of filter structures**
  - Software tested using MATLAB 2007b and SynDSP 3.6
- ◆ **The tool works only for the following Simulink models**
  - SynDSP models
  - Single input, single output models
  - Models that use adders and multipliers, no control structures
- ◆ **Usage instructions**
  - Unzip the the .rar files all into a single directory (e.g. E:\Tool)
  - Start MATLAB
  - Make E:\Tool your home directory
  - The folder E:\Tool\models has all the relevant SynDSP models
  - On the command line type: ArchTran_GUI
- ◆ **Check the readme file in E:\Tool\docs for instructions**
- ◆ **Examples shown in the next few slides**

12.11

Slide 12.11

We can formalize the process of architecture selection by automatically generating all feasible solutions within the given area, power and performance constraints. A system designer only needs to create a simple direct-mapped architecture using the Simulink fixed-point library. This information is used to extract a data-flow graph, which is used in scheduling and retiming routines. The output is a set of synthesis-ready architecture solutions. These architectures can be mapped onto any Simulink hardware library. This example shows mapping of a second-order digital filter onto the Synplify DSP library from Synplicity and resulting architectures with different levels of folding. Control logic that facilitates dataflow is also automatically generated. This way, we can quickly explore many architectures and choose the one that best meets our area, power, and performance constraints.

Using GUI for Transformation

- ◆ **Direct mapped DFG (Simulink/SynDSP model)**
- ◆ **Use GUI to open Simulink model from drop down menu**

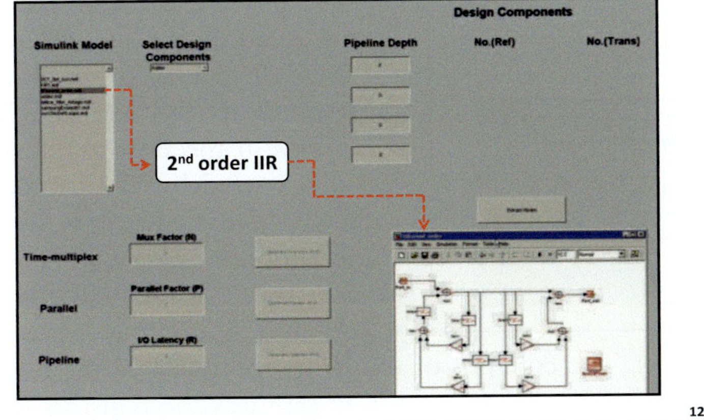

12.12

Slide 12.12

The entire process of architectural transformations using the CAD algorithms described in Chap. 11 has been embedded in a graphical user interface (GUI) within the MATLAB environment. This was done to facilitate architectural optimization from direct-mapped Simulink models. A snapshot of the GUI framework is shown. A drop-down menu allows the user to select the desired Simulink model to be transformed. In the figure a 2nd-order IIR filter has been selected, and on the right is the corresponding Simulink model for it.

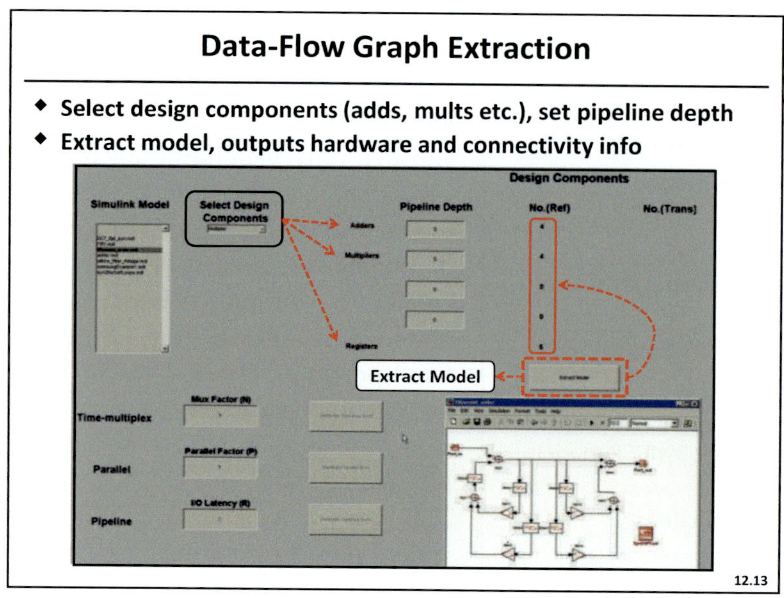

Slide 12.13

Following model selection, the user must choose the components and macros present in the model. These components can be simple datapath units like adders and multipliers or more complex units like radix-2 butterfly or CORDIC. The selected component names appear on the right side of the GUI (indicated by the arrows in the slide). If the macros or datapath units are to be pipelined implementations, then the user can set the pipeline depth of these components in the dialog box adjacent to their names. Once this information is entered, the tool is ready to extract the data-flow-graph netlist from the Simulink model. Clicking the "extract model" button accomplishes this task. The GUI displays the number of design components of each type (adders and multipliers in this example) present in the reference model ("**No.(Ref)**" field shown in the red box).

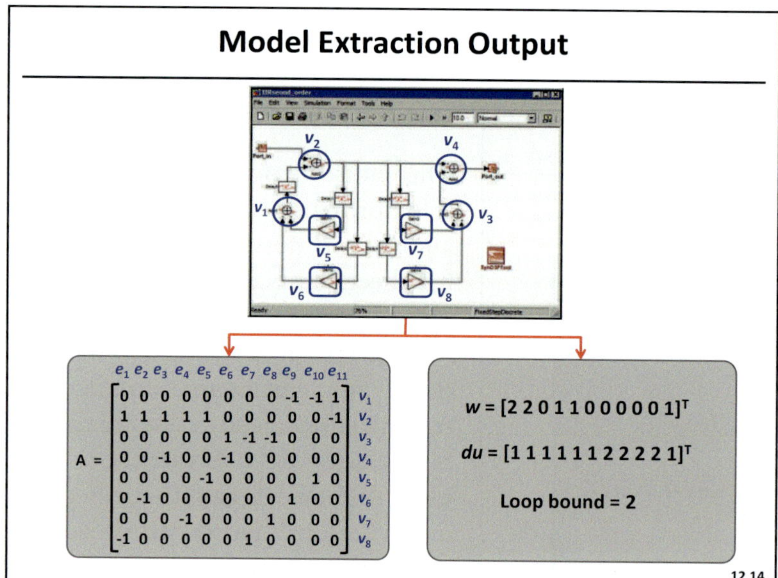

$$A = \begin{bmatrix} e_1 & e_2 & e_3 & e_4 & e_5 & e_6 & e_7 & e_8 & e_9 & e_{10} & e_{11} \\ 0 & 0 & 0 & 0 & 0 & 0 & 0 & -1 & -1 & 1 & \\ 1 & 1 & 1 & 1 & 1 & 0 & 0 & 0 & 0 & 0 & -1 \\ 0 & 0 & 0 & 0 & 0 & 1 & -1 & -1 & 0 & 0 & 0 \\ 0 & 0 & -1 & 0 & 0 & -1 & 0 & 0 & 0 & 0 & 0 \\ 0 & 0 & 0 & 0 & -1 & 0 & 0 & 0 & 0 & 1 & 0 \\ 0 & -1 & 0 & 0 & 0 & 0 & 0 & 1 & 0 & 0 & 0 \\ 0 & 0 & 0 & -1 & 0 & 0 & 1 & 0 & 0 & 0 & 0 \\ -1 & 0 & 0 & 0 & 0 & 0 & 1 & 0 & 0 & 0 & 0 \end{bmatrix} \begin{matrix} v_1 \\ v_2 \\ v_3 \\ v_4 \\ v_5 \\ v_6 \\ v_7 \\ v_8 \end{matrix}$$

$$w = [2\,2\,0\,1\,1\,0\,0\,0\,0\,0\,1]^T$$

$$du = [1\,1\,1\,1\,1\,1\,2\,2\,2\,2\,1]^T$$

Loop bound = 2

Slide 12.14

The slide shows the netlist information generated after model extraction. The tool generates the netlist in the form of the incidence matrix A, the weight matrix w, the pipeline matrix du and the loop bound given by the slowest loop in the DFG. The rows of the A matrix correspond to the connectivity of the nodes in the flow-graph of the Simulink model. The nodes v_i shown in the figure represent the computational elements in the model. The 2nd-order IIR filter has 4 add and 4 multiply operations, each corresponding to a node in the A matrix. The columns of the A matrix represent the dataflow edges in the model, as explained in Chap. 10. The weight vector w captures the registers on the dataflow edges and has the same number of columns as the matrix A. The pipeline vector du stores the number of pipeline stages in each node while loop bound computes the minimum possible delay of the slowest loop in the model.

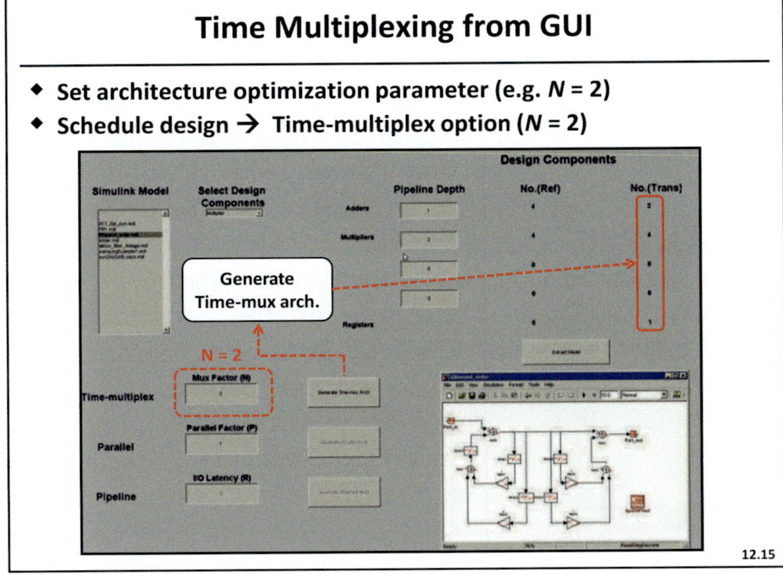

Slide 12.15

After the netlist extraction phase, the data-flow-graph model can be transformed to various other functionally equivalent architectures using the algorithms described in Chap. 11. The GUI environment supports these transformations via a push-button flow. Examples of these with the 2nd-order IIR filter as the baseline architecture will be shown next. In this slide we see the time-multiplexing option turned on with a folding factor of $N=2$. Once the Mux factor (N) is entered in the dialog box, the "Generate Time-mux arch." option is enabled. Pushing this button will schedule the DFG in $N=2$ clock cycles in this example. The GUI displays the number of design components in the transformed architecture after time-multiplexing ("**No.(Trans)**" field shown in the *red box*).

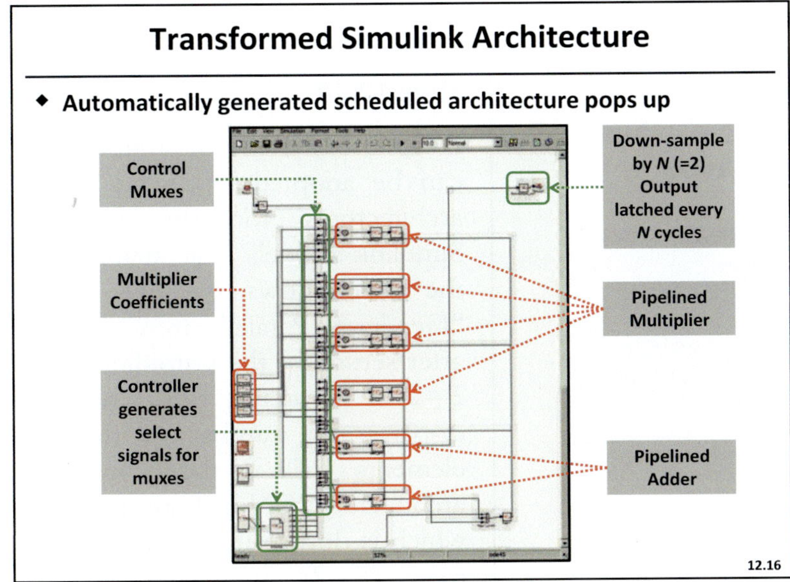

Slide 12.16

This slide shows a snapshot of the architecture generated after time-multiplexing the 2nd-order IIR by a factor $N=2$. The Simulink model including all control units and pipeline registers for the transformed architecture is generated automatically. The control unit consists of the muxes which route the input signals to the datapath resources and a finite state machine based controller, which generates the select signals for the input multiplexors. The datapath units are pipelined by the user-defined pipeline stages for each component. The tool also extracts the multiplier coefficients from the reference DFG and routes them as predefined constants to the multiply units. Since the original DFG is folded by a factor of N, the transformed model produces an output every N cycles, requiring a down-sample-by-N block at the output.

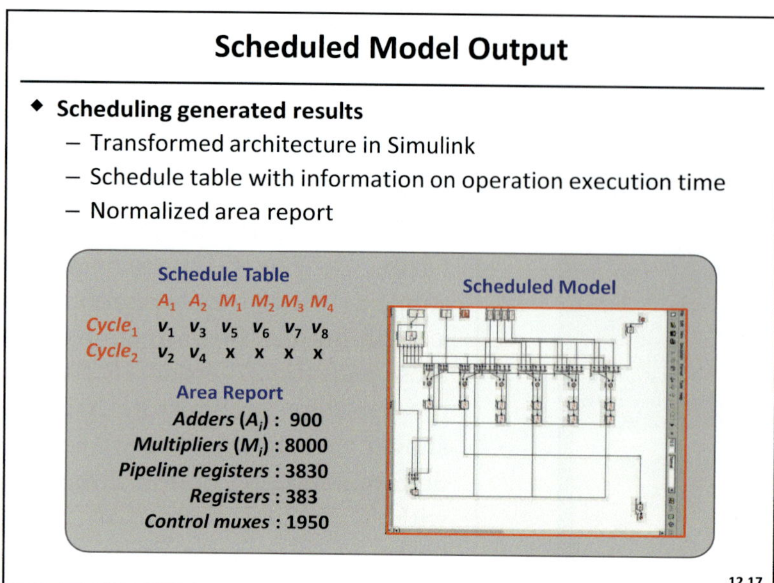

Slide 12.17

The schedule table, which is the result of the list-scheduling algorithm employed by the tool, is output along with the transformed architecture. The schedule table follows the same format as was described in Chap. 11. The columns in the table represent resource units, while the rows refer to the operational clock cycle ($N=$ 2 cycles in this example). Each row lists out the operations scheduled in a particular clock cycle, while also stating the resource unit executing the operation. The extent to which the resource units are utilized every cycle gives a rough measure of the quality of the schedule. It is desirable that maximum resource units are operational every cycle, since this leads to lower number of resource units and consequently less area. The tool uses area estimates from a generic 90nm CMOS library to compute the area of the datapath elements, registers and control units in the transformed architecture.

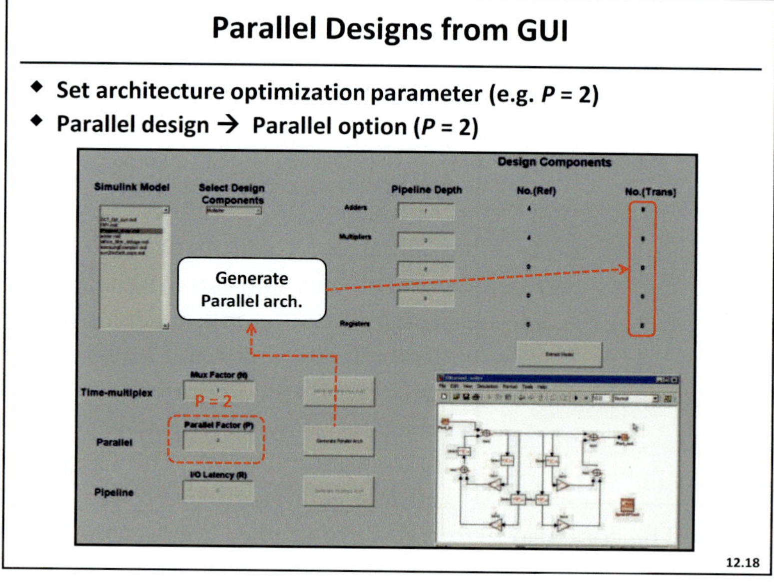

Slide 12.18

A flow similar to time-multiplexing can be adopted to create parallel architectures for the reference Simulink model. The snapshot in the slide shows the use of the "Generate Parallel arch." button after setting of the unfolding factor $P=2$ in the dialog box. The GUI displays the number of datapath elements in the unfolded architecture ("**No.(Trans)**" field), which is double the number in this case ($P=2$) of those in the reference one ("**No.(Ref)**" field).

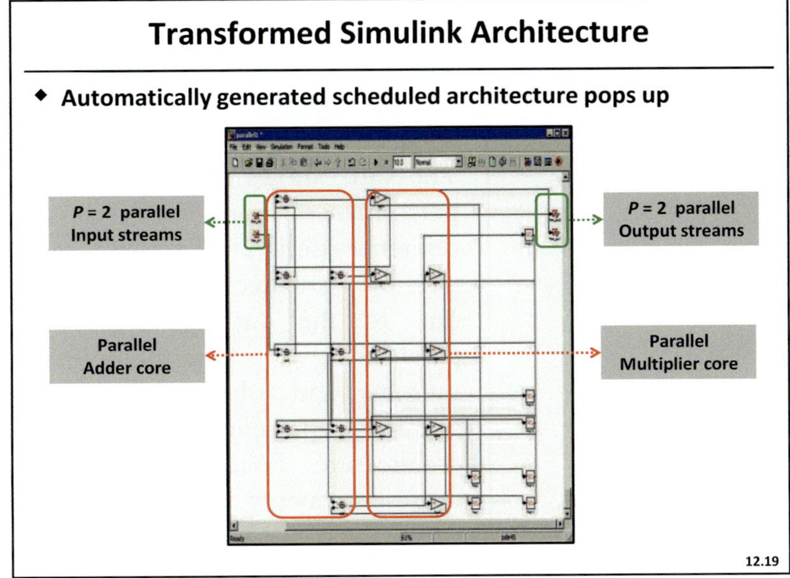

Slide 12.19

The parallelized Simulink model is automatically generated by the tool, an example of it being shown in this slide for the 2nd-order IIR filter. The parallel model has P input streams of data arriving at a certain clock rate and P output data streams going out at the same rate. In order to double the data rate, the output streams of data can be time-multiplexed to create a single output channel.

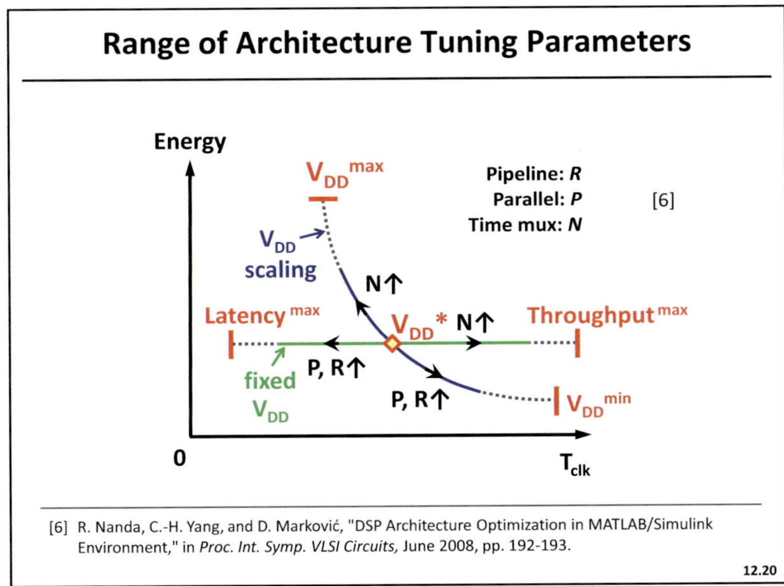

Slide 12.20

Architecture optimization is performed through a custom graphical interface as shown on this slide [6]. In this example, a 16-tap FIR filter is selected from a library of Simulink models. Parameters, such as pipeline depth for adders and multipliers, can be specified as well as the number of adders and multipliers (15 and 16, respectively, in this example). Based on the extracted DFG model of the reference design, and architectural parameters N, P, R, a user can choose to generate various architectural realizations. The parameters N, P, and R are calculated based on system specs and hardware estimates for the reference design.

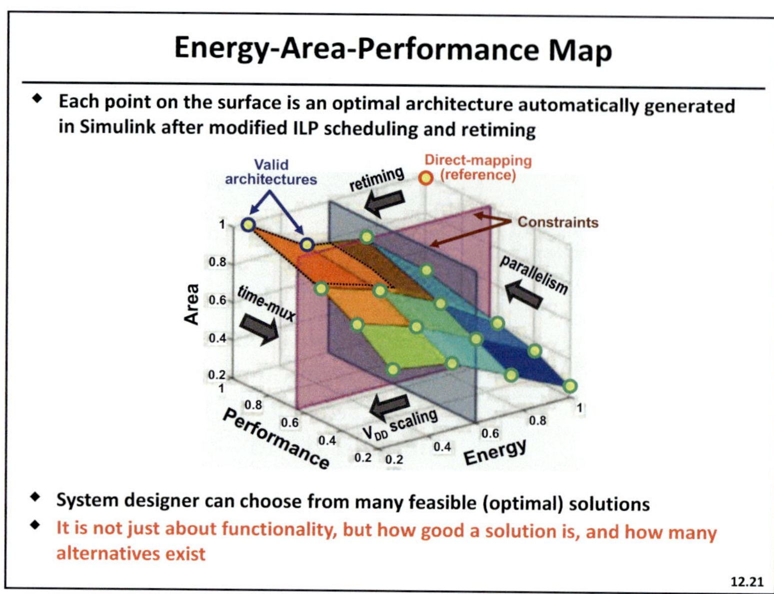

Slide 12.21

The result of the flow from Slide 12.8 is a design mapped into the energy-area-performance space. The slide shows energy, area, and performance normalized to the reference design. There are many possible architectural realizations due to the varying degrees of parallelism, time-multiplexing, retiming, and voltage scaling. Each point represents a unique architecture. Solid vertical planes represent energy and performance constraints. For example, if energy lower than 0.6 and performance better than 0.65 are required, there are two valid architectures that meet these constraints. The flow does not just give a set of all possibilities; it also provides a quantitative comparison of alternative solutions. The designer can then select the one that best meets system specifications.

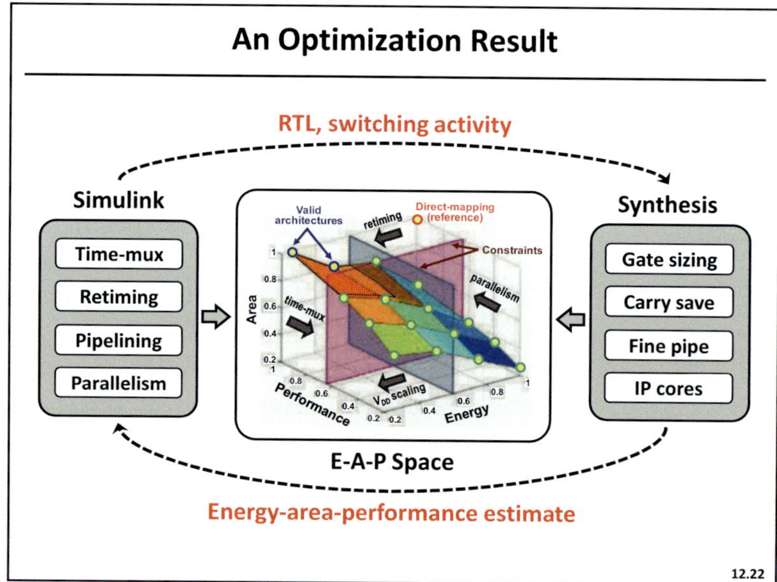

Slide 12.22

High-level architectural techniques such as parallelism, pipelining, time-multiplexing, and retiming can be combined with low-level circuit tuning; which includes gate sizing, fine-grain pipelining, and dedicated IP cores or special arithmetic (such as carry save). The combined effects of architecture and circuit parameters gives the most optimal solution since additional degrees of freedom can be explored to further improve energy, area, and performance metrics. High-level techniques are explored first, followed by circuit-level refinements.

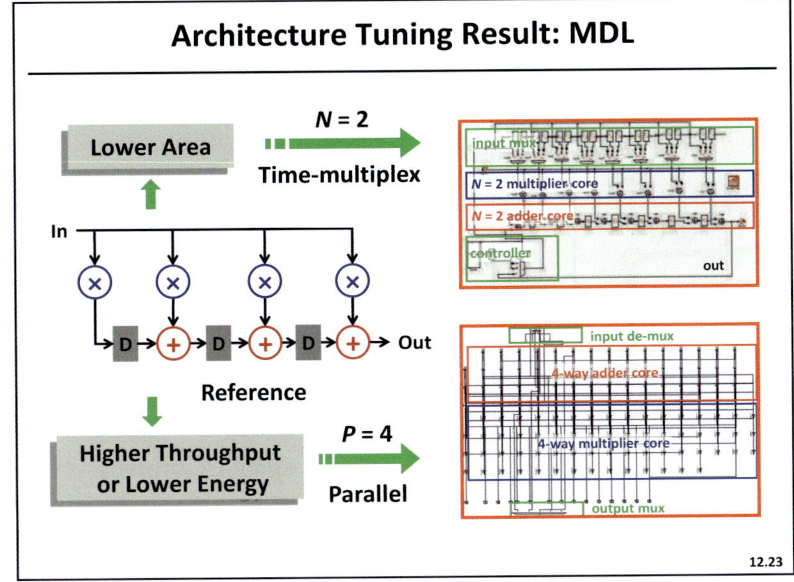

Slide 12.23

A typical output of the optimizer in the Simulink MDL format is shown on this slide. Starting from a direct-mapped filter (4 taps shown for brevity), we can vary level of time-multiplexing and parallelism to explore various architectures. Logic for input and output conditioning is automatically created. Each of these designs is bit- and cycle-equivalent to the reference design. The architectures can be synthesized through the backend flow and mapped to the energy-area-performance space previously described.

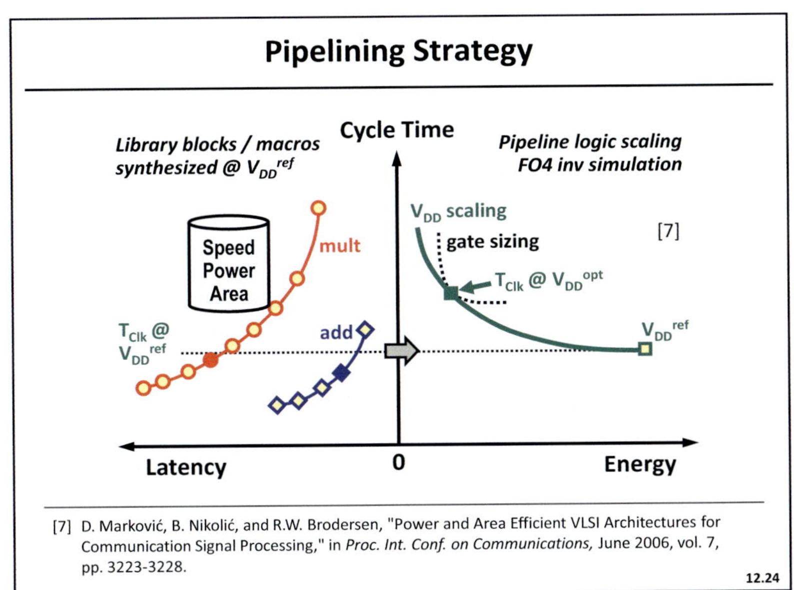

[7] D. Marković, B. Nikolić, and R.W. Brodersen, "Power and Area Efficient VLSI Architectures for Communication Signal Processing," in *Proc. Int. Conf. on Communications*, June 2006, vol. 7, pp. 3223-3228.

Slide 12.24

This slide shows basic information needed for high-level estimates: energy-delay tradeoff for pipeline logic and latency vs. cycle-time tradeoff for a logic block (such as adder or multiplier) [7]. The shape of the energy-delay line is estimated by transistor-level simulations of simple digital logic. Circuit parameters need to be balanced at target operating voltage, but we translate timing constraints to the reference voltage dictated by standard-cell libraries. Initial logic depth for design blocks is estimated from latency - cycle time tradeoff to provide balanced pipelines and ease retiming. This information is used in the architecture selection.

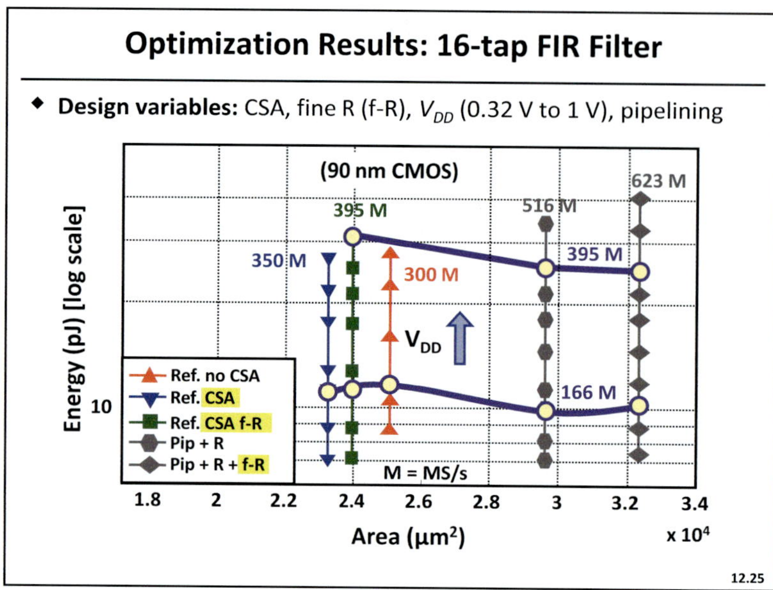

Slide 12.25

The optimization flow described in Slide 12.8 was verified on a 16-tap FIR filter. Transformations like scheduling, retiming and parallelism were applied to the filter; integrated with supply voltage scaling and micro-architectural techniques, like the usage of carry-save arithmetic. The result was an array of optimized architectures, each unique in the energy-area-performance space. Comparison of these architectures has been made in the graph with contour lines connecting architectures, which have the same max throughput. Vertical lines in the slide indicate the variation of delay and power of the architectures with supply voltage scaling.

We first look at the effect of carry-save optimization on the reference architecture. The reference architecture without carry-save arithmetic (CSA) consumes larger area and is slower compared to the design that employs CSA optimization (rightmost green line). To achieve the same reference throughput (set at 100 MS/s for all architectures during logic synthesis), the architecture without CSA must upsize its gates or use complex adder structures like carry-look-ahead; which increases the area and switched capacitance leading to an increase in energy consumption as well. The CSA-optimized architecture (leftmost *green line*) still performs better in terms of achievable throughput; highlighting the effectiveness of CSA. All synthesis estimates are based on a general-purpose 90-nm CMOS technology library.

Following CSA optimization the design is retimed to further improve the throughput (central green line). From the graph we can see that retiming improves the achievable throughput from 350 MS/s to 395 MS/s (a 13% increase). This is accompanied by a small area increase (3.5%), which can be attributed to extra register insertion during retiming. Pipelining in feed-forward systems also results in considerable throughput enhancement. This is illustrated in the leftmost red line of the graph. Pipelining is accompanied by an increased area and I/O latency. The results from logic synthesis show a 30% throughput improvement from 395 MS/s to 516 MS/s The area increases by 22% due to extra register insertion as shown in the figure. Retiming with pipelining during logic synthesis does fine-grain pipelining inside the multipliers to balance logic depth across the design (rightmost red line). This step significantly improves the throughput to 623 MS/s (20% increase).

Scheduling the filter results in area reduction by about 20% ($N=2$) compared to the retimed reference architecture and throughput degradation by 40% (leftmost blue line). The area reduction is small for the filter, since the number of taps in the design is small and the increased area of registers and multiplexers in the control units offsets the decrease in adder and multiplier area. Retiming the scheduled architecture results in a 12% improvement in throughput (rightmost blue line) but also a 5% increase in area due to increased register count.

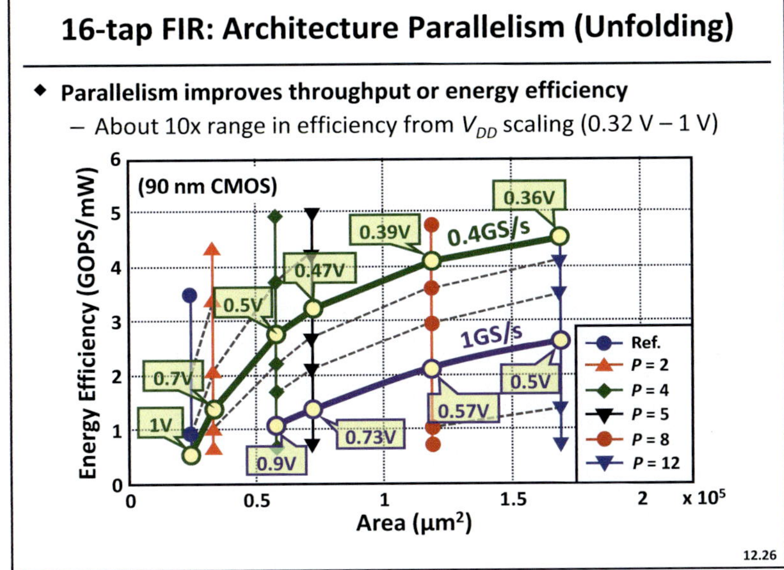

Slide 12.26

The unfolding algorithm was applied to the 16-tap FIR to generate the architectures shown in the graph. Parallelism P was varied from 2 to 12 to generate an array of architectures that exhibit a broad range of throughput and energy efficiencies. The throughput varies from 40 MS/s to 3.4 GS/s while the energy efficiency ranges from 0.5 GOPS to 5 GOPS. It is possible to improve the energy efficiency significantly with continued voltage scaling if sufficient delay slack is available. Scaling of the supply voltage has been done in 90-nm technology in the range of 1 V to 0.32 V. The graph shows a clear tradeoff between energy/throughput and area. The final choice of architecture will ultimately depend on the throughput constraints, available area and power budget.

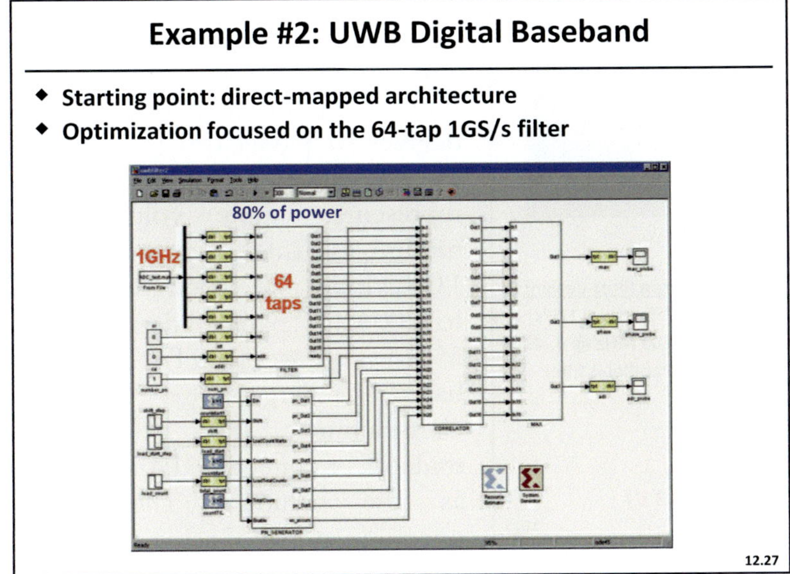

Slide 12.27

Another example is an ultra-wideband (UWB) digital baseband filter. The filter takes inputs sampled at 1 GHz and represents 80% of power of the direct-mapped design. The focus is therefore on power minimization in the filter. The 1 GHz stream is divided into five 200 MHz input streams to meet the speed requirements of the digital I/O pads.

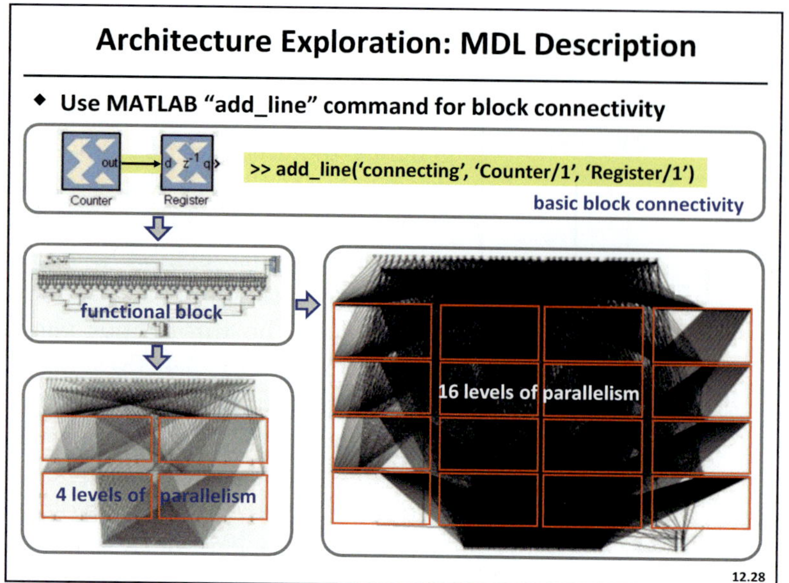

Slide 12.28

Transformation of the 64-tap filter architecture is equally simple as the 16-tap example that was previously shown. Matrix-based DFG design description is used to construct block-level model for the design using the "add_line" function in MATLAB. The function specifies connectivity between the input and output ports as described in this slide. Construction of a functional block is shown on the left-middle plot. This block can then be abstracted to construct architectures with varying level of parallelism. Four and sixteen levels of parallelism are shown. Signal interleaving at the output is also automatically constructed.

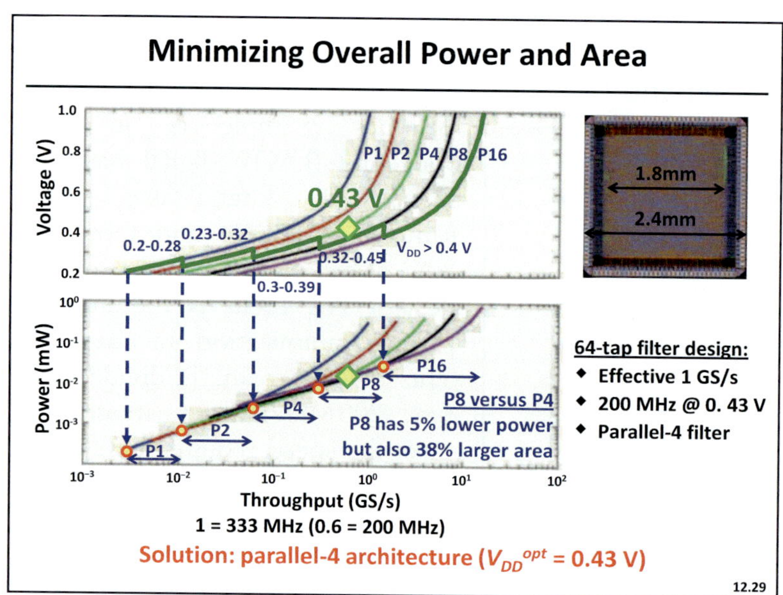

Slide 12.29

The results show the supply voltage and estimated total power as a function of throughput for varying degrees of parallelism (P1 to P16) in the filter design. Thick line represents supply voltage that minimizes total power. For a 1 GS/s input rate, which is equivalent to 200 MHz parallel data, P8 has 5% lower power than P4, but it also has 38% larger area than P4. Considering the power-area tradeoff, architecture P4 is chosen as the solution. The design operating at 0.43 V achieved overall 68% power reduction compared to direct-mapped P1 design operating at 0.68 V. Die photo of the UWB design is shown. The next step is to verify chip functionality.

FPGA Based Chip Verification

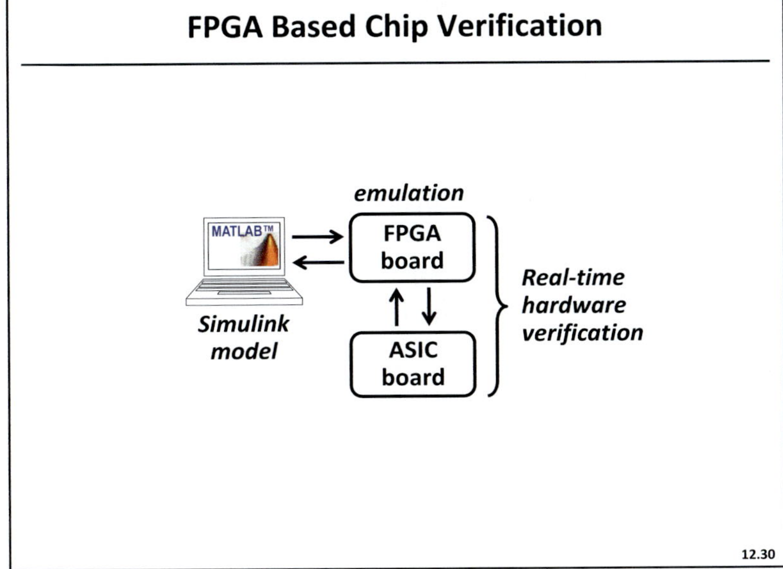

Slide 12.30

Functional verification of ASICs is also done within MATLAB/Simulink framework that is used for functional simulations and architecture design. Since the Simulink model can be mapped to both FPGA and ASIC, an FPGA board can be used to facilitate real-time I/O verification. The FPGA can emulate the functionality in real-time and compare results online with the ASIC. It can also host test vectors and capture ASIC output for offline analysis.

Hardware Test Bench Support

- **Approach: use Simulink test bench (TB) for ASIC verification** [8]
    - Develop custom interface blocks (I/O)
    - Place I/O and ASIC RTL into TB model

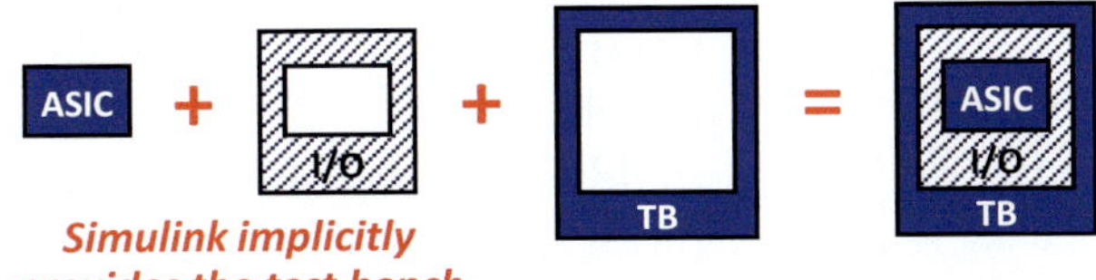

- **Additional requirements from the FPGA test platform**
    - As general purpose as possible (large memories, fast I/O)
    - Use embedded CPU to provide high-level interface to FPGA

[8] D. Marković *et al.*, "ASIC Design and Verification in an FPGA Environment," in *Proc. Custom Integrated Circuits Conf.*, Sep. 2007, pp. 737-740.

12.31

Slide 12.31

In order to verify chip functionality, the idea is to utilize Simulink test bench that's implicitly available from algorithm development [8]. We therefore need emulation interface between the test bench and emulation ASIC, so Simulink can fully control the electrical interface between the FPGA and ASIC boards. Simulink library is therefore extended with the required I/O functionality.

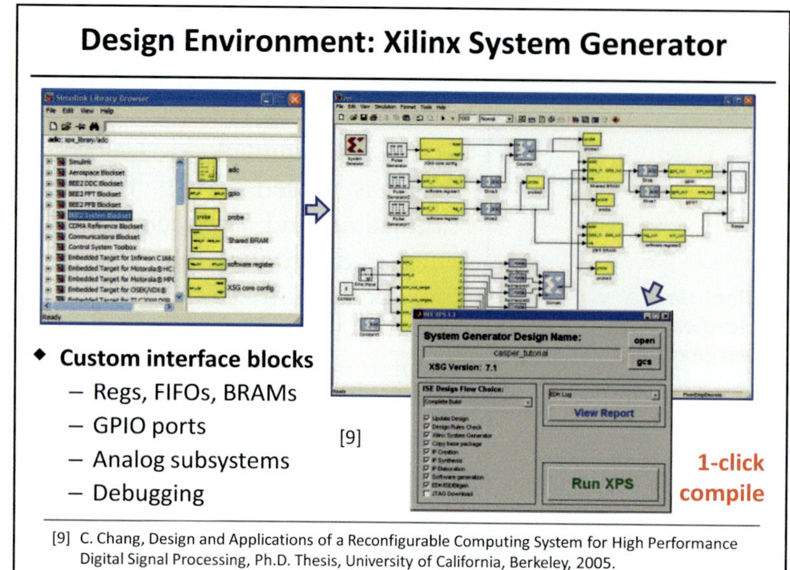

Slide 12.32

The yellow-block library is used for chip verification. It is good for dataflow designs such as DSP. The library is called the BEE2 system blockset [9], designed at the Berkeley Wireless Research Center, and now maintained by international collaboration ([www.casper.berkeley.edu]). The I/O debugging functionality is provided by several types of blocks. These include software / hardware interfaces such as registers, FIFOs, block memories; external general purpose I/O port interfaces (these two types are primarily used for ASIC verification); there are also A/D and D/A interfaces for external analog components; and software-controlled debugging resources. This library is accompanied with custom scripts that automate FPGA backend flow to abstract away FPGA specific interface and debugging details. From a user standpoint, it is push-of-a-button flow that generates configuration file for the FPGA.

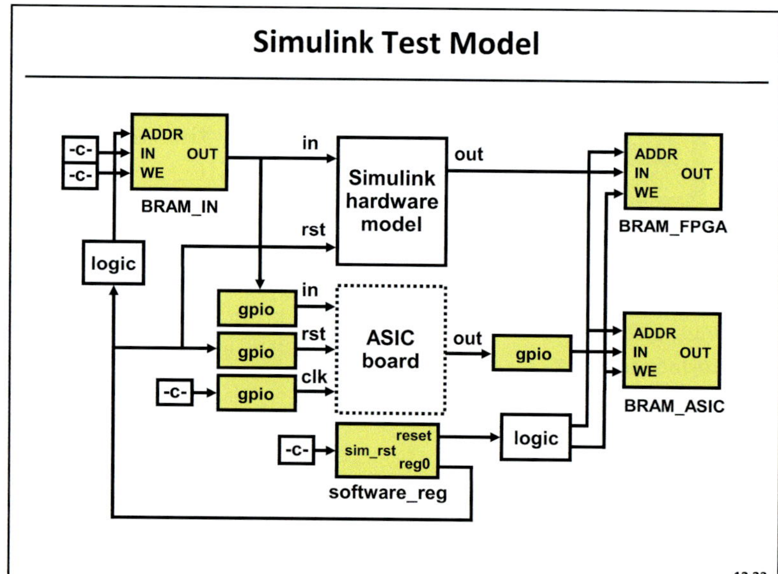

Slide 12.33

This slide shows a typical I/O library usage in a Simulink test bench model. The block in the middle is the Simulink hardware model and the ASIC functionality is described with Simulink blocks. Block in the dashed lines indicates actual ASIC board, which is driven by the FGPA. The yellow blocks are the interface between two hardware platforms; they reside on the FGPA. The GPIO blocks define mapping between ASIC I/O and the GPIO headers on the FPGA board. The software register block allows single 32-bit word transfer between MATLAB and FPGA board. This block is used to configure control bits that manage ASIC verification. The ASIC clock is provided by the FPGA, which can generate the clock internally or synchronize to an external source. Input test vectors are stored in block RAM memory. Block RAMs are also used to store results of FPGA emulation (BRAM_FPGA) and sampled outputs from the ASIC board (BRAM_ASIC). So, the yellow block interface is programmed on the FPGA board and controlled from MATLAB environment.

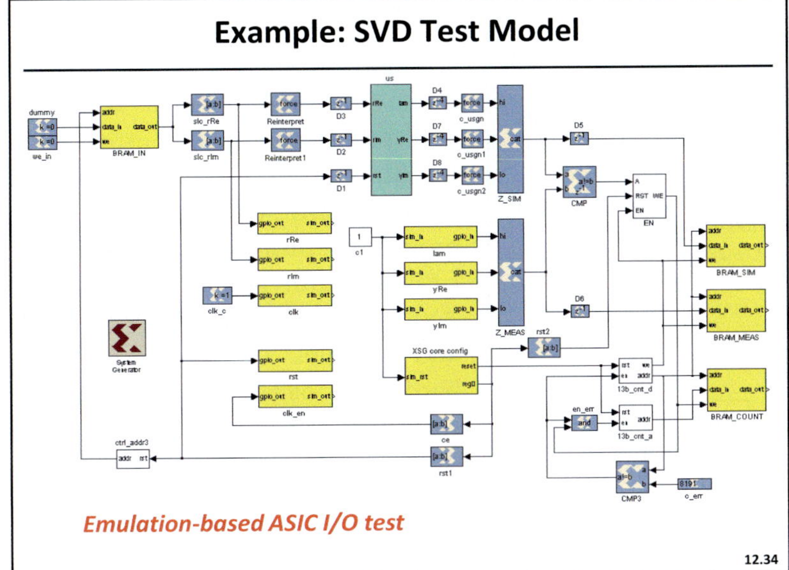

Slide 12.34

Here is an example of Simulink test bench model used for I/O verification of the singular value decomposition (SVD) chip. ASIC is modeled with the blue SVD block, with many layers of hierarchy underneath. The yellow blocks are the interface between the FPGA and the ASIC. The white blocks are simple controllers, built from counters and logic gates, to manage memory access. Inputs are taken from the input block RAM and fed into both the ASIC board and its equivalent description on the FGPA. Outputs of the FPGA and ASIC are stored in output block RAMs. Finally we can use another block RAM for debugging purposes to locate the samples where eventual mismatch has occurred.

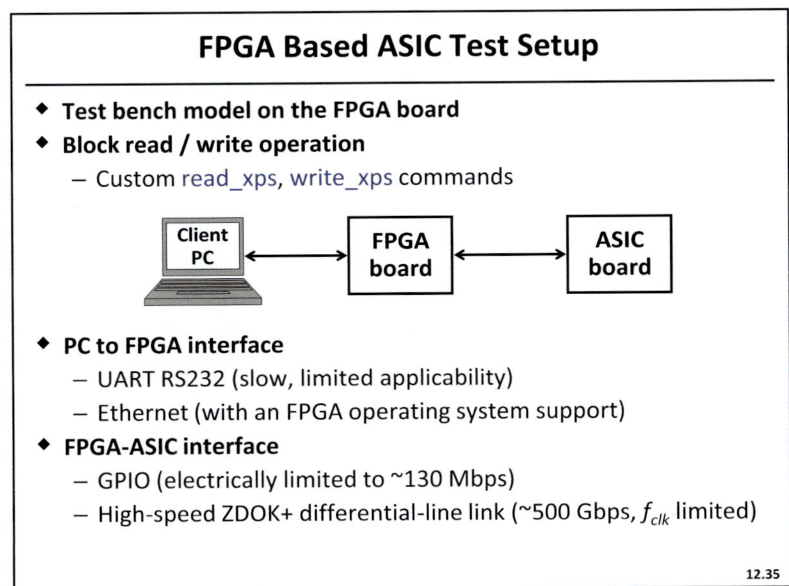

Slide 12.35

This slide illustrates the hardware test setup. The client PC has MATLAB/Simulink and custom BEE Platform Studio flow featuring I/O library and custom routines for data access. Test vectors are managed using custom "read_xps" and "write_xps" software routines that exchange data between FPGA block RAMs (BRAMs) and MATLAB. The PC-FPGA interface can be realized using serial port or Ethernet. The FPGA-ASIC interface can use general-purpose I/O (GPIO) connectors or high-speed differential-line ZDOK+ connectors.

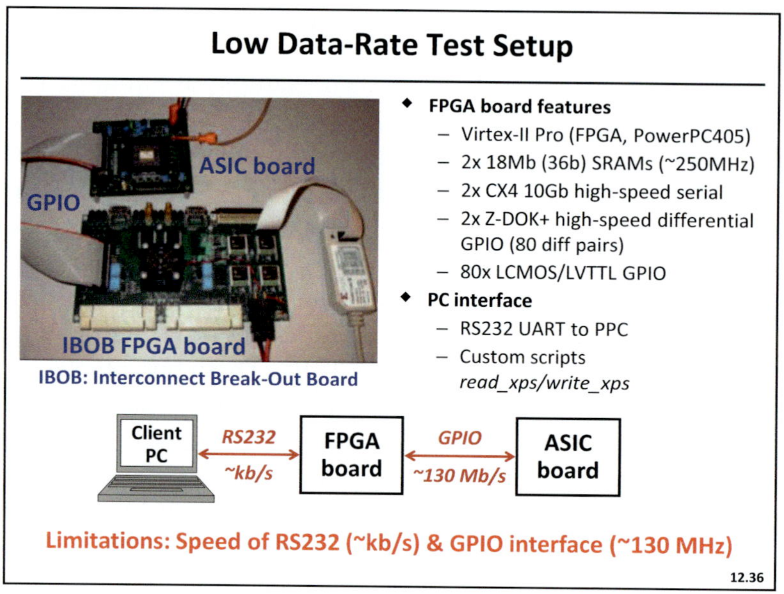

Slide 12.36

A low-data-rate test setup is shown here. Virtex-II FPGA board is accessed from the PC using serial RS232 link and connects to the ASIC board using GPIO. This model is programmed onto the FPGA board, which stimulates the ASIC over general purpose I/Os and samples outputs from both hardware boards. Block RAMs and software registers are controlled through the serial port. This setup is convenient for at-speed verification of the ASIC that operate below 130 MHz and do not require large amounts of data. Key limitations of this setup are low speed of the serial port (~kb/s) and relatively slow GPIO interface that is electrically limited to ~130 Mb/s.

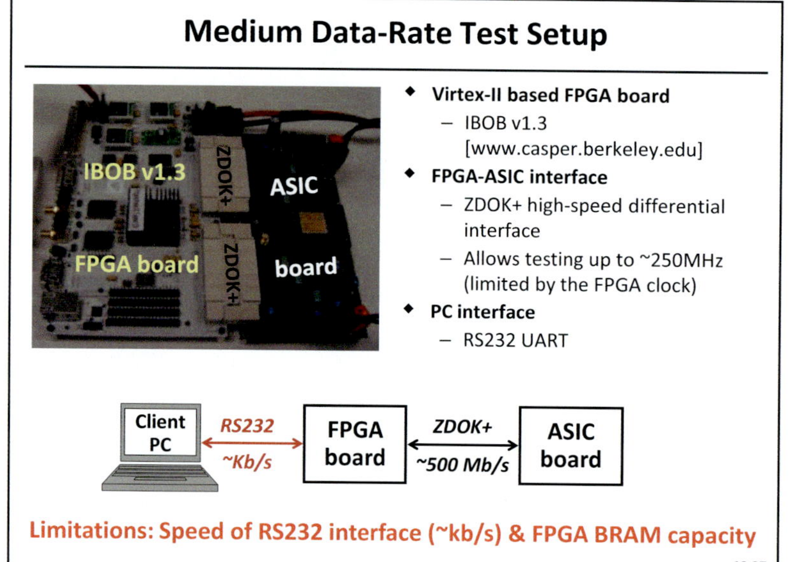

Slide 12.37

Shown is an example of a medium-data-rate test setup. The I/O speed between the FPGA and ASIC boards can be improved with differential pair connectors such as those used in the Xilinx personality module. With ZDOK+ links projected to work in the multi-GS/s range, the FPGA-ASIC bandwidth is now limited by the FPGA clock and/or speed of ASIC I/O ports. This setup is based on advanced Virtex-II board from the UC Berkeley Radio Astronomy Group (CASPER). The board is called IBOB (Interface Break-out Board). Limitations of this setup are serial PC-FPGA interface and limited capacity of BRAM that is used to store test vectors.

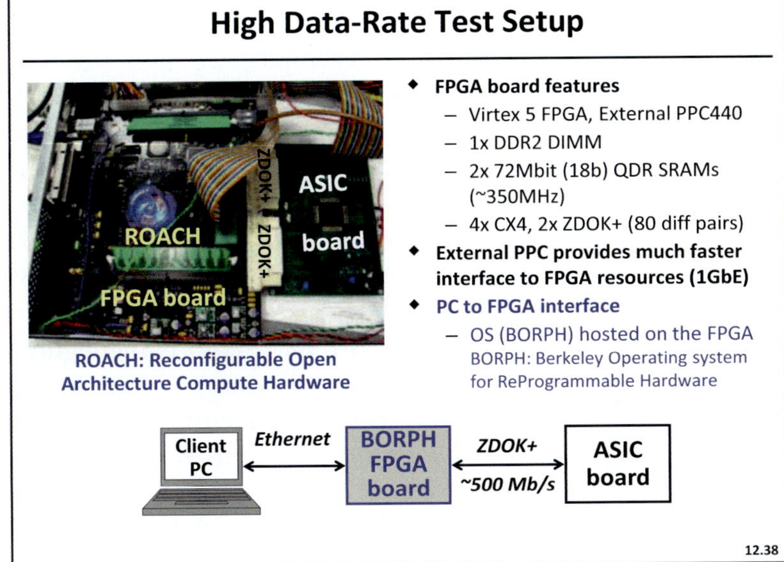

Slide 12.38

To address the speed bottleneck of the serial port, Ethernet link between the user PC and the FPGA board is used. The Ethernet link between the PC and FPGA is already a standard feature on most commercial parts. Shown here is another board developed by the CASPER team, called ROACH (Reconfigurable Open Architecture Computing Hardware). The board is based on a Virtex-5 FPGA chip. Furthermore, the client PC functionality can be pushed into the FPGA board with operating system support.

Slide 12.39

Custom operating system, BORPH (Berkeley Operating system for ReProgrammable Hardware) extends a standard Linux system with integrated kernel support for FPGA resources [9]. The BEE2 flow produces BORPH object files instead of usual FPGA configuration "bit" files. This allows users to execute hardware processes on FPGA resources the same way they run software on conventional machines. BORPH also allows access to the general Unix file system, which enables test bench to access the same test-vector files as the top-level Simulink. The OS is supported by BEE2/3 and ROACH boards, and has limited support on IBOB boards.

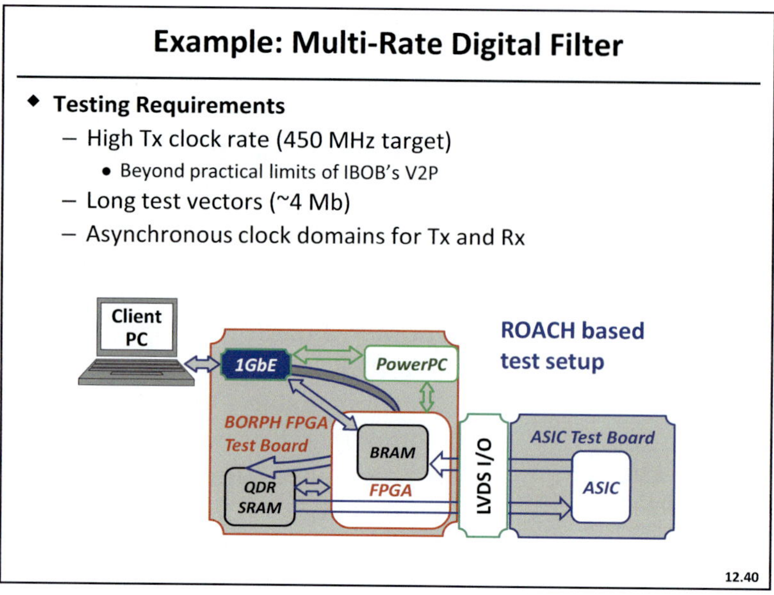

Slide 12.40

As an example, we illustrate test setup for high-speed digital filter. This system requires clock of 400 MHz, which is beyond the capability of Virtex-II based boards. The chip also requires long test vectors, over 4 Mb, and the use of asynchronous clock domains. ROACH-based setup is highlighted in the diagram. High-speed ZDOK+ connectors are used between the ASIC and FPGA boards. Quad-data-rate (QDR) SRAM on the FPGA board is used to support longer test vectors. The ROACH board is controlled by the PC using 1 GbE Ethernet port. Optionally, a PowerPC based control using BORPH could be used.

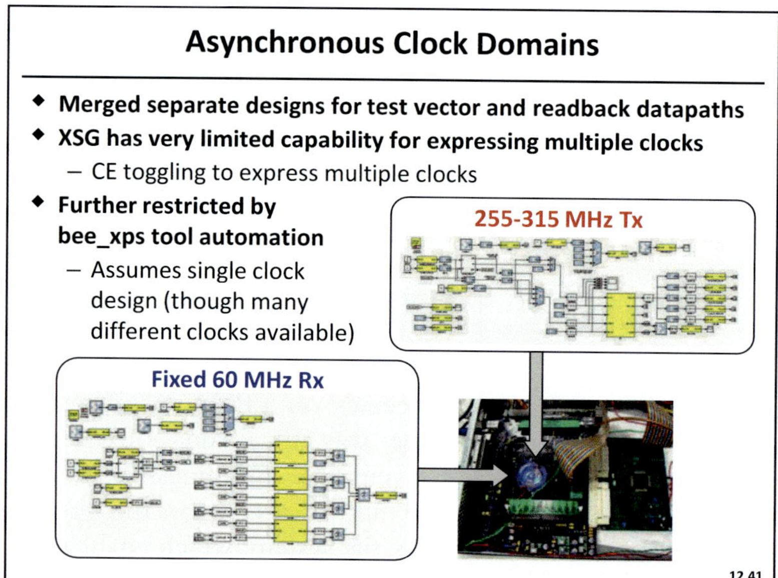

Slide 12.41

This setup was used for real-time FPGA verification up to 330 MHz (limited by the FPGA clock). The ASIC board shown here is a high-speed digital front-end DSP designed to operate with I/O rates of up to 450 MHz (limited by the speed of the chip I/O pads).

Support for expressing asynchronous clock domains is limited compared to what is physically possible in the FPGA. Asynchronous clock domains are required when the ASIC does not have equal data rates at the input and output, for example in decimation. Earlier versions of XSG handled this by running in the fastest clock domain while toggling clock enable (CE) for slower domains. While newer versions of XSG can infer clock generation, the end-to-end toolflow was built on a single-clock assumption. Therefore, this test setup was composed of two independent modules representing the two different clock speeds, which were merged into a single end-design.

Results and Limitations

- **Results**
  - Test up to 315 MHz w/ loadable vectors in QDR;
    up to 340 MHz with pre-compiled vectors in ROMs
  - 55 dB SNR @ 20 MHz bandwidth

- **Limitations**
  - DDR output FF critical path @ 340 MHz (clock out)
  - QDR SRAM bus interface critical path @ 315 MHz
  - Output clock jitter?
  - LVDS receivers usually only 400-500 Mbps
    - OK for data, not good for faster clocks
    - Get LVDS I/O cells?

Slide 12.42

The full shared-memory testing infrastructure allowed testing the chip up to 315 MHz. This included mapping one of the 36 Mb QDR SRAMs onto the PowerPC bus to enable software access for loading arbitrary test vectors at runtime. At 315 MHz, the critical paths in the control logic of the bus attachment presented a speed bottleneck.

This speed barrier was overcome by pre-loading test vectors into the FPGA bitstream using ROMs. Trading off the flexibility of runtime-loadable test vectors pushed the maximum speed of chip test up to 340 MHz. At this point, there existed a physical critical path in the FPGA for generating a 340 MHz clock for the test chip.

FPGA Based ASIC Verification: Summary

- **The trend is towards fully embedding logic analysis on FPGA, including OS support for remote access**

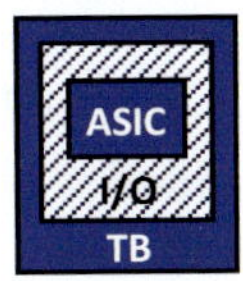

	ASIC	I/O	TB	
Simulation	Simulink	Simulink	Simulink	Pure SW Simulation
	HDL	Simulink	Simulink	Simulink ModelSim co-simulation
Emulation	FPGA	HIL tools	Simulink	Hardware-in-the-loop simulation
	FPGA	FPGA	FPGA	Pure FPGA emulation
ASIC I/O Test	FPGA & ASIC	FPGA	Custom SW	Testvectors outside FPGA
	FPGA & ASIC	FPGA	FPGA	Testvectors inside FPGA

Slide 12.43

Various verification strategies are summarized in this slide. We look at ASIC emulation, I/O interface, and test bench.

Simulation of design built from Simulink blocks is straightforward, but can be quite slow. The alternative is Simulink ModelSim co-simulation, with HDL description of the ASIC.

Mapping this HDL onto FPGA and using hardware-in-the-loop tools greatly improves the verification speed, but is limited by the speed of PC-to-FPGA interface. Pure hardware emulation is the best solution, because it can fully utilize processing capability of the FPGA.

After mapping the final design onto ASIC, we close the loop by using FPGA for the I/O test. The idea is to move towards test setup fully embedded on the FPGA that includes local operating system and remote access support.

Further Extensions

- **Design recommendations**
  - Send source-synchronous clock with returned data
  - Send synchronization information with returned data
    - "Vector warning" or frame start, data valid
- **KATCP: communication protocol interfacing to BORPH**
  - Can be implemented over TCP telnet connection
  - Libraries and clients for C, Python
  - KATCP MATLAB client (replaces read_xps, write_xps)
    - Can program FPGA from directly from MATLAB – no more JTAG cable!
    - Provides byte-level read/write granularity
    - Increases speed from ~Kb/s to ~Mb/s
      (Room for improvement; currently high protocol overhead)
- **Towards streaming**
  - Transition to TCP/IP-based protocols facilitates streaming
  - Ethernet streaming w/o going through shared memory

12.44

Slide 12.44

Currently, data rates in and out of the FPGA are limited due to the use of what is essentially a monitor and control interface. Without having to resort to custom electrical interfaces or pushing up to 10GbEthernet standards, several steps can be taken in order to push up to the bandwidths needed for testing.

A support for long test vectors (~10 Mb) at moderate data rates (~Mb/s) is required by some applications. The current infrastructure on an IBOB would only allow for loading of a single test vector at a time, with each load occurring at ~10 Kb/s over an RS232 link.

A UDP (Uniform Datagram Protocol)-based link to the same software interface is also available. Test vectors would still be loaded and used one at a time, but the time required for each would be decreased by about three orders of magnitude as the link bandwidth increases to about ~10 Mb/s. This would enable moderate-bandwidth streaming with a software-managed handshaking protocol.

Fully-streaming test vectors beyond these data rates would require hardware support that is not available in the current generations of boards. Newer boards that have an Ethernet interface connected directly to the FPGA, not just the PowerPC, would allow TCP/IP-based streaming into the FPGA, bypassing the PowerPC software stack

Summary

- **MATLAB/Simulink is an environment for algorithm modeling and optimized hardware implementation**
  - Bit-true cycle-accurate model can be used for functional verification and mapping to FPGA/ASIC hardware
  - The environment is suited for automated architecture exploration using high-level scheduling and retiming
  - Test vectors used in algorithm development can also be used for functional verification of fabricated ASIC
- **Enhancements to traditional FPGA-based verification**
  - Operating system can be hosted on an FPGA for remote access and software-like execution of hardware processes
  - Test vectors can be hosted on FPGA for real-time data streaming (for data-limited or high-performance applications)

12.45

Slide 12.45

MATLAB/Simulink environment for algorithm modeling and hardware implementation was discussed. The environment captures bit-true cycle-accurate behavior and can be used for FPGA and ASIC implementations. Hardware description allows for rapid prototyping using FPGAs. The unified design environment can be used for wordlength optimization, architecture transformations and logic verificaiton. Leveraging hardware equivalency between FPGA and ASIC, an FPGA can host logic analysis for I/O verification of fabricated ASICs. Further refinements include operating system support for remote access and real-time data streaming. The use of the design environment presented in Chaps. 9, 10, 11, and 12 will be exemplified on several examples in Chaps. 13, 14, 15, and 16.

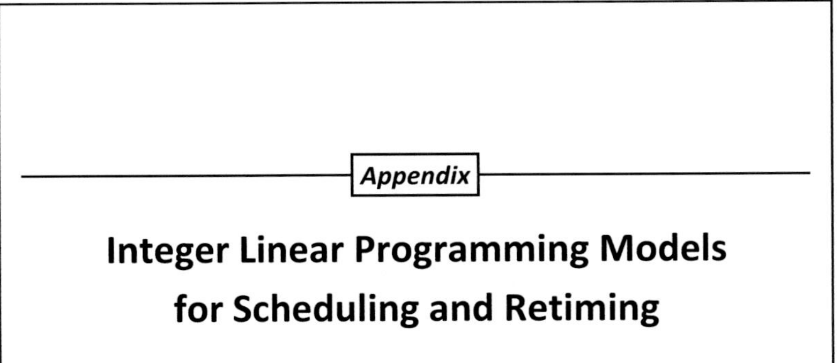

Slide 12.46

Next several slides briefly show ILP formulation of scheduling and retiming. Retiming step is the bottleneck in CPU runtime for complex algorithms. Several approaches for retiming are compared in terms of runtime and optimality of results.

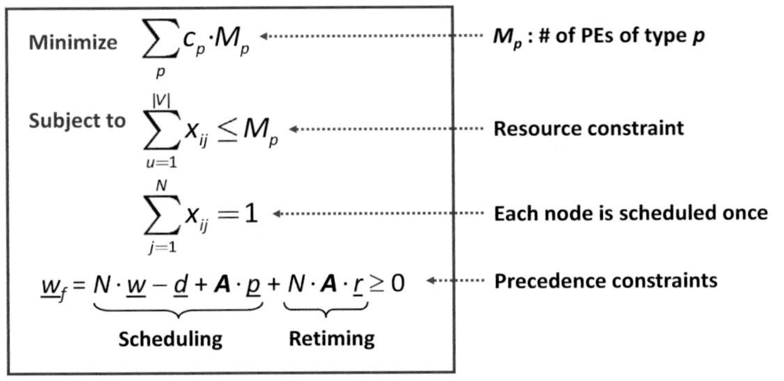

Slide 12.47

Scheduling and retiming routines are the essence of architectural transformations, as described in Chap. 11. By employing scheduling and retiming, we can do parallelism, time-multiplexing (folding), and pipelining. This slide shows a traditional model for scheduling and retiming based on ILP formulation.

The objective is to minimize cost function $\Sigma c_p \cdot M_p$, where M_p is the number of processing elements of type p and c_p is the normalized cost of the processing element M_p. For example, processing element of type 1 can be an adder; processing element of type 2 can be a multiplier; in which case M_1 is the number of adders, M_2 is the number of multipliers, etc. If the normalization is done with respect to the adders, c_1 is 1, c_2 can be 10 to account for the higher area cost of the multipliers. Constraints in the optimization are that the number of processes of type p executed in any cycle cannot exceed available resources (M_p) for that process, and that each node x_{uj} of the algorithm is scheduled once during N cycles. To ensure correct functionality, precedence constraints have to be maintained as described by the last equation. By setting the retiming vector to 0, the optimization reduces to scheduling and does not yield optimal solution. If the retiming vector is non-zero, the ILP formulation works with a number of unbounded integers and results in exponential run-time, which is impractical for large designs. Therefore, an alternate problem formulation has to be specified to ensure feasible run-time and optimal result.

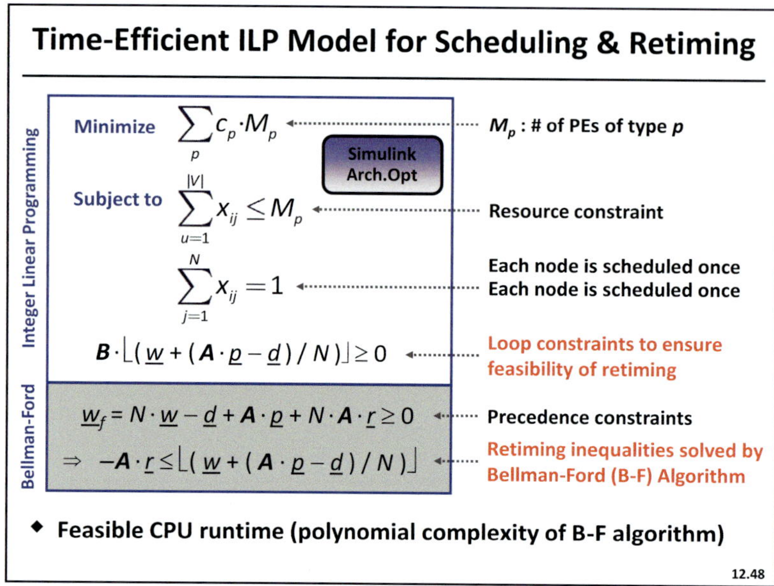

Slide 12.48

Improved scheduling formulation is shown here. The algorithm separates scheduling and retiming routines in order to avoid simultaneously solving a large number of integer variables (for both scheduling and retiming). Additional constraint is placed in scheduling to enable optimal retiming after the scheduling step. The additional constraint specifies that all loops should have non-zero latency, which allows proper movement of registers after scheduling. The retiming inequalities are then solved using polynomial-complexity Bellman-Ford algorithm. By decoupling scheduling and retiming tasks, we can achieve optimal results with feasible run-time.

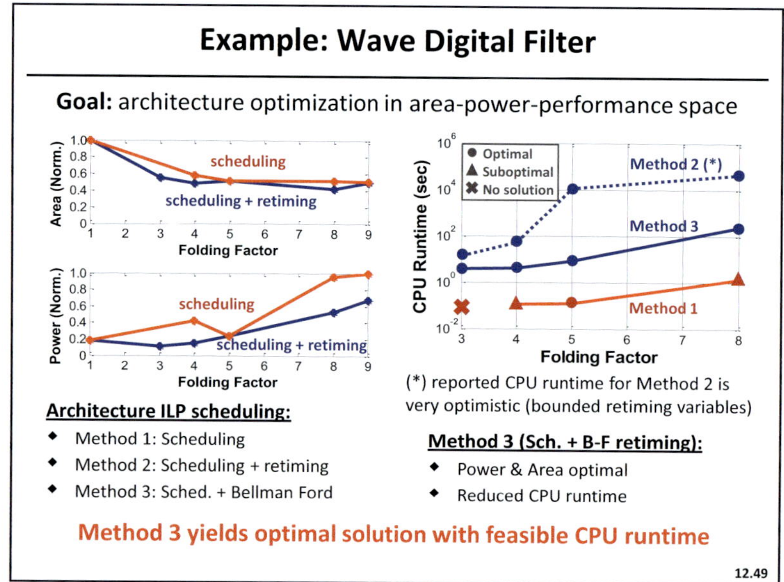

Slide 12.49

Scheduling and retiming solutions are compared on this slide for the case of a wave digital filter. Normalized area, power and CPU runtime are shown for varying folding factors. It can be seen that scheduling and retiming (methods 2 and 3) always improve the results in comparison to a scheduling-only approach (method 1). In some cases, scheduling does not even find a solution (folding factors 2, 3, 6, and 7). Methods 2 and 3 achieve optimal solutions, but they vary in run-time. Method 2 is a traditional

ILP approach that simultaneously solves scheduling and retiming problems. To make a best-case comparison, we bound the retiming variables in the close proximity (within ± 1) of the solution. Due to the increased number of variables, method 2 still takes very long time. In method 3, we use unbounded retiming variables, but due to separation of scheduling and retiming, a significantly reduced CPU runtime is achieved. The retiming variables are decoupled from the ILP and appear in the Bellman-Ford algorithm (polynomial time complexity) once ILP completes scheduling.

References

- W.R. Davis *et al.*, "A Design Environment for High Throughput, Low Power Dedicated Signal Processing Systems," *IEEE J. Solid-State Circuits*, vol. 37, no. 3, pp. 420-431, Mar. 2002.

- K. Kuusilinna, *et al.*, "Real Time System-on-a-Chip Emulation," in *Winning the SoC Revolution,* by H. Chang, G. Martin, Kluwer Academic Publishers, 2003.

- D. Marković, A Power/Area Optimal Approach to VLSI Signal Processing, Ph.D. Thesis, University of California, Berkeley, 2006.

- R. Nanda, DSP Architecture Optimization in Matlab/Simulink Environment, M.S. Thesis, University of California, Los Angeles, 2008.

- R. Nanda, C.-H. Yang, and D. Marković, "DSP Architecture Optimization in Matlab/Simulink Environment," in *Proc. Int. Symp. VLSI Circuits,* June 2008, pp. 192-193.

- D. Marković, B. Nikolić, and R.W. Brodersen, "Power and Area Efficient VLSI Architectures for Communication Signal Processing," in *Proc. Int. Conf. Communications,* June 2006, vol. 7, pp. 3223-3228.

- D. Marković, *et al.*, "ASIC Design and Verification in an FPGA Environment," in *Proc. Custom Integrated Circuits Conf.,* Sept. 2007, pp. 737-740.

- C. Chang, Design and Applications of a Reconfigurable Computing System for High Performance Digital Signal Processing, Ph.D. Thesis, University of California, Berkeley, 2005.

- H. So, A. Tkachenko, and R.W. Brodersen, "A Unified Hardware/Software Runtime Environment for FPGA-Based Reconfigurable Computers using BORPH," in *Proc. Int. Conf. Hardware/Software Codesign and System Synthesis,* 2008, pp. 259-264.

Part IV

Design Examples: GHz to kHz

Multi-GHz Radio DSP

Chapter 13

with Rashmi Nanda
University of California, Los Angeles

Slide 13.1

This chapter discusses DSP techniques used to design digitally intensive front ends for radio systems. Conventional radio architectures use analog components and RF filters that do not scale well with technology, and have poor tunability required for supporting multiple modes of operation. Digital CMOS scales well in power, area, and speed with each new generation, and can be easily programmed to support multiple modes of operation. In this chapter, we will discuss techniques used to "digitize" radio front ends for standards like LTE and WiMAX. Signal processing and architectural techniques combined will be demonstrated to enable operation across all required channels.

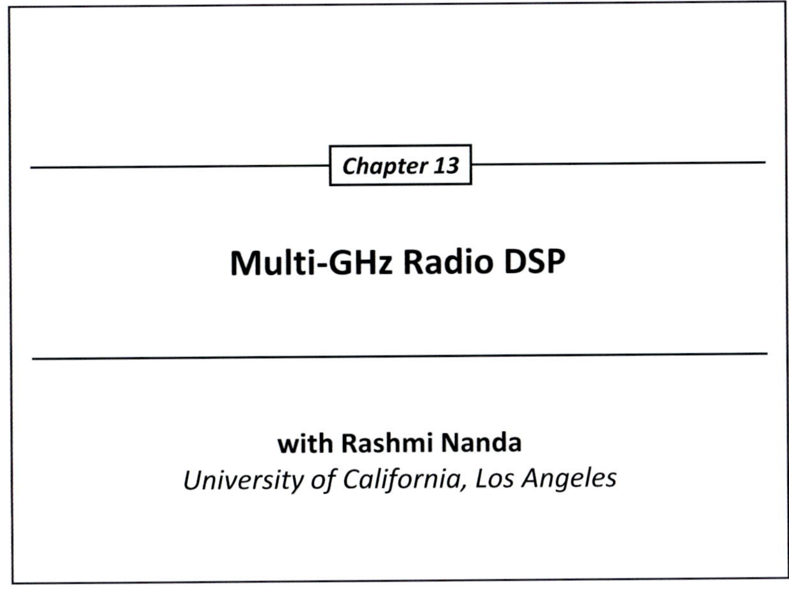

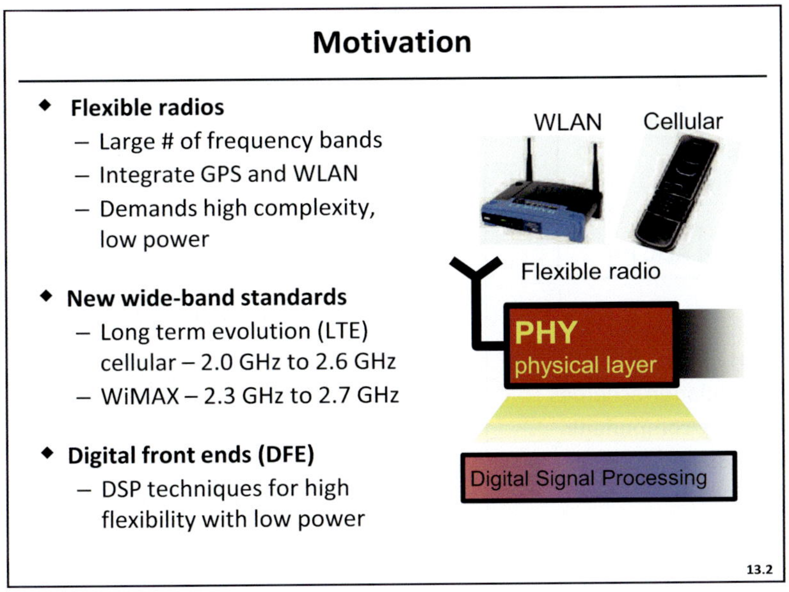

Slide 13.2

Cellular and wireless LAN devices are becoming more and more complex every day. Users want seamless Internet connectivity and GPS services on the go. Data channels in emerging standards like LTE and WiMAX are spread across widely varying frequency bands. The earlier practice was to build a separate front end for each of these bands. But this leads to excess area overhead and redundancy. In this scenario, a single device capable of transmitting and receiving data in a wide range of frequency bands becomes an attractive solution. The idea is to enable a radio system capable of capturing data at different RF frequencies and signal bandwidths, and adapt to different standards through external tuning parameters. Circuit designers will not have to re-design the front end for new RF carriers or bandwidth. Compared to its dedicated counterpart, there will be overhead associated with designing such flexible front ends. Hence, the main objective is to implement a tunable radio with minimal cost by maximizing digital signal processing techniques. This chapter takes a look at this problem and discusses solutions to enable reconfigurable radios.

D. Marković and R.W. Brodersen, *DSP Architecture Design Essentials*, Electrical Engineering Essentials,
DOI 10.1007/978-1-4419-9660-2_13, © Springer Science+Business Media New York 2012

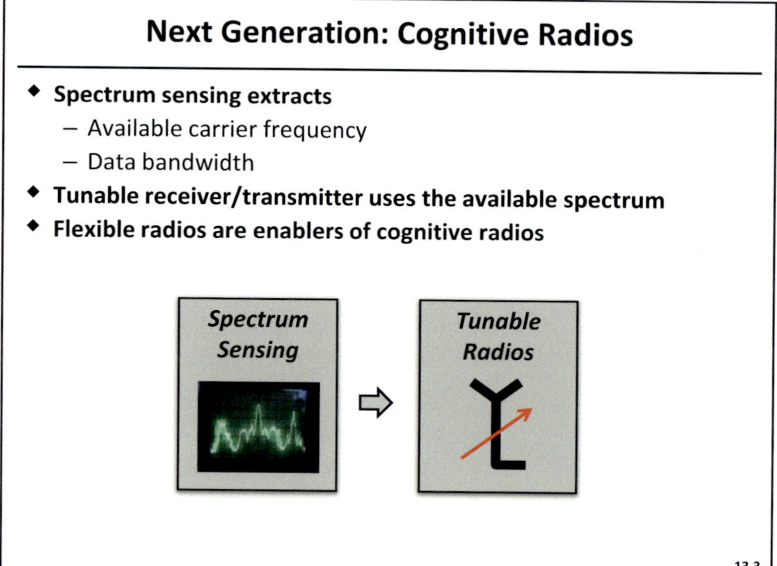

Slide 13.3

Flexibility in radio circuits will enable the deployment of cognitive radios in future wireless devices. Cognitive radios allow secondary users to utilize the unused spectrum of the primary user. The spectrum-sensing phase in cognitive radios detects the available carrier frequencies and signal bandwidths not presently being utilized by the primary user. The tunable radio frontend then makes use of this information to send and receive data in this available bandwidth. A low cost digital frontend will enable easy reconfiguration of transmit and receive RF carriers as well as the signal bandwidths.

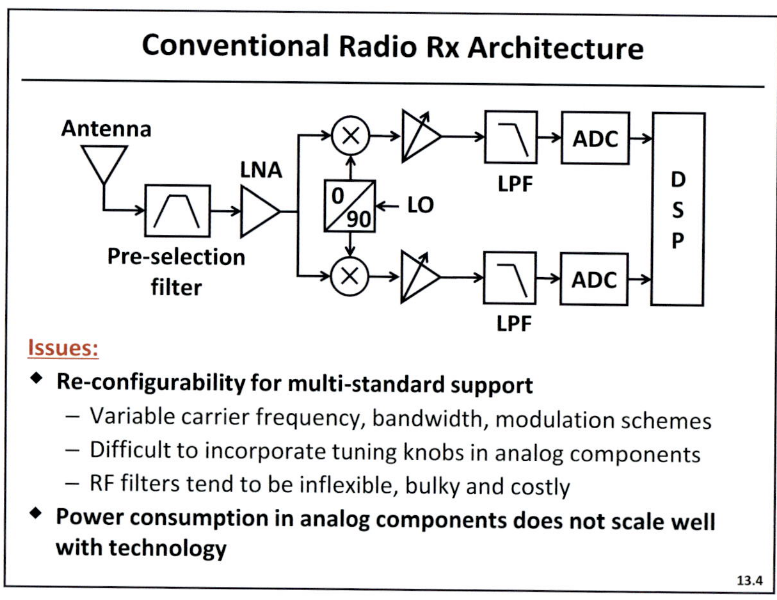

Slide 13.4

The traditional architecture of radio receiver frontends has RF and analog components to support the signal conditioning circuits. The RF signal received from the antenna is filtered, amplified, and down-converted before being digitized by the ADC and sent to the baseband blocks, as shown in the receiver chain in the figure. The RF and analog components in the system work well when operating at a single carrier frequency and known signal bandwidth. The problem arises when the same design has to support multi-mode functionality. The front end needs programmable blocks if multiple modes are to be supported in a single integrated system. Incorporating programmability in RF or analog blocks is very challenging. Designers face several issues while ensuring acceptable performance across wide range of frequency bands. The RF filter blocks are a big bottleneck due to their inflexibility and huge design cost. Also, power consumption in analog blocks does not scale well with scaled technology, making the design process considerably more difficult for new generations of CMOS.

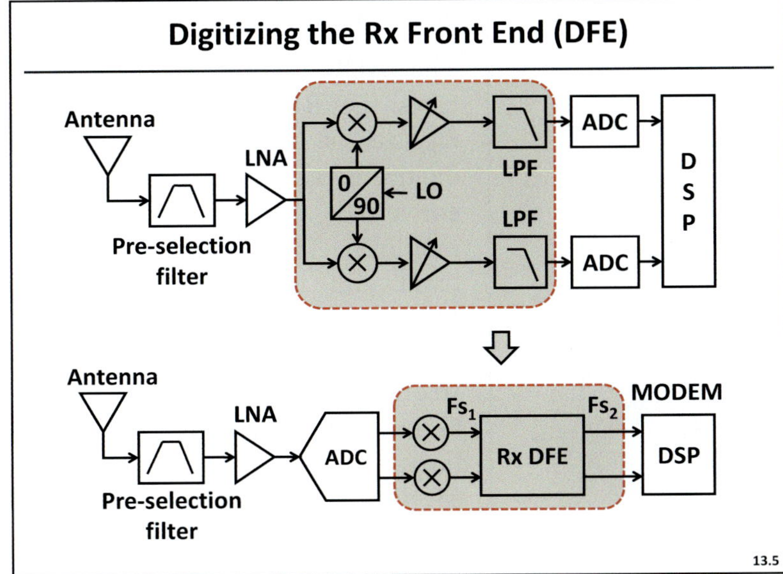

Slide 13.5

The increasing speed of digital CMOS with each new technology generation, led to the idea of moving more and more of the signal conditioning circuits to the digital domain. In the most ideal scenario, the receiver digital front-end (DFE) will enable direct digitization of the RF signal after low-noise amplification in the receiver chain, as shown in the bottom figure in the slide. The digital signal is then down-converted and filtered before being sent to the baseband modem. In this approach, almost all the signal conditioning has been moved to the mixed-signal/digital domain.

DFE: Benefits and Challenges

* **Benefits:**
  - Easy programmability
  - Small area and low power of the DSP components
* **Challenges:**
  - More complex ADC, which has to work at GHz speeds [1]
  - Some DSP blocks have to process GHz-rate signals
* **Some existing solutions:**
  - Intermediate-frequency ADC, analog filtering and digitization [2]
  - Discrete-time signal processing for signal conditioning [3]

[1] N. Beilleau *et al.*, "A 1.3V 26mW 3.2GS/s Undersampled LC Bandpass ΣΔ ADC for a SDR ISM-band Receiver in 130nm CMOS," in *Proc. Radio Frequency Integrated Circuits Symp.*, June 2009, pp. 383-386.
[2] G. Hueber *et al.*, "An Adaptive Multi- Mode RF Front-End for Cellular Terminals, " in *Proc. Radio Frequency Integrated Circuits RFIC Symp.*, June 2008, pp. 25-28.
[3] R. Bagheri *et al.*, "An 800MHz to 5GHz Software-Defined Radio Receiver in 90nm CMOS," in *Proc. Int. Solid-State Circuits Conf.*, Feb. 2006, pp. 480-481.

13.6

Slide 13.6

The main benefits of this idea are easy programmability, small area and low power of the DSP components, which replace the RF and analog blocks. But in doing so, we have pushed a large part of the computational complexity into the ADC, which must now digitize the incoming RF signal at GHz speeds (f_{s1}) [1], while also ensuring sufficient dynamic range necessary for wideband digitization. Also, some of the DSP blocks in the DFE chain must now process signals at GHz sample rates. To mitigate the difficulties of high-speed, high dynamic range ADC design, other intermediate realizations for digital front ends have been proposed in literature. In [2], the ADC design constraints were relaxed by first down-converting the received signal to an intermediate frequency, followed by analog filtering and subsequent digitization at 104 Ms/s. In [3], the authors use discrete-time signal processing approaches for signal conditioning. We take a look at the design challenges and advantages associated with adopting a wholly digital approach in radio receiver design. This will enable us to understand challenges associated with high-speed DSP applications.

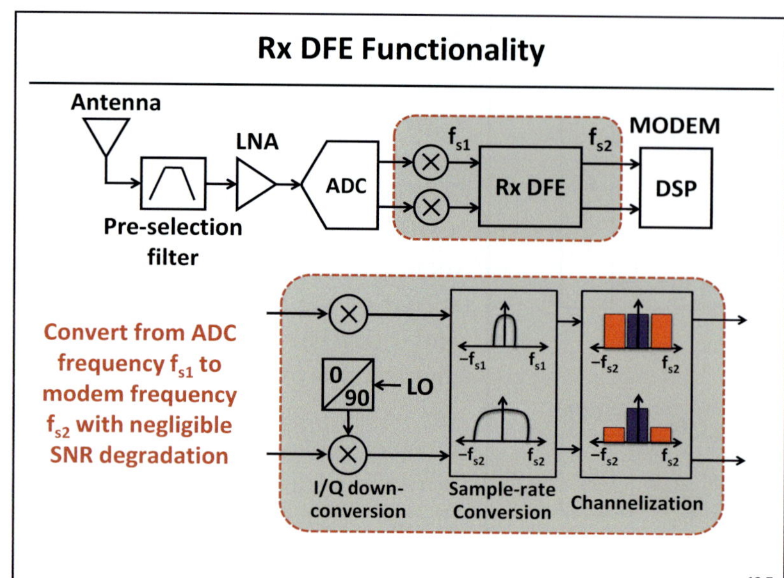

Slide 13.7

The ADC in the receiver chain takes the RF signal as input and digitizes it at frequency f_{s1}. The digitized signal is first down-converted using a mixer/digital multiplier. This baseband signal has bandwidth in the order of 0–20 MHz for the current cellular and WLAN standards. The sampling frequency f_{s1}, however, is in the GHz range, making the signal heavily over-sampled. The MODEM at the end of the receiver chain accepts signals at frequency f_{s2}, typically in the MHz range, its value being dictated by the standard. The Rx DFE blocks must down-sample the baseband signal from sample rate f_{s1} to the MODEM sampling rate of f_{s2}. This down-sampling operation must be performed with negligible SNR degradation. Also, for full flexibility, the DFE must be able to down-sample from arbitrary frequency f_{s1} to any baseband MODEM frequency f_{s2}. This processing must be achieved with power dissipation and area comparable to or less than the traditional RF receivers built with analog components.

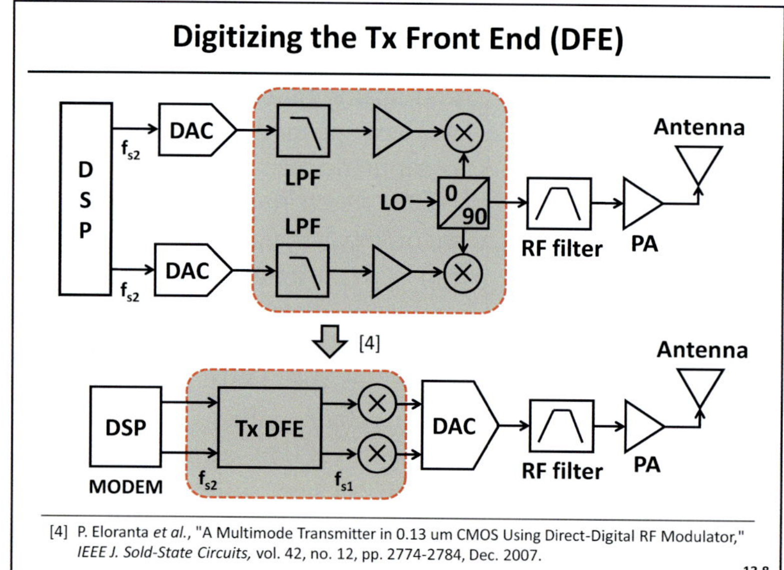

Slide 13.8

Up to now we have talked about the digitization of the receiver front-end chain. The same concept applies to the transmitter chain as well. The top figure in the slide shows the conventional transmitter architecture. The low-frequency baseband signal from the MODEM is converted to an analog signal using a digital-to-analog converter. This signal is then low-pass filtered, amplified, and up-converted to the RF carrier frequency using analog and RF blocks. Incorporating flexibility with minimum power and area overhead in the analog and RF blocks, again, prove to be a bottleneck in this implementation. The bottom figure in the slide shows the proposed Tx DFE implementation [4]. Here, the signal conditioning circuits have been moved to the digital domain with the objective of incorporating flexibility in the design components, and also to avoid problems of linearity, mismatch, and dynamic range, commonly associated with analog circuits.

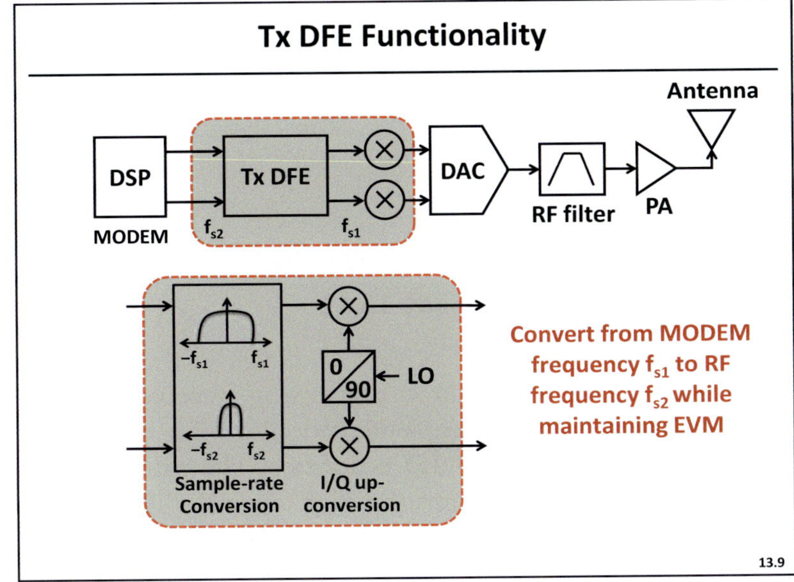

Slide 13.9

The Tx DFE up-samples the incoming baseband signal from rate f_{s2} (MODEM sampling rate) to the higher sampling rate f_{s1} at the D/A input. Over-sampling the signal is necessary for more than one reason. Firstly, the spectral images at multiples of the baseband sampling frequency can be suppressed by digital low-pass filtering after over-sampling. Secondly, the quantization noise spreads across a larger spectrum after over-sampling, with the amplitude reducing by 3 dB for every doubling of sampling frequency, provided the number of digital bits in the data stream is not reduced. Noise amplitude reduction is necessary to satisfy the power emission mask requirement in the out-of-band channels. The degree of over-sampling is determined by the extent to which quantization noise power levels need to be suppressed.

Over-sampling, however, is not without its caveats. A large part of the complexity is now pushed to the D/A converter, which must convert digital bits to analog signals at GHz rates. The Tx DFE structure must provide up-sampling and low-pass filtering of spectral images at small power and area overhead, to allow larger power headroom for the D/A converter. To ensure full flexibility, the DFE chain must up-sample signals from any MODEM frequency f_{s2} to any D/A frequency f_{s1}. Also, this up-sampling should be done with an acceptable value of error vector magnitude (EVM) of the transmitted signal.

Example: LTE & WiMAX Requirements

- Wide range of RF carrier frequencies
- Support for multiple data bandwidths & MODEM frequencies

Uplink (LTE) 1.92-2.4 GHz
Downlink (LTE) 2.1-2.4 GHz
WiMAX 2.3-2.69 GHz

Uplink
Downlink

MODEM Frequencies

Channel BW [MHz]	Sampling Frequency [MHz]	
	LTE	WiMAX
1.25	1.92	-
2.5	3.84	-
5	7.68	5.6
7	-	8
8.75	-	10
10	15.36	11.2
15	23.04	-
20	30.72	22.4

13.10

Slide 13.10

The slide shows examples of flexibility requirements for present-day cellular and WLAN standards like long-term evolution (LTE) and the wireless MAN (WiMAX 802.16). The RF carrier frequencies can be anywhere between 1.92–2.69 GHz in the uplink (transmitter chain) and 2.11–2.69 GHz in the downlink (receiver chain). The signal bandwidth ranges between 1.25–20 MHz. The RF carrier frequencies determine the bandwidth of the direct conversion ADC, and consequently the value of f_{s1}, the ADC sampling frequency. The signal bandwidth determines the MODEM sampling frequencies (f_{i2}) for both DFEs. The table shows corresponding MODEM sampling frequency for different signal bandwidths for the LTE and WiMAX standards. The MODEM sampling frequencies are slightly greater than the channel bandwidth (Nyquist rate). This is due to the presence of extra bits for guard bands and headers, which increase the actual signal bandwidth. In the next slides we look at the design challenges associated with the design of digital front-end receivers and transmitters.

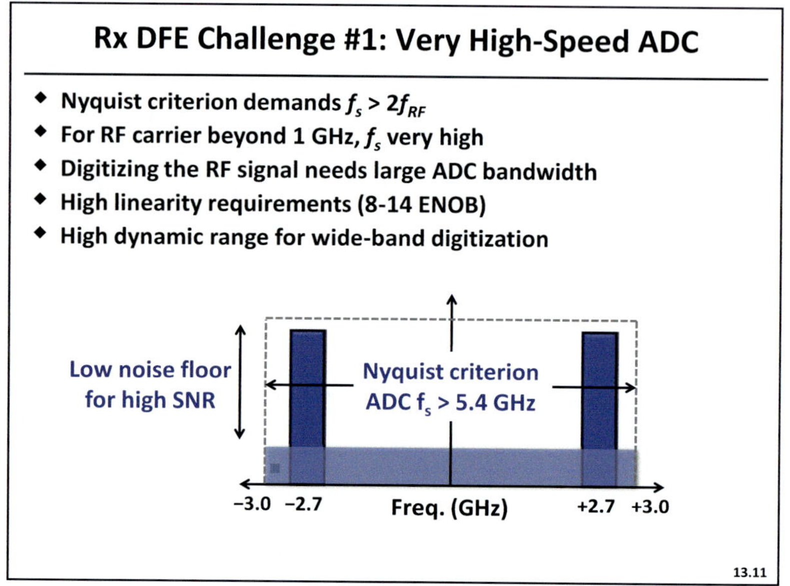

Slide 13.11

The first challenge in the design of direct-conversion receivers is the implementation of the high speed A/D converter. The sampling rate must be high since the RF signal is centered at a GHz carrier ranging between 2 and 2.7 GHz for LTE/WiMAX. According to Nyquist criterion, a sampling frequency of 4 to 5.4 GHz is required for direct RF signal digitization. For example, the slide illustrates an ADC bandwidth of 4 GHz required to digitize a signal centered at 2 GHz. A point to note is that the signal bandwidth is much smaller than the RF carrier frequency (in the order of MHz). Nevertheless, if Nyquist criterion is to be followed then the sampling frequency will be dictated by the RF carrier frequency and not the signal bandwidth. The second challenge stems from the SNR specifications of greater than 50 dB for current standards. This imposes a linearity requirement in the range of 8–14 effective number of bits on the ADC. The ADC design is therefore constrained by two difficult specifications of high sampling rate as well as high linearity.

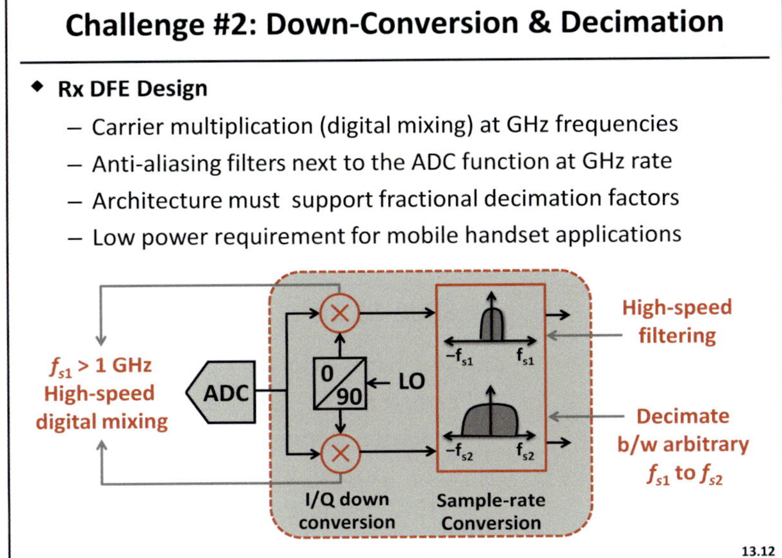

Slide 13.12

After the ADC digitizes the incoming RF signal at frequency f_{s1}, the digital signal must be down-converted from the RF carrier to the baseband. This has to be done by a mixer/digital multiplier. Digital multiplication at GHz rate is practically infeasible or extremely power hungry even for short wordlengths (4–5 bits). The down-sampling/decimation filters in the DFE must process the high-speed incoming data at frequency f_{s1}. Another problem lies in enabling the processing of such high-speed data with minimal power and area overhead, which is mandatory if the DFE is to be migrated to mobile handset type of applications. Since the choice of f_{s1} and f_{s2} is arbitrary in both the Tx and Rx chain, it will often be the case that the sample rate change factor (f_{s1}/f_{s2}) will be fractional. Supporting fractional rate change factors becomes an additional complexity in the DFE design.

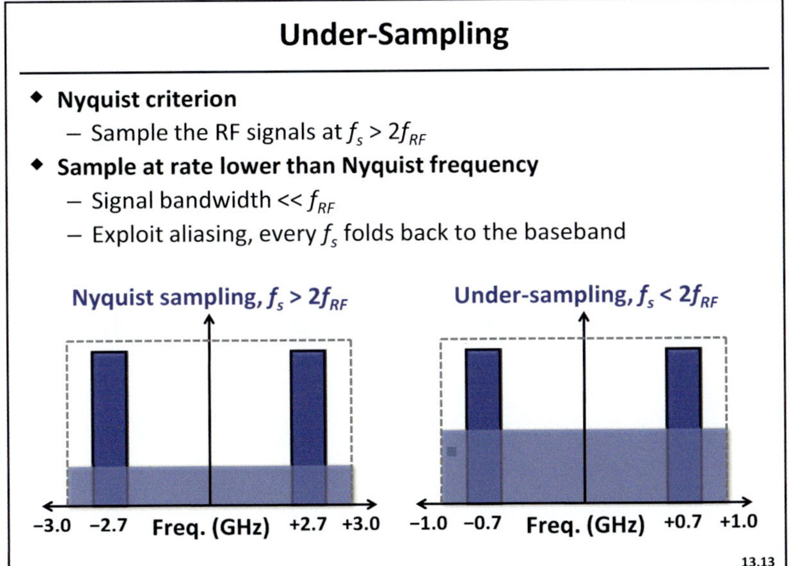

Slide 13.13

Processing of digital samples at frequency f_{s1} in the ADC and the DSP blocks is a primary bottleneck in the receiver chain. The timing constraints on the entire DFE chain can reduce significantly, if f_{s1} can be lowered through optimization techniques. This reduction in sample rate, however, must not adversely affect the signal-to-noise ratio. One such technique is under-sampling or sub-Nyquist sampling. Under-sampling exploits the fact that the signal bandwidth in cellular and WLAN signals is orders of magnitude lower than the RF carrier frequency. Hence, even if we sample the RF signal at frequencies lesser than the Nyquist value of $2f_{RF}$, a replica of the original signal can be constructed through aliasing. When a continuous-time signal is sampled at rate f_s, then post sampling, analog-domain multiples of frequency band f_s overlap in the digital domain. In the example shown here, an RF signal centered at 2.7 GHz is sampled at 2 GHz. Segments of spectrum in the frequency band of 1 to 3 GHz and −1 to −3 GHz fold back into the −1 to +1 GHz sampling band. We get a replica of the original signal at 0.7 GHz and −0.7 GHz. This phenomenon is referred to as aliasing. If there is no interference in the −1 to +1 GHz spectrum range, then the signal replica at 0.7 GHz is uncorrupted. Band-pass filtering before the A/D conversion can ensure this. The number of bits in the ADC determines the total noise power, and this noise (ideally white) spreads uniformly over the entire sampling frequency band. Lowering the sampling frequency results in quantization noise spread in a smaller frequency band, as shown in the figure on the right, which has higher noise floor levels. Hence, the extent of under-sampling is restricted by the allowed in-band noise power level.

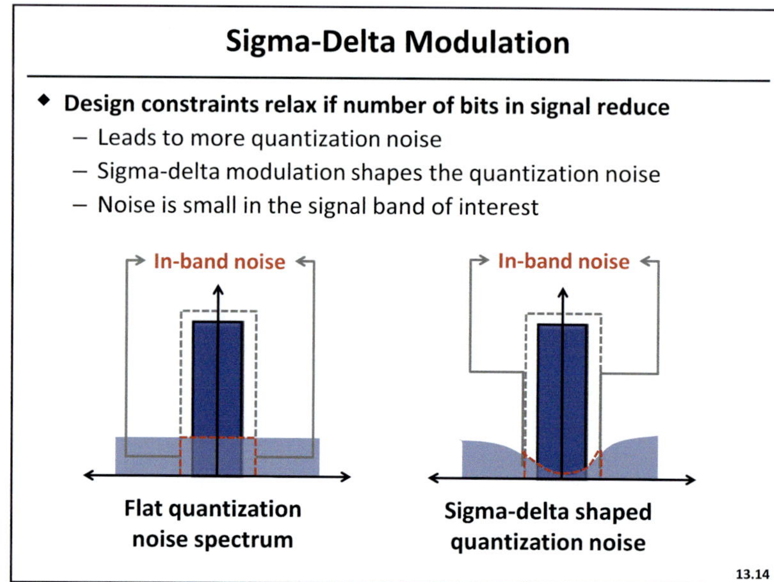

Slide 13.14

We have now looked at the critical constraints of bandwidth and linearity imposed on the ADC. We also looked at ways to reduce the ADC sample rate. But achieving linearity specifications of 50–60 dB is still difficult over a uniform ADC bandwidth spanning several hundred MHz. For smaller bandwidth signals, a popular approach is to reduce the noise in the band of interest through over-sampling and sigma-delta modulation. Over-sampling is a technique used to push down the quantization noise floor in the ADC bandwidth. The total quantization noise power of the ADC remains unchanged if the number of bits in the ADC is fixed. With higher sampling frequency, the same noise power is spread over a larger frequency band, pushing down the noise amplitude levels. After filtering, only the noise content in the band of interest contributes to the SNR, everything outside is attenuated. Sigma-delta modulation is a further step towards reducing the noise power in the signal band of interest. With this modulation, the quantization noise is shaped in a manner that pushes the noise out of the band of interest leading to a higher SNR. The figure on the right shows an example of this noise shaping.

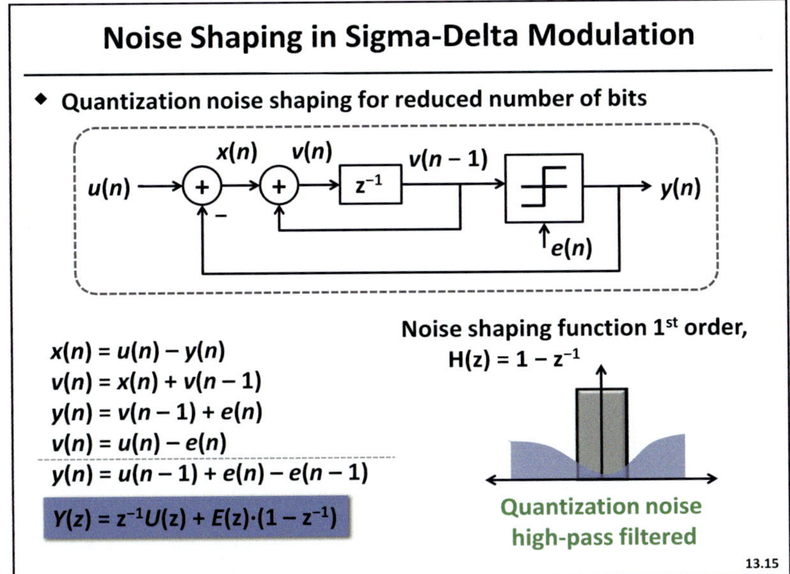

Slide 13.15

The slide shows an example of first-order sigma-delta noise shaping. The noise shaping is implemented by high-pass filtering the quantization noise $E(z)$, while the incoming signal $U(z)$ is unaltered. The equations in the slide illustrate the process for first-order filtering with transfer function $H(z) = (1 - z^{-1})$. Changing this filter transfer function can increase the extent of noise shaping. Common practice is to use higher-order IIR filters (2nd to 4th) in the feedback loop to reduce the noise power in the band of interest.

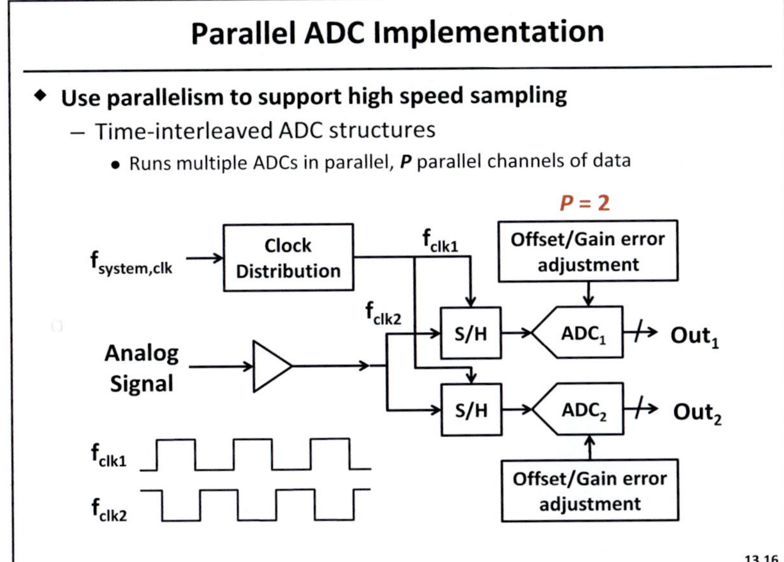

Slide 13.16

If the value of f_{s1} is quite large even after under-sampling, then further optimization is needed to enable high-throughput signal processing. One such technique is parallelism, which will be utilized several times in the course of DFE design. Time-interleaved ADC is the common term used for parallel ADCs. In this case, the incoming continuous-time signal is processed by N ADCs running in parallel and operating at sampling frequency f_s/N, f_s being the sampling frequency of the complete ADC. Time interleaving uses multiple ADCs functioning with time-shifted clocks, such that the N parallel ADCs generate a chunk of N continuous samples of the incoming signal. Adjacent ADC blocks operate on clocks that are time shifted by $1/f_s$. For example, in the figure, for a 2-way parallel ADC, the system clock at rate f_s is split into two clocks, f_{clk1} and f_{clk2}, which are time-shifted by $1/f_s$, and with equal frequency of $f_s/2$. Both clocks independently sample the input signal for A/D conversion. The output of the ADC are 2 parallel channels of data that generate 2 digital samples of the input at rate $f_s/2$, making the overall sampling frequency of the system equal to f_s. Although this is a very attractive technique to increase the sample rate of any ADC, the method has its shortcomings. One problem is the timing jitter between multiple clocks that can cause sampling errors. To avoid this, the use of a small number (2–4) of parallel channels is recommended. With fewer channels, the number of clock domains is less, and the timing jitter is easier to compensate through calibration mechanisms after A/D conversion.

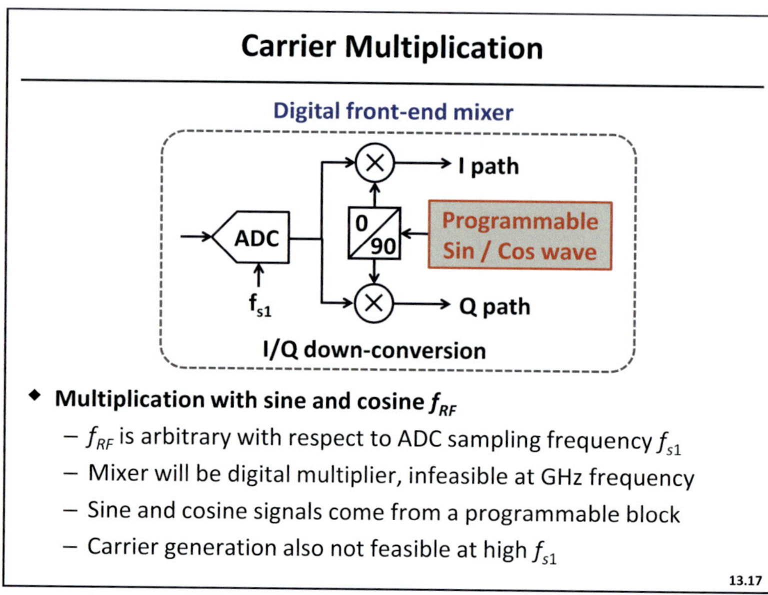

Slide 13.17

Once the incoming RF signal is digitized, the next task is to down-convert it to the baseband. For this purpose, we require a digital mixer. The mixer has two components, a multiplier and a frequency synthesizer. If the ADC sampling frequency f_{s1} is in the GHz range, then the multiply operation becomes infeasible. Digital sine/cosine signal generation is usually implemented through look-up tables or CORDIC units. These blocks also cannot support GHz sampling rates. Use of parallel channels of data through time-interleaved ADCs is one workaround for this problem. N parallel channels make the throughput per channel equal to f_{s1}/N. This, however, does not quite solve the problem, since to ensure timing feasibility of carrier multiplication, the value of N will have to be very large leading to ADC timing jitter issues, discussed previously.

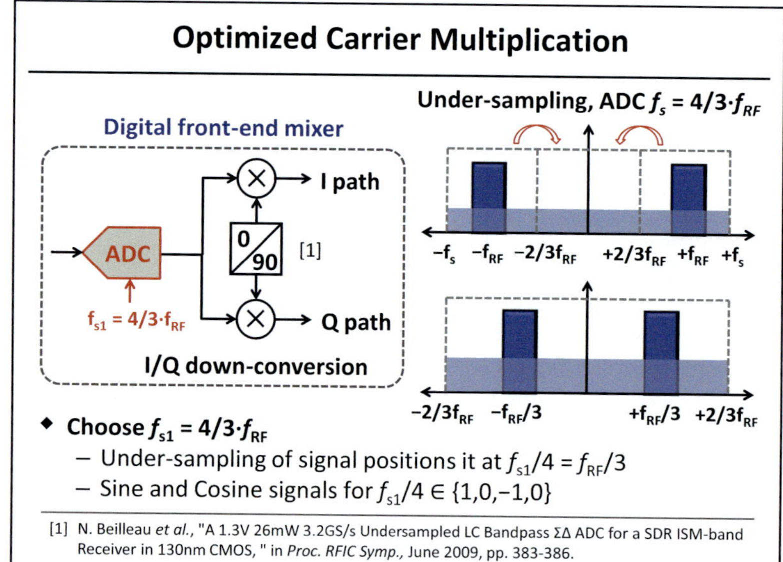

Slide 13.18

A solution to the carrier multiplication problem is to make the ADC sampling frequency programmable. If the value of f_{s1} is a function of f_{RF}, then we can ensure that after under-sampling the replica signal is judiciously positioned, so that we avoid any carrier multiplication after all. For example, in the figure shown in the slide, $f_{s1} = (4/3) \cdot f_{RF}$. After under-sampling, the replica signal is created at frequency $f_{RF}/3$ that corresponds to the $\pi/2$ position in the digital domain ($f_{s1}/2$ being the π position) [1]. The mixing process reduces to multiplication with $\sin(\pi/(2n))$ and $\cos(\pi/(2n))$, both of which are elements of the set $\{1, 0, -1, 0\}$, when n is an integer. Hence, implementing the mixer becomes trivial, requiring no look up tables or CORDIC units. Similarly $f_{s1} = f_{RF}$ can also be used, in which case the replica signal is created at the baseband. In this case the ADC sampling and carrier mixing will be done simultaneously; but two ADCs will be necessary in this case to generate separate *I* and *Q* data streams. A programmable PLL will be required for both these implementations in order to tune the sampling clock of the ADC depending on the received carrier frequency.

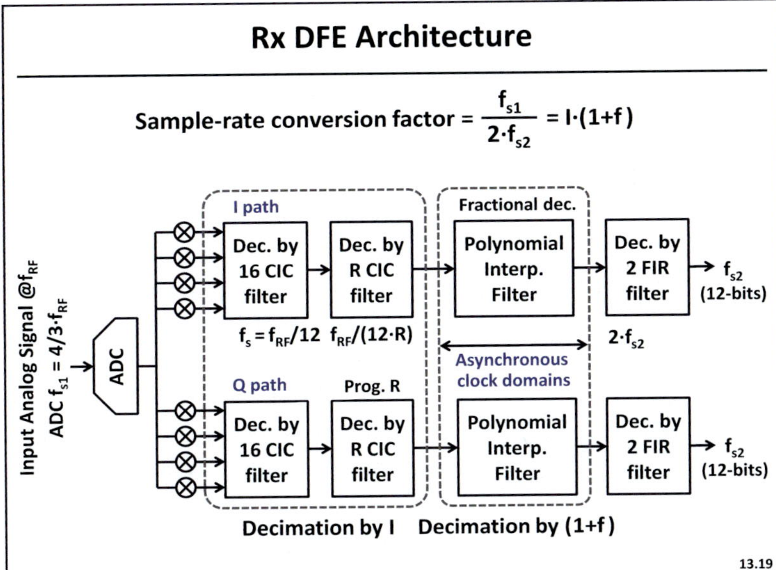

Slide 13.19

After incorporating all the optimization techniques discussed, a possible implementation of the Rx DFE is shown in the figure. The receiver takes in the RF signal and under-samples it at frequency $4 \cdot f_{RF}/3$. The ADC is time-interleaved with four channels of data. Under-sampling positions the replica signal at $\pi/2$ after digitization, making the mixing process with cosine and sine waves trivial. This is followed by decimation by 16 through a cascade-integrated-comb (CIC) filter, which brings the sampling frequency down to $f_{RF}/12$. The decimation by R CIC block is the first programmable block in the system. This block takes care of the down-conversion by integer factor R (variable value set by user). Following this block is the fractional sample rate conversion block, which is implemented using a polynomial interpolation filter. The fractional rate change factor is user specified. The polynomial interpolation filter has to hand-off data between two asynchronous clock domains. The final block is the decimation-by-2 filter, which is a low-pass FIR. In the next few slides we will discuss the DSP blocks shown in this system.

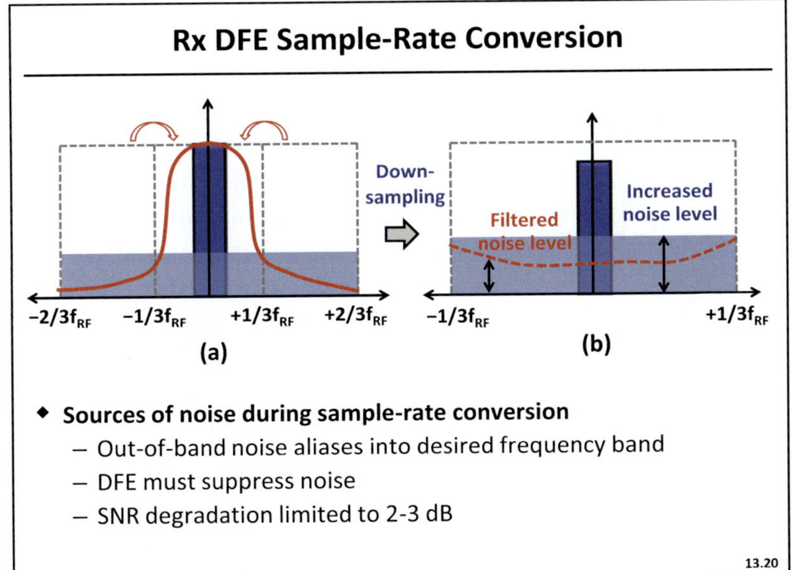

- **Sources of noise during sample-rate conversion**
  - Out-of-band noise aliases into desired frequency band
  - DFE must suppress noise
  - SNR degradation limited to 2-3 dB

Slide 13.20

We earlier saw the phenomenon of aliasing, when a continuous time signal is under-sampled. The same concept is applicable when a discrete signal sampled at frequency f_{s1} is down-sampled to a lower frequency f_{s2}. Any noise/interference outside the band $(-f_{s2}/2, f_{s2}/2)$ folds back into the baseband with an increased noise level, as shown in Fig. (b). This aliased noise can degrade the SNR significantly. The DFE must attenuate the out-of-band noise through low-pass filtering (shown with the *red* curve in Fig. (a)) before down-sampling. The low-pass filtering reduces the noise level shown by the dashed curve in Fig. (b). The SNR degradation from the output of the ADC to the input of the MODEM should be limited to within 2–3 dB, to maximize the SNR at the input of the MODEM. One way to suppress this noise is through CIC filtering, which is attractive due to its simple structure and low-cost implementation.

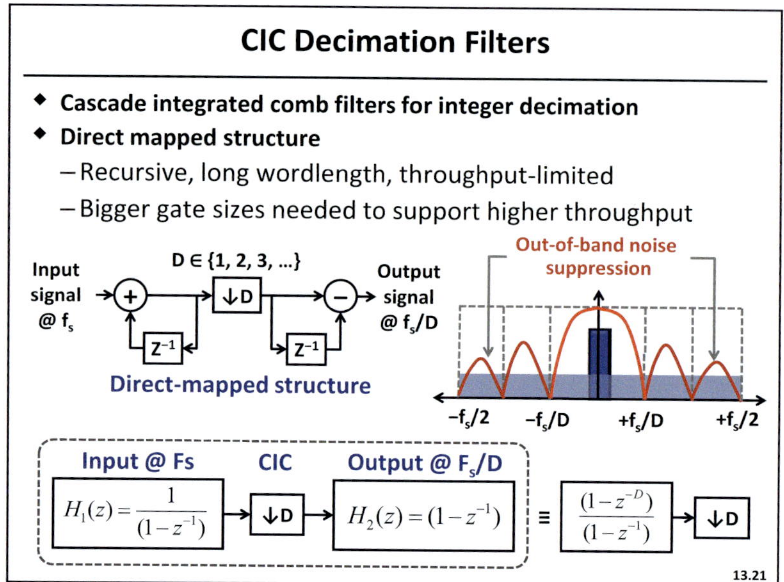

$$H_1(z) = \frac{1}{(1-z^{-1})} \quad \downarrow D \quad H_2(z) = (1-z^{-1}) \quad \equiv \quad \frac{(1-z^{-D})}{(1-z^{-1})} \quad \downarrow D$$

Slide 13.21

The slide shows an implementation of a CIC filter used for down-sampling by a factor of D. The structure has an integrator followed by a down-sampler and differentiator. Frequency response of this filter is shown on the left. The response ensures that the attenuation is small in the band of interest spanning from $-f_s/D$ to f_s/D. The out-of-band noise and interference lies in the band f_s/D to $f_s/2$ and $-f_s/D$ to $-f_s/2$. The filter attenuates the signal in the out-of-band region. Increasing the number of CIC filter sections increases the out-of-band attenuation. The integrated transfer function of the CIC filter for a single section and decimation factor D is shown at the bottom of the slide. Characteristics of higher-order CIC filtering are discussed next.

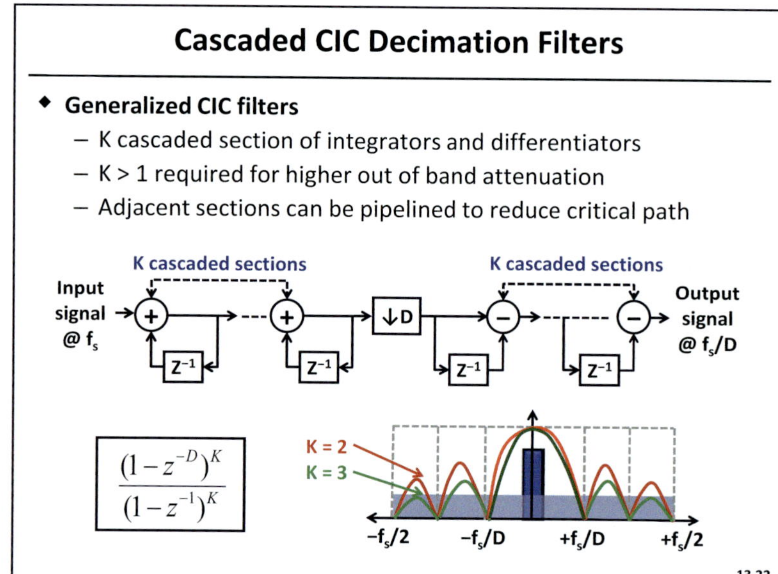

$$\frac{(1-z^{-D})^K}{(1-z^{-1})^K}$$

Slide 13.22

The previous slide shows a single integrator and differentiator section in the CIC. An increase in attenuation can be obtained with additional sections. Although the implementation of the filter looks quite simple, there still remain a few design challenges. Firstly, the integrator is a recursive structure that often requires long wordlengths. This creates a primary bottleneck with respect to the maximum throughput (f_s) of the filter. In our system (Slide 13.19), a CIC filter is placed just after the ADC, where data streams can have throughput up to several hundred MHz. A second drawback of the feedback integrator structure is its lack of support for parallel streams of data input. Since we would like to use time-interleaved ADCs, the CIC should be able to take the parallel streams of data as input. In the next slide we will see how the CIC transfer function can be transformed to solve both of these problems.

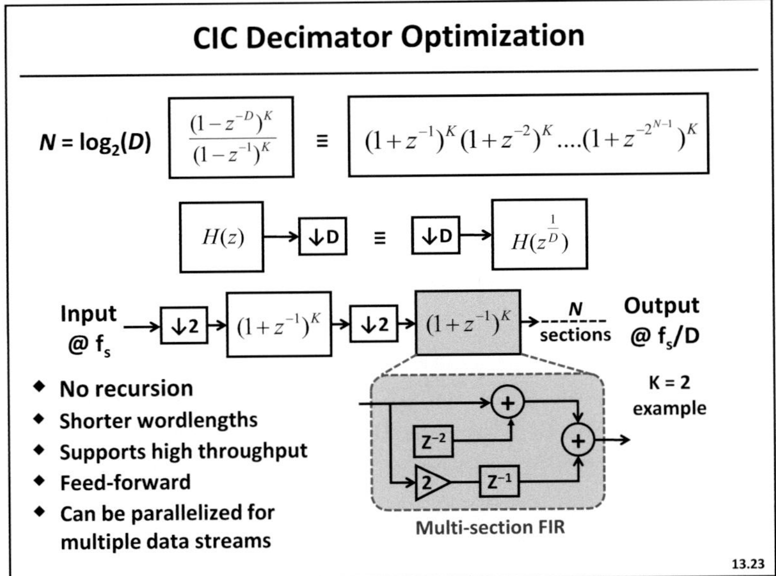

Slide 13.23

The CIC filter transfer function, when expanded, is an FIR or feed-forward function as shown in the first equation on this slide. When the decimation factor D is of the form a^x, the transfer function can be expressed as a cascade of x units decimating by a factor of a. In the example shown in the slide, the transfer function for decimation by 2^N is realized using a cascade of decimation-by-2 structures. The number of such structures is equal to N or $\log_2(D)$ where $D=2^N$. The decimation-by-2 block is simple to implement requiring multiply-by-2 (shift operation) and additions when the original CIC has 2 sections. Also, the feed-forward nature of the design results in smaller wordlengths, which makes higher throughput possible. An additional advantage of the feed-forward structure is its support for parallel streams of data coming from a time-interleaved ADC. So, we get the desired filter response along with a design that is amenable to high-throughput specifications. It may look like this implementation adds extra area and power overhead due to the multiple sections in cascade, however it should be noted that the successive sections work at lowered frequencies (every section decimates the sampling frequency by 2) and reduced wordlengths. The overall structure, in reality, is power and area efficient.

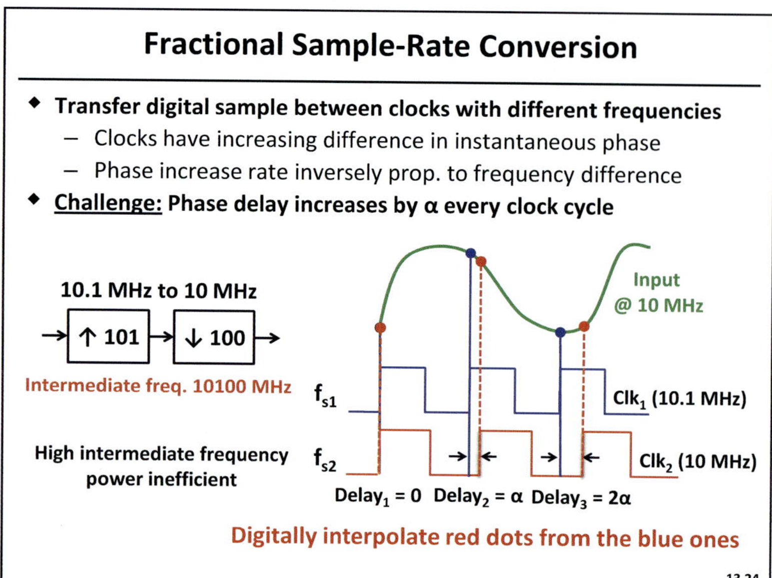

Slide 13.24

Although integer sample-rate conversion can be done using CIC or FIR filters as shown earlier, these filters cannot support fractional decimation. Owing to the flexibility requirement in the DFE, the system must support fractional sample-rate conversion. There a couple of ways to implement such rate conversions. One way is to express the fractional rate conversion factor as a rational number p/q. The signal is first up-sampled by a factor of q and then down-sampled by a factor of p to obtain the final rate conversion factor. For example, to down-convert from 10.1 MHz to 10 MHz, the traditional scheme is to first up-sample the signal by a factor of 101 and then down-sample by 100, as shown in the figure on the left. But this would mean that some logic components would have to function at an intermediate frequency of 10.1 GHz, which can be infeasible. To avoid this problem, digital interpolation techniques are used to reconstruct samples of the signal at the output clock rate given the samples at the input clock rate. The figure shows digital interpolation of the output samples (*red dots*), given the input samples (*red dots*). Since both clocks have different frequencies, the phase delay between their positive edges changes every cycle by a difference of alpha (α), which is given by the difference in the time-period of both clocks. This value of alpha is used to re-construct the output samples from the input ones, through use of interpolation polynomials. The accuracy of this re-construction depends on the order of interpolation. Typically, third-order is sufficient for 2–3 dB of noise figure specifications. To minimize area and power overhead, fractional sample-rate conversion should be done, as far as possible, at slow input clock rates.

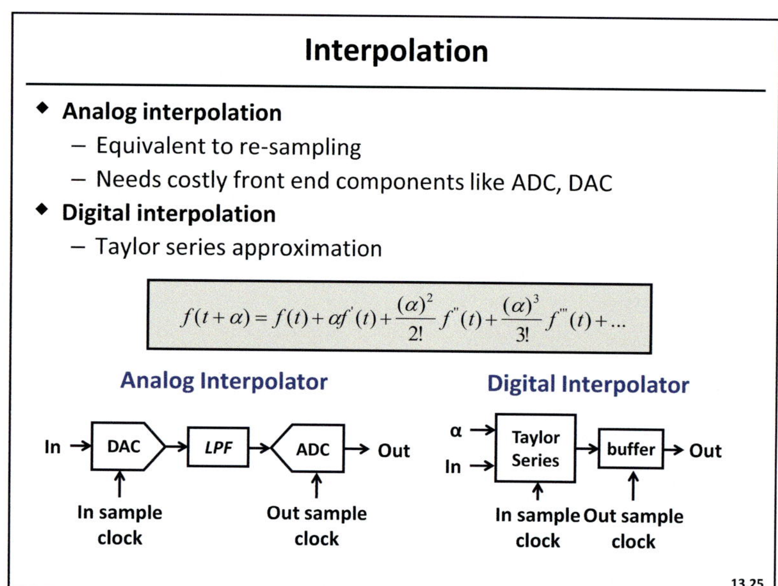

$$f(t+\alpha) = f(t) + \alpha f'(t) + \frac{(\alpha)^2}{2!} f''(t) + \frac{(\alpha)^3}{3!} f'''(t) + \ldots$$

Slide 13.25

The interpolation process is equivalent to re-sampling. Given a discrete signal at a certain input clock f_{in}, analog interpolation would involve a digital to analog conversion and re-sampling at output clock f_{out} using an ADC. This method uses costly mixed-signal components, and ideally we would like to pursue an all-digital alternative. As mentioned in the previous slide, it is possible to construct the digital samples at clock f_{out}, given the input samples at clock f_{in} and the time difference alpha (α) between both clocks. The process is equivalent to an implementation of Taylor's series, where the value of the signal at time $\tau + \alpha$ is obtained by knowing the value of the signal at time τ, and the value of the signal derivatives at time τ. This slide shows the Taylor's series expansion of a function at $f(\tau + \alpha)$. The value of $f(\tau)$ is already available in the incoming input sample. The remaining unknowns are the derivatives of the signal at time τ.

Implementing Taylor Series

- **Third-order truncation of the series is typically sufficient**
- **Implementing differentiators**
  - Implement $D_1(w)$ using an FIR filter
  - Ideal response needs infinite # of taps
  - Truncate FIR → poor response at high frequencies

$$f'(t) = \frac{df}{dt} = \frac{dF(s)}{dt} = sF(s) = jwF(w) = D_1(w)F(w)$$

Slide 13.26

Signal derivatives can be obtained by constructing an FIR transfer function, which emulates the frequency response of the derivative function. For example, the transfer function for the first derivative in the Laplace domain is $s \cdot F(s)$. This corresponds to a ramp-like frequency response $D_1(w)$. This ideal response will require a large number of taps, if implemented using a time-domain FIR type function. An optimization can be performed at this stage, if the signal of interest is expected to be band-limited. For example, if the signal is band-limited in the range of $-2\pi/3$ to $+2\pi/3$ ($-f_s/3$ to $+f_s/3$), then the function can emulate the ideal derivative function in this region, and attenuate the signals outside this band, as shown in the right figure. This will not corrupt the signal of interest, with the added benefit of attenuating the interference/noise outside the signal band. The modified frequency response can be obtained with surprisingly small number of taps using an FIR function, leading to a low-cost power-efficient implementation of the interpolator.

The Farrow Structure

- **8 taps in the FIR to approximate differentiator $D_i(w)$**
- **The upper useful frequency is 0.7π**
- **C_0 is an all-pass filter in the band of interest**
- **Structure can be re-configured easily by controlling parameter α**

$$y_i = \frac{f^i(nT)}{i!}$$

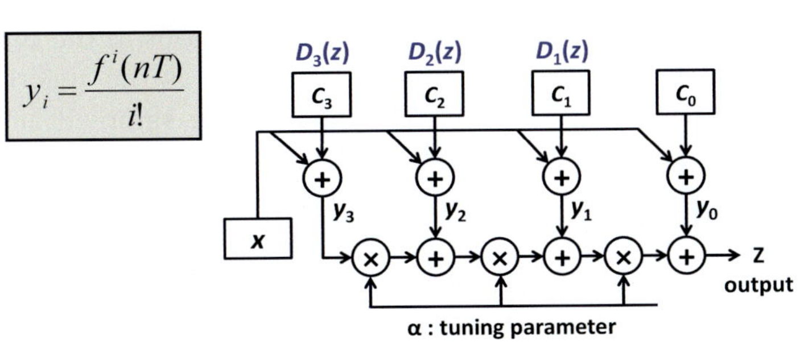

Slide 13.27

The slide shows a third-order Taylor's series interpolator. The computation uses the first three derivative functions $D_1(w)$, $D_2(w)$ and $D_3(w)$. Following the approximation techniques described in the previous slide, the three functions were implemented using 8-tap FIRs. The functions emulate the ideal derivative response up to 0.7π, which will be referred to as the upper useful frequency. Signals outside the band of -0.7π to 0.7π will be attenuated due to the attenuating characteristics of the approximate transfer functions. The most attractive property of this structure is its ability to interpolate between arbitrary input and output clocks. For different sets of input and output clocks, the only change is in the difference between their time-periods alpha (α). Since α is an external input, it can be easily programmed, making the interpolator very flexible. The user need only make sure that the signal of interest lies within the upper useful frequency band of $(-0.7\pi, 0.7\pi)$, to be able to use the interpolator.

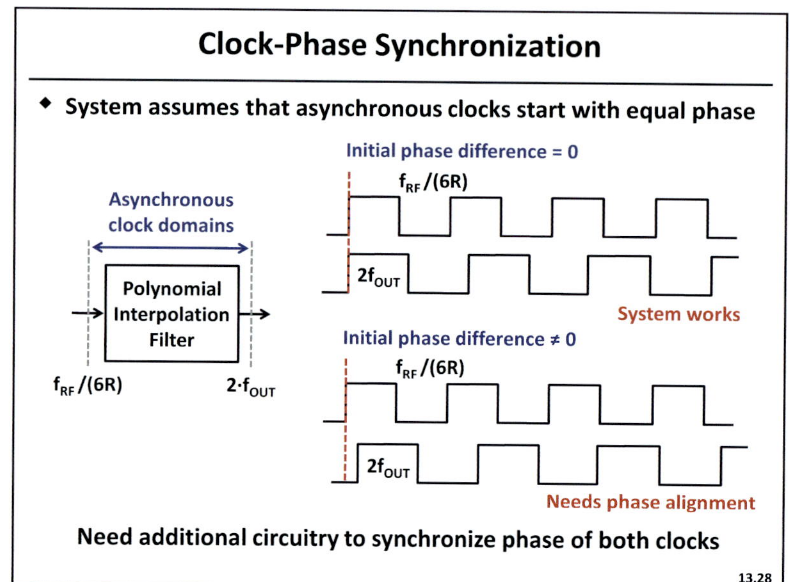

Slide 13.28

The unknown initial phase difference between the input and output clock becomes a problem for the interpolation process. If both clocks are synchronized so that their first positive edges are aligned at $t=0$, as shown in the top timing diagram, then the interpolation method works. But if there is an initial phase difference between the two clocks at time $t=0$, then the subsequent phase difference will be the sum of the initial phase difference and α. Additional phase detection circuit will be required to detect the initial phase. The phase detector can be implemented by tracking the positive-edge transitions of both clocks. At some instant the faster clock has two rising edges within one cycle of the slower clock. The interpolator calibrates the initial phase by detecting this event.

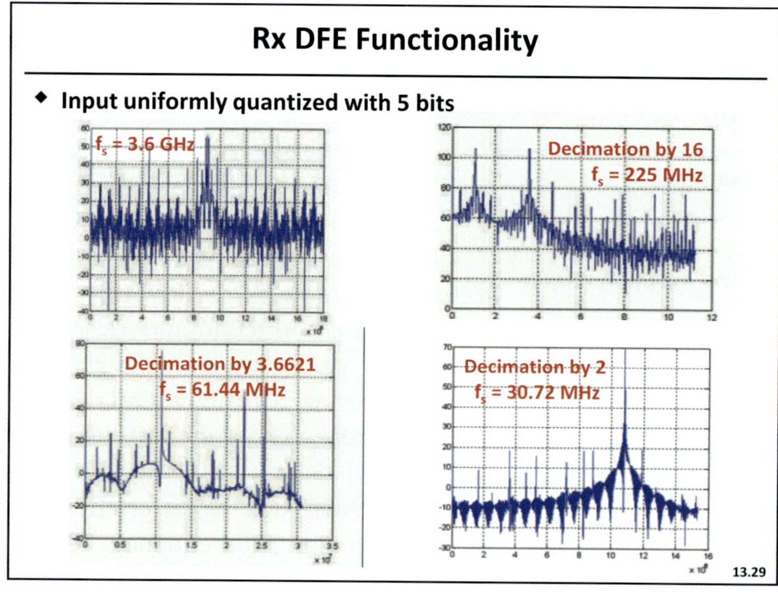

Slide 13.29

Now that we have looked at all the components of the block diagram for the Rx DFE shown in Slide 13.19, we can take a look at the expected functionality of this Rx DFE. The slide shows the frequency spectrum after various stages of computations in the DFE. The top-left figure is the spectrum of the analog input sampled at 3.6 GHz with a 5-bit ADC. The ADC input consists of two discrete tones (the tones are only a few MHz apart and cannot be distinguished in the top-left figure). The ADC noise spectrum is non-white, with several spurious tones scattered across the entire band. The top-right figure shows the signal after decimation by 16 and at a sample rate of 225 MHz. The two input tones are discernable in the top-right figure, with the noise beyond 225 MHz filtered out by the CIC. This is followed by two stages of decimation, integer decimation-by-3 using CIC filters and fractional decimation by 1.2207. The overall decimation factor is the product 3.6621, which brings the sample rate down to 61.44 MHz. The resulting spectrum is shown in the bottom-left figure. We can still see the two tones in the picture. The bottom-right figure shows the output of the final decimation-by-2 unit that filters out the second tone and leaves behind the signal sampled at 30.72 MHz. This completes our discussion on the Rx DFE design and we move to Tx DFE design challenges in the next slide.

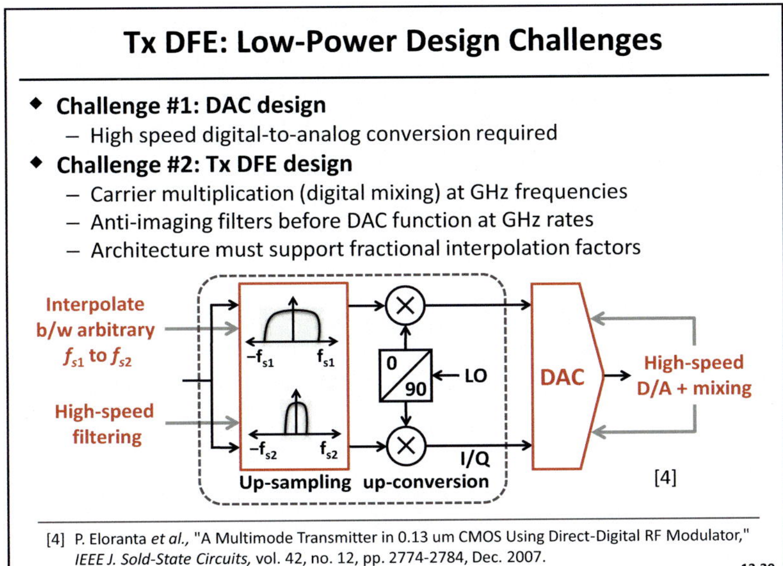

Slide 13.30

The Tx DFE has similar implementation challenges as the Rx. The figure shows an example of a Tx DFE chain. The baseband signal from the MODEM is up-sampled and mixed with the RF carrier in the digital domain, before being sent to a D/A converter. High-speed digital-to-analog conversion is required at the end of the DFE chain. The digital signal has to be over-sampled heavily, to lower noise floor levels, as discussed earlier. This would mean that anti-imaging filters and digital mixers in the Tx DFE chain would have to work at enormously high speeds. Implementing such high-throughput signal processing and data up-conversion becomes infeasible without opportunistic use of architectural and signal-processing optimizations. DRFC techniques [4] allow RF carrier multiplication to be integrated in the D/A converter architecture. The D/A converter power has to be optimized to lower the power consumption associated with digitally intensive Tx architectures.

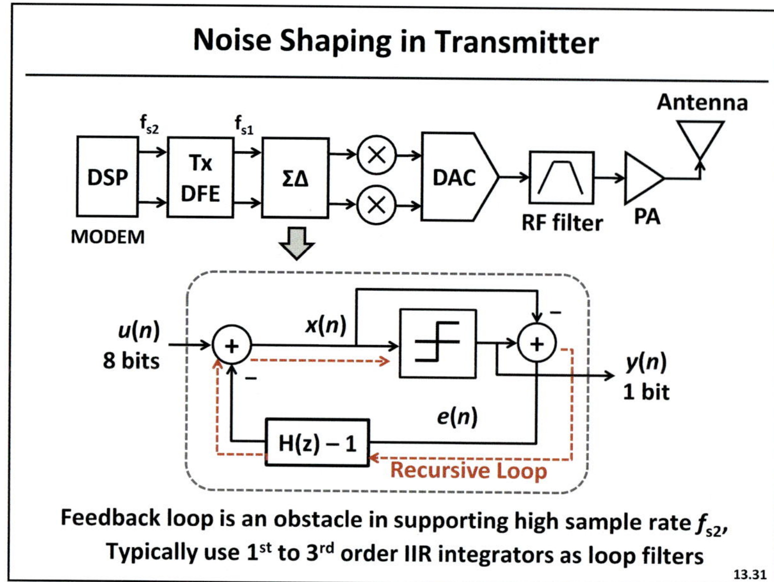

Slide 13.31

The power consumption of the D/A conversion block is directly proportional to the number of bits in the incoming sample. Sigma-delta modulation can be used at the end of the transmitter chain before up-conversion, to reduce the total number of bits going into the D/A converter. We saw earlier that sigma-delta modulation reduces the noise in the band of interest by high-pass filtering the quantization noise. This technique reduces the number of output bits for a fixed value of in-band noise power, since the noise introduced due to quantization is shaped outside the region of interest.

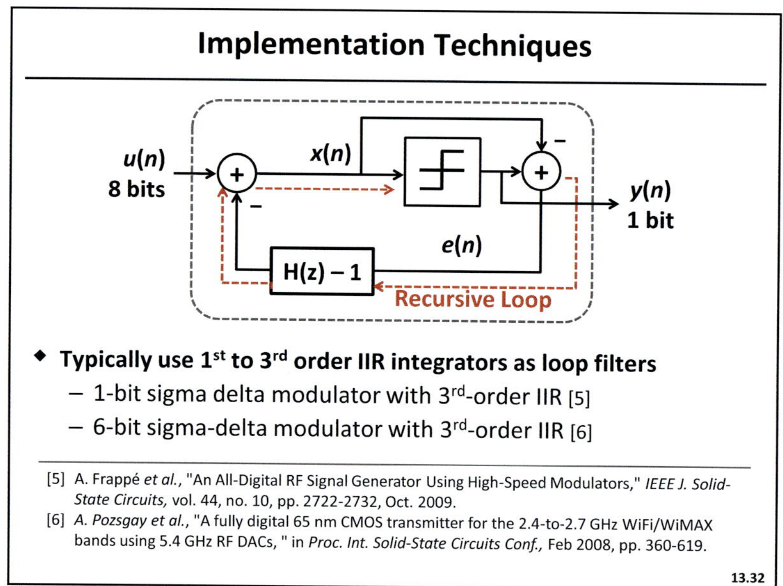

Slide 13.32

The implementation proposed in [5] uses a 1-bit sigma-delta modulator with third-order IIR filtering in the feedback loop. A second approach [6] uses 6-bit sigma-delta modulator with 3rd-order IIR noise shaping. The high-speed sigma-delta noise shaping filters are the main challenge in the implementation of these modulators.

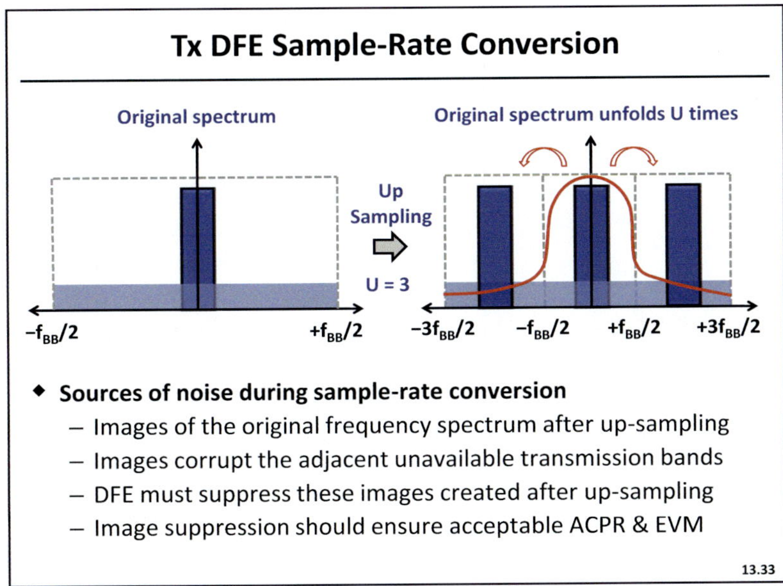

Slide 13.33

The up-sampling process results in images of the original signals at multiples of the original sampling frequency. In the figure, up-sampling by 3 leads to the 3 images shown in the spectrum on the right. While transmitting a signal, the Tx must not transmit power above a maximum level in the spectrum adjacent to its own band, since the adjacent bands belong to a different user. The adjacent-channel power ratio (ACPR) metric measures the ratio of the signal power in the channel of transmission and the power leaked in the adjacent channel. This ratio should be higher than the specifications set by the standard. Hence, images of the original spectrum must be attenuated through low-pass filtering before signal transmission. The common way is to do this through FIR and CIC filtering. The filtering process must not degrade the error vector magnitude (EVM) of the transmit constellation. The EVM represents the aggregate of the difference between the ideal and received vectors of the signal constellation. High quantization noise and finite wordlength effects during filtering can lower the EVM of the transmitted signal.

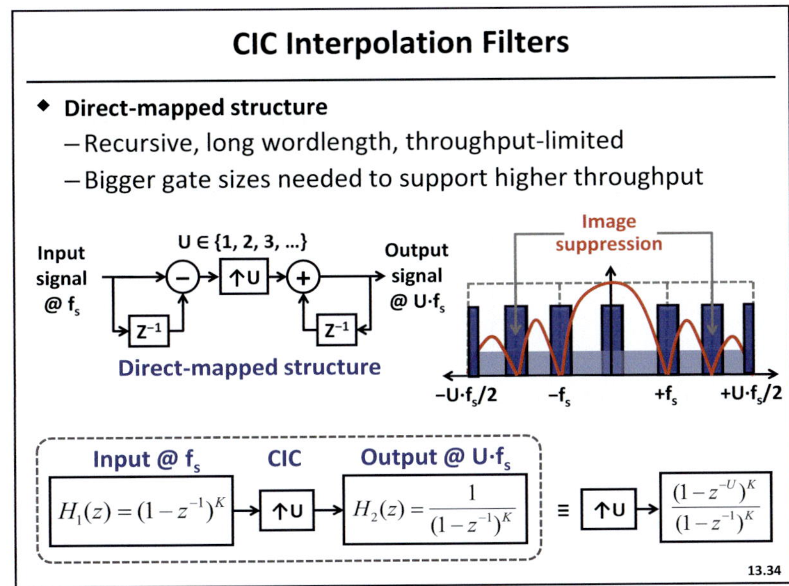

In the CIC block diagram, the transfer functions are:

$$H_1(z) = (1 - z^{-1})^K \quad \uparrow U \quad H_2(z) = \frac{1}{(1 - z^{-1})^K} \quad \equiv \quad \uparrow U \quad \frac{(1 - z^{-U})^K}{(1 - z^{-1})^K}$$

Slide 13.34

CIC filters are used for suppressing the images created after up-sampling. The CIC transfer function is shown on the right in the slide. The nulls of the transfer function lie at multiples of the sampling frequency f_s, which are the center of the images formed after up-sampling. The traditional recursive implementation of these filters suffers from long wordlengths and low throughput due to the feedback integrator. Optimization of these structures follows similar lines of feed-forward implementation as in the Rx.

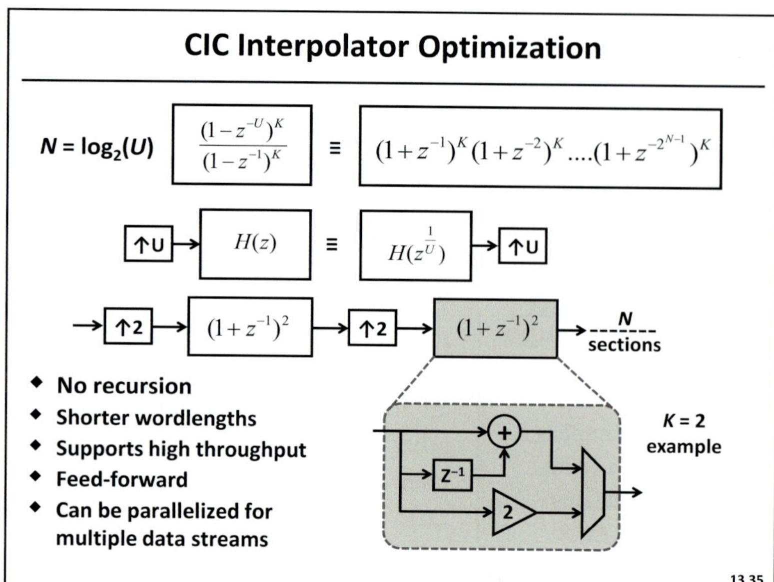

Slide 13.35

CIC filters used for up-sampling can also be expressed as feed-forward transfer functions, as shown in the first equation in the slide. A cascade of x up-sampling filters can implement the modified transfer function, when the up-sample factor U equals a^x. The example architecture shows up-sampling by 2^N. For a cascade of two sections ($K=2$), the filter architecture can be implemented using shifts and adds. This architecture avoids long wordlengths and is able to support parallel streams of output data as well as higher throughput per output stream.

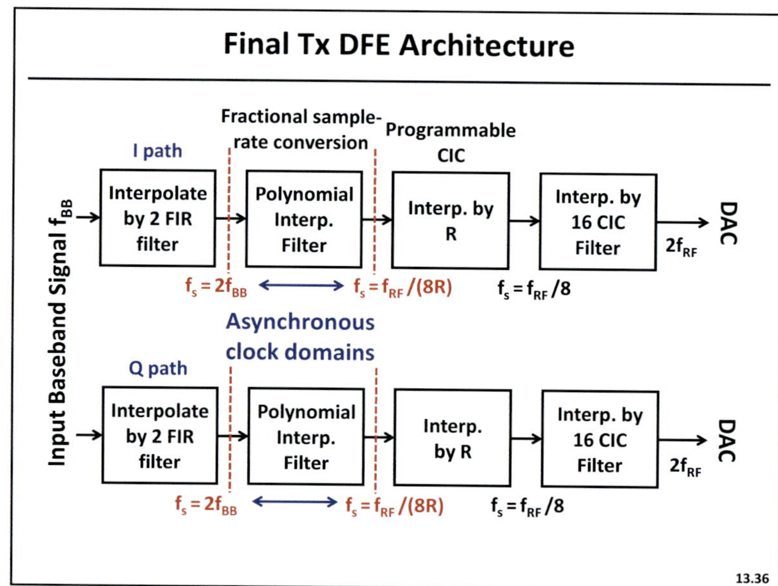

Slide 13.36

The transmit DFE structure shown in this slide can be regarded as the twin of the Rx DFE (from Slide 13.19) with the signal flow reversed. The techniques used for Tx optimization are similar to those used in the Rx architecture. The only difference is that the sampling frequency is higher at the output end as opposed to the input end in the Rx and all the DSP units up-sampling/interpolation instead of decimation. The polynomial interpolation filter described for the Rx DFE design can be used in the Tx architecture as well for data handoff between asynchronous clock domains.

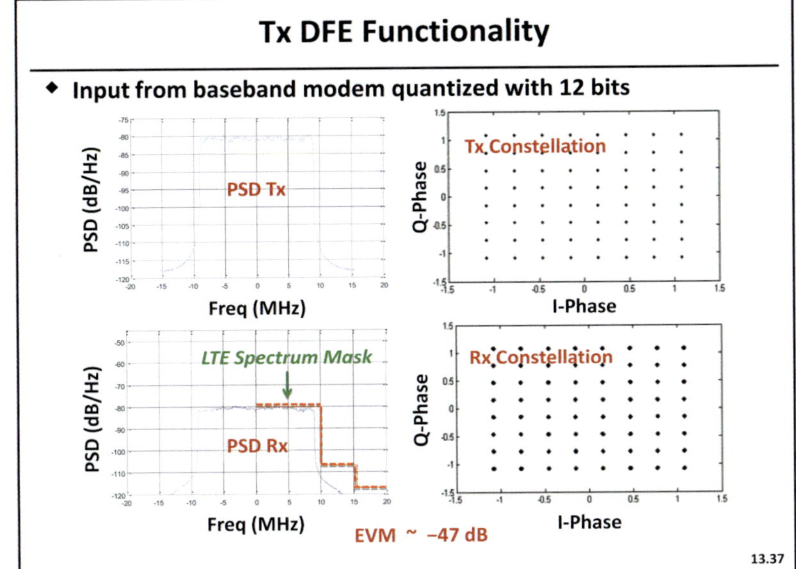

Slide 13.37

Simulation results for Tx DFE are shown in this slide. The input signal was at the baseband frequency of 30.72 MHz and up-converted to 5.4 GHz. The received signal PSD shows that the output signal is well within the spectrum mask and ACPR requirement. The measure EVM (error vector magnitude) was about −47 dB for the received spectrum.

Summary

- **Digital front-end implementation**
  - Avoids analog non-linearity
  - Power and area scales with technology

- **Challenges**
  - High dynamic range ADC in receiver
  - High throughput at Rx input and Tx output
  - Minimum SNR degradation of received signal
  - Quantization noise and image suppression at Tx output
  - Fractional sample-rate conversion

- **Optimization**
  - Sigma-delta modulation
  - Feed-forward CIC filtering
  - Farrow structure for fractional rate conversion

13.38

Slide 13.38

In summary, this chapter talked about implementation of receiver and transmitter signal conditioning circuits in a mostly digital manner. The main advantages associated with this approach lie in avoiding non-linear analog components and utilizing benefits of technology scaling with new generations of digital CMOS. Digitizing the receiver radio chain is an ongoing topic of research, mainly due to the requirement of high-speed, high dynamic-range mixed-signal components. Sigma-delta modulation and time interleaving are some techniques that could make such designs feasible. DSP challenges and optimization techniques like use of feed-forward CIC filters, and polynomial interpolators were also discussed in detail to handle the filtering, down/up-sampling and fractional interpolation in the transceiver chain.

References

- N. Beilleau *et al.*, "A 1.3V 26mW 3.2GS/s Undersampled LC Bandpass $\Sigma\Delta$ ADC for a SDR ISM-band Receiver in 130nm CMOS," in *Proc. Radio Frequency Integrated Circuits Symp.*, June 2009, pp. 383-386.

- G. Hueber *et al.*, "An Adaptive Multi-Mode RF Front-End for Cellular Terminals," in *Proc. Radio Frequency Integrated Circuits Symp.*, June 2008, pp. 25-28.

- R. Bagheri *et al.*, "An 800MHz to 5GHz Software-Defined Radio Receiver in 90nm CMOS," in *Proc. IEEE Int. Solid-State Circuits Conf.*, Feb. 2006, pp. 480-481.

- P. Eloranta *et al.*, "A Multimode Transmitter in 0.13 um CMOS Using Direct-Digital RF Modulator," *IEEE J. Sold-State Circuits*, vol. 42, no. 12, pp. 2774-2784, Dec. 2007.

- A. Frappé *et al.*, "An All-Digital RF Signal Generator Using High-Speed Modulators," *IEEE J. Solid-State Circuits*, vol. 44, no. 10, pp. 2722-2732, Oct. 2009.

- A. Pozsgay *et al.*, "A Fully Digital 65 nm CMOS Transmitter for the 2.4-to-2.7 GHz WiFi/WiMAX Bands using 5.4 GHz RF DACs," in *Proc. Int. Solid-State Circuits Conf.*, Feb 2008, pp. 360–619.

| Chapter 14 |

MHz-rate Multi-Antenna Decoders:
Dedicated SVD Chip Example

with Borivoje Nikolić
University of California, Berkeley
and Chia-Hsiang Yang
National Chiao Tung University, Taiwan

Slide 14.1

This chapter will demonstrate hardware realization of multidimensional signal processing. The emphasis is on managing design complexity and minimizing power and area for complex signal processing algorithms. As an example, adaptive algorithm for singular value decomposition will be used. Power and area efficiency derived from this example will also be used as a reference for flexibility considerations in Chap. 15.

Introduction to Chapters 14 & 15

- **Goal:** develop universal MIMO radio that works with multiple signal bands / users

- **Challenges**
  - Integration of complex multi-antenna (MIMO) algorithms
  - Flexibility to varying operating conditions (single-band)
  - Flexibility to support processing of multiple signal bands

- **Implication:** DSP for distributed / cooperative MIMO systems

flexibility to multiple signal bands.

Slide 14.2

The goal of the next two chapters is to present design techniques that can be used to build a universal MIMO (abbreviation of multi-input multi-output) radio architecture for multiple signal bands or multiple users. This flexible radio architecture can be used to support various standards ranging from wireless LAN to cellular devices. The design challenges are how to integrate complex MIMO signal processing, how to provide flexibility to various operating conditions, and how to extend the

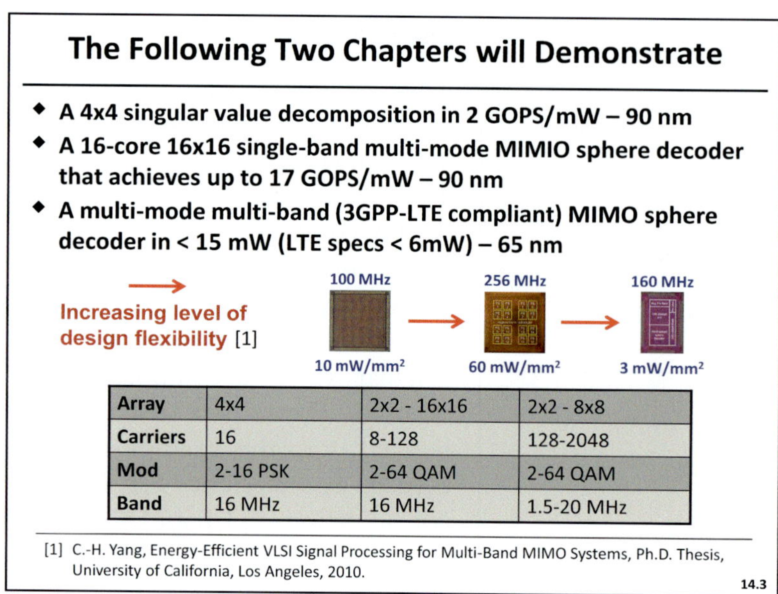

Array	4x4	2x2 - 16x16	2x2 - 8x8
Carriers	16	8-128	128-2048
Mod	2-16 PSK	2-64 QAM	2-64 QAM
Band	16 MHz	16 MHz	1.5-20 MHz

Slide 14.3

In the following two chapters, we will demonstrate design techniques for dealing with complexity and flexibility [1]. This chapter will describe chip realization of a MIMO singular value decomposition (SVD) for fixed antenna array size, to demonstrate design complexity. In the next chapter, two MIMO sphere decoder chips will demonstrate varying levels of flexibility. The first sphere decoder chip will demonstrate the flexibility in antenna array size, modulation scheme, search method, and number of sub-carriers (single-band). The second sphere decoder chip will extend the flexibility to support multiple signal bands and support both hard/soft outputs. As shown in the table, the flexibility includes antenna array size (array sizes) from 2×2 to 8×8, number of sub-carriers from 128 to 2048, modulation scheme 2-64QAM, and multiple bands from 1.25 to 20 MHz.

Slide 14.4

We will start with background on MIMO communication. Diversity-multiplexing tradeoff in MIMO channels will be introduced and illustrated on several common algorithms. After covering the algorithm basics, architecture design techniques for implementing algorithm kernels will be described, with emphasis on energy and area minimization. The use of design techniques will be illustrated on dedicated MIMO SVD chip.

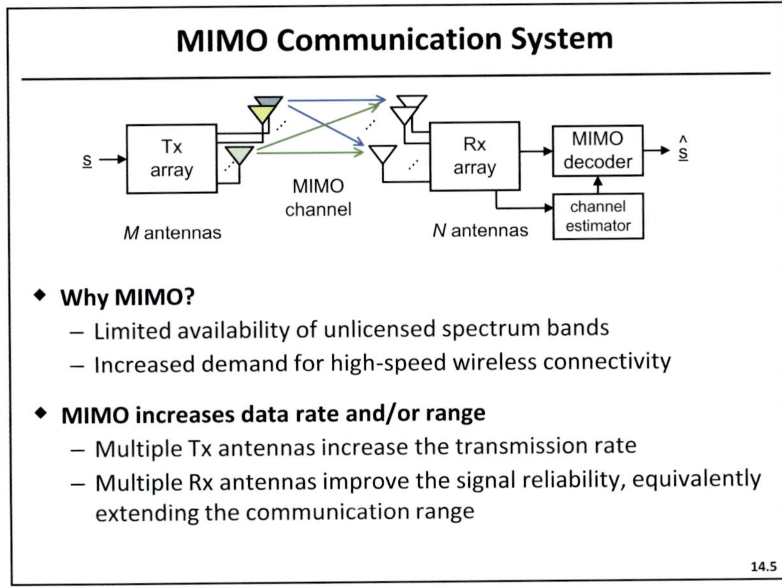

Slide 14.5

Multi-input multi-output (or MIMO) communication systems use multiple transmit and receive antennas for data transmission and reception. The function of the MIMO decoder is to constructively combine the received signals to improve the system performance. MIMO communication is inspired by the limited availability of spectrum resources. The spectral efficiency has to be improved to satisfy the increased demand for high-speed wireless links by using MIMO technology. MIMO systems can increase data rate and/or communication range. In principle, multiple transmit antennas increase the transmission rate, and multiple receive antennas improve signal robustness, thereby extending the communication range. The DSP challenge is the MIMO decoder, hence it is the focus of this chapter.

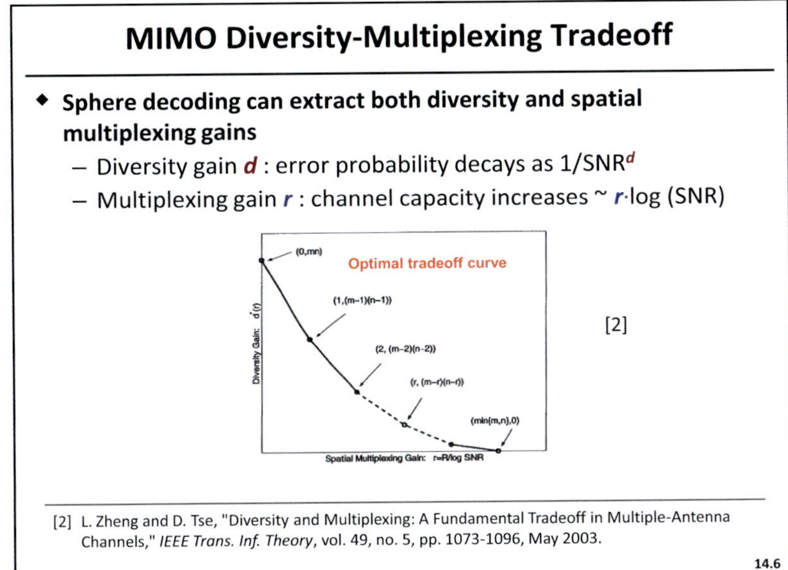

Slide 14.6

Given a MIMO system, there is a fundamental tradeoff between diversity and spatial multiplexing algorithms [2], as shown in this slide. MIMO technology is used to improve the reliability of a wireless link through increased diversity or to increase the channel capacity through spatial multiplexing. The diversity gain d is characterized by decreasing error probability as $1/SNR^d$. Lower BER or higher diversity gain improves the path loss and thereby increases the range. The spatial multiplexing gain r is characterized by increasing channel capacity proportional to r·log(SNR). Higher spatial multiplexing gain, for a fixed SNR, supports higher transmission rate per unit bandwidth. Both gains can be improved using a larger antenna array, but for a given antenna array size, there is a fundamental tradeoff between these two gains.

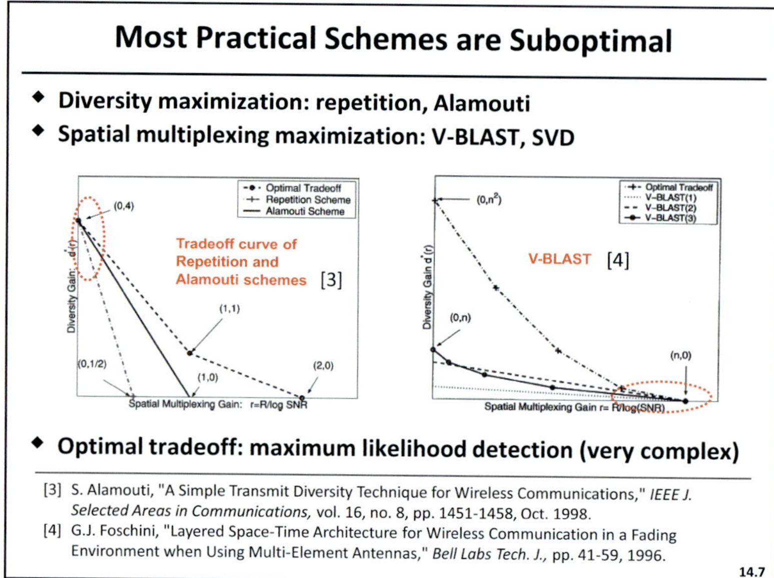

Slide 14.7

Diversity algorithms, including repetition and Alamouti schemes [3], can only achieve the optimal diversity gain (which is relevant to the transmission range). In contrast, Spatial-multiplexing algorithms such as V-BLAST algorithm [4], can only achieve the optimal spatial multiplexing gain (which is related to the transmission rate). Then the question is how to span the entire tradeoff curve to unify these point-wise solutions, and how can we do it in hardware?

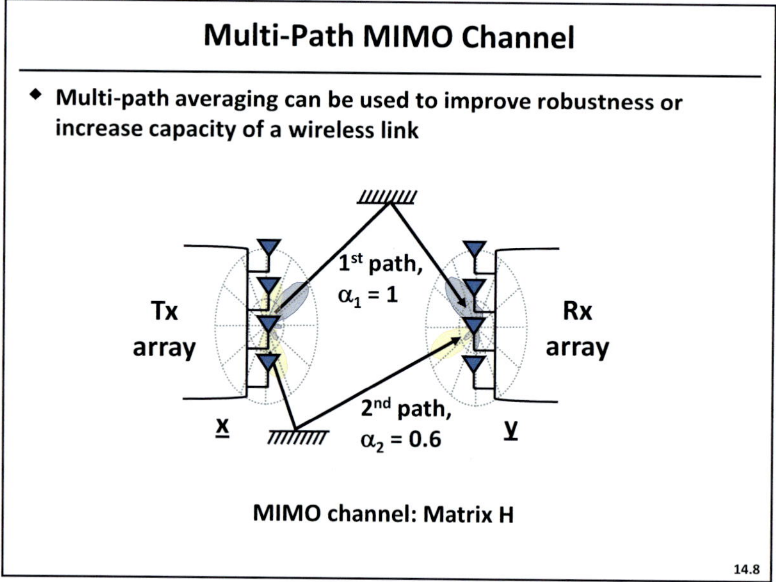

Slide 14.8

This slide illustrates multi-path wireless channel with multiple transmit and multiple receive antennas. MIMO technology can be used to improve robustness or increase capacity of a wireless link. Link robustness is improved by multi-path averaging as shown in this illustration. The number of averaging paths can be artificially increased by sending the same signal over multiple antennas. MIMO systems can also improve capacity, which is done by spatially localizing transmission beams, so that independent data streams can be sent over transmit antennas.

In a MIMO system, channel is a complex matrix H formed of transfer functions between individual antenna pairs. Vectors x and y are Tx and Rx symbols, respectively. Given x and y, the question is how to estimate gains of these spatial sub-channels.

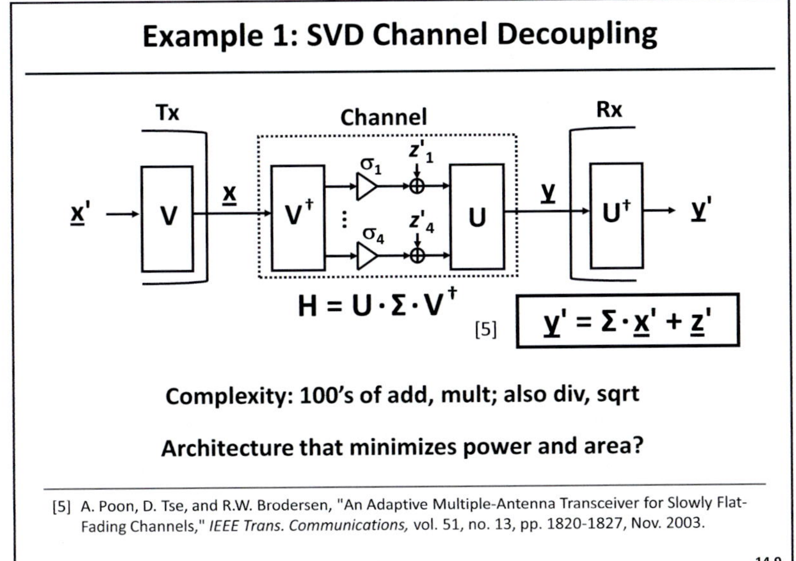

Slide 14.9

Singular value decomposition is an optimal way to extract spatial multiplexing gains [5]. Channel matrix H is a product of U, Σ, and and V, where U and V are unitary, and Σ is a diagonal matrix. With partial channel knowledge at the transmitter, we can project modulated symbols onto V matrix, essentially sending signals along eigen-modes of the fading channel. If we post-process received data by rotating y along U matrix, we can fully orthogonalize the channel between x' and y'. Then, we can send independent data streams through spatial sub-channels, which gains are described with Σ matrix.

This algorithm involves hundreds of adders and multipliers, and also dividers and square roots. This is well beyond the complexity of an FFT or Viterbi unit. Later in this chapter, we will illustrate design strategy for implementing the SVD algorithm.

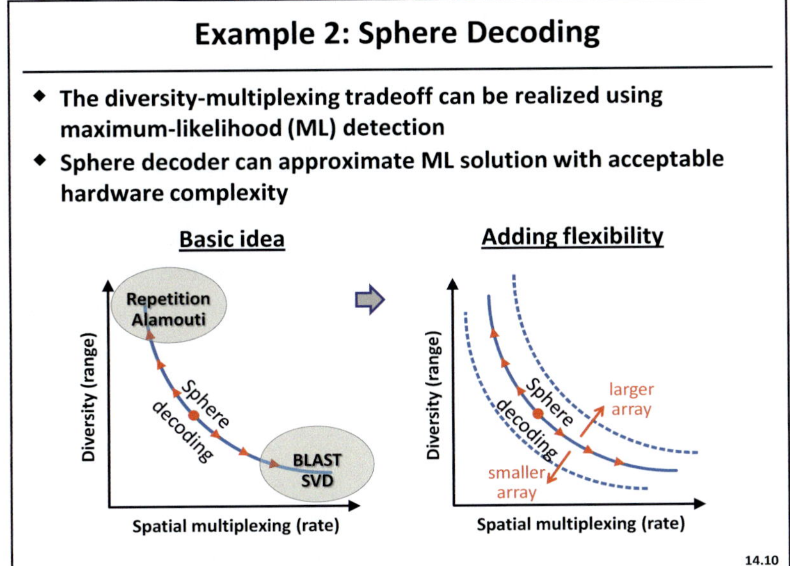

Slide 14.10

SVD is just one of the points on the optimal diversity-multiplexing tradeoff curve; the point which maximizes spatial multiplexing as shown on the left plot. Theoretically, optimal diversity-multiplexing can be achieved with maximum likelihood (ML) detection. Practically, ML is very complex and infeasible for large antenna-array size. A promising alternative to ML is the sphere decoding algorithm. It can closely achieve the maximum likelihood (ML) detection performance with several orders of magnitude lower computational complexity (polynomial vs. exponential). This way, sphere decoder can be used to extract both diversity and spatial multiplexing gains in a computationally efficient way.

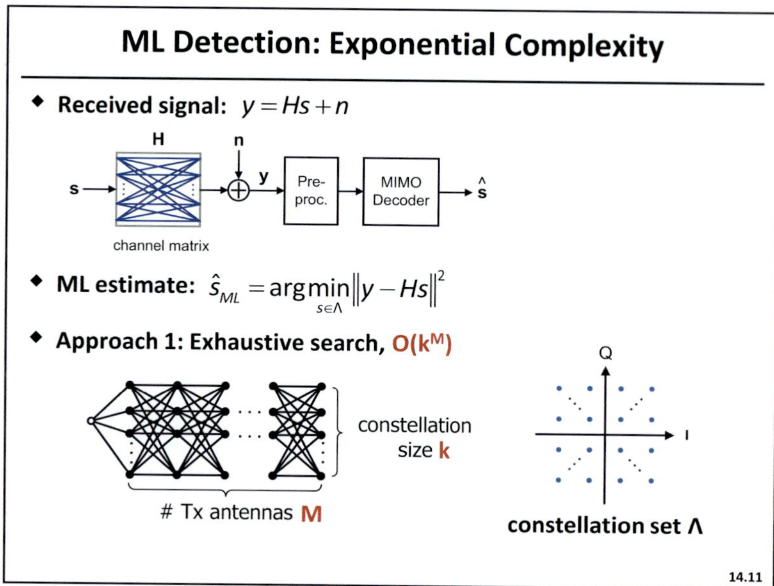

Slide 14.11

Mathematically, we can formulate the received signal y as $Hs + n$, where H is the channel matrix describing the fading gains between any two transmit and receive antennas. s is the transmit signal, and n is AWGN. Theoretically, the maximum-likelihood estimate is optimal in terms of bit error rate performance. It is achieved by minimizing the Euclidean distance of $y - Hs$, where s is drawn from a constellation set Λ. Straightforward approach is to use an exhaustive search. As shown here, each node represents one constellation point. The trellis records the decoded symbols. For M Tx antennas, we have to enumerate all possible solutions, which have the complexity k^M, and find the best one. Since the complexity is exponential, it's not feasible for practical implementation.

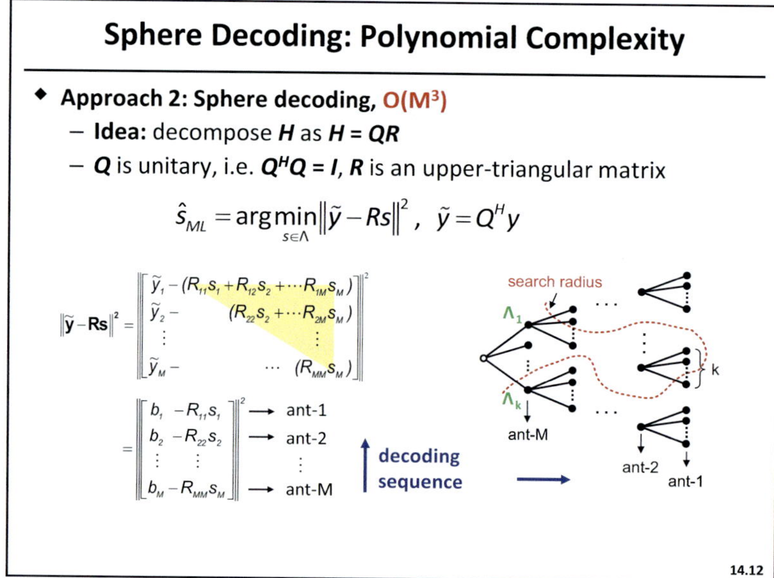

Slide 14.12

Another approach is the sphere decoding algorithm. We decompose the channel matrix H into $Q \cdot R$, where Q is unitary and R is upper-triangular. After the matrix transformation, the ML estimate can be written in another form, which minimizes the Euclidean distance $\tilde{y} - Rs$, where $\tilde{y} = Q^H \cdot y$. By using the unique structure of R, we decode s from antenna M first, and use the decoded symbol to decode the next one, and so on. The decoding process can be modeled as tree search. Each node represents one constellation point and search path records the decoded sequence. Since the Euclidean distance is non-decreasing as the increase of the search depth, we can discard the branches if the partial Euclidean distance is already larger than the search radius.

Complexity Comparison

- **Maximum likelihood: $O(k^M)$**

- **Sphere decoding: $O(M^3)$**

- **Example: $k = 64$ (64-QAM), $M = 16$ (16 antennas)**

  - Sphere decoding has 10^{25} lower complexity!

- **Reduced complexity allows for higher-order MIMO systems**

14.13

Slide 14.13

On average, the complexity of the sphere decoding algorithm is around cubic in the number of antennas, $O(M^3)$. We can see a significant complexity reduction for 16×16 64-QAM system in high SNR regime. This reduced complexity means one could consider higher-order MIMO systems and use extra degrees of freedom that are made possible by a more efficient hardware.

Hardware Emulation Results

- **Comparable BER performance of 4×4, 8×8, and 16×16, with different throughput given a fixed bandwidth**
- **Repetition coding by a factor 2 reduces the throughput by 2×, but improves BER performance**
- **An 8×8 system with repetition coding by 2 outperforms the ML 4×4 system performance by 5dB**

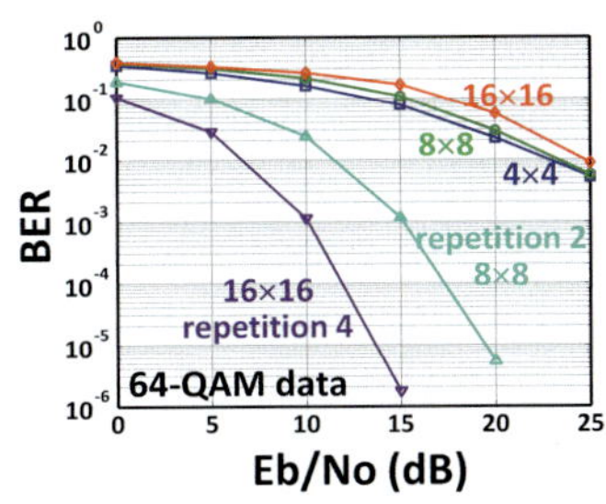

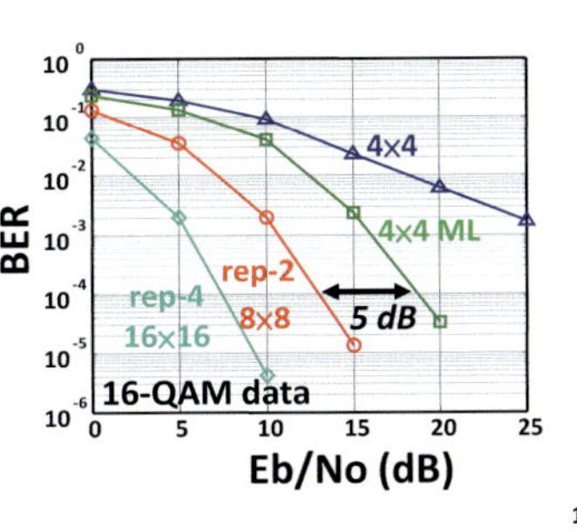

14.14

Slide 14.14

The sphere decoding algorithm is mapped onto FPGA to compute BER vs. SNR plots. The x-axis is the E_b/N_0, and the y-axis is the bit-error rate. For the left plot, we see that the performance of 4×4, 8×8, and 16×16 is comparable, but the throughput is different given a fixed bandwidth. For example, the throughput of 8×8 is twice faster than 4×4. For the second plot, the throughput of 4×4, 8×8 with repetition coding by 2, and 16×16 with repetition coding by 4 is the same, but the BER performance is improved significantly. One interesting observation is that the performance of the 8×8 system with repetition coding by 2 has outperformed the 4×4 system with the ML performance by 5dB. This was made possible with extra diversity gain achieved by repetition.

Outline

- **MIMO communication background**
  - Diversity-multiplexing tradeoff
  - Singular value decomposition
  - Sphere decoding algorithm

- ⇨ **Architecture design techniques**
  - Design challenges and solutions
  - Multidimensional data processing
  - Energy and area optimization

- **Chip 1:**
  - Single-mode single-band
  - 4x4 MIMO SVD chip

14.15

Slide 14.15

Implementation of the SVD algorithm is discussed next. We will show key architectural techniques for dealing with multiple frequency sub-carriers and large algorithm complexity. The focus will be on area and energy minimization.

Adaptive Blind-Tracking SVD

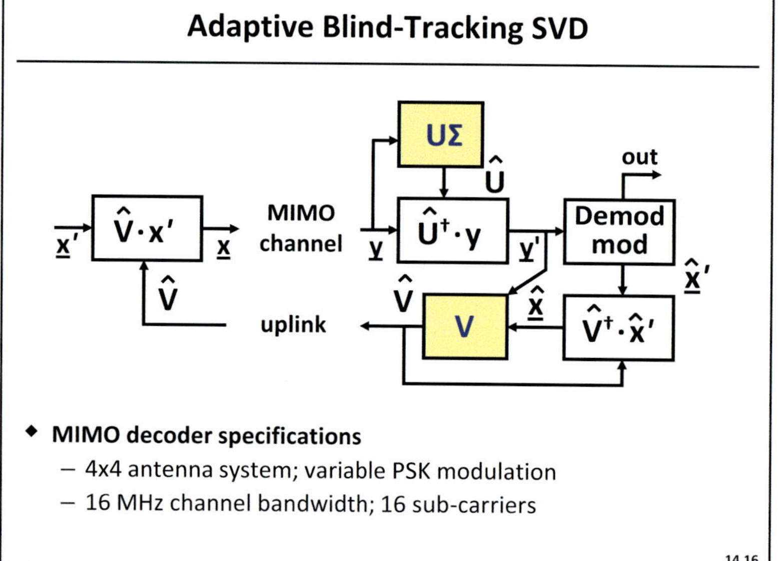

- **MIMO decoder specifications**
  - 4x4 antenna system; variable PSK modulation
  - 16 MHz channel bandwidth; 16 sub-carriers

14.16

Slide 14.16

This slide shows a block diagram of an adaptive blind-tracking SVD algorithm. The core of the SVD algorithm are the UΣ and V blocks, which estimate corresponding matrices. Hat symbol is used to indicate estimates. Channel decoupling is done at the receiver. As long as there is a sizable number of received symbols within a fraction of the coherence time, the receiver can estimate U and Σ from the received data alone. Tracking of V matrix is based on decision-directed estimates of the transmitted symbols. V matrix is periodically sent to the transmitter through the feedback channel.

We derive MIMO decoder specifications from the following system: a 4×4 antenna system that uses variable PSK modulation, 16 MHz channel bandwidth and 16 sub-carriers. In this chapter, we will illustrate implementation of the UΣ algorithm and rotation along U matrix, which has over 80% of complexity of the entire SVD.

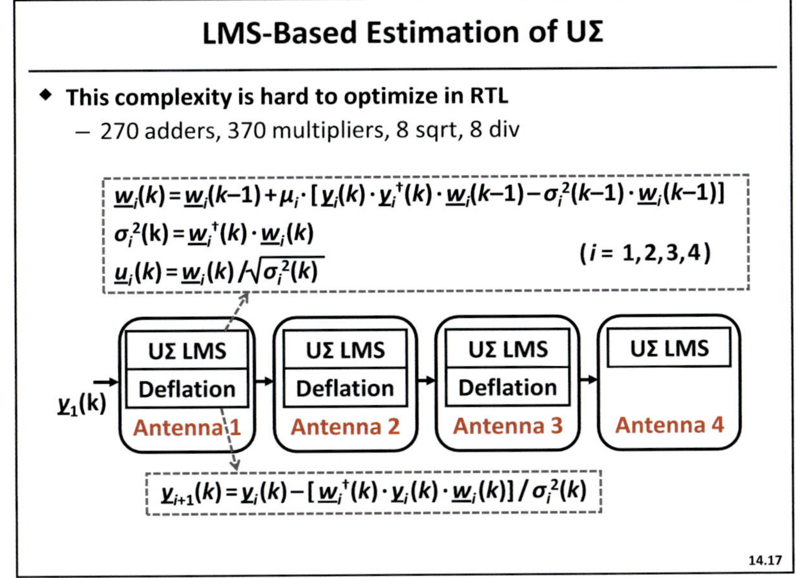

$$\underline{w}_i(k) = \underline{w}_i(k-1) + \mu_i \cdot [\underline{y}_i(k) \cdot \underline{y}_i^{\dagger}(k) \cdot \underline{w}_i(k-1) - \sigma_i^2(k-1) \cdot \underline{w}_i(k-1)]$$
$$\sigma_i^2(k) = \underline{w}_i^{\dagger}(k) \cdot \underline{w}_i(k)$$
$$\underline{u}_i(k) = \underline{w}_i(k) / \sqrt{\sigma_i^2(k)} \qquad (i = 1,2,3,4)$$

$$\underline{y}_1(k)$$

$$\underline{y}_{i+1}(k) = \underline{y}_i(k) - [\underline{w}_i^{\dagger}(k) \cdot \underline{y}_i(k) \cdot \underline{w}_i(k)] / \sigma_i^2(k)$$

Slide 14.17

The UΣ block performs sequential estimation of the eigenpairs, eigenvectors and eigenvalues, using adaptive MMSE algorithm. Traditionally, this kind of tracking is done by looking at the eigenpairs of the autocorrelation matrix. The algorithm shown here uses vector-based arithmetic with additional square root and division, which greatly reduces implementation complexity.

These equations are meant to show the level of complexity we work with. On top, we estimate components of U and Σ matrices using adaptive LMS-based tracking algorithm. The algorithm also uses adaptive step size, computed from the estimated gains in different spatial channels. Then we have square root and division implemented using Newton-Raphson iterative formulas. The recursive operation also means nested feedback loops.

Overall, this is about 300 adders, 400 multipliers, and 10 square roots and dividers. This kind of complexity is hard to optimize at the RTL level and chip designers typically don't like to work with equations. The question is, how do we turn the equations into silicon?

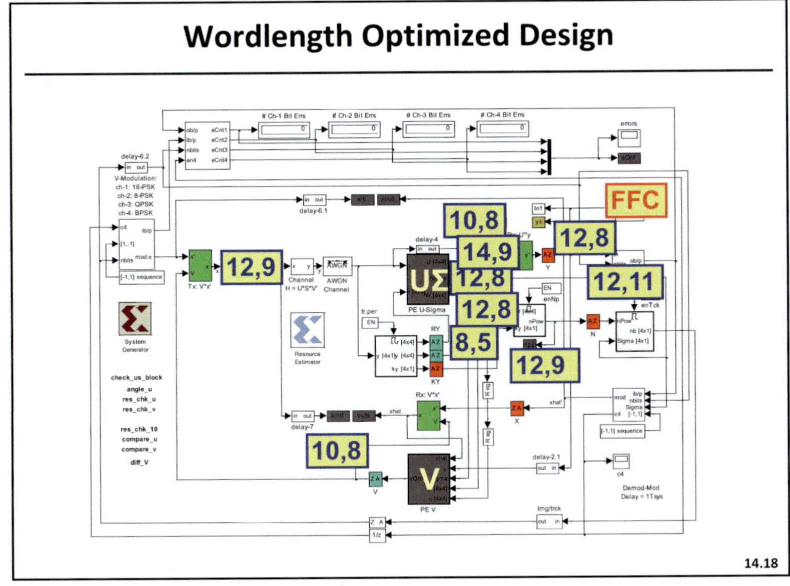

Slide 14.18

This slide shows Simulink model of a MIMO transceiver. With this graphical timed data-flow model, we can evaluate the SVD algorithm in a realistic closed-loop environment. The lines between the blocks carry wordlength information.

The first step in design optimization is wordlength optimization, which is done using the floating-to-fix point conversion tool (FFC) described in Chap. 10. The goal is to minimize hardware utilization subject to user-specified MSE error at the output due to quantization. The tool does range detection for integer bits and uses perturbation theory to determine fractional bits. Shown here are total and fractional bits at the top level. The optimization is performed hierarchically due to memory constraints and long simulation time.

The next step is to go down the hierarchy and optimize what's inside the blocks.

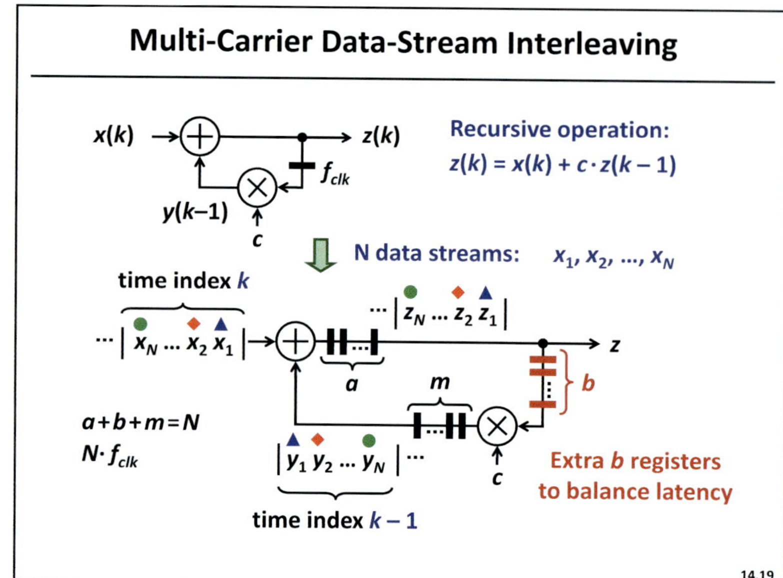

Slide 14.19

Data-stream interleaving is applied to reduce area of the implementation. Recursive operation is the underlying principle in the LMS-based tracking of eigenmodes, so we analyze simple case here to illustrate the concept. Top diagram implements the recursive operation on the right. The output $z(k)$ is a sum of current input and delayed and scaled version of previous output. Clock frequency corresponds to the sample time. This simple model assumes ideal *multiply* and *add* blocks with zero latency.

So, we refine the model by adding appropriate latency at the output of the *multiply* and *add* blocks. Then we can take this as an opportunity to interleave multiple streams of data and reduce area compared to the case of parallel realization. This is directly applicable to multiple carriers corresponding to narrowband sub-channels.

If the number of carriers N exceeds the latency required from arithmetic blocks, we add balancing registers. We have to up-sample computation by N and time-interleave incoming data. Data stream interleaving is applicable to parallel execution of independent data streams.

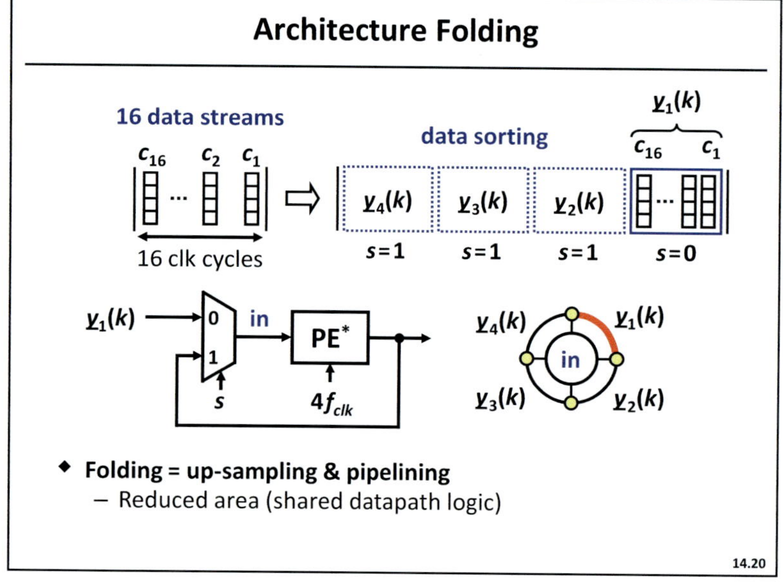

Slide 14.20

For time-serial ordering, we use folding. PE* operation performs a recursive operation (* indicated recursion). We can take output of the PE* block and fold it over in time back to its input or select incoming data stream y_1 using the life-chart on the right. The 16 sub-carriers, each carrying a vector of real and imaginary data, are sorted in time and space to occupy 16 consecutive clock cycles to allow folding over antennas.

Both interleaving and folding introduce pipeline registers to memorize internal states, but share pipeline logic to save overall area.

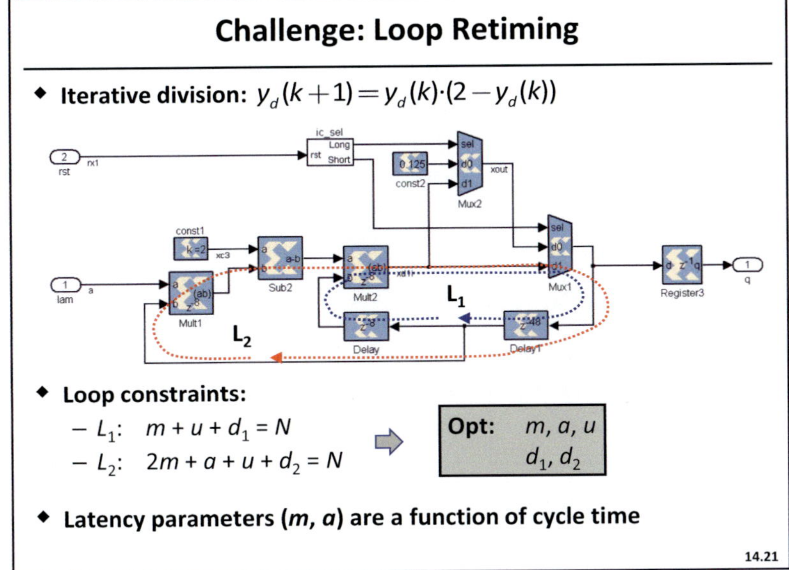

Slide 14.21

One of the key challenges in implementing recursive algorithms is loop retiming. This example shows loop retiming for an iterative divider that was used to implement $1/\sigma_i(k)$ in the formulas from Slide 14.17. Using the DFG representation and simply going around each of these loops; we identify the number of latencies in the multiplier, adder, multiplexer, and then add balancing delays (d_1, d_2) so that the loop latency equals N, the number of sub-carriers. It may seem that this is an ill-conditioned system because there are more degrees of freedom than constraints, but that is not the case. Multiplier and adder latency are both a function of cycle time.

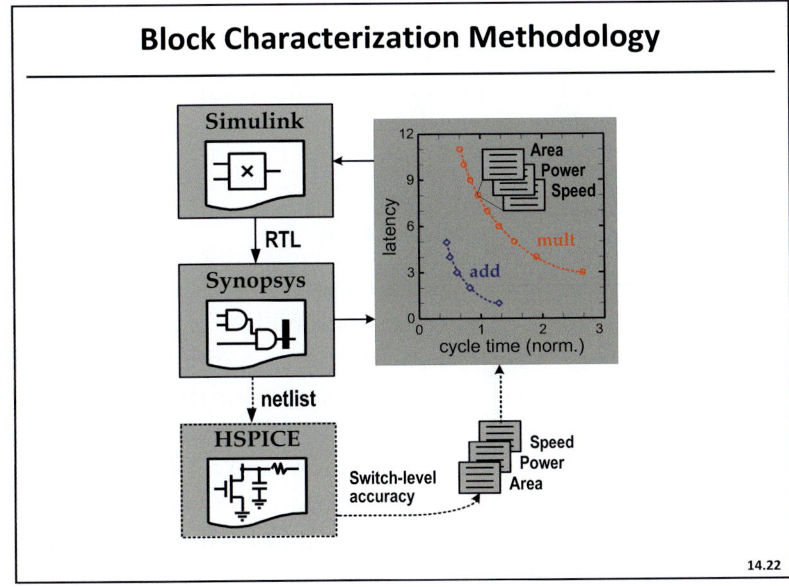

Slide 14.22

In order to determine proper latency in the multiplier and adder blocks, their latency is characterized as a function of cycle time. It is expected that the multiplier has longer latency than the adder due to larger complexity. For the same cycle time we can exactly determine how much add and multiply latency we need to specify in our implementation. The latencies are obtained using the characterization flow shown on the left. We thus augment Simulink blocks with library cards for area, power, and speed of the building blocks.

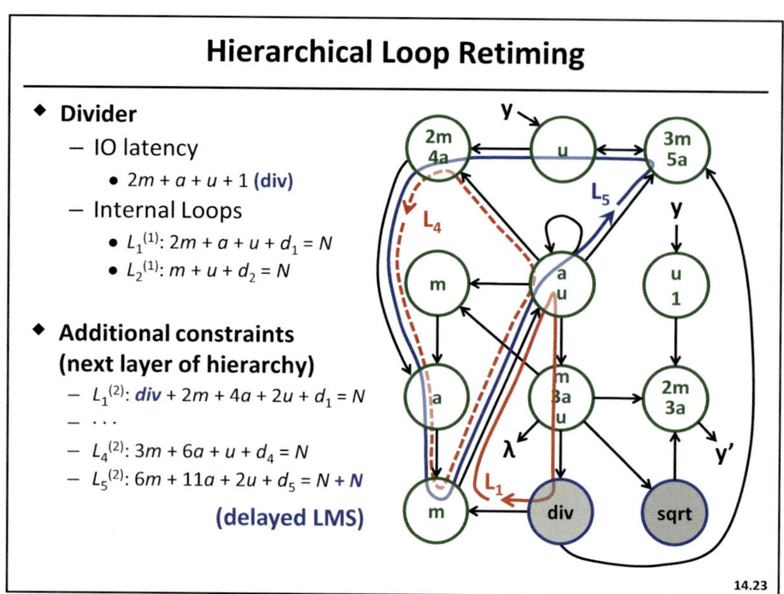

Slide 14.23

This loop retiming approach can be hierarchically extended to an arbitrary level of complexity. This DFG shows the entire UΣ block, which has five nested feedback loops. We use the divider and square root blocks from lower level of hierarchy as nodes in the top-level DFG. Each of the lower hierarchical blocks brings information about latency of primary inputs to primary outputs, and internal loops. In this example, the internal loops $L_1^{(1)}$ and $L_2^{(1)}$ are shown on the left. The superscript indicates the level of hierarchy. Additional latency constraints are specified for each loop at the next level of hierarchy, level 2 in this example. Loops $L_1^{(2)}$, $L_4^{(2)}$, and $L_5^{(2)}$ are shown. Another point to note is that we can leverage the use of delayed LMS by allowing extra sample period (N clock cycles) to relax constraints on the most critical loop. This was algorithmically possible in the adaptive SVD.

After setting the number of registers at the top level, there is no need for these registers to cross the loops during circuit implementation. We ran top-level retiming on the square root and divide circuits, and compared following two approaches: (1) retiming of designs with pre-determined block latencies as described in Slides 14.21, and 14.22, (2) retiming of flattened top-level design with latencies. The first approach took 15 minutes and the second approach took 45 minutes. When the same comparison was ran for the UΣ block, hierarchical approach with pre-determined retiming took 100 minutes, while flat top-level retiming did not converge after 40 hours! This clearly illustrates the importance of top-level hierarchical retiming approach.

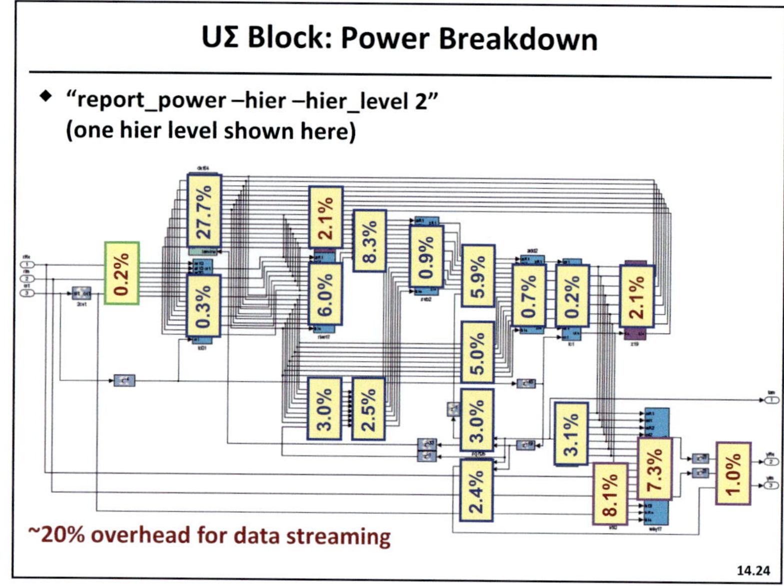

Slide 14.24

We can use synthesis estimates for various hierarchical blocks and feed them back into Simulink. Here are power numbers for the UΣ blocks. About 80% of power is used for computations and 20% for data manipulation to facilitate interleaving and folding. The blocks do not have the same numerical complexity or data activity, so power numbers vary from 0.2% to 27.7%. We can back-annotate each of these values early in the design process and estimate required power for various system components. From this Simulink description, we can map to FPGA or ASIC.

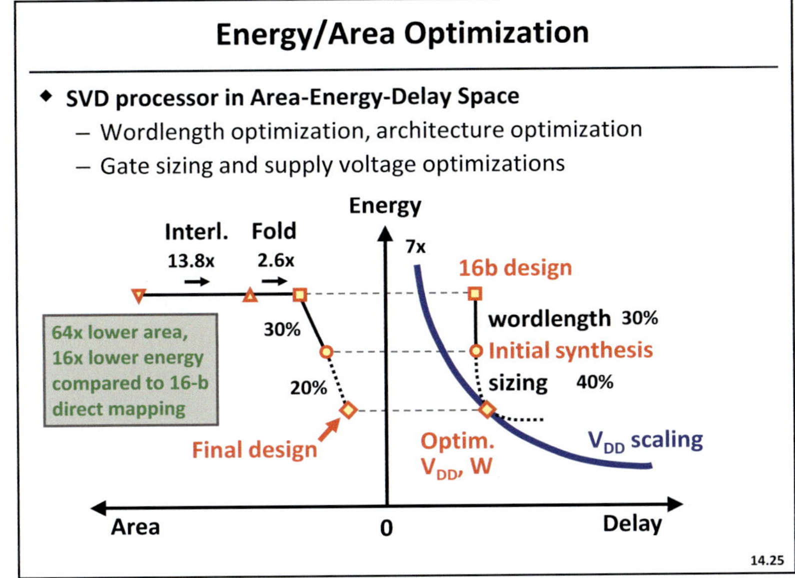

Slide 14.25

This is the architecture we implemented. Interleaving by 16 (16 sub-carriers) and folding by 4 (4 antennas) combined reduce area by 36 times compared to direct-mapped parallel implementation. In the energy-delay space, this architecture corresponds to the energy-delay sensitivity for a pipeline stage of 0.8, targeting 0.4 V operation. The architecture is then optimized as follows:

Starting from a 16-bit realization of the algorithm, we apply wordlength optimization for a 30% reduction in energy and area.

The next step is logic synthesis where we need to incorporate gate sizing and supply voltage optimizations. From circuit-level optimization results, we know that sizing is the most effective at small incremental delays compared to the minimum delay. Therefore we synthesize the design with 20% slack and perform incremental compilation to utilize benefits of sizing for a 40% reduction in energy and a 20% reduction in area of the standard-cell implementation. Standard cells are characterized for 1V supply, so we translate timing specifications to that voltage. At the optimal V_{DD} and W, energy-delay curves of sizing and V_{DD} are tangent, corresponding to equal sensitivity.

Compared to the 16-bit direct-mapped parallel realization with gates optimized for speed, the total area reduction of the final design is 64 times and the total energy reduction is 16 times. Major techniques for energy reduction are supply voltage scaling and gate sizing.

Summary of Design Techniques

- ◆ **Technique** → **Impact**
  - Wordlength opt → 30% Area ↓ (compared to 16-bit)
  - Gate sizing opt → 15% Area ↓
  - Voltage scaling → 7x Power ↓
  - Loop retiming → (together with gate sizing)
  - Interleaving → 16x Throughput ↑
  - Folding → 3x Area ↓

- ◆ **Net result**
  - Energy efficiency: 2.1 GOPS/mW
  - Area efficiency: 20 GOPS/mm² (90 nm CMOS, OP = 12-bit add)

14.26

---

explorations in Chap. 15.

Slide 14.26

Here is the summary of all design techniques and their impact on energy and area. Main techniques for minimizing energy are wordlength reduction and gate sizing (both reduce the switching capacitance), and voltage scaling. Area is primarily minimized by interleaving and folding. Overall, 2 Giga additions per second per mW (GOPS/mW) of energy efficiency is achieved with $20\,\mathrm{GOPS/mm^2}$ of integration density in 90 nm CMOS technology. These numbers will serve as reference for flexibility

---

Outline

- ◆ **MIMO communication background**
  - Diversity-multiplexing tradeoff
  - Singular value decomposition
  - Sphere decoding algorithm

- ◆ **Architecture design techniques**
  - Design challenges and solutions
  - Multidimensional data processing
  - Energy and area optimization

- ⇨ ◆ **Chip 1:**
  - Single-mode single-band
  - 4x4 MIMO SVD chip

14.27

Slide 14.27

Next, chip measurement results will be shown. The chip implements dedicated algorithm optimized for reduced power and area. The algorithm works with single frequency band and does not have flexibility for adjusting antenna-array size.

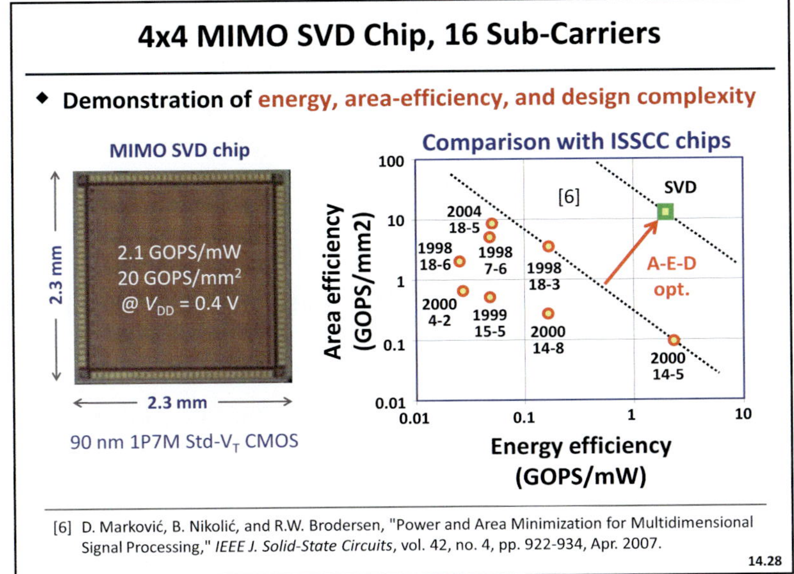

Slide 14.28

The chip implements an adaptive 4×4 singular value decomposition in a standard-Vt 90 nm CMOS process. The core area is $3.5\,\text{mm}^2$, and the total chip area with I/O pads is $5.1\,\text{mm}^2$. The chip is optimized for 0.4 V. It runs at 100 MHz and executes 70 GOPS 12-bit equivalent add operations, consuming 34 mW of power in full activity mode with random input data. The resulting power/energy efficiency is 2.1 GOPS/mW. Area efficiency is 20 GOPS/mm². This 100 MHz operation is measured over 9 die samples in a range of 385 to 425 mV.

Due to the use of optimization techniques for simultaneous area and power minimization, the chip achieves considerable improvement in area and energy efficiency as compared to representative chips from the ISSCC conference [6]. The numbers next to the dots indicate year of publication and paper number. All designs were normalized to a 90 nm process for fair comparison. The SVD chip is therefore a demonstration of achieving high energy and area efficiencies for a complex algorithm. The next challenge is to add flexibility for multiple operation modes with minimal degradation of energy and area efficiency. This will be discussed in Chap. 15.

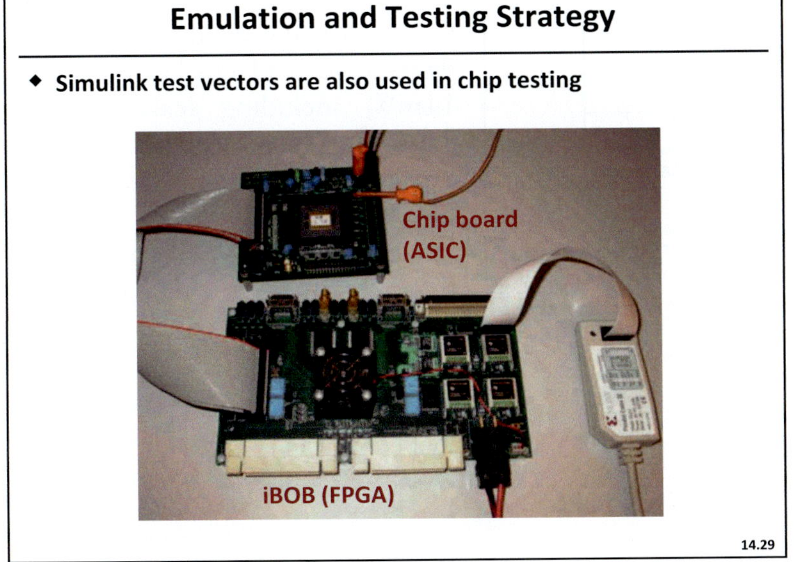

Slide 14.29

Simulink was used for design entry and architecture optimization. It is also used for chip testing. We export test vectors from Simulink, program the design onto FPGA, and stimulate the ASIC over general-purpose I/Os. The results are compared on the FPGA in real time. We can bring in external clock or use internally generated clock from the FPGA board.

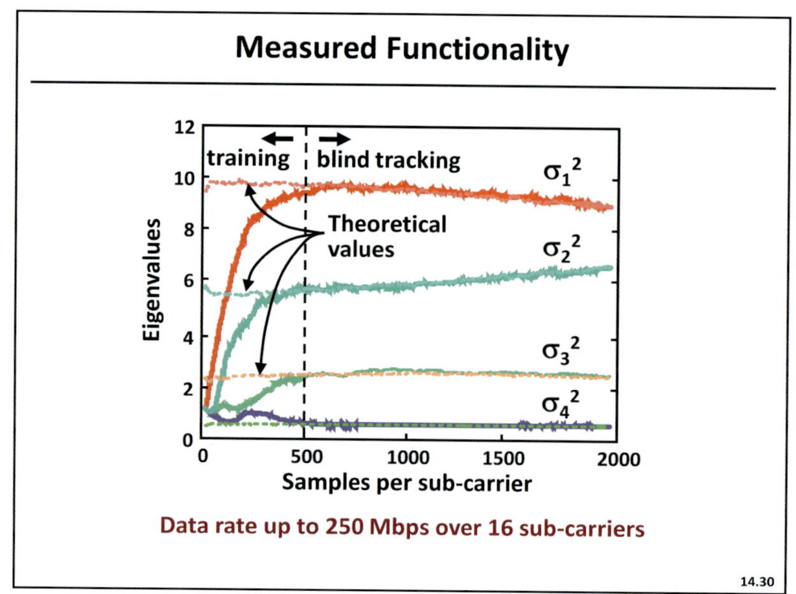

Slide 14.30

This is the result of functional verification. The plot shows tracking of eigenvalues over time, for one sub-carrier. These eigenvalues are the gains of different spatial sub-channels.

After the reset, the chip is trained with a stream of identity matrices and then it switches to blind tracking mode. Shown are measured and theoretical values to illustrate tracking performance of the algorithm. Although the algorithm is constrained with constant-amplitude modulation, we are still able to achieve 250 Mbps over 16 sub-carriers using adaptive PSK modulation.

SVD Chip: Summary of Measured Results [7]

Silicon Technology	90 nm 1P7M std-Vt CMOS
Total Power Dissipation	34 mW
Leakage / Clocking	4 mW / 14 mW
UΣ / Deflation	20 mW / 14 mW
Active Chip Area	3.5 mm²
Power Density	10 mW/mm²
Opt V_{DD} / f_{clk}	0.4 V / 100 MHz
Min V_{DD} / f_{clk}	0.25 V / 10 MHz
Max Throughput	250 Mbps / 16 carriers

[7]   D. Marković, R.W. Brodersen, and B. Nikolić, "A 70GOPS 34mW Multi-Carrier MIMO Chip in 3.5mm²," in *Proc. Int. Symp. VLSI Circuits,* June 2006, pp. 196-197.

14.31

Slide 14.31

This table is a summary of measured results reported in [7].

Optimal supply voltage is 0.4 V with 100 MHz clock, but the chip is functional at 255 mV, running with a 10 MHz clock. The leakage power is 12% of the total power in the worst case, and clocking power is 14 mW, including leakage. With 3.5 mm² of core area, the achieved power density is 10 mW/mm². Maximal throughput is 250 Mbps using 16 frequency sub-channels.

Summary and Conclusion

♦ **Summary**
- Maximum likelihood detection can extract optimal diversity and spatial multiplexing gains from a multi-antenna channel
- Singular value decomposition maximizes spatial multiplexing
- Sphere decoder is a practical alternative to maximum likelihood
- Design techniques for managing design complexity while minimizing power and area have been discussed

♦ **Conclusion**
- High-level retiming is critical for optimized realization of complex recursive algorithms
- Dedicated complex algorithms in 90 nm technology can achieve
  - 2 GOPS/mW
  - 20 GOPS/mm²

14.32

Slide 14.32

This chapter presented optimal diversity-spatial multiplexing tradeoff in multi-antenna channels and discussed algorithms that can extract diversity and multiplexing gains. The optimal tradeoff is achieved using maximum likelihood (ML) detection, but ML is infeasible due to numerical complexity. Singular value decomposition can maximize spatial multiplexing (data rate), but does not have flexibility to increase diversity. Sphere decoder algorithm emerges as a practical alternative to ML.

Implementation techniques for a dedicated 4×4 SVD algorithm have been demonstrated. It was shown that high-level retiming of recursive loops is crucial for timing closure during logic synthesis. Using architectural and circuit techniques for power and area minimization, it was shown that complex algorithms in a 90 nm technology can achieve energy efficiency of 2.1 GOPS/mW and area efficiency of 20 GOPS/mm². Next, it is interesting to investigate the energy and area cost of adding flexibility for multi-mode operation.

References

- C.-H. Yang, Energy-Efficient VLSI Signal Processing for Multi-Band MIMO Systems, Ph.D. Thesis, University of California, Los Angeles, 2010.

- L. Zheng and D. Tse, "Diversity and Multiplexing: A Fundamental Tradeoff in Multiple-Antenna Channels," *IEEE Trans. Information Theory,* vol. 49, no. 5, pp. 1073–1096, May 2003.

- S. Alamouti, "A Simple Transmit Diversity Technique for Wireless Communications," *IEEE J. Selected Areas in Communications,* vol. 16, no. 8, pp. 1451–1458, Oct. 1998.

- G.J. Foschini, "Layered Space-Time Architecture for Wireless Communication in a Fading Environment when Using Multi-Element Antennas," *Bell Labs Tech. J.,* vol. 1, no. 2, pp. 41–59, 1996.

- A. Poon, D. Tse, and R.W. Brodersen, "An Adaptive Multiple-Antenna Transceiver for Slowly Flat-Fading Channels," *IEEE Trans. Communications,* vol. 51, no. 13, pp. 1820–1827, Nov. 2003.

- D. Marković, B. Nikolić, and R.W. Brodersen, "Power and Area Minimization for Multidimensional Signal Processing," *IEEE J. Solid-State Circuits,* vol. 42, no. 4, pp. 922-934, April 2007.

- D. Marković, R.W. Brodersen, and B. Nikolić, "A 70GOPS 34mW Multi-Carrier MIMO Chip in 3.5mm²," in *Proc. Int. Symp. VLSI Circuits,* June 2006, pp. 196-197.

Additional References

- S. Santhanam *et al.*, "A Low-cost 300 MHz RISC CPU with Attached Media Processor," in *Proc. IEEE Int. Solid-State Circuits Conf.*, Feb. 1998, pp. 298–299. (paper 18.6)

- J. Williams *et al.*, "A 3.2 GOPS Multiprocessor DSP for Communication Applications," in *Proc. IEEE Int. Solid-State Circuits Conf.*, Feb. 2000, pp. 70–71. (paper 4.2)

- T. Ikenaga and T. Ogura, "A Fully-parallel 1Mb CAM LSI for Realtime Pixel-parallel Image Processing," in *Proc. IEEE Int. Solid-State Circuits Conf.*, Feb. 1999, pp. 264–265. (paper 15.5)

- M. Wosnitza *et al.*, "A High Precision 1024-point FFT Processor for 2-D Convolution," in *Proc. IEEE Int. Solid-State Circuits Conf.*, Feb. 1998, pp. 118–119. (paper 7.6)

- F. Arakawa *et al.*, "An Embedded Processor Core for Consumer Appliances with 2.8 GFLOPS and 36 M Polygons/s FPU," in *Proc. IEEE Int. Solid-State Circuits Conf.*, Feb. 2004, pp. 334–335. (paper 18.5)

- M. Strik *et al.*, "Heterogeneous Multi-processor for the Management of Real-time Video and Graphics Streams," in *Proc. IEEE Int. Solid-State Circuits Conf.*, Feb. 2000, pp. 244–245. (paper 14.8)

- H. Igura *et al.*, "An 800 MOPS 110 mW 1.5 V Parallel DSP for Mobile Multimedia Processing," in *Proc. IEEE Int. Solid-State Circuits Conf.*, Feb. 1998, pp. 292–293. (paper 18.3)

- P. Mosch *et al.*, "A 720 uW 50 MOPs 1 V DSP for a Hearing Aid Chip Set," in *Proc. IEEE Int. Solid-State Circuits Conf.*, Feb. 2000, pp. 238–239. (paper 14.5)

Slide 15.1

This chapter discusses design techniques for dealing with design flexibility, in addition to complexity that was discussed in the previous chapter. Design techniques for managing adjustable number or antennas, modulations, number of sub-carriers and search algorithms will be presented. Multi-core architecture, based on scalable processing element will be described. At the end, flexibility for multi-band operation will be discussed, with emphasis on flexible FFT that operates over many signal bands. Area and energy cost of the added flexibility will be analyzed.

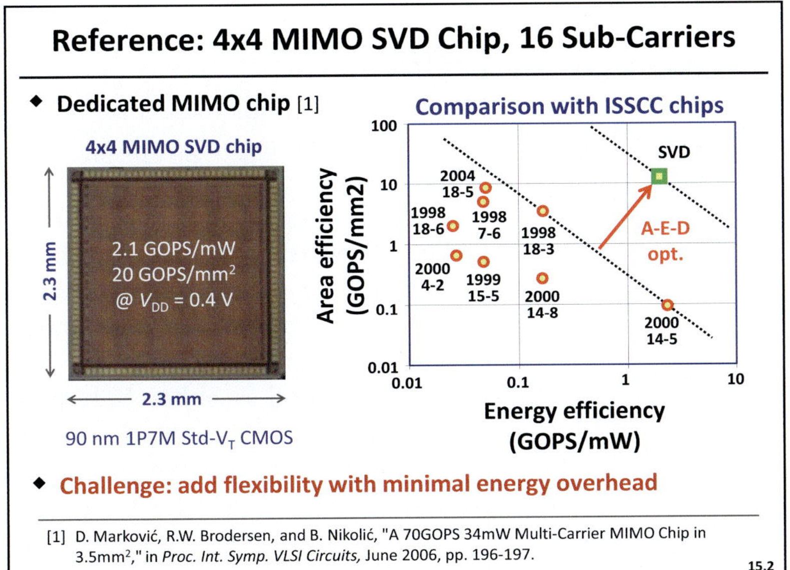

Slide 15.2

We use a MIMO SVD chip as a starting point in our study of the energy and area cost of hardware flexibility. This chip implements singular value decomposition for MIMO channels [1]. It supports 4×4 MIMO systems and 16 sub-carriers. This chip achieves a 2.1 GOPS/mW energy-efficiency and 20 GOPS/mm² area efficiency at 0.4 V. Compared with the baseband and multimedia processors published in ISSCC, this chip achieves the highest efficiency considering both area and energy.

Now the challenge is how to add flexibility without compromising efficiency? Design flexibility is the focus of this chapter.

D. Marković and R.W. Brodersen, *DSP Architecture Design Essentials*, Electrical Engineering Essentials, DOI 10.1007/978-1-4419-9660-2_15, © Springer Science+Business Media New York 2012

Outline

 ◆ **Scalable decoder architecture**
- Design challenges and solutions
- Scalable PE architecture
- Hardware complexity reduction

◆ **Chip 1:**
- Multi-mode single-band
- 16x16 MIMO SD Chip
- Energy and area efficiency

◆ **Chip 2:**
- Multi-mode multi-band
- 8x8 MIMO SD + FFT Chip
- Flexibility Cost

15.3

Slide 15.3

The chapter is structured as follows. Design concepts for handling flexibility for multiple operating modes with respect to the number of antennas, modulations and carriers will be described first. Scalable processing element (PE) will be presented. The PE will be used in two sphere-decoder chips to demonstrate varying levels of flexibility. The first chip will demonstrate multi-PE architecture for multi-mode single-band operation that supports up to 16×16 antennas. The second chip will extend the flexibility to multiple frequency bands. It features flexible FFT operation and also soft-output generation for iterative decoding.

Representative Sphere Decoders

Reference	Array size	Modulation	Search method	$N_{carriers}$
Shabany, ISSCC'09	4×4	64-QAM	K-best	1
Knagge, SIPS'06	8×8	QPSK	K-best	1
Guo, JSAC'05	4×4	16-QAM	K-best	1
Burg, JSSC'05 Garrett, JSSC'04	4×4	16-QAM	Depth-first	1

◆ **Design challenges**
- Antenna array size: constrained by complexity (N_{mult})
- Constellation size: only fixed constellation size considered
- Search method: no flexibility for both K-best and depth-first
- Number of sub-carriers: only single-carrier systems considered

◆ **We will next discuss techniques that address these challenges**

15.4

Slide 15.4

In hardware implementation of the sphere decoding algorithm, focus has mainly been on fixed antenna array size and modulation scheme. The supported antenna-array size is restricted to 8×8 and mostly 4×4, and the supported modulation scheme is up to 64-QAM. The search method is either K-best or depth-first. In addition, most designs consider only single-carrier systems. The key challenges are the support of larger antenna-array sizes, higher-order modulation schemes and flexibility. In this chapter, we will demonstrate solutions to these challenges by minimizing hardware complexity so that we can increase the supported antenna-array size, modulation scheme and add flexibility.

Numerical Strength Reduction

- **Multiplier size is the key factor for complexity reduction**
- **Two equivalent representations**

$$\hat{s}_{ML} = \arg\min \left\| R(s - s_{ZF}) \right\|^2$$

$$\hat{s}_{ML} = \arg\min \left\| Q^H y - Rs \right\|^2$$

- **The latter is a better choice from hardware perspective**
  - <u>Idea</u>: s_{ZF} and $Q^H y$ can be precomputed
  - Wordlength of s (3-bit real/imag part) is usually shorter than s_{ZF}
    → separate terms → multipliers with reduced wordlength/area

Area and delay reduction due to numerical strength reduction

WL of s_{ZF}	6	8	10	12
WL(R) = 12, Area/delay	0.5/0.83	0.38/0.75	0.3/0.68	0.25/0.63
WL(R) = 16, Area/delay	0.5/0.68	0.38/0.63	0.3/0.58	0.25/0.54

15.5

Slide 15.5

We start by reducing the size of the multipliers, because multiplier size is the key factor for complexity reduction and allows for increase in the antenna-array size. We consider two equivalent representations. At the first glance, the former has one multiplication while the latter has two. However, a careful investigation shows the latter is a better choice from hardware perspective. First, s_{ZF} and $Q^H y$ can be precomputed. Hence, they have negligible impact on the total number of operations. Second, the wordlength of s is usually shorter than s_{ZF}. Therefore, separating terms results in multipliers with reduced wordlength, thus reducing the area and delay. The area reduction is at least 50% and the delay reduction also reaches 50% for larger wordlength of R and s_{ZF}.

#1: Multiplier Simplification

- **Hardware area is dominated by complex multipliers**
  - 8.5× reduction in the number of multipliers (folding: 136 → 16)
  - 7× reduction in multiplier size
    - Choice of mathematical expression: $16 \times 16 \rightarrow 16 \times 4$
    - Gray encoding, only odd numbers used: $16 \times 4 \rightarrow 16 \times 3$
    - Wordlength optimization: $16 \times 3 \rightarrow 12 \times 3$

Overall **40%** lower multiplier area

Re{s}/Im{s}	2's complement	Gray code	operation
7	0111	100	8−1
5	0101	101	4+1
3	0011	111	4−1
1	0001	110	1
−1	1111	010	1 × −1
−3	1101	011	3 × −1
−5	1011	001	5 × −1
−7	1001	000	7 × −1

Final multiplier design [2]

[2] C.-H. Yang and D. Marković, "A Flexible DSP Architecture for MIMO Sphere Decoding," *IEEE Trans. Circuits & Systems I*, vol. 56, no. 10, pp. 2301-2314, Oct. 2009.

15.6

Slide 15.6

The first step is to simplify complex multipliers since complex multipliers used for Euclidean-distance calculation dominate the hardware area. An 8.5 times reduction is achieved by using folding technique. The size of the multipliers also affects hardware area directly. A seven times area reduction is achieved by choosing a simpler representation, using Gray encoding to compactly represent the constellation, and optimizing wordlength of datapath. We can further reduce the multiplier area by using shift-and-add operations, replacing the negative operators with inverters and leaving the carry-in bits in the succeeding adder tree. The final multiplier design is shown here, which has only one adder and few inverters, and multiplexers [2]. With this design, a 40% area reduction is achieved compared to the conventional implementation.

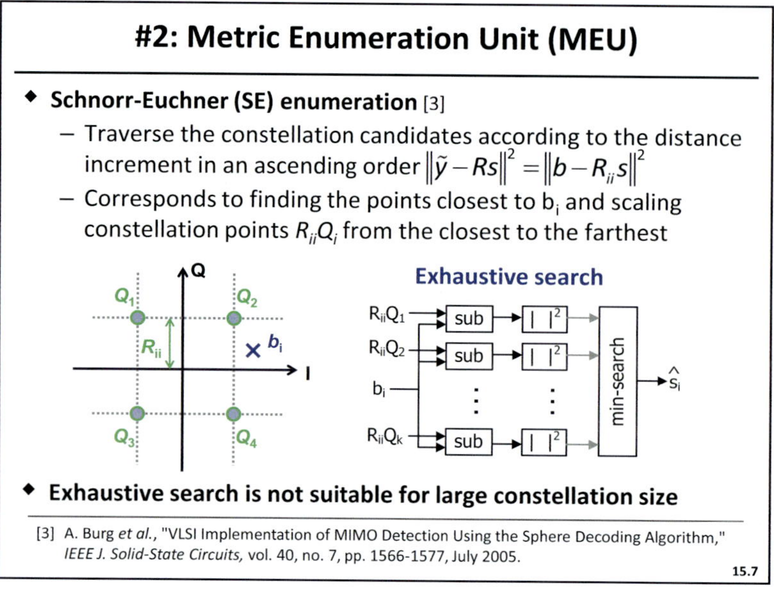

Slide 15.7

The second challenge is how to support larger modulation sizes. To improve BER performance, the Schnorr-Euchner (SE) enumeration technique states that we have to traverse the constellation points according to the distance increment in an ascending order for each transmit antenna. The basic idea of this algorithm is using locally optimal solution to speed up the search of globally optimal solution. In the constellation plane, we have to traverse the point closest to b_i and scale constellation points $R_{ii}Q_i$ from the closest to the farthest [3]. In this example, Q_2 should be visited first, and Q_4 second. Straightforward implementation is to calculate the distance of all possible constellation points and find the point with the minimum distance as the first candidate. The hardware cost grows quickly as the modulation size increases.

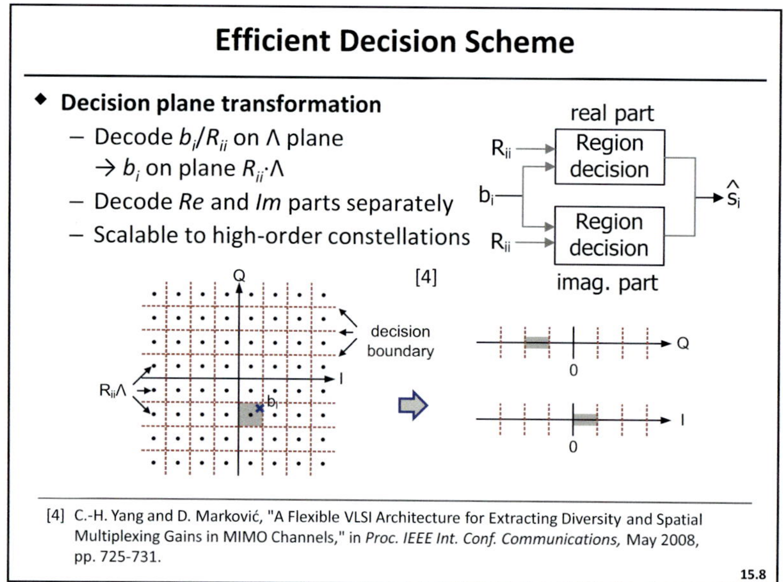

Slide 15.8

Efficient decision scheme is shown here. The goal is to enumerate possible constellation points according to the distance increment in an ascending order. In the constellation plane, for example, we first need to examine the point that is the closest between $R_{ii}Q_i$ and b_i. Region partition search is based on deciding the closest point by deciding the region in which b_i lies [4]. This method is feasible because real part and imaginary part can be decoded separately as shown in the gray highlights. The next question is how to enumerate the remaining points?

Efficient Candidate Enumeration

- **8 surrounding constellation points divided into 2 subsets:**
  - 1-bit error and 2-bit errors if Gray coding is used
- **The 2nd closest point decided by the decision boundaries**
- **The remaining points decided by the search direction**

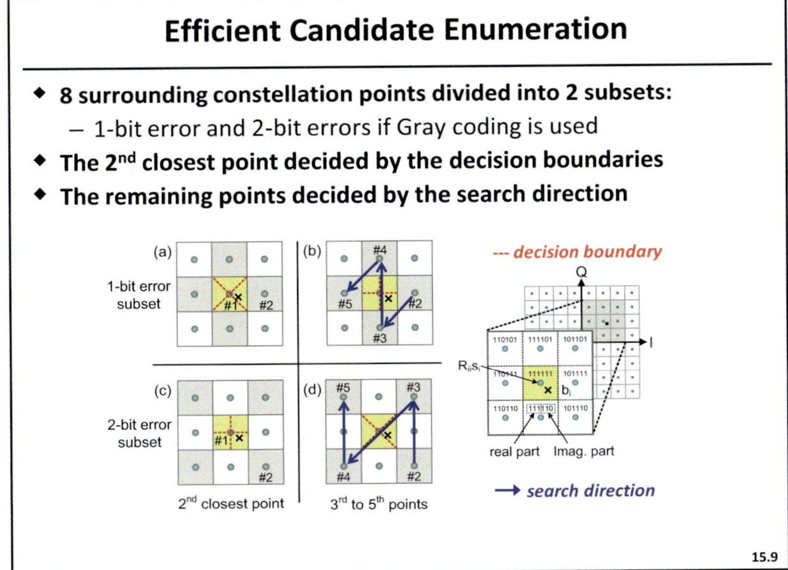

Slide 15.9

Geometric relationship can be used instead of sorting algorithm to simplify calculation and reduce hardware cost. Again, we use decision boundary to decide the 2nd search point and remaining points. Eight surrounding constellation points are divided into 2 subsets first: 1-bit error and 2-bit errors using Gray coding. Take the 2-bit error subset as an example: the 2nd point is decided by checking the sign of the real part and imaginary part of $(b_i - R_{ii}s_i)$, which has four combinations. After examining the 2nd point, the 3rd to 5th points are decided by the search direction, which is decided by another decision boundary. The same idea is used for 1-bit error subset.

#3: Search Algorithm

- **Conventional search algorithms**
  - Depth-first: starts from the root of the tree and explores as far as possible along each branch until a leaf node is found
  - K-best: keeps only K branches with the smallest partial Euclidean distance (PED) at each level

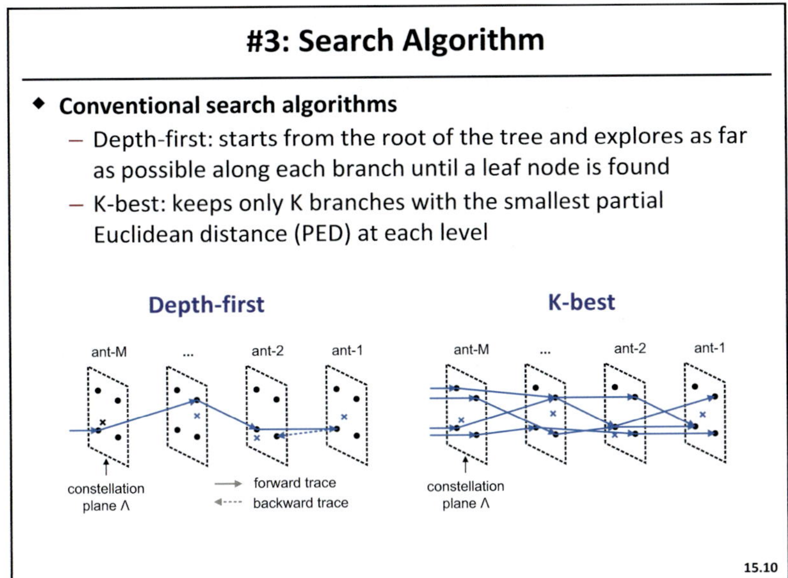

Slide 15.10

The third step is to resolve limitation of the conventional search algorithms: depth-first and K-best. Depth-first starts the search from the root of the tree and explores as far as possible along each branch until a leaf node is found or the search is outside the search radius. It can examine all possible solutions, but the throughput is low. K-best approximates the breath-first search by keeping K branches with the smallest partial Euclidean distance at each level and finding the best one at the last stage. Since only forward search is allowed, error performance is limited.

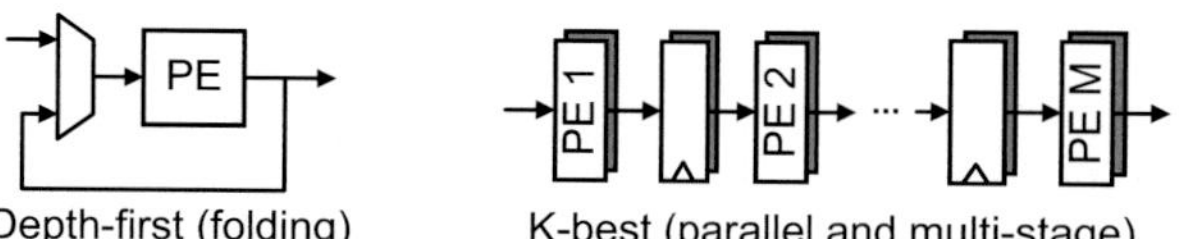

Architecture Tradeoffs

- **Basic architecture**

- **Practical considerations**

Algorithm	Area	Throughput	Latency	Radius shrinking
Depth-first	Small	Variable	Long	Yes
K-best	Large	Constant	Short	No

- **Algorithmic considerations**
  - Depth-first can reach ML given enough cycles
  - K-best cannot reach ML, it can approach it

15.11

Slide 15.11

The basic architectures for these two search methods are shown here. Depth-first uses folding architecture and K-best uses multi-stage architecture. In essence, depth-first allows back-trace search, but K-best searches forward only. The advantages of depth-first include smaller area and better performance. In contrast, K-best provides constant throughput and shorter latency. Combining these advantages is quite challenging.

Metric Calculation

- **Metric Calculation Unit (MCU) computes** $\sum_{j=i+1}^{M} R_{ii} s_i$

- **Bi-directional shift register chain is embedded to support back trace and forward trace**

- **Area-efficient storage of *R* matrix coefficients: off-diagonal terms organized into a square memory**

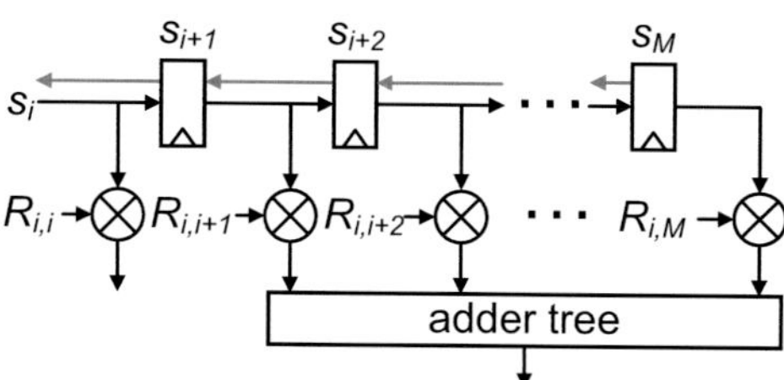

15.12

Slide 15.12

Metric calculation unit has to compute summation of $R_{ii}s_i$. We use an FIR-like architecture to facilitate metric calculation. Because the trace goes back up by one layer instead of random jump, a bi-directional shift register chain is used to support back trace and forward trace. Coefficients of **R** matrix can be stored in an area-efficient way, because **R** is an upper-triangular matrix. The real and imaginary parts can be organized into a square memory without storing 0s.

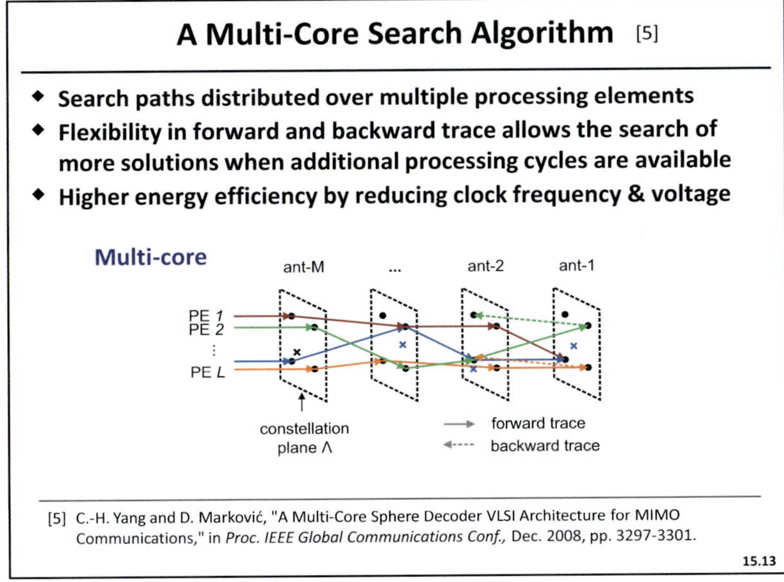

Slide 15.13

Given the ability to use multiple processing elements concurrently, we consider a multi-core search algorithm from [5]. It distributes search paths over multiple processing elements. Since each PE has the flexibility in forward and backward search, the whole search space can be examined when additional processing cycles are available. When a higher throughput is not required, we can increase the energy efficiency by reducing clock frequency and supply voltage due to the distribution of computations.

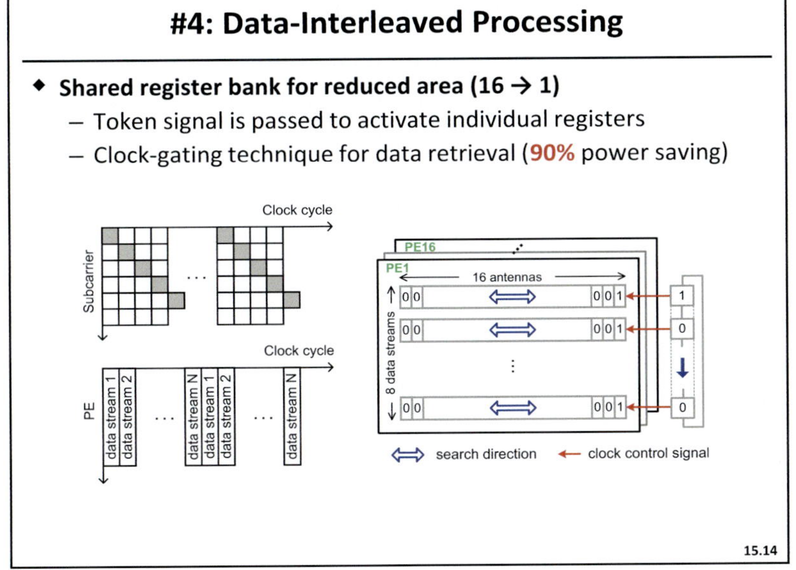

Slide 15.14

To extend the processing period for signal processing across PEs, multiple data streams or sub-carriers are interleaved into each PE sequentially. As shown here, only one sub-carrier is processed in one clock cycle, but it is processed over all PEs. For area saving, the input register bank can be shared by all PEs, so we don't need to build 16 copies for 16 PEs. For power saving, we can disable inactive registers for other sub-carriers using clock-gating technique since only one sub-carrier is processed in one clock cycle.

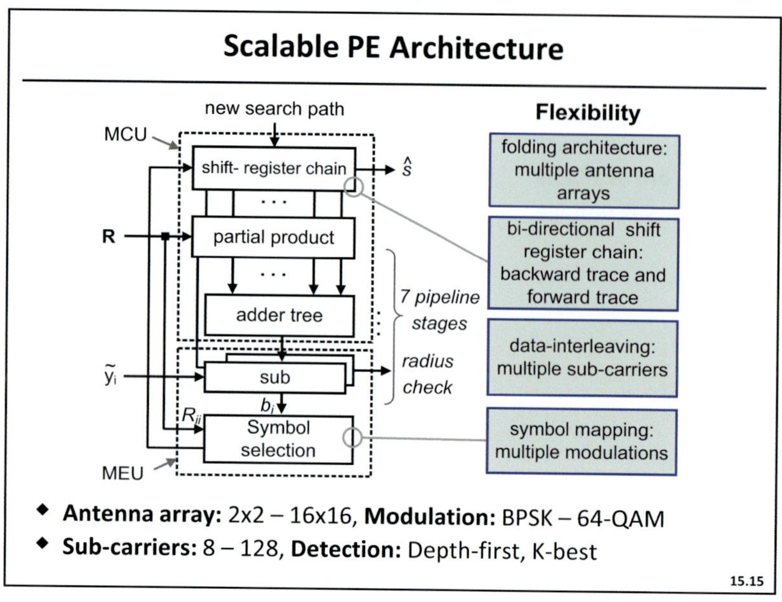

Slide 15.15

The block diagram of one PE is shown here. Each PE is scalable to support varying antenna arrays, multiple modulations, and multiple data streams through signal processing and circuit techniques. Flexibility in antenna-array size is supported through hardware reuse. Simplified boundary-decision scheme is used to support multiple modulation schemes. Bi-directional shift registers support flexible search method. Data-interleaving technique is used to support multiple sub-carriers. Flexibility is added with a small area overhead.

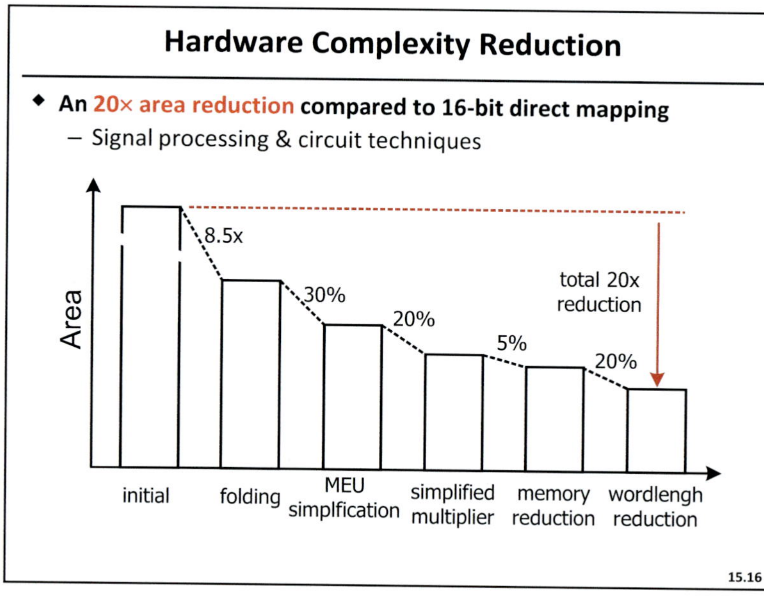

Slide 15.16

Overall, a 20 times area reduction is achieved compared to 16-bit direct-mapped architecture using signal processing and circuit techniques listed here. Main area impact comes from architecture folding (an 8.5x reduction). Simplified metric enumeration gives additional 30% area reduction, simplified multiplier further reduces overall area by 20% and so does the wordlength reduction step.

Outline

- **Scalable decoder architecture**
  - Design challenges and solutions
  - Scalable PE architecture
  - Hardware complexity reduction
- ➡ **Chip 1:**
  - Multi-mode single-band
  - 16x16 MIMO SD Chip
  - Energy and area efficiency
- **Chip 2:**
  - Multi-mode multi-band
  - 8x8 MIMO SD + FFT Chip
  - Flexibility Cost

15.17

Slide 15.17

The scalable processing element from Slide 15.15 will now be used for multi-PE architecture to demonstrate hard-output multi-mode single-band MIMO sphere decoding. The focus will be on energy and area efficiency analysis with respect to added design flexibility. 16×16 MIMO chip will be used as design example for the flexibility analysis.

Chip 2: Multi-Core Sphere Decoder, Single-Band [6]

- **Flexibility**
  - $2\times2 - 16\times16$ antennas
  - BPSK – 64QAM
  - K-best / depth-first
  - 8-128 sub-carriers
- **Supply voltage**
  - **Core:** 0.32 V – 1.2 V
  - **I/O:** 1.2 V / 2.5 V
- **Silicon area**
  - **Die:** 8.88 mm^2
  - **Core:** 5.29 mm^2
  - **PE:** 0.24 mm^2

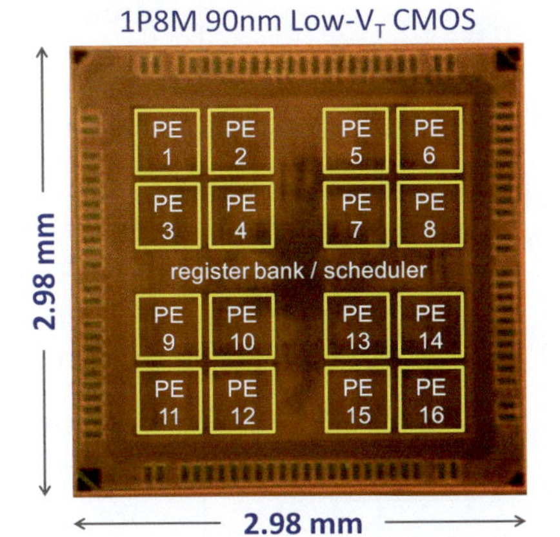

[6] C.-H. Yang and D. Marković, "A 2.89mW 50GOPS 16x16 16-Core MIMO Sphere Decoder in 90nm CMOS," in *Proc. IEEE European Solid-State Circuits Conf.*, Sep. 2009, pp. 344-348.

15.18

Slide 15.18

Here is the chip micrograph of multi-core single-band sphere decoder [6]. This chip is fabricated in a standard-V$_T$ 1P8M 90 nm CMOS process. It can support the flexibility summarized on the left. The core supply voltage is tunable from 0.32 V to 1.2 V according to the operation mode. The overall die area is 8.88 mm^2, within which the core area is 5.29 mm^2; each PE has area of 0.24 mm^2. Register banks and scheduler facilitate multi-core operation.

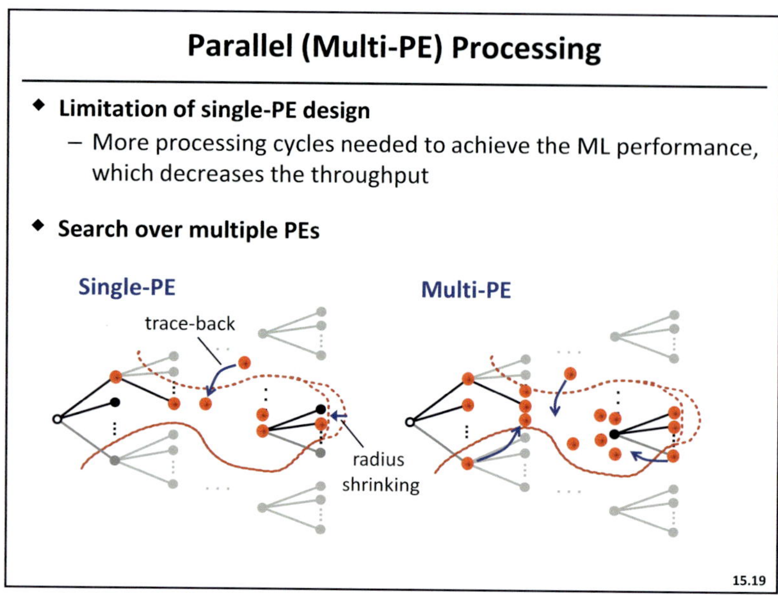

Slide 15.19

More PEs are used to improve BER performance or throughput. In the single-PE case, one node is examined each cycle. When the node is outside the search radius, backward-search is required. When a solution with a smaller Euclidean distance is found, the search radius can be shrunk to reduce the search space. The search process continues until all the possible nodes are examined if there is no timing constraint. In the multiple-PE case, we put multiple PEs to search at the same time. Therefore, more nodes are examined each cycle. Like the single-PE, the search paths will be bounded in the search radius for all PEs. Since multiple PEs are able to find the better solutions earlier, the search radius can be shrunk faster. Again, the search process continues until all the possible nodes are examined if there is no timing constraint.

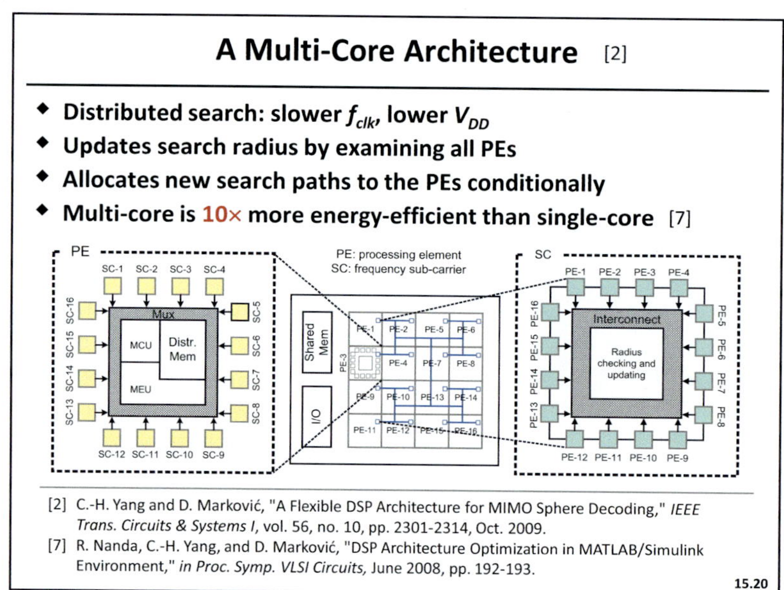

Slide 15.20

By connecting multiple PEs, a new search path can be loaded directly when a previously assigned search tree is examined [2]. Search radius is updated through interconnect network when a solution with a smaller Euclidean distance is found.

Search paths are distributed over 16 processing elements. The multi-core architecture provides a 10x higher energy efficiency than the single-core architecture by reducing clock frequency and supply voltage while keeping the same throughput [7]. On the other hand, operating at the same clock frequency, it can provide a 16x higher throughput. By connecting multiple PEs, the search radius is updated through the interconnect network. It updates the search radius by checking the Euclidean distance of all PEs to speed up search process. It also supports multiple sub-carriers. Multiple sub-carriers are interleaved into PEs through hardware sharing.

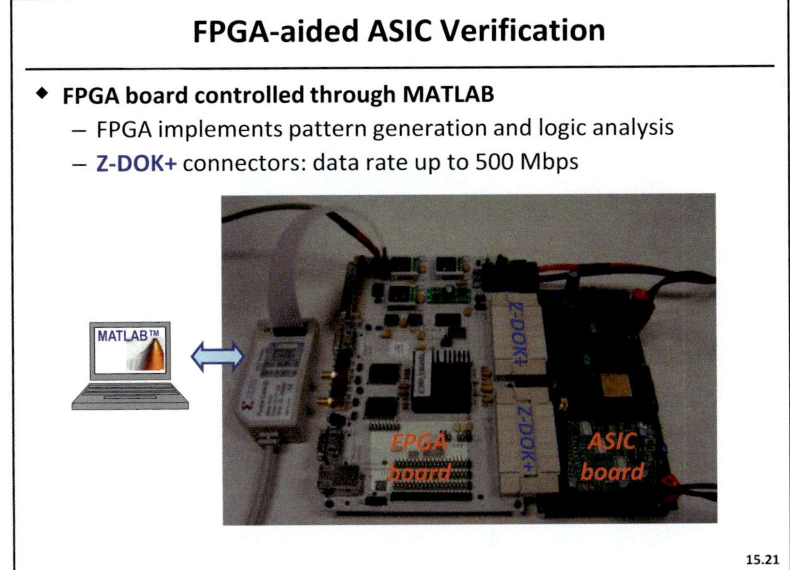

Slide 15.21

To test the chip, we use an FPGA board to implement pattern generation and logic analysis. The functionality of the FPGA is built in a Simulink graphical environment and controlled through MATLAB commands. Two high-speed Z-DOK+ connectors provide data rate up to 500 Mbps. The test data are stored in the on-board block RAM and fed into the ASIC board. The outputs of the ASIC board are captured and stored in the block RAM, and transferred to PC for further analysis. The real-time operation and easy programmability simplify the testing process and data analysis.

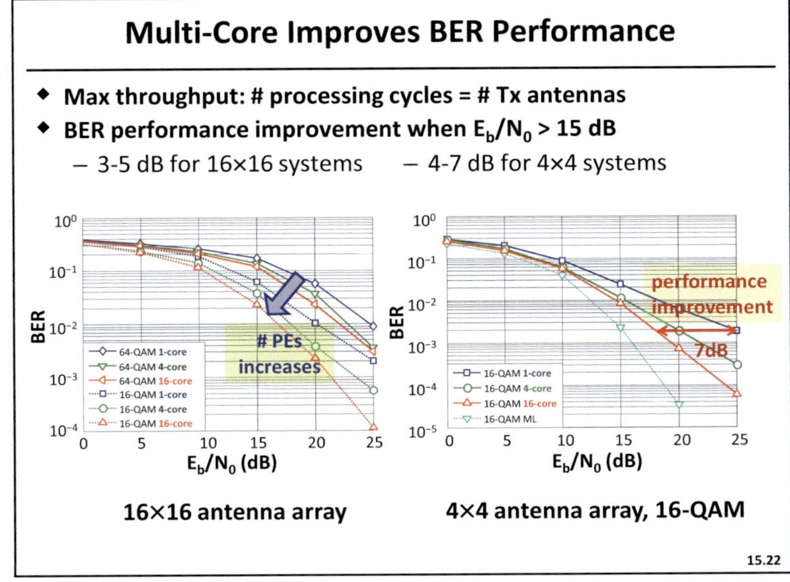

Slide 15.22

Here are the results of BER performance. We constrain the number of processing cycles to be equal to the number of transmit antennas for the maximum throughput. For 16×16 array size, the solid line represents the 64-QAM modulation and the dashed line represents the 16-QAM modulation. As we increase the number of PEs, the BER performance can be improved. A 3-5 dB improvement is observed for 16×16 systems. For 4×4 array size, 16 PEs provide a 7 dB improvement over the 1-PE architecture. Note that the performance gap between the *red-line* and the ML detection comes from the constraint on the number of processing cycles. The circuit performance is shown next.

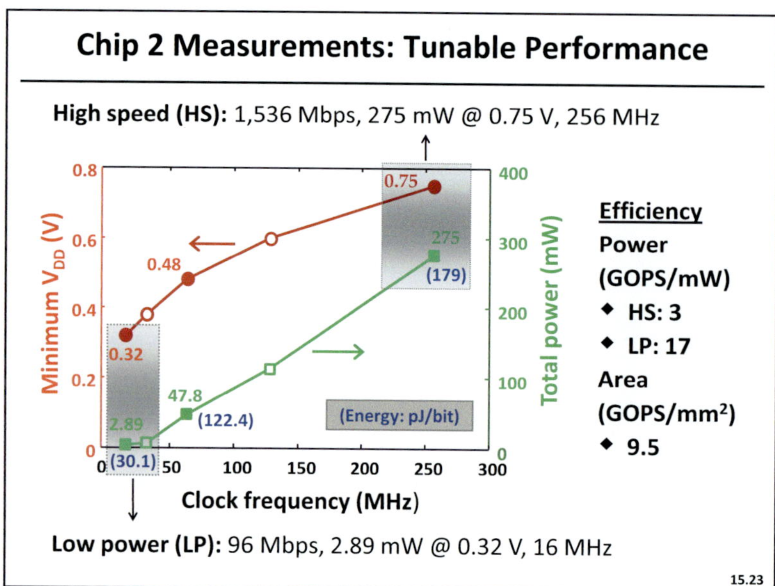

Slide 15.23

The clock frequency of the chip is set to 16, 32, 64, 128, and 256 MHz for a bandwidth of 16 MHz. Power consumption for these operation points ranges from 2.89 mW at 0.32 V to 275 mW at 0.75 V, which corresponds to the energy efficiency of 30.1 pJ/bit to 179 pJ/bit.

Chip 2: Performance Summary

	Modulation	BPSK – 64QAM
Flexibility	Data streams	8 – 128
	Array size	2×2 – 16×16
	Algorithm	K-best/depth-first
	Mode (complex/real)	C
	Process (nm)	90
	Clock freq. (MHz)	16 – 256
	Power (mW)	2.89/0.32V – 275/0.75V
	Energy (pJ/bit)	30.1 – 179
	Bandwidth (MHz)	16
	Throughput (bps/Hz)	12 – 96

15.24

Slide 15.24

Compared with the state-of-the-art chip, the proposed chip has up to 8x higher throughput/BW, in which 4 times comes from the antenna array size and 2 times comes from complex-valued signal processing. This chip also has 6.6x lower energy/bit, which mainly comes from hardware reduction, clock-gating technique and voltage scaling. It also has higher flexibility in terms of antenna array size, modulation scheme, search algorithm, and number of sub-carriers with a small hardware overheard.

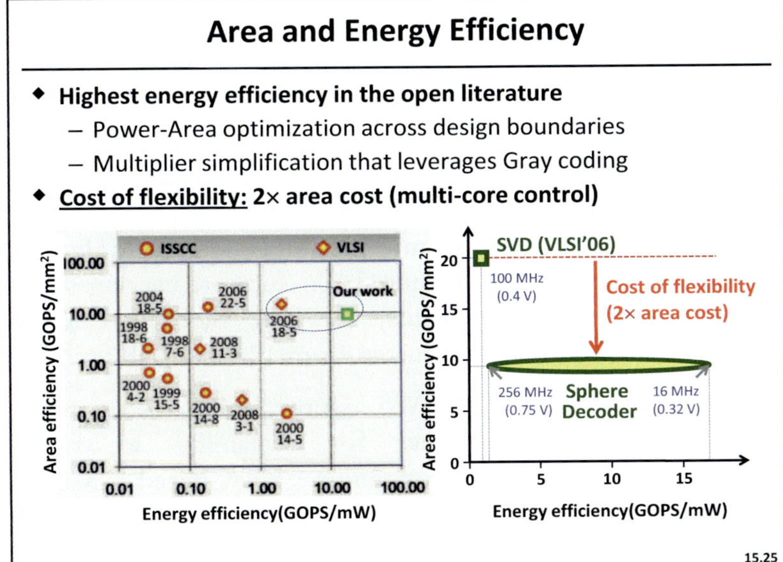

Slide 15.25

If we compare the MIMO decoder chip with the SVD chip from Chap. 14, we find that the sphere decoder has even better energy efficiency, as shown on the right. This is mainly due to multiplier simplification, leveraging Gray code based arithmetic, clock gating, and aggressive voltage scaling. The area efficiency of the sphere decoder is reduced by 2x due to interconnect and multi-core control overhead.

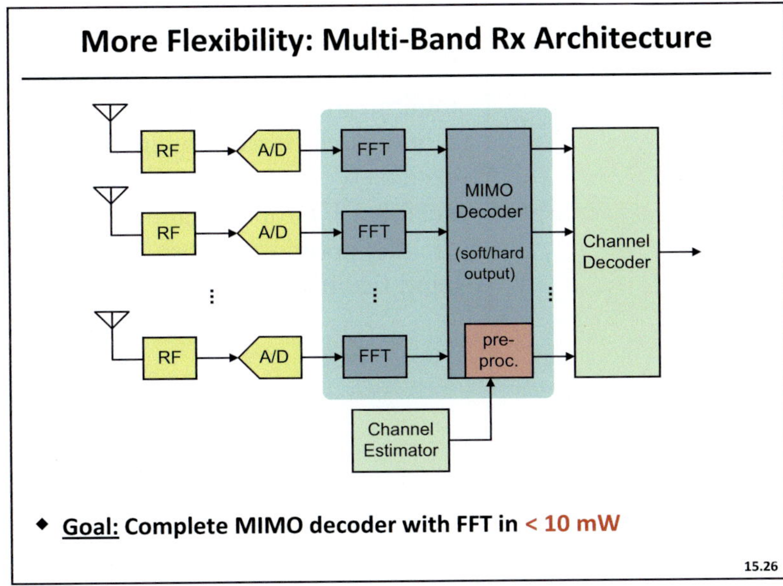

Slide 15.26

We next consider extending the flexibility to multiple signal bands. The receiver architecture is shown in this slide. The signals captured by multiple receive antennas are digitized through ADCs and then converted back to frequency domain through FFT. The modulated data streams carried over the narrow-band sub-carriers are constructively combined through the MIMO decoder. Soft outputs are generated for advanced error-correction signal processing. The design target is to support multiple signal bands (1.25 MHz to 20 MHz) with power consumption less than 10 mW.

Reference Design Specifications

LTE Specs		
Bandwidth (MHz)	1.25, 2.5, 5, 10, 15, 20
FFT size	128, 256, 512, 1024, 1536, 2048
Antenna configuration	1×1, 2×2, 3×2, 4×2
Modulation	QPSK, 16QAM, 64QAM

* **Design considerations**
  - Make 8x8 array: towards distributed MIMO
  - Leverage hard-output multi-core SD from chip 1 (re-design for 8x8 antenna array)
  - Include flexible FFT (128-2048 points) front-end block
  - Include soft-output generation
* **What is the cost of multi-mode multi-band flexibility?**

15.27

Slide 15.27

The 3GPP-LTE is chosen as an application driver in this design. The specifications of 3GPP-LTE systems related to digital baseband are shown in the table. Signal bands from $1.25\,\mathrm{MHz}$ to $20\,\mathrm{MHz}$ are required. To support varying signal bands, the FFT block has to be adjusted from 128 to 2,048 points.

We assume up to 8×8 antenna array for scalability to future cooperative relay-based MIMO systems. In this work, we will leverage hard-output sphere decoder architecture scaled down to work with 8×8 antennas, include flexible FFT and add soft-output generation.

Outline

* **Scalable decoder architecture**
  - Design challenges and solutions
  - Scalable PE architecture
  - Hardware complexity reduction
* **Chip 1:**
  - Multi-mode single-band
  - 16x16 MIMO SD Chip
  - Energy and area efficiency
* **Chip 2:**
  - Multi-mode multi-band
  - 8x8 MIMO SD + FFT Chip
  - Flexibility Cost

15.28

Slide 15.28

We will next present power-area minimization techniques for multi-mode multi-band sphere decoder design. We will mainly focus on the flexible FFT module and evaluation of the additional cost for multi-band flexibility.

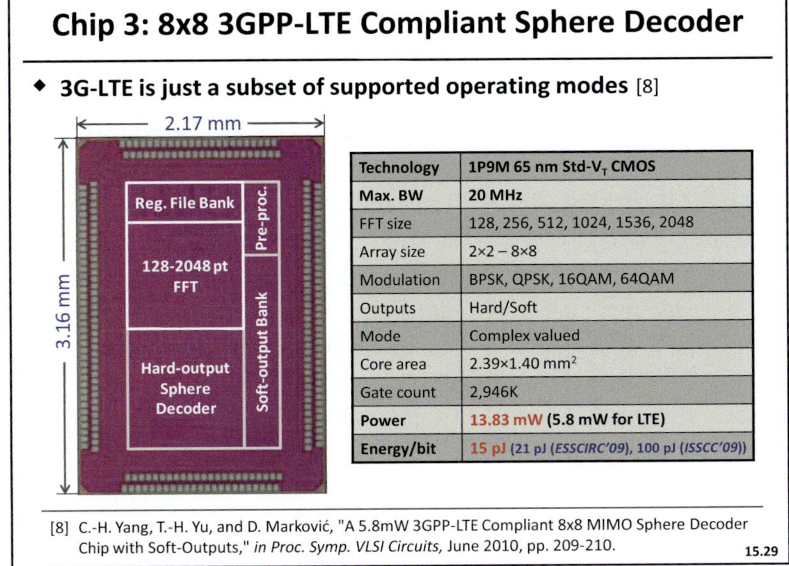

Technology	1P9M 65 nm Std-V_T CMOS
Max. BW	20 MHz
FFT size	128, 256, 512, 1024, 1536, 2048
Array size	2×2 – 8×8
Modulation	BPSK, QPSK, 16QAM, 64QAM
Outputs	Hard/Soft
Mode	Complex valued
Core area	2.39×1.40 mm²
Gate count	2,946K
Power	13.83 mW (5.8 mW for LTE)
Energy/bit	15 pJ (21 pJ (ESSCIRC'09), 100 pJ (ISSCC'09))

[8] C.-H. Yang, T.-H. Yu, and D. Marković, "A 5.8mW 3GPP-LTE Compliant 8x8 MIMO Sphere Decoder Chip with Soft-Outputs," *in Proc. Symp. VLSI Circuits,* June 2010, pp. 209-210.

15.29

Slide 15.29

Hard-output MMO sphere decoder kernel from the first chip is redesigned to accommodate increased number of sub-carriers as required by LTE channelization. Soft outputs are supported through a low-power clock-gated register bank (to calculate Log-likelihood ratio). Power consumption listed here is for 20 MHz band in the 64-QAM mode. The chip dissipates 5.8 mW for the LTE standard, 13.83 mW for full 8×8 array with soft-outputs [8]. The hard-output sphere decoder kernel achieves E/bit of 15 pJ/bit (18.7GOPS/mW) and outperforms prior work. The chip micrograph and summary are shown on this slide. The chip fully supports LTE and has added flexibility for systems with larger array size or cooperative MIMO processing on smaller sub-arrays.

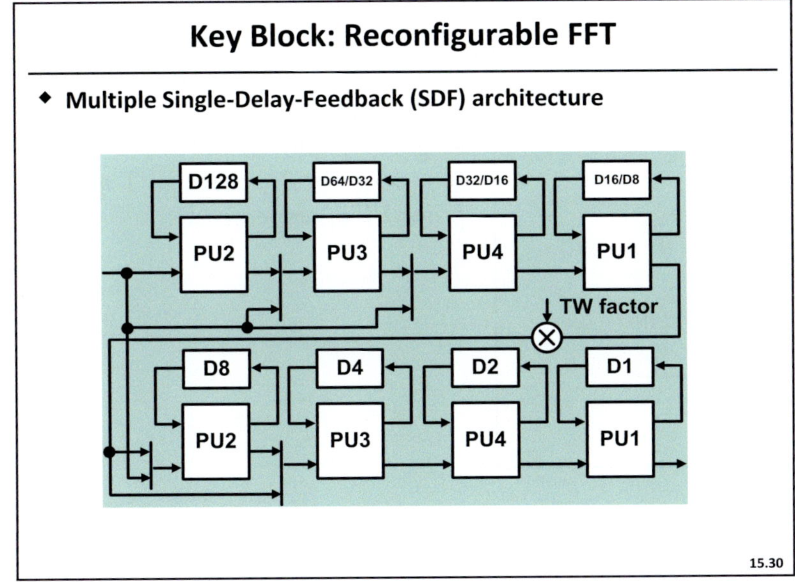

15.30

Slide 15.30

Reconfigurable FFT block is implemented by several small processing units (PUs), as shown in this slide. The FFT block is scalable to support 128 to 2,048 points by changing datapath inter-connection. Multi-path single-delay-feedback (SDF) architecture provides high utilization for varying FFT sizes. Unused PUs and delay lines are clock-gated for power saving. Twiddle (TW) factors are generated by trigonometric approximation instead of fetching coefficients from ROMs for area reduction. To support varying FFT sizes, we adopt a single-path-delay-feedback (SDF) architecture.

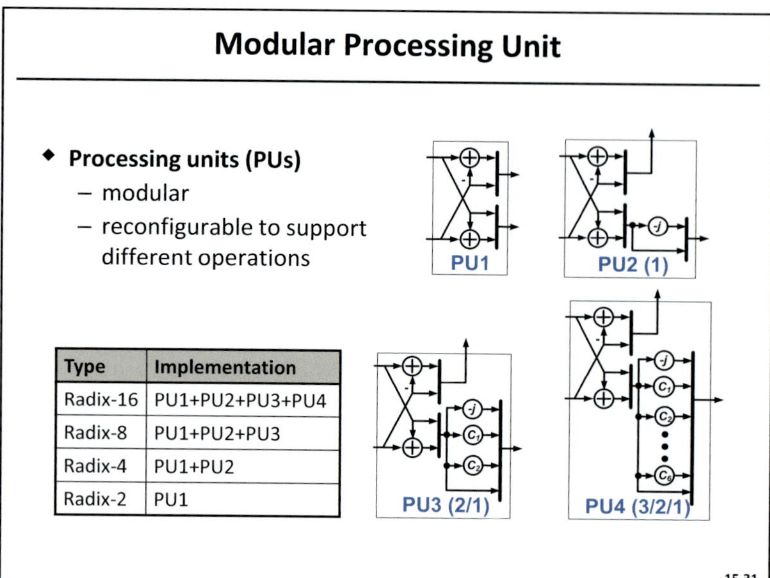

Type	Implementation
Radix-16	PU1+PU2+PU3+PU4
Radix-8	PU1+PU2+PU3
Radix-4	PU1+PU2
Radix-2	PU1

Slide 15.31

The PUs are modular to support different radix representations, allowing power-area tradeoffs. We take the highest radix to be 16. This requires four processing units, as the building blocks shown on the right. Higher-order processing elements can be reconfigured to lower-order ones. For example, PU4 can be simplified to PU3, PU2, or PU1; PU3 can be simplified to PU2 or PU1, etc. This modularity will be examined in architectural exploration for each FFT size.

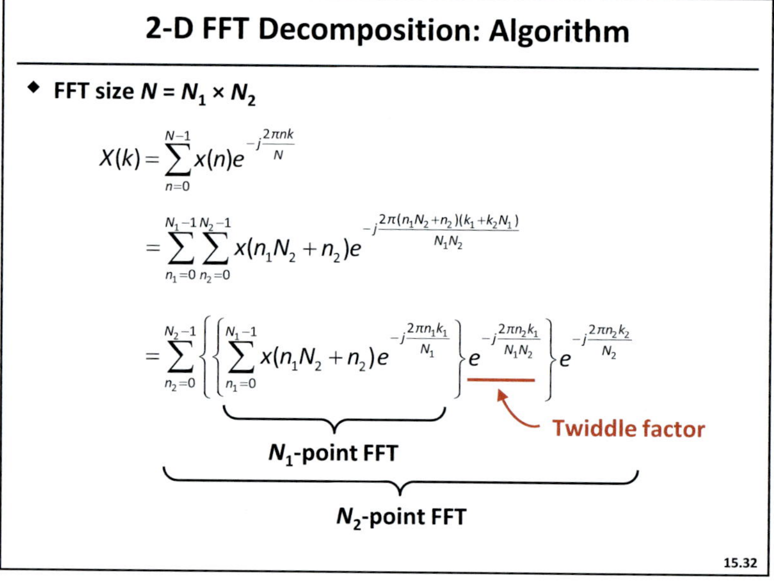

$$X(k) = \sum_{n=0}^{N-1} x(n)e^{-j\frac{2\pi nk}{N}}$$

$$= \sum_{n_1=0}^{N_1-1}\sum_{n_2=0}^{N_2-1} x(n_1 N_2 + n_2)e^{-j\frac{2\pi(n_1 N_2 + n_2)(k_1 + k_2 N_1)}{N_1 N_2}}$$

$$= \sum_{n_2=0}^{N_2-1}\left\{\left\{\sum_{n_1=0}^{N_1-1} x(n_1 N_2 + n_2)e^{-j\frac{2\pi n_1 k_1}{N_1}}\right\}e^{-j\frac{2\pi n_2 k_1}{N_1 N_2}}\right\}e^{-j\frac{2\pi n_2 k_2}{N_2}}$$

Slide 15.32

With the 2-D decomposition, we can re-formulate the FFT operation as shown in the equations on this slide. $k = k_1 \times k_2$ is the FFT bin index and $N = N_1 \times N_2$ is the FFT length. k_1 and k_2 are the sub-FFT bin index for N_1-point FFT and N_2-point FFT, respectively.

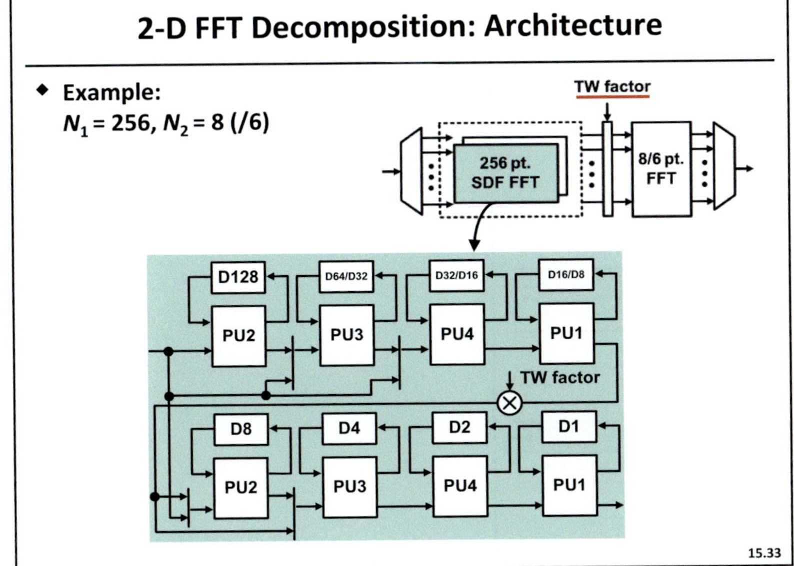

Slide 15.33

Each N_1-point FFT can be operated at N_2-times lower sampling rate in order to maintain the throughput while further reducing the supply voltage to save power. In our application $N_2 = 8$ for all but 1,536-point design where $N_2 = 6$. The FFT is built with several smaller processing units, so we can change the supported FFT size by changing the connection.

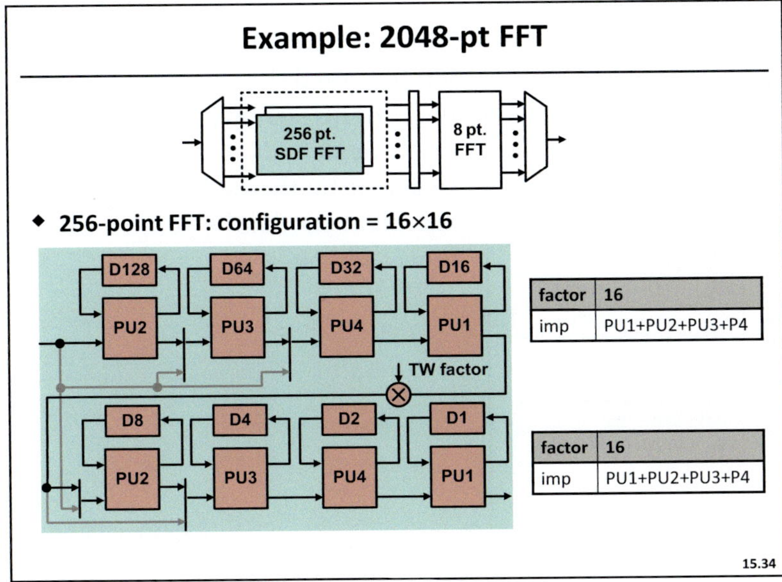

Slide 15.34

2,048-point FFT is made with $N_1 = 256$ and $N_2 = 8$. The 256-point FFT is decomposed into two stages with radix-16 factorization. Radix-16 is implemented by concatenating PU1, PU2, PU3, and PU4, as shown in Slide 15.31. Delay-lines associated with PUs are configured to adjust signal delay properly.

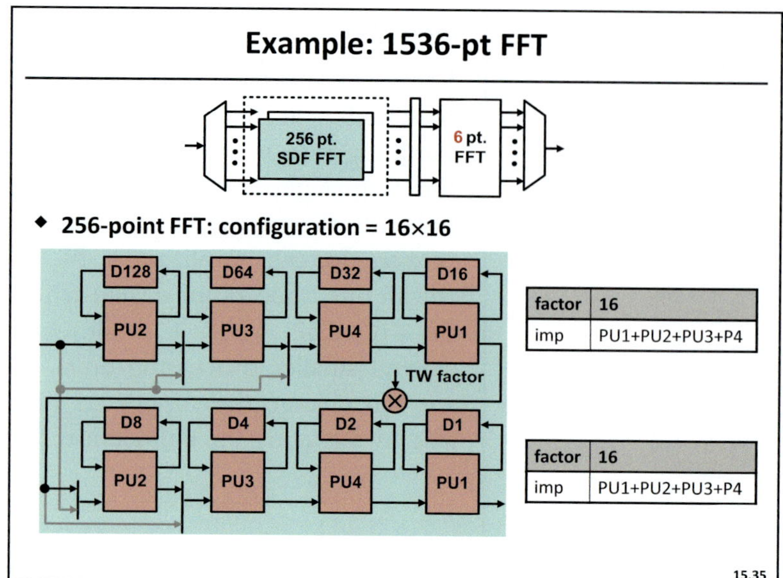

factor	16
imp	PU1+PU2+PU3+P4

factor	16
imp	PU1+PU2+PU3+P4

Slide 15.35

1,536-point design shares the same front-end processing with 2,048-point design and uses $N_2 = 6$ in the back-end.

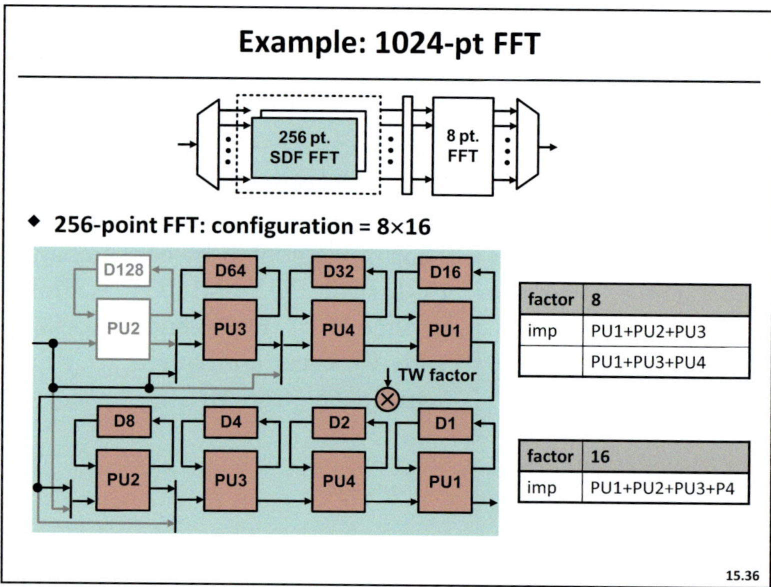

factor	8
imp	PU1+PU2+PU3
	PU1+PU3+PU4

factor	16
imp	PU1+PU2+PU3+P4

Slide 15.36

The 1,024-point FFT is implemented by disabling PU2 at the input (shown in gray) to implement the radix-8 stage followed by the radix-16 stage. The idle PU2 module is turned off for power saving.

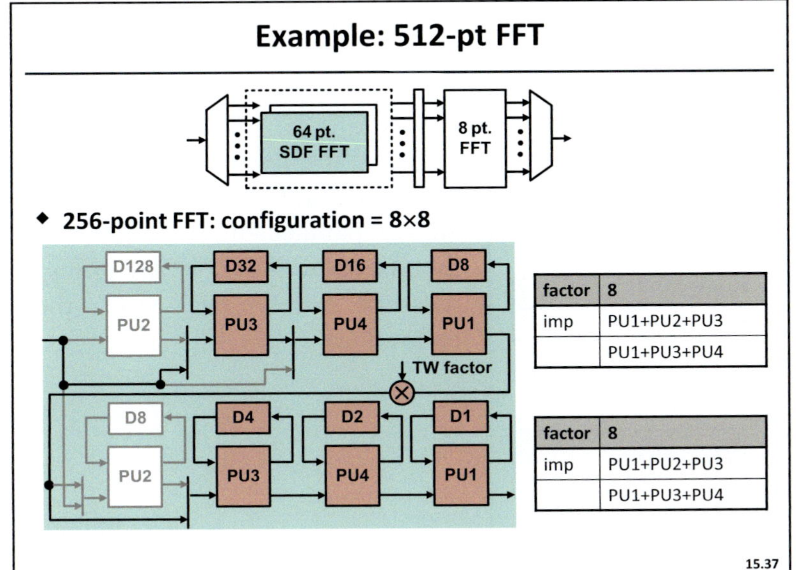

factor	8
imp	PU1+PU2+PU3
	PU1+PU3+PU4

factor	8
imp	PU1+PU2+PU3
	PU1+PU3+PU4

Slide 15.37

512-point FFT can be implemented with two radix-8 stages by disabling the two PU2 front units shown in gray.

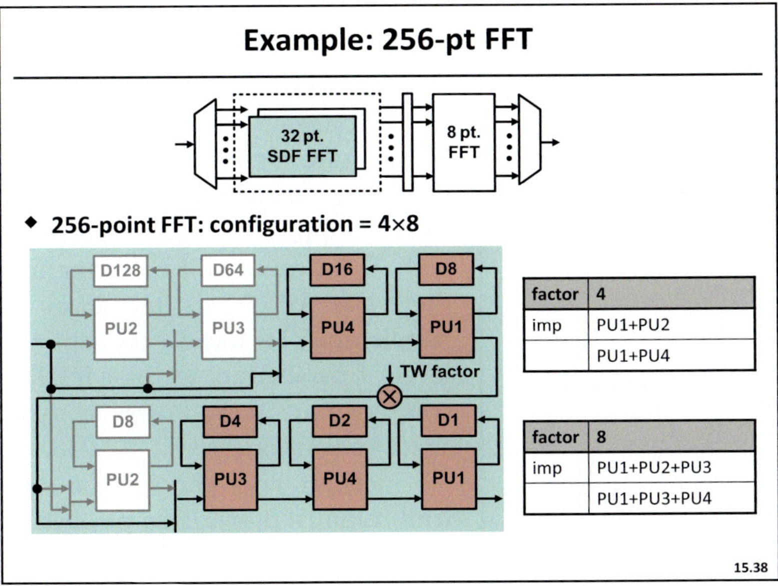

factor	4
imp	PU1+PU2
	PU1+PU4

factor	8
imp	PU1+PU2+PU3
	PU1+PU3+PU4

Slide 15.38

256-point design is implemented with the radix-4 stage followed by the radix-8 stage by disabling the units shown in gray.

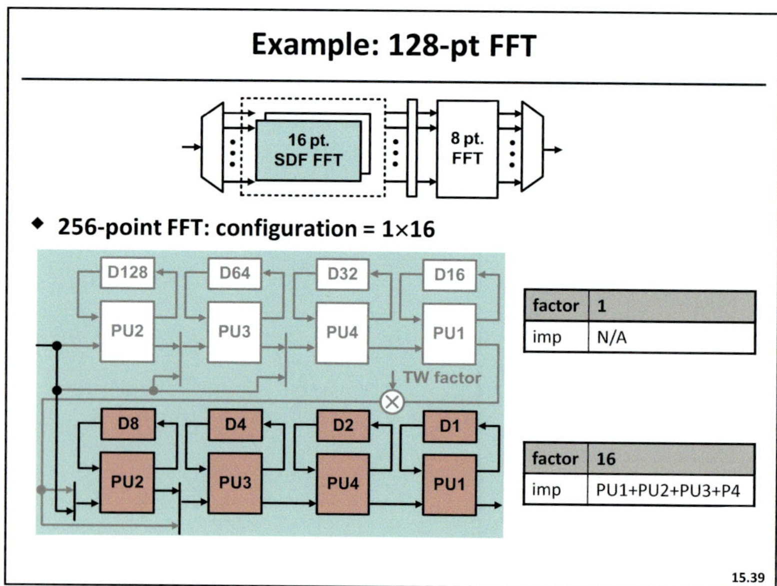

factor	1
imp	N/A

factor	16
imp	PU1+PU2+PU3+P4

Slide 15.39

Finally, 128-point FFT that uses radix-16 in the first stage is shown.

Two key design techniques for FFT area and power minimization are parallelism and FFT factorization. Next, we are going to illustrate the use of these techniques.

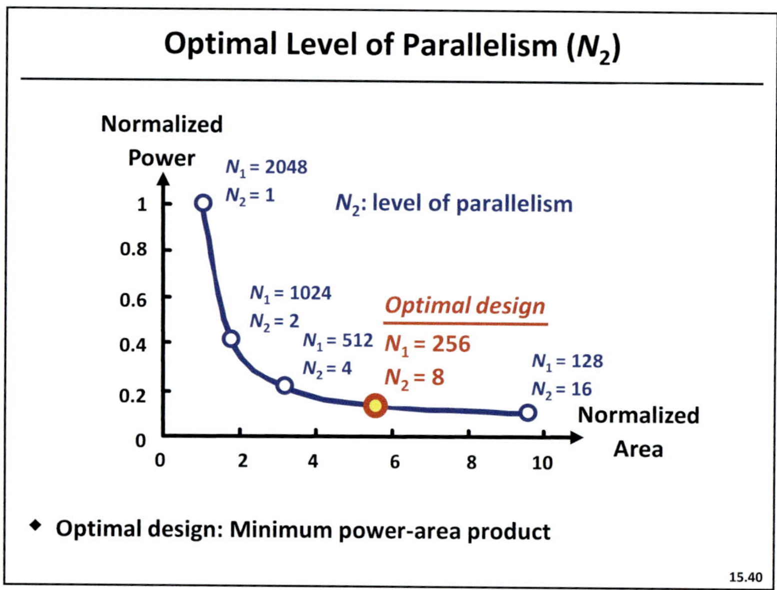

Slide 15.40

Parallelism is used to reduce power consumption by taking advantage of voltage scaling. Parallel architecture allows for slower operating frequency and therefore lower supply voltage. Combined with the proposed FFT structure, the value of N_2 is the level of parallelism. To decide the optimal level of parallelism, we plot feasible combinations in the Power vs. Area space (normalized to $N_1 = 2,048$, $N_2 = 1$ design) and choose the one with minimal power-area product as the optimal design. According to our analysis, $N_2 = 4$ and $N_2 = 8$ have similar power-area products. We choose 8-way parallel design to lower the power consumption.

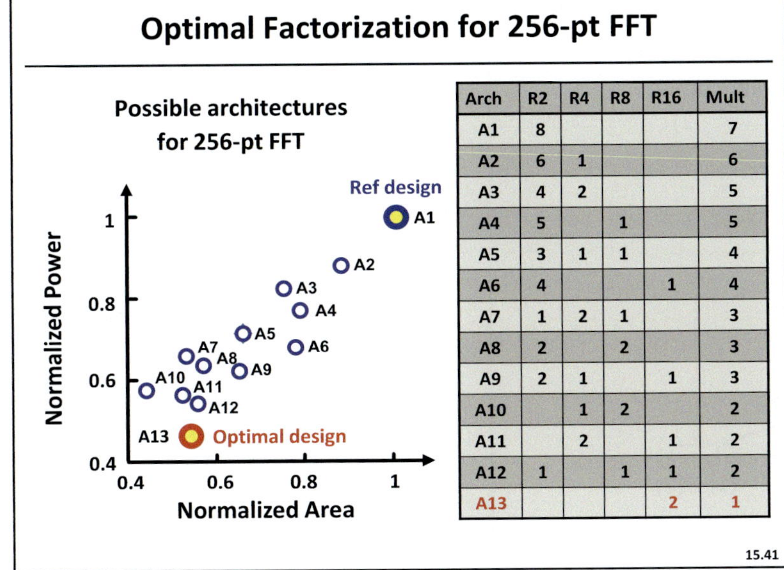

Arch	R2	R4	R8	R16	Mult
A1	8				7
A2	6	1			6
A3	4	2			5
A4	5		1		5
A5	3	1	1		4
A6	4			1	4
A7	1	2	1		3
A8	2		2		3
A9	2	1		1	3
A10		1	2		2
A11		2		1	2
A12	1		1	1	2
A13				2	1

Slide 15.41

Next, we optimize each N_1-point FFT by the processing unit factorization. Take a 256-point FFT as the design example. There are 13 possible radix factorizations, as illustrated in the Table. Each processing unit and the concatenated twiddle factor multiplier have different areas and switching activities, allowing power and area tradeoff. The plot shows power and area among the possible architectures for a 256-point FFT, normalized to the radix-2 case. Power-area product reduction of 75 % can be achieved by using 2 radix-16 processing elements (A13) as the building block.

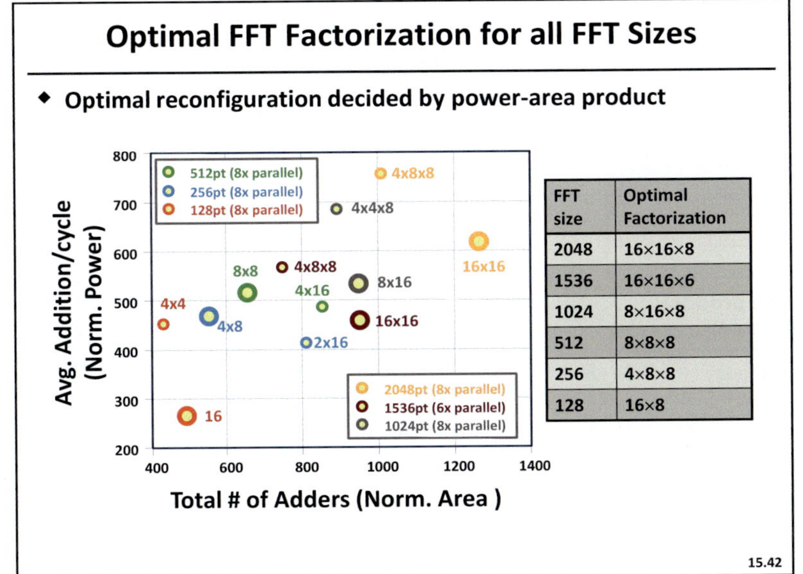

FFT size	Optimal Factorization
2048	16×16×8
1536	16×16×6
1024	8×16×8
512	8×8×8
256	4×8×8
128	16×8

Slide 15.42

Here, we analyze different radix factorizations for the 128–2,048-point FFT design. The optimal reconfiguration is decided by the architecture with the minimum power-area product. For 128-point FFT, 8-times parallel with radix-16 is the optimal choice. Other combinations are not shown here due to their poor performance. Using this methodology, we can decide the optimal reconfiguration for all required FFT sizes. All the big dots represent the optimal factorization for various FFT sizes.

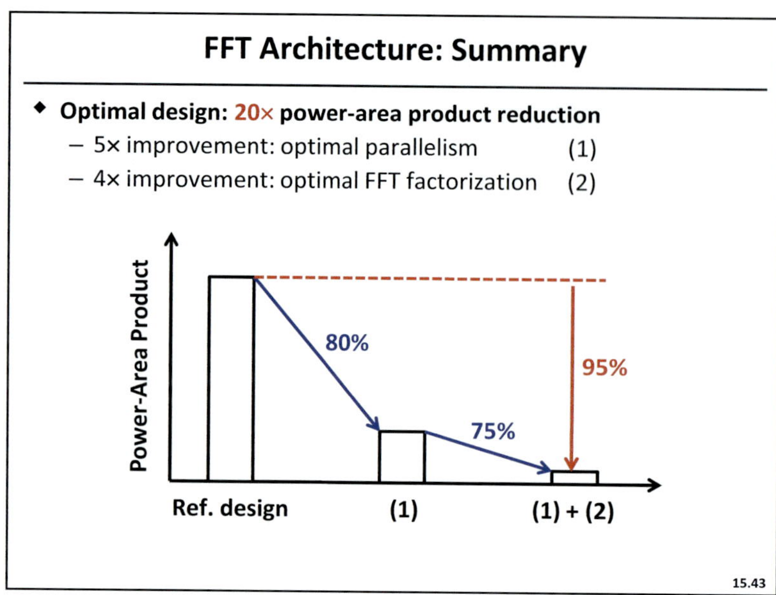

Slide 15.43

Combining the 2-D FFT decomposition and processing-unit factorization, a 20-times power-area product reduction can be achieved compared to the single delay feedback radix-2 1,024-point FFT reference design. To decide the optimal level of parallelism, we enumerate and analyze all possible architectures and plot them in the power vs. area space. An 8× parallelism has the minimum power-area product considering voltage scaling. Overall, a 95% (20x) power-area reduction is achieved compared to the reference design, where 5× comes from architecture parallelism and 4× comes from radix factorization.

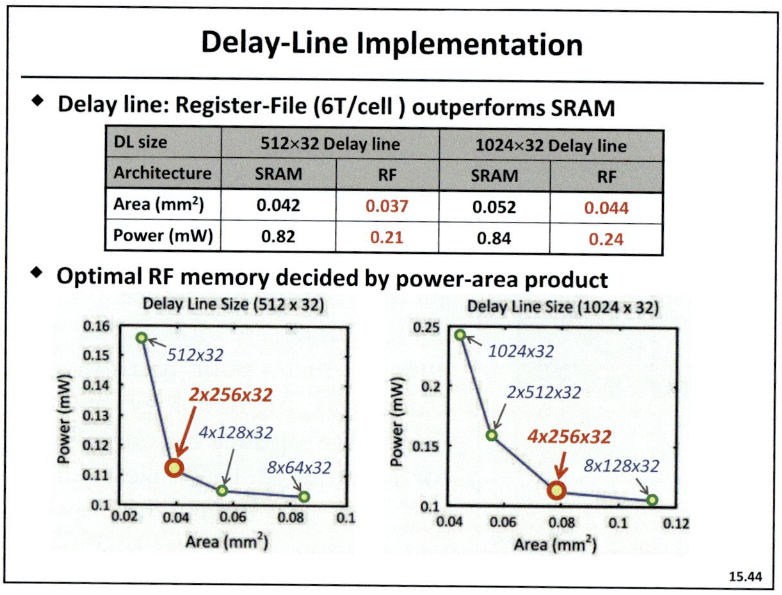

DL size	512×32 Delay line		1024×32 Delay line	
Architecture	SRAM	RF	SRAM	RF
Area (mm²)	0.042	0.037	0.052	0.044
Power (mW)	0.82	0.21	0.84	0.24

Slide 15.44

For the delay line implementation, we compare register-file (RF) and SRAM-based delay lines. According to the analysis, for a larger size delay line, RF has smaller area and less power consumption even though the cell area of RF is bigger. But even for the same length delay line, there are several different memory partitions. Again, we plot all possible combinations in the power-area space and choose the design with the minimum power-area product. 2×256 is optimal for length 512, and 4×256 is optimal for length 1024.

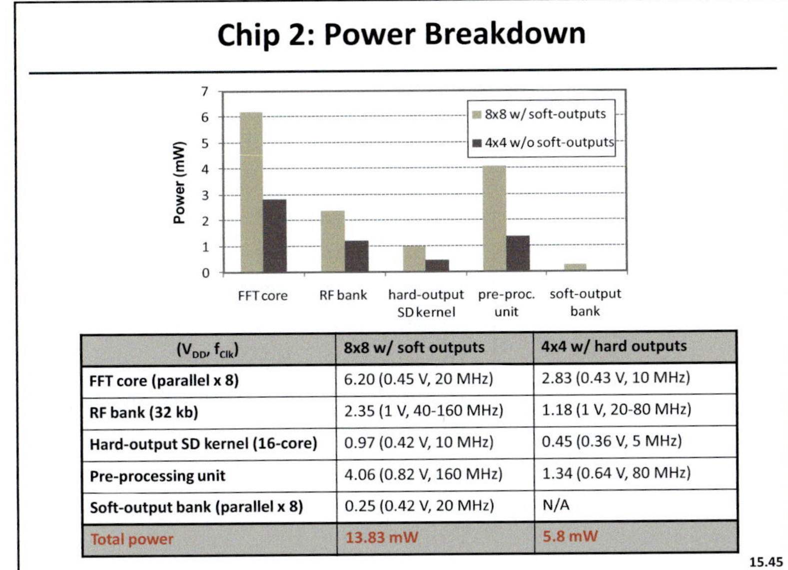

(V_{DD}, f_{Clk})	8x8 w/ soft outputs	4x4 w/ hard outputs
FFT core (parallel x 8)	6.20 (0.45 V, 20 MHz)	2.83 (0.43 V, 10 MHz)
RF bank (32 kb)	2.35 (1 V, 40-160 MHz)	1.18 (1 V, 20-80 MHz)
Hard-output SD kernel (16-core)	0.97 (0.42 V, 10 MHz)	0.45 (0.36 V, 5 MHz)
Pre-processing unit	4.06 (0.82 V, 160 MHz)	1.34 (0.64 V, 80 MHz)
Soft-output bank (parallel x 8)	0.25 (0.42 V, 20 MHz)	N/A
Total power	13.83 mW	5.8 mW

Slide 15.45

Power breakdown of the chip is shown here. Since these five blocks have individual power supplies, we can measure their own powers. Operating for 8×8 antenna array with soft outputs, the chip dissipates 13.83 mW. Operating for 4×4 mode with only hard outputs, the chip dissipates 5.8 mW of power.

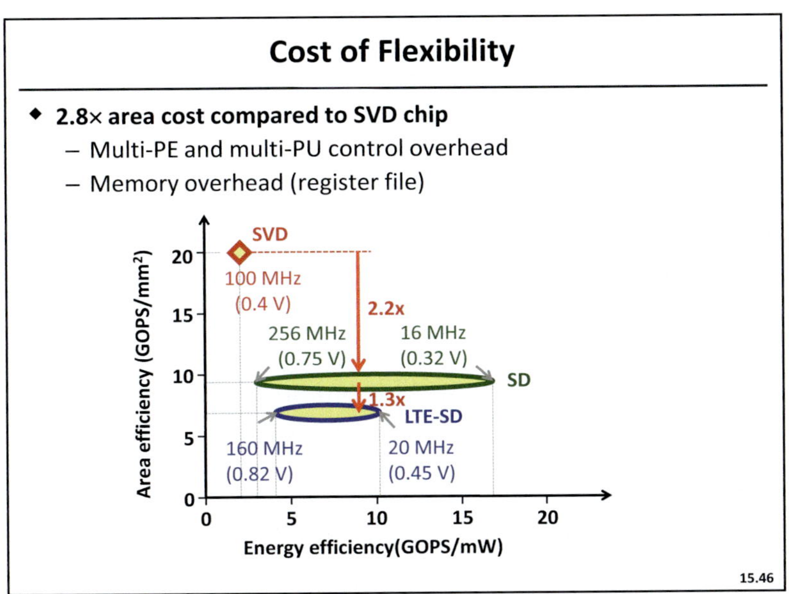

Slide 15.46

Let's see the cost of flexibility. Area and energy efficiency of SVD chip (chip1), SD chip (chip 2), and LTE-SD chip (chip3) are compared. The area efficiency of LTE-SD chip is dropped by 2.8× compared to the SVD chip, and 1.3× compared to the previous SD chip. The area-efficiency reduction comes from the control and interconnect overhead of multi-PE, multi-PU, and also from memory overhead for register file. The energy efficiency of SD chip and LTE-SD chip is 1.5×–8.5× higher than that of the SVD chip. The improved energy efficiency is attributed to optimization across design boundaries, arithmetic simplification, voltage scaling and clock-gating techniques.

Results

♦ **A 4x4 SVD achieves 2 GOPS/mW & 20 GOPS/mm² (90 nm CMOS)**

♦ **Multi-mode single-band sphere decoder is even more efficient**
  - Scalable PE allows wide range of performance tuning
    • Low power: 2.89 mW & 96 Mbps @ 0.32 V (17 GOPS/mW)
    • High speed: 275 mW & 1,536 Mbps @ 0.75 V (3 GOPS/mW)
  - Multi-core (PE) architecture provides
    • Higher power efficiency: 6x for same throughput
    • Improved BER performance: 3-5 dB
    • Higher throughput: 16x

♦ **Multi-mode multi-band flexibility is achieved through flexible FFT**
  - 3GPP-LTE compliant 8×8 MIMO decoder can be implemented in
    < 6mW in a 65 nm CMOS

15.47

Slide 15.47

In conclusion, the flexible SD and LTE-SD chips are able to provide $1.5\times - 8.5\times$ higher energy efficiency than the dedicated SVD chip. 16-core sphere decoder architecture provides $6\times$ higher energy efficiency for same throughput or $16\times$ higher throughput for same operating frequency compared to single-core architecture. Optimal reconfigurable FFT provides a 20 times power-area product reduction from algorithm, architecture and circuit design. We optimize this FFT through FFT factorization, PU reconfiguration, memory partition and delay-line implementation. Combined with clock-gating technique, multi-voltage design, and aggressive voltage scaling, this chip consumes less than 6 mW for LTE standard in a 65 nm CMOS technology.

Summary

♦ **Sphere decoding is a practical ML approaching algorithm**
♦ **Implementations for large antenna array, constellation, and number of carriers is challenging**
  - Multipliers are simplified by exploiting Gray-coded modulation and wordlength reduction
  - Decision plane partitioning simplifies metric enumeration
  - Multi-core search allows for increased energy efficiency or improved throughput
♦ **Flexible processing element for multi-mode/band operation**
  - Supports antennas 2x2 to 16x16, modulations BPSK to 64QAM, 8 to 128 sub-carriers, and K-best/depth-first search
  - Multi-mode operation is supported with 2x reduction in area efficiency due to overhead to operate multi-core architecture
  - Multi-band LTE operation incurs another 1.3x overhead

15.48

Slide 15.48

Hardware realization of MHz-rate sphere decoding algorithm is presented in this chapter. Sphere decoding can approach maximum likelihood (ML) detection with feasible computational complexity, which makes it attractive for practical realization. Scaling the algorithm to higher number of antennas, modulations, and number of frequency sub-carriers is challenging. The chapter discussed simplifications in multiplier implementation that allows extension to large antenna arrays (16×16), decision plane partitioning for metric enumeration, and multi-core search for improved energy efficiency and performance. Flexible processing element is used in a multi-core architecture to demonstrate multi-mode and multi-band operation with minimal overhead in area efficiency as compared to dedicated MIMO SVD chip.

References

- C.-H. Yang and D. Marković, "A Multi-Core Sphere Decoder VLSI Architecture for MIMO Communications," in *Proc. IEEE Global Communications Conf.,* Dec. 2008, pp. 3297-3301.

- C.-H. Yang and D. Marković, "A 2.89mW 50GOPS 16x16 16-Core MIMO Sphere Decoder in 90nm CMOS," in *Proc. IEEE Eur. Solid-State Circuits Conf.,* Sept. 2009, pp. 344-348.

- R. Nanda, C.-H. Yang, and D. Marković, "DSP Architecture Optimization in Matlab/Simulink Environment," in *Proc. Symp. VLSI Circuits,* June 2008, pp. 192-193.

- C.-H. Yang, T.-H. Yu, and D. Marković, "A 5.8mW 3GPP-LTE Compliant 8x8 MIMO Sphere Decoder Chip with Soft-Outputs," in *Proc. Symp. VLSI Circuits,* June 2010, pp. 209-210.

- C.-H. Yang, T.-H. Yu, and D. Marković, "Power and Area Minimization of Reconfigurable FFT Processors: A 3GPP-LTE Example," *IEEE J. Solid-State Circuits*, vol. 47, no.3, pp. 757-768, Mar. 2012.

| | Chapter 16 | |

kHz-Rate Neural Processors

with Sarah Gibson and Vaibhav Karkare
University of California, Los Angeles

This chapter presents a design example of a kHz-rate neural processor. A brief introduction to kHz design will be provided, followed by an introduction to neural spike sorting. Several spike-sorting algorithms will be reviewed. Lastly, the design of a 130-μW, 64-channel spike-sorting DSP chip will be presented.

Need for kHz Processors

- **Conventional DSPs for communications operate at a few hundred MHz and consume tens to hundreds of mW of power**

- **Signals in biomedical and geological applications have bandwidths of ~ 10 to 20 kHz**
  - These applications are severely power constrained (desired power consumption is tens of μW)
    - Implantable neural devices are power density limited (<< 800 μW/mm^2)
    - Seismic sensor nodes are limited by battery life (desired battery life of ~ 5 – 10 years)

- **There is a need for energy-efficient implementation of kHz-rate DSPs to process signals for these applications**

- **In this chapter, we describe the design of a 64-channel spike-sorting DSP chip as an illustration of design techniques for kHz-rate DSP processors**

16.2

The designs we analyzed thus far operate at few tens to hundreds of MHz and consume a few tens to hundreds of milliwatts of power. The sample rate requirement for these applications roughly tracks the speed of the underlying technology. There are many applications that dictate sample rates much below the speed of technology such as those found in medical implants or seismic sensors. Medical implants in particular impose very tight power density limits to avoid tissue overheating.

This chapter will therefore address implementation issues related to kHz-rate processors with stringent power density requirements (10–100x below communication DSP processors).

D. Marković and R.W. Brodersen, *DSP Architecture Design Essentials*, Electrical Engineering Essentials,
DOI 10.1007/978-1-4419-9660-2_16, © Springer Science+Business Media New York 2012

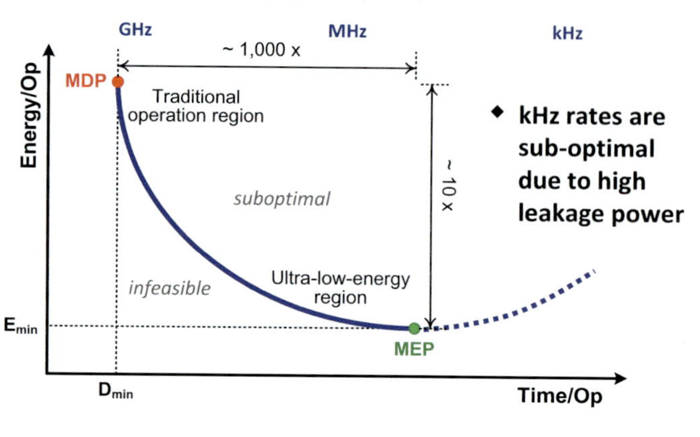

Slide 16.3

This slide reviews the energy-delay tradeoff in combinational logic. The solid line indicates the optimal tradeoff that is bounded by the minimum-delay point (MDP) and the minimum-energy point (MEP). All designs below the line are infeasible and all designs above the line are suboptimal. The discrepancy between the MDP and the MEP is about one order of magnitude in energy (for example, scaling V_{DD} from 1.0 V to 0.3 V gives an order of magnitude reduction in energy per operation) and about three orders of magnitude in speed. The dotted line shows kHz region, beyond MEP, where the energy is limited by leakage. Designing for the kHz sample rates thus requires circuit methods for aggressive leakage minimization as well as architectural techniques that minimize leakage through reduced area. The use of these techniques will be illustrated in this chapter.

Slide 16.4

Electrophysiology is the technique of recording electrical signals from the brain using implantable microelectrodes. As shown in this figure, the voltage recorded by a microelectrode is the resulting sum of the activity from multiple neurons surrounding the electrode. In many cases, it is important to know which action potentials, or "spikes", come from which neuron. The process of assigning each action potential to its originating neuron so that information can be decoded is known as "spike sorting." This process is also used in basic science experiments seeking to understand how the brain processes information as well as in medical applications like brain-machine interfaces (BMIs), which are often controlled by single neurons.

The Spike-Sorting Process

1. **Spike Detection:** Separating spikes from noise

2. **Alignment: Aligning detected spikes to a common reference**

3. **Feature Extraction: Transforming spikes into a certain set of features (e.g. principal components)**

4. **Dimensionality Reduction: Choosing which features to use in clustering**

5. **Clustering: Classifying spikes into different groups (neurons) based on extracted features**

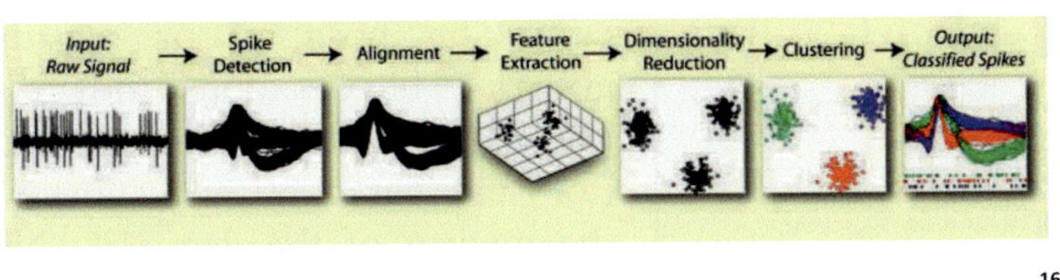

16.5

Slide 16.5

This slide provides background on neural spike sorting. Spike sorting is generally performed in five successive steps: spike detection, spike alignment, feature extraction (FE), dimensionality reduction, and clustering. Spike detection is simply the process of separating the actual spikes from the background noise. In alignment, each spike is aligned to a common point, such as the maximum value or the maximum derivative. In feature extraction, spikes are transformed into a certain feature space that is designed to separate groups of spikes more easily than can be done in the original space (i.e., the time domain). Dimensionality reduction is choosing which features to retain for clustering and which to discard. It is often performed after feature extraction to improve the accuracy of, and to reduce the complexity of, subsequent clustering. Finally, clustering is the process of classifying spikes into different groups (i.e., neurons) based on the extracted features. The final result of spike sorting is a list of spike times for each neuron in the recording. This information can be represented in a graph called a "raster plot", shown on the bottom of the right-most subplot of this figure.

Algorithm Overview

* **Spike detection**
  - Absolute-value threshold
  - Nonlinear energy operator (NEO)
  - Stationary wavelet transform product (SWTP)

* **Feature extraction**
  - Principal component analysis (PCA)
  - Discrete wavelet transform (DWT)
  - Discrete derivatives (DD)
  - Integral transform (IT)

16.6

Slide 16.6

Now we will review a number of spike-detection and feature-extraction algorithms. All spike detection methods involve first pre-emphasizing the spike and then applying a threshold to the waveform. The method of pre-emphasis and the method of threshold calculation are presented for each of the following spike-detection algorithms: absolute-value thresholding, the nonlinear energy operator (NEO), and the stationary wavelet transform product (SWTP). Then, four different feature-extraction algorithms will be described: principal component analysis (PCA), the discrete wavelet transform (DWT), discrete derivatives (DD), and the integral transform (IT).

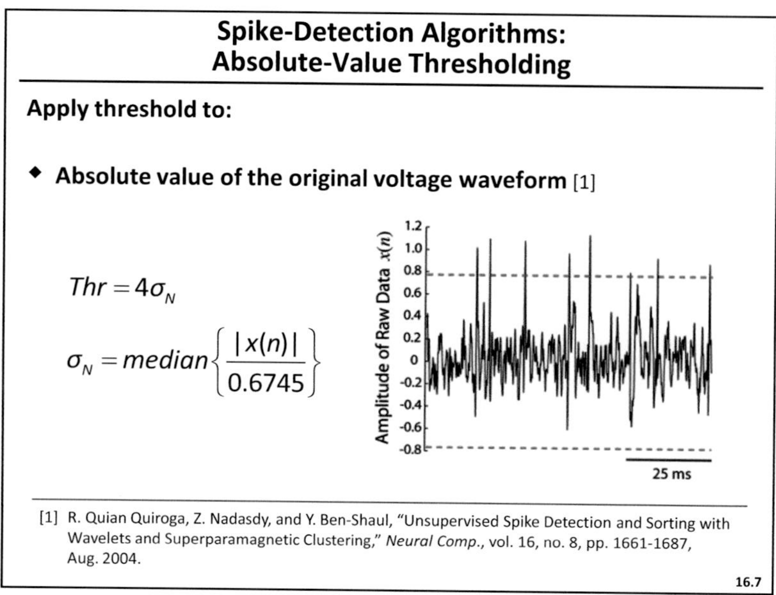

Slide 16.7

One simple, commonly used detection method is to apply a threshold to the voltage of the waveform. This threshold can be applied to either the raw waveform or to the absolute value of the waveform. Applying a threshold to the absolute value of the signal is more intuitive, since spikes can either be positive- or negative-going. The equations shown for the calculation of the threshold (Thr) are based on an estimate of the median of the data [1]. Note that this figure shows $\pm Thr$ (red dashed lines) applied to the original waveform rather than $+Thr$ applied to the absolute value of the waveform.

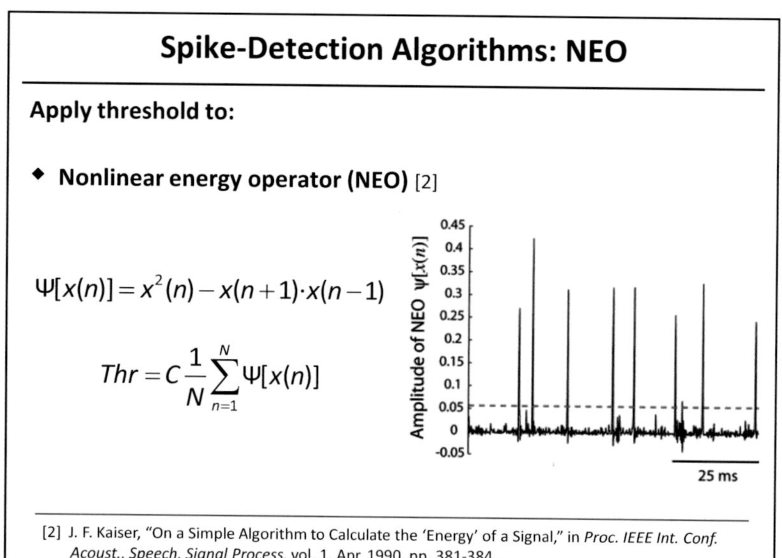

Slide 16.8

The nonlinear energy operator (NEO) [2] is another method of pre-emphasizing the spikes. The NEO is large only when the signal is both high in power (i.e., $x^2(n)$ is large) and high in frequency (i.e., $x(n)$ is large while $x(n+1)$ and $x(n-1)$ are small). Since a spike by definition is characterized by localized high frequencies and an increase in instantaneous energy, this method has an obvious advantage over methods that look only at an increase in signal energy or amplitude without regarding the frequency. Similarly to the method in [1], the threshold Thr was automatically set to a scaled version of the mean of the NEO. It can be seen from this figure that the NEO has greatly emphasized the spikes compared to the figure in the previous slide.

Spike-Detection Algorithms: SWTP

Apply threshold to Stationary Wavelet Transform Product (SWTP)

1. Calculate SWT at 5 consecutive dyadic scales:
$$W(2^j, n), \; j = 1, \ldots, 5$$

2. Find the scale $2^{j,\max}$ with the largest sum of absolute values:
$$j_{max} = \underset{j}{\arg\max}\left(\sum_{n=1}^{N} |W(2^j, n)| \right)$$

3. Calculate point-wise product between SWTs at this & the two previous scales:
$$P(N) = \prod_{j=j_{max}-2}^{j_{max}} |W(2^j, n)|$$

4. Smooth with Bartlett window $w(n)$

5. Threshold: $\quad Thr = C\dfrac{1}{N}\sum_{n=1}^{N} w(n) * P(n)$

16.9

Slide 16.9

The Discrete Wavelet Transform (DWT), originally presented in [4], is ideally suited for the detection of signals in noise (e.g., edge detection, speech detection). Recently, it has also been applied to spike detection. The stationary wavelet transform product (SWTP) is a variation on the DWT presented in [3], as outlined in this slide.

Spike-Detection Algorithms: SWTP

◆ **Apply threshold to Stationary Wavelet Transform Product** [3]

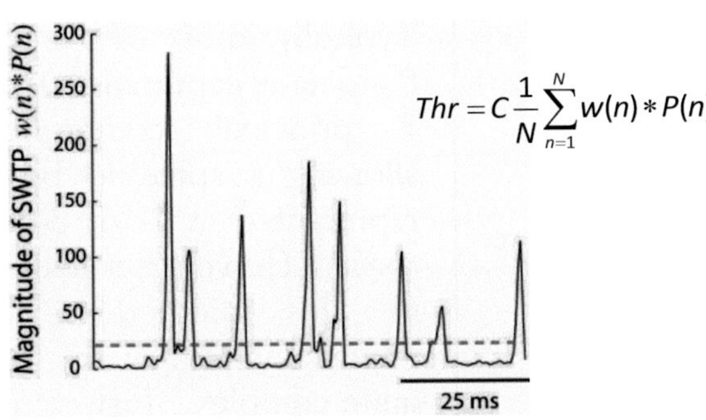

[3] K.H. Kim and S.J. Kim, "A Wavelet-based Method for Action Potential Detection from Extracellular Neural Signal Recording with Low Signal-to-noise Ratio," *IEEE Trans. Biomed. Eng.*, vol. 50, no. 8, pp. 999-1011, Aug. 2003.

16.10

Slide 16.10

This figure shows an example of the SWTP signal and the calculated threshold.

Feature-Extraction Algorithms: PCA

- **Projects data onto an orthogonal set of basis vectors such that the first coordinate (called the first principal component) represents the direction of largest variance**

- **Algorithm:**
  1. Calculate covariance matrix of data (spikes) (N-by-N).
  2. Calculate eigenvectors ("principal components") of covariance matrix (N 1-by-N vectors).
  3. For each principal component (i = 1,...,N), calculate the ith "score" as the scalar product of the data point (spike) and the ith principal component.

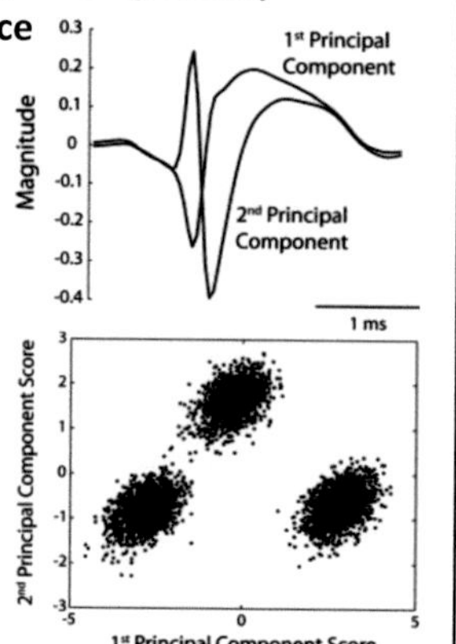

Slide 16.11

Principal Component Analysis (PCA) has become a benchmark FE method in neural signal processing. In PCA, an orthogonal basis ("principal components" or PCs) is calculated for the data that captures the directions in the data with the largest variation. Each spike is expressed as a series of PC coefficients. (These PC coefficients or scores, shown in the scatter plot, are then used in subsequent clustering.) The PCs are found by performing eigenvalue decomposition of the covariance matrix of the data; in fact, the PCs are the eigenvectors themselves.

Feature-Extraction Algorithms: PCA

- **Pros:**
  - Efficient (coding): can represent spike in 2 or 3 PCs

- **Cons:**
  - Inefficient (implementation): hard to perform eigendecomposition in hardware
  - Only finds independent axes if the data is Gaussian
  - There is no guarantee that the directions of maximum variance will contain good features for discrimination

16.12

Slide 16.12

Besides being a well understood method, PCA is often used because it is an efficient way of coding data. Typically, most of the variance in the data is captured in the first 2 or 3 principal components, thus allowing a spike to be accurately represented in 2- or 3-dimensional space. However, a main drawback to this method is that the implementation of the algorithm is quite complex. Just computing the principal component scores for each spike takes over 5,000 additions and over 5,000 multiplications (assuming 72 samples per spike), not to mention the even greater complexity required to perform the initial eigenvalue decomposition to calculate the principal components themselves. Another drawback of PCA is that it achieves the greatest performance when the underlying data is from a unimodal Gaussian distribution, which may or may not be the case for neural data. If the data is non-Gaussian, or multimodal Gaussian, then the basis vectors will not be independent, but only uncorrelated. Furthermore, there is no guarantee that the direction of maximum variance will correspond to features that help discriminate between groups. For example, consider a feature taken from a unimodal distribution with a high variance. Now consider a feature taken from a multimodal distribution with a lower variance. Clearly the feature with the multimodal distribution will allow for the discrimination between groups, even though the overall variance of that feature could be lower.

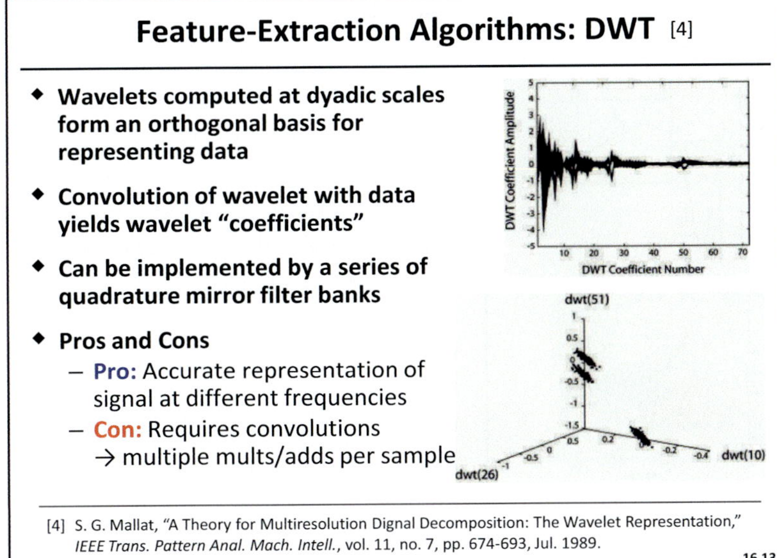

Slide 16.13

The discrete wavelet transform has been proposed for FE by [4]. The DWT should be a good method for FE since it is a multi-resolution technique that provides good time resolution at high frequencies and good frequency resolution at low frequencies. The DWT is also appealing because it can be implemented using a series of filter banks, keeping the complexity relatively low. However, depending on the choice of mother wavelet and on the number of wavelet scales used, the number of computations required to compute the wavelet coefficients for each spike can become high.

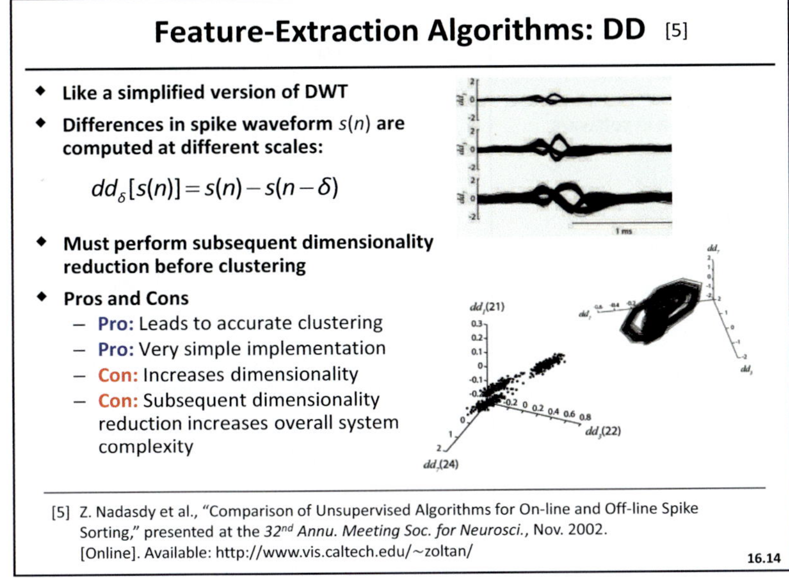

$$dd_\delta[s(n)] = s(n) - s(n-\delta)$$

Slide 16.14

Discrete Derivatives (DD) is a method similar to the DWT but simpler [5]. In this method, discrete derivatives are computed by calculating the slope at each sample point, over a number of different time scales. A derivative with time scale δ means taking the difference between spike sample at time n and spike sample that is δ samples apart.

The benefits of this method are that its performance is very close to that of DWT while being much less computationally expensive. The main drawback, however, is that it increases the dimensionality of the data, thus making subsequent dimensionality reduction even more critical.

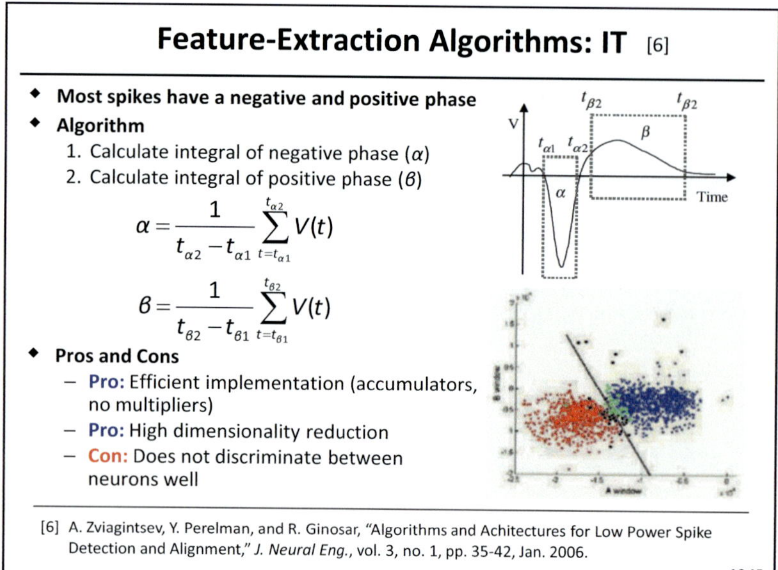

$$\alpha = \frac{1}{t_{\alpha 2} - t_{\alpha 1}} \sum_{t=t_{\alpha 1}}^{t_{\alpha 2}} V(t)$$

$$\beta = \frac{1}{t_{\beta 2} - t_{\beta 1}} \sum_{t=t_{\beta 1}}^{t_{\beta 2}} V(t)$$

Slide 16.15

The last feature-extraction method considered here is the integral transform (IT) [6]. As shown in the top figure, most spikes have both a positive and a negative phase. In this method, the integral of each phase is calculated, which yields two features per spike. This method seems ideally suited for hardware implementation, since it can be implemented very simply and since it inherently achieves such a high level of dimensionality reduction (thereby reducing the complexity of subsequent clustering). The main drawback, however, is that it has been shown not to perform well. That is because these features tend not to be good discriminators for different neurons.

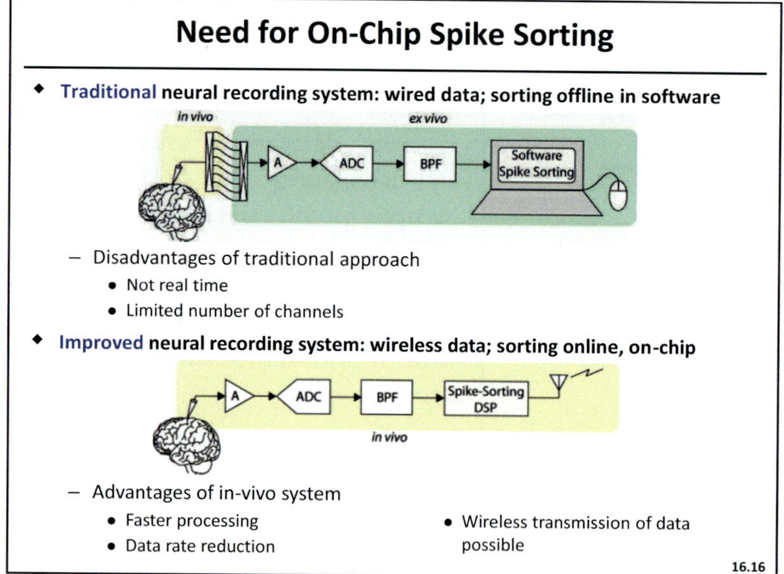

Slide 16.16

In a traditional neural recording system, unamplified raw data is sent outside the body through wires. Spike sorting of this data is performed offline, in software. This setup for neural recording has several disadvantages. It precludes real-time processing of data, and can only provide support for a limited number of channels. It also restricts freedom of movement of the subject, and the wires increase the risk of infection. Spike sorting on a chip implanted inside the skull solves these problems. It provides the fast, real-time processing that is necessary for brain-machine interfaces. It also provides output data-rate reduction, which enables wireless transmission of data for a large number of channels. Thus, there is a clear need for the development of spike-sorting DSP chips.

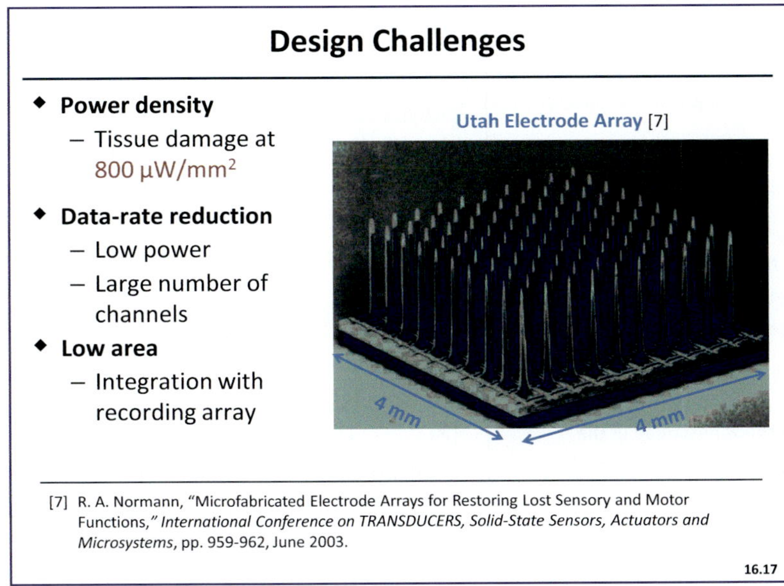

Slide 16.17

There are a number of design constraints for a spike-sorting DSP chip. The most important constraint for an implantable neural recording device is power density. A power density of $800\,\mu\mathrm{W/mm}^2$ has been shown to damage brain cells, so the power density must be significantly lower than this limit. Output data rate reduction is another important design constraint. Low output data rates imply low power for wireless transmission. Thus, a higher number of channels can be supported for a given power budget. The spike-sorting DSP should also have a low area to allow for integration of the DSP, along with the analog front end and RF circuits, to the base of the recording electrode. This figure shows a 100-channel recording electrode array, which was developed at the University of Utah [7] and which has a base area of $16\,\mathrm{mm}^2$.

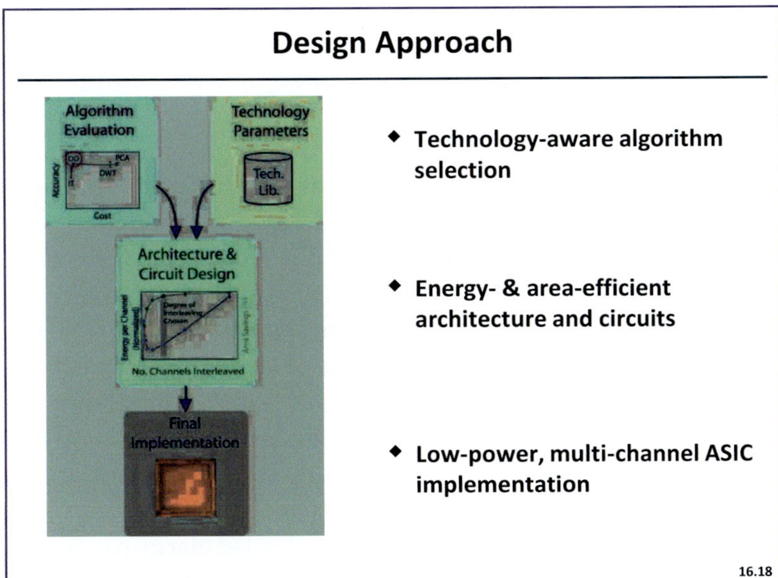

Slide 16.18

In the following design example, we have used a technology-driven algorithm-architecture selection process to implement the spike-sorting DSP. Complexity-performance tradeoffs of several algorithms have been analyzed in order to select the most hardware friendly spike-sorting algorithms [8]. Energy- and area-efficient choices have been made in the architecture and circuit design process that lead to a low-power implementation for the multichannel spike-sorting ASIC [9]. The following few slides will elaborate on each of these design steps.

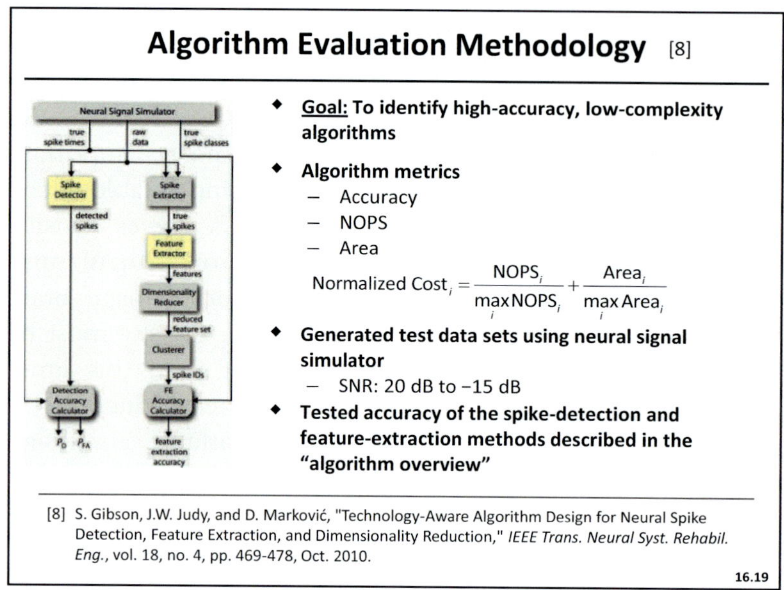

$$\text{Normalized Cost}_i = \frac{NOPS_i}{\max_i NOPS_i} + \frac{Area_i}{\max_i Area_i}$$

Slide 16.19

Several algorithms have been published in literature for each of the spike-sorting steps described earlier. However, there is no agreement as to which of these algorithms are the best-suited for hardware implementation. To elucidate this issue, we analyzed the complexity-performance tradeoffs of several spike-sorting algorithms [8]. Probability of detection, probability of false alarm, and classification accuracy were used to evaluate the algorithm performance. The figure on this slide shows the signal-processing flow used to calculate the accuracy of each of the detection and feature-extraction (FE) algorithms considered. Algorithm complexity was evaluated in terms of the number of operations per second required for the algorithm and the estimated area (for 90-nm CMOS). These two metrics were then combined into a "normalized cost" metric for each algorithm.

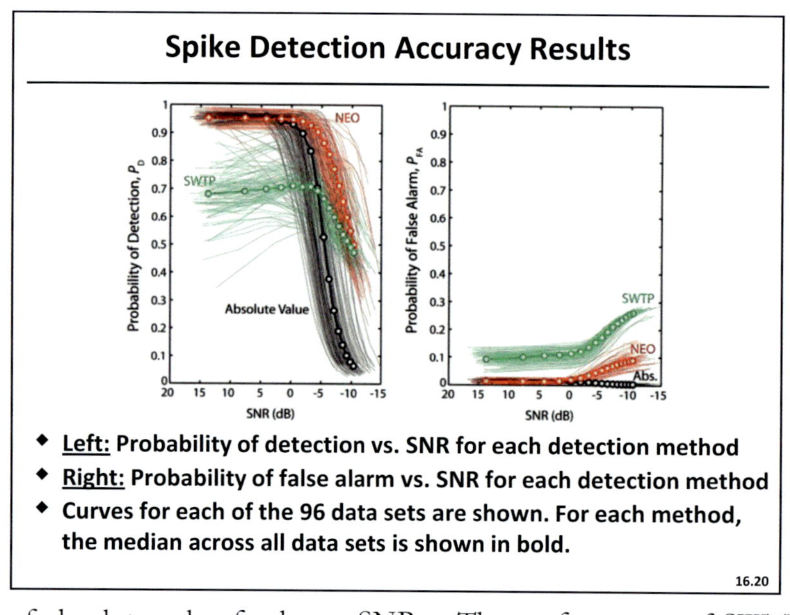

Slide 16.20

This figure shows both the probability of detection versus SNR (*left*) and the probability of false alarm versus SNR (*right*) for each detection method that was investigated. The absolute-value method has a low probability of false alarm across SNRs, but the probability of detection falls off for low SNRs. The probability of detection for NEO also falls off for low SNRs, but the drop-off occurs later and is less drastic. The probability of false alarm for NEO, however, is slightly higher than that of absolute-value for lower SNRs. The performance of SWTP is generally poorer than that of the other two methods.

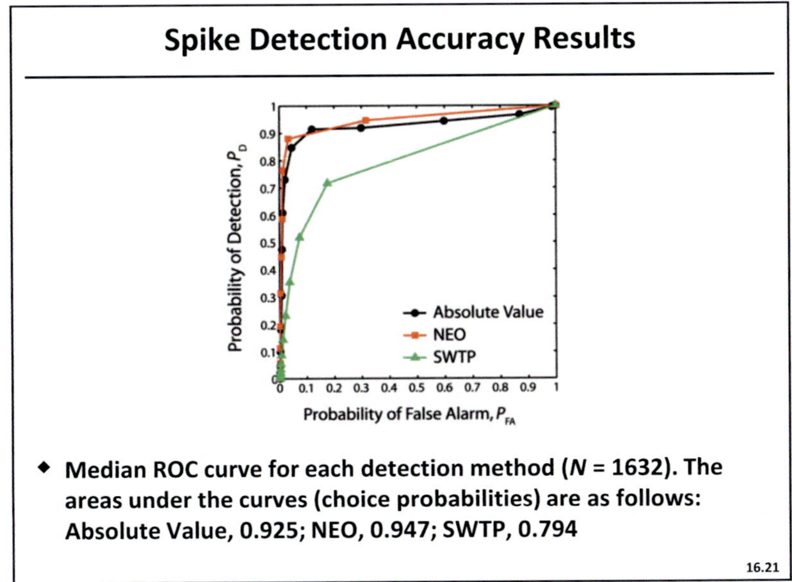

- **Median ROC curve for each detection method ($N = 1632$). The areas under the curves (choice probabilities) are as follows: Absolute Value, 0.925; NEO, 0.947; SWTP, 0.794**

16.21

Slide 16.21

Receiver Operating Characteristic (ROC) curves were used to evaluate the performance of the various spike-detection algorithms. For a given method, the ROC curve was generated by first performing the appropriate pre-emphasis and then systematically varying the threshold on the pre-emphasized signal from very low (the minimum value of the pre-emphasized signal) to very high (the maximum value of the pre-emphasized signal). At each threshold value, spikes were detected and P_D and P_{FA} were calculated in order to form the ROC curve. The area under the ROC curve (also called the "choice probability") represents the probability that an ideal observer will correctly classify an event in a two-alternative forced-choice task. Thus, a higher choice probability corresponds to a better detection method.

This figure shows the median ROC curve for all data sets and noise levels ($N=1632$) for each spike-detection method. It is again clear from this figure that the SWTP method is inferior to the other two methods. However, the curves corresponding to the absolute value and NEO methods are quite close, so it is necessary to consider the choice probability for each method in order to determine which of these methods is better. Since NEO has the highest choice probability, it is the most accurate of the three spike-detection methods.

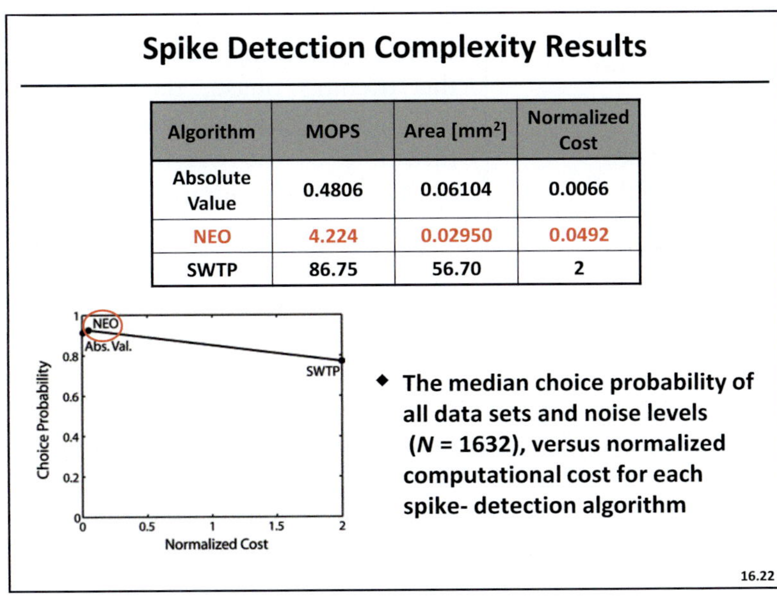

Algorithm	MOPS	Area [mm²]	Normalized Cost
Absolute Value	0.4806	0.06104	0.0066
NEO	4.224	0.02950	0.0492
SWTP	86.75	56.70	2

- **The median choice probability of all data sets and noise levels ($N = 1632$), versus normalized computational cost for each spike- detection algorithm**

16.22

Slide 16.22

This table shows the estimated MOPS, area, and normalized cost for each of the spike-detection algorithms considered. The absolute-value method is shown to have the lowest overall cost of the three methods, with NEO in a close second. The figure beneath the table shows a plot of median choice probability versus the normalized cost for each algorithm. The best choice for hardware implementation would lie in the high-accuracy, low-cost (upper, left) corner. Thus, it is clear that NEO is best-suited for hardware implementation.

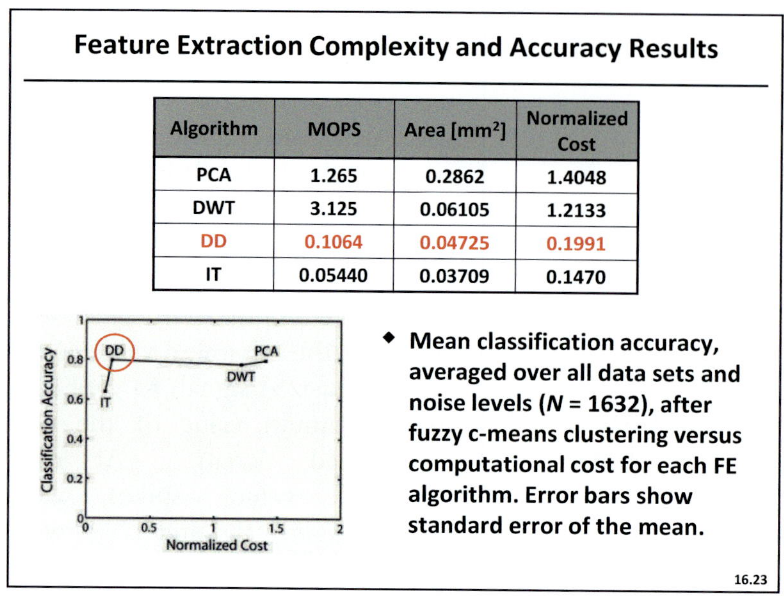

Algorithm	MOPS	Area [mm²]	Normalized Cost
PCA	1.265	0.2862	1.4048
DWT	3.125	0.06105	1.2133
DD	0.1064	0.04725	0.1991
IT	0.05440	0.03709	0.1470

Slide 16.23

This table shows the same results as before but for the feature-extraction methods. The IT method is shown to have the lowest overall cost of the three methods, with DD in a close second. The figure beneath the table shows the mean classification accuracy versus normalized cost for each feature-extraction algorithm. Again, the best choice for hardware implementation would lie in the high-accuracy, low-cost (upper, left) corner. Thus, it is clear that DD is best-suited feature-extraction algorithm for hardware implementation.

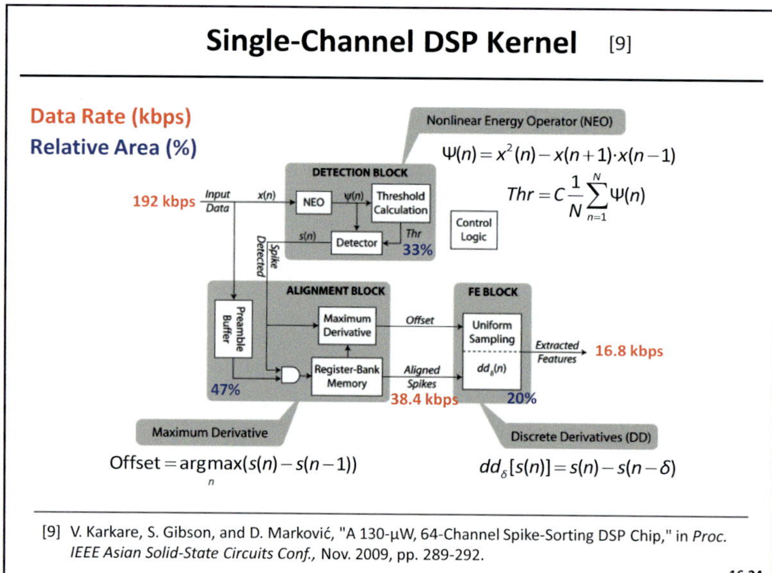

[9] V. Karkare, S. Gibson, and D. Marković, "A 130-μW, 64-Channel Spike-Sorting DSP Chip," in *Proc. IEEE Asian Solid-State Circuits Conf.,* Nov. 2009, pp. 289-292.

Slide 16.24

We chose NEO for detection, alignment to the maximum derivative, and discrete derivatives for feature extraction. This figure shows the block diagram of a single-channel spike-sorting DSP core [9]. The NEO block calculates $\psi(n)$ for incoming data samples. In the training phase, the threshold calculation module calculates the threshold as the weighted average of $\psi(n)$ for the input data. This threshold is used by the detector to signal spike detection. The preamble buffer saves a sliding window of the input data samples. Upon threshold crossing, this preamble is saved along with the following samples into the register bank memory as the detected spike. The maximum derivative block calculates the offset required to align the spikes to the point of maximum derivative. The FE block then calculates the uniformly sampled discrete derivative coefficients for the aligned spike. The 8-bit input data, which enters the module at a rate of 192 kbps, is converted to aligned spikes, which have a rate of 38.4 kbps. The extracted features have a data rate of 16.8 kbps, thus providing a 91.25% reduction in the data rate. The detection and alignment modules occupy 33% and 47% of the total $0.1\,mm^2$ area of a single-channel DSP core. The FE module occupies the remaining 20%. It is interesting to note that the single-channel DSP core is a register-dominated design with 50% of the area occupied by registers. The low operating frequency and register dominance makes the design of a spike-sorting DSP different from the conventional DSP ASICs, which tend to be high-frequency and logic-dominated systems.

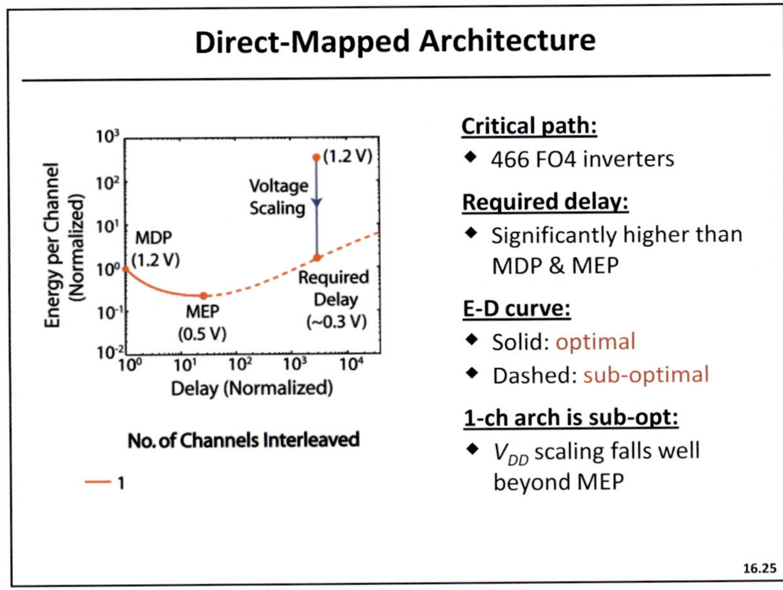

Slide 16.25

Now that we are familiar with the spike-sorting DSP and its features, let us analyze the design from an energy-delay perspective. This plot shows the normalized energy per channel versus the normalized delay for the spike-sorting DSP core [9]. The DSP has a critical-path delay equivalent to 466 FO4 inverters at the nominal supply voltage. This implies that the design at the minimum-delay point (MDP) is 2000 times faster than the application delay requirement. The E-D curve shown by the red line assumes operation at the maximum possible frequency at each voltage. However, since the application delay is fixed, there is no reward for early computation, as the circuit continues to leak for the remainder of the clock cycle. Operating the DSP at the nominal supply voltage would thus put us at the high energy point labeled 1.2V, where the DSP is heavily leakage-dominated. In order to reduce the energy consumed, we used supply-voltage scaling to bring the design from the high energy point at 1.2V to a much lower energy at 0.3V. However, mere supply-voltage scaling for a single-channel DSP puts us at a sub-optimal point beyond the minimum-energy point (MEP) for the design. To bring the DSP to a desirable operating point between the minimum-delay and minimum-energy points, we chose to interleave the single-channel architecture.

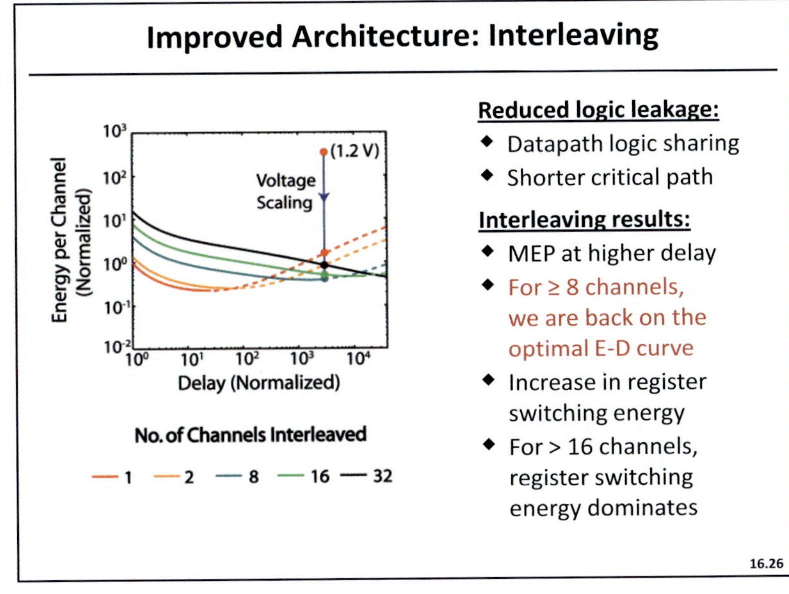

Slide 16.26

Interleaving allows us to use the same logic hardware for multiple channels, thereby reducing the logic leakage energy consumed per channel. Interleaving also reduces the depth of the datapath and pushes the MEP to a higher delay, thus bringing the MEP closer to the application delay requirement. The reduction in the logic leakage energy and the shift to a better operating point cause the energy per channel at application delay to decrease with higher degrees of interleaving. Beyond 8-channel interleaving, the design operates between the minimum-delay and the minimum-energy points. However, the number of registers in the design remains constant with respect to the number of channels interleaved. The switching energy of these registers increases with interleaving. Therefore, as more than 16 channels are interleaved, the energy per channel starts to increase due to an increase in register switching energy [9].

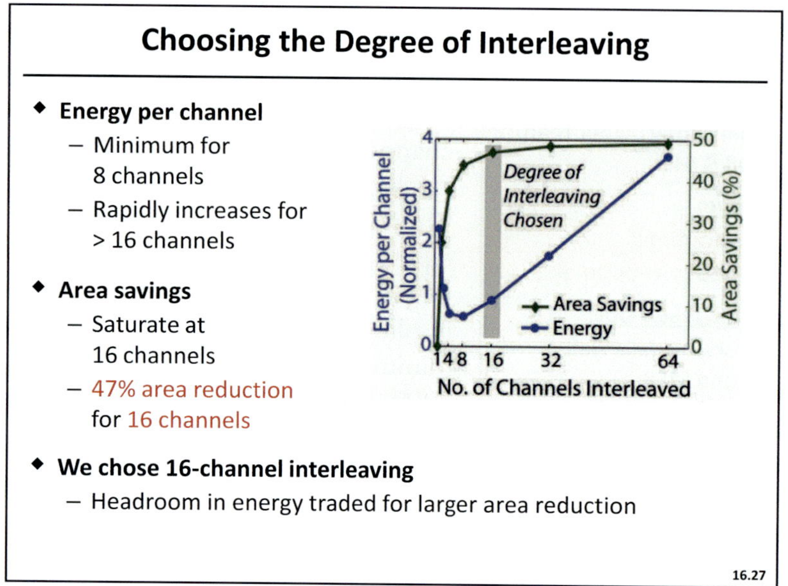

Slide 16.27

This curve shows the energy per channel at the application delay versus the number of channels interleaved [9]. It can be seen that the energy is at a minimum at 8-channel interleaving and increases significantly beyond 16-channel interleaving. In addition to reduction in energy per channel, interleaving also allows us to reduce area for the DSP, since the logic is shared between multiple channels. The area occupied by registers, however, remains constant. At high degrees of interleaving, the register area dominates over the logic area. This causes the area savings to saturate at 16-channel interleaving, which offers an area reduction of 47%. Since we expect to be well within the power density limit, we chose to trade headroom in energy in favor of increased area reduction. We, therefore, chose to implement the 64-channel DSP with a modular architecture consisting of four 16-channel interleaved cores.

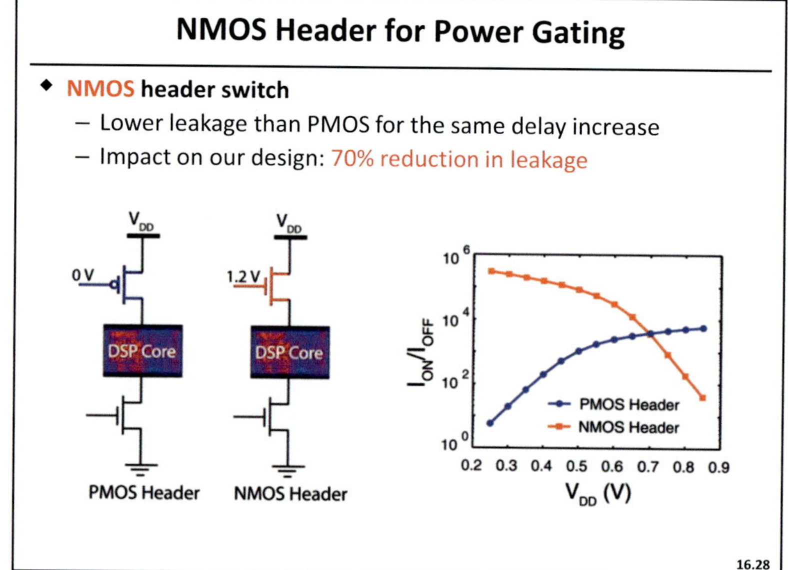

Slide 16.28

The modularity in the architecture allows us to power gate the inactive cores to reduce the leakage. In conventional designs, we need a PMOS header at the V_{DD}, since the source voltage of an NMOS can only go up to $V_{DD} - V_{TN}$. However, in our design the supply voltage is 0.3 V, but the voltage on the sleep signal at the gate can be pushed up to 1.2 V. Therefore it is possible to use an NMOS header for gating the V_{DD} rail. The NMOS sleep transistor has a V_{GS} greater than the core V_{DD} in the active mode, while maintaining negative V_{GS} in the sleep mode. The high overdrive in the active mode implies that we need a much narrower NMOS device for a given on-state current requirement than the corresponding PMOS. The narrow NMOS device leads to a lower leakage in the sleep mode. This plot shows the ratio of the maximum on-state versus off-state current for the NMOS and PMOS headers at different supply voltages. It is seen that for voltages lower than 0.7 V, the NMOS header has a better I_{ON}/I_{OFF} ratio than has the PMOS header. In this chip, we used NMOS devices for gating the V_{DD} and the GND rail [9].

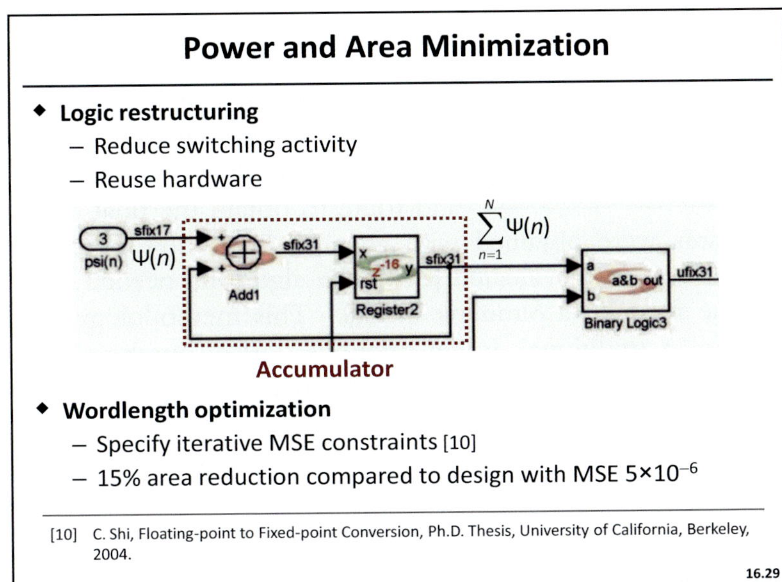

Slide 16.29

We also used logic restructuring and wordlength optimization to further reduce the power and area of the DSP. The logic was designed so as to avoid redundant signal switching. For instance, consider the circuit shown here for the accumulation of $\psi(n)$ for the threshold calculation in the training phase. The output of the accumulation node is gated such that the required division for averaging only happens once at the end of the training period, thereby avoiding the redundant switching as $\psi(n)$ is being accumulated. We also exploited opportunities for hardware sharing. For example, the circuit for the calculation of $\psi(n)$ is shared between the training and detection phases. Wordlength optimization was performed using an automated wordlength optimization tool [10]. Iteratively increasing constraints were specified on the mean squared error (MSE) at the signal of interest until detection or classification errors occur. For instance, it was determined that a wordlength of 31 bits is needed at the accumulation node for $\psi(n)$ to avoid detection errors. Wordlength optimization offers 15% area reduction compared to a design with a fixed MSE of 5×10^{-6} (which is equal to the input MSE).

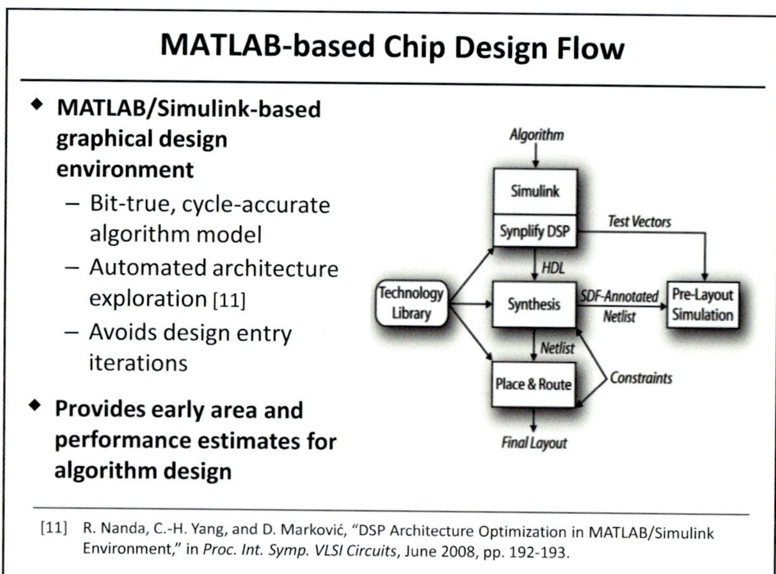

Slide 16.30

We used a MATLAB/Simulink-based graphical environment to design the spike-sorting DSP core. The algorithm was implemented in Simulink, which provided a bit-true, cycle-accurate representation of the design. The Synplify DSP blockset was used to auto-generate the HDL from the Simulink model. The HDL was synthesized and the output of the delay-annotated netlist was verified with respect to the output of the Simulink model. The netlist was also used for place-and-route to obtain the final layout for the DSP core. Since synthesis estimates were obtained with an automated flow from the Simulink model, various architecture options could be evaluated [11]. The algorithm needed to be entered only once in the design flow in the form of a Simulink model. This methodology thus avoided the design re-entry that is common in traditional design methods. Also, numbers from technology characterization could be used with the Simulink model to obtain area and performance estimates early in design phase.

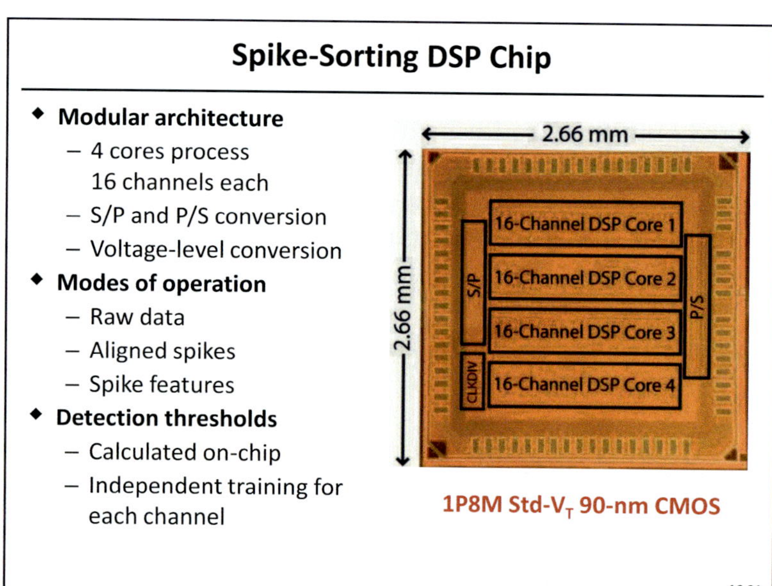

Slide 16.31

This slide shows the die photo of the spike-sorting DSP chip. As mentioned earlier, the 64-channel DSP has a modular architecture consisting of four cores that process 16 channels each. All cores except core 1 are power-gated. The modularity in the architecture allows for easy extension of the design to support a higher number of channels. Input data for 64 channels arrives at a clock rate of 1.6 MHz and is split into four streams of 400 kHz each using a serial-parallel (S/P) converter. At the output, the data streams from the four cores are combined by the parallel-serial (P/S) converter to form a 64-channel-interleaved output stream. Since we use a reduced voltage at the core, a level converter with cross-coupled PMOS devices was designed to provide a 1.2-V swing at the I/O pads. The chip supports three output modes to output raw data, aligned spikes, or spike features. The detection thresholds are calculated on-chip, as opposed to many previous spike-sorting DSPs that require the detection threshold to be specified by the user. The training phase for threshold calculation can be initiated independently for each channel. Hence, a subset of noisy channels can be trained more often than the rest without affecting the detection output on the other channels.

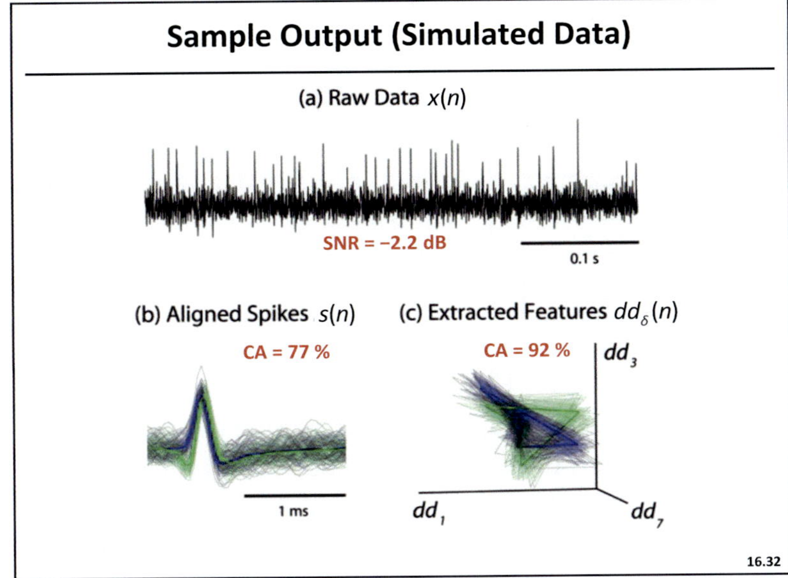

Slide 16.32

This figure shows a sample output for simulated neural data having an SNR of $-2.2\,$dB. Part (a) shows a snapshot of the raw data, part (b) the aligned spikes, and part (c) the uniformly sampled DD coefficients. In parts (b) and (c), samples have been color-coded according to the correct classification of that sample. That is, for every spike in the time-domain (b) and for every DD coefficient in the feature domain (c), the sample is colored green if it originates from neuron 1 or purple if it originates from neuron 2. It is known a priori that the simulated neural data has two neurons which, as seen in (b), are difficult to distinguish in the time domain. However, these classes separate nicely in the discrete-derivative domain (c). The classification accuracy of this example is 77% when time-domain spikes are clustered. When discrete derivative coefficients are used instead, the classification accuracy is improved to 92%.

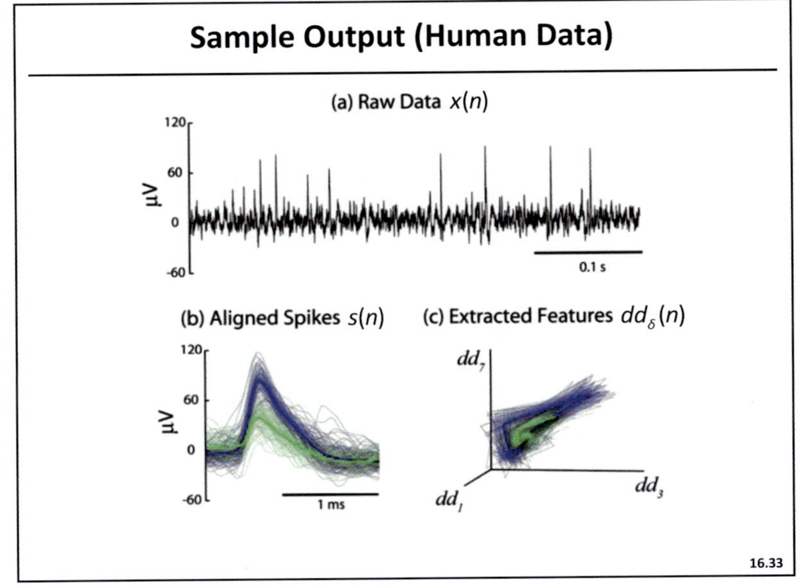

Slide 16.33

This figure shows a sample output of the chip when human data is processed. The raw data is recorded using one of the nine 40-μm-diameter electrodes positioned in the hippocampal formation of a human epilepsy patient at UCLA. Part (a) shows the raw data, part (b) the aligned spikes, and part (c) the FE coefficients. In this case, the samples in (b) and (c) have been colored according to the results of clustering.

Chip Performance

- **Power**
  - 130 µW for 64 channels
  - 52 µW for 16 channels
- **Area**
  - Die: 7.07 mm^2
  - Core: 4 mm^2
- **Classification accuracy**
  - Over 90 % for SNR > 0 dB

Technology	1P8M 90-nm CMOS
Core V$_{DD}$	0.55 V
Gate count	650 k
Clock domains	0.4 MHz, 1.6 MHz
Power	2 µW/channel
Data reduction	91.25 %
No. of channels	16, 32, 48, 64

SNR	−2.2 dB	Median
P$_D$	86 %	87 %
P$_{FA}$	1 %	5 %
Class. accuracy	92 %	77 %

Power density: 30 µW/mm^2

16.34

Slide 16.34

This table shows a summary of the chip, built in a 90-nm CMOS process. The DSP core designed for 0.3 V operation could be verified in silicon for a minimum core voltage of 0.55 V. The chip consists of 650 k gates and uses two clocks of 0.4 MHz and 1.6 MHz. The slower clock is derived on-chip from the faster clock. The power consumption of the chip is 2 µW/channel when processing all 64 channels. Data-rate reduction of 91.25 % is obtained as input data at a rate of 11.71 Mbps is converted to spike features at 1.02 Mbps. The chip can be used to process 16, 32, 48, or 64 channels at a time. The accuracy of the chip for the simulated neural data illustrated earlier is given by a probability of detection of 86 %, a probability of false alarm of 1 %, and a classification accuracy of 92 %. The corresponding median numbers calculated for the SNR range of −15 dB to 20 dB are 87 %, 5 %, and 77 %, respectively. The total power consumption is 130 µW for 64-channel operation. When only one core is active, the chip consumes 52 µW for processing 16 channels. The die occupies an area of 7.07 mm^2 with a core area of 4 mm^2. A classification accuracy of over 90 % is obtained for all simulated data sets with positive SNRs. The average power density of the cores is 30 µW/mm^2.

Summary

- **Spike sorting classifies spikes to their putative neurons**
  - It involves detection, alignment, feature extraction, and classification steps, starting with a 20-30kS/s data
  - Each step reduces data rate for wireless transmission
    - Feature extraction reduces data rate by 11x
  - Using spike features instead of aligned spikes improves classification accuracy

- **Real-time neural spike sorting works at speeds (30 kHz) far below the speed of GHz-rate CMOS technology**
  - Low data rates result in leakage-dominated designs
  - Extensive power gating is required to reduce power

16.35

Slide 16.35

Neural spike sorting is an example of a kHz-rate application where the speed of technology far exceeds application requirements. This necessitates different architectural and circuit solutions. Architecture based on heavy interleaving is used to reduce area and leakage. At the circuit level, supply voltage scaling down to the sub-threshold regime and additional power gating can be used to reduce leakage power.

References

- R. Quian Quiroga, Z. Nadasdy, and Y. Ben-Shaul, "Unsupervised Spike Detection and Sorting with Wavelets and Superparamagnetic Clustering," *Neural Comp.*, vol. 16, no. 8, pp. 1661–1687, Aug. 2004.

- J. F. Kaiser, "On a Simple Algorithm to Calculate the 'Energy' of a Signal," in *Proc. IEEE Int. Conf. Acoust., Speech, Signal Processing*, Apr. 1990, vol. 1, pp. 381–384.

- K.H. Kim and S.J. Kim, "A Wavelet-based Method for Action Potential Detection from Extracellular Neural Signal Recording with Low Signal-to-noise Ratio," *IEEE Trans. Biomed. Eng.*, vol. 50, no. 8, pp. 999–1011, Aug. 2003.

- S.G. Mallat, "A Theory for Multiresolution Signal Decomposition: The Wavelet Representation," *IEEE Trans. Pattern Anal. Mach. Intell.*, vol. 11, no. 7, pp. 674–693, July 1989.

- Z. Nadasdy et al., "Comparison of Unsupervised Algorithms for On-line and Off-line Spike Sorting," *presented at the 32nd Annual Meeting Soc. for Neurosci.*, Nov. 2002. [Online]. Available: http://www.vis.caltech.edu/~zoltan

- A. Zviagintsev, Y. Perelman, and R. Ginosar, "Algorithms and Architectures for Low Power Spike Detection and Alignment," *J. Neural Eng.*, vol. 3, no. 1, pp. 35–42, Jan. 2006.

- R.A. Normann, "Microfabricated Electrode Arrays for Restoring Lost Sensory and Motor Functions," in *Proc. International Conference on TRANSDUCERS, Solid-State Sensors, Actuators and Microsystems*, pp. 959-962, June 2003.

- S. Gibson, J.W. Judy, and D. Marković, "Technology-Aware Algorithm Design for Neural Spike Detection, Feature Extraction, and Dimensionality Reduction," *IEEE Trans. Neural Syst. Rehabil. Eng.*, vol. 18, no. 4, pp. 469-478, Oct. 2010.

- V. Karkare, S. Gibson, and D. Marković, "A 130 uW, 64-Channel Spike-Sorting DSP Chip," in *Proc. IEEE Asian Solid-State Circuits Conf.*, Nov. 2009, pp. 289-292.

- C. Shi, Floating-point to Fixed-point Conversion, Ph.D. Thesis, University of California, Berkeley, 2004.

- R. Nanda, C.-H. Yang, and D. Marković, "DSP Architecture Optimization in Matlab/Simulink Environment," in *Proc. Int. Symp. VLSI Circuits*, June 2008, pp. 192-193.

Additional References

- M. Abeles and M.H. Goldstein Jr., "Multispike Train Analysis," *Proceedings of the IEEE*, vol. 65, no. 5, pp. 762–773, May 1977.

- R.J. Brychta *et al.*, "Wavelet Methods for Spike Detection in Mouse Renal Sympathetic Nerve Activity," *IEEE Trans. Biomed. Eng.*, vol. 54, no. 1, pp. 82–93, Jan. 2007.

- S. Gibson, J.W. Judy, and D. Marković, "Comparison of Spike-Sorting Algorithms for Future Hardware Implementation," in *Proc. Int. IEEE Engineering in Medicine and Biology Conf.*, Aug. 2008, pp. 5015-5020.

- E. Hulata *et al.*, "Detection and Sorting of Neural Spikes using Wavelet Packets," *Phys. Rev. Lett.*, vol. 85, no. 21, pp. 4637–4640, Nov. 2000.

- K.H. Kim and S.J. Kim, "Neural Spike Sorting under Nearly 0-db Signal-to-noise Ratio using Nonlinear Energy Operator and Artificial Neural-network Classifier," *IEEE Trans. Biomed. Eng.,* vol. 47, no. 10, pp. 1406–1411, Oct. 2000.

- M.S. Lewicki. "A Review of Methods for Spike Sorting: The Detection and Classification of Neural Action Potentials," *Network,* vol. 9, R53-R78, 1998.

- S. Mukhopadhyay and G. Ray, "A New Interpretation of Nonlinear Energy Operator and its Efficacy in Spike Detection," *IEEE Trans. Biomed. Eng.,* vol. 45, no. 2, pp. 180–187, Feb. 1998.

- I. Obeid and P.D. Wolf, "Evaluation of Spike-detection Algorithms for a Brain-machine Interface Application," *IEEE Trans. Biomed. Eng.,* vol. 51, no. 6, pp. 905–911, Jun. 2004.

- A. Zviagintsev, Y. Perelman, and R. Ginosar, "Low-power Architectures for Spike Sorting," in *Proc. Int. IEEE EMBS Conf. Neural Eng.,* Mar. 2005, pp. 162–165.

Slide 17.1

The material in this book shows core ideas of DSP architecture design. Techniques presented can aid in extracting performance, energy and area efficiency in future applications. New ideas will also have to emerge to solve future problems. In the future, we feel, the main design problem will be how to achieve hardware flexibility and energy efficiency simultaneously.

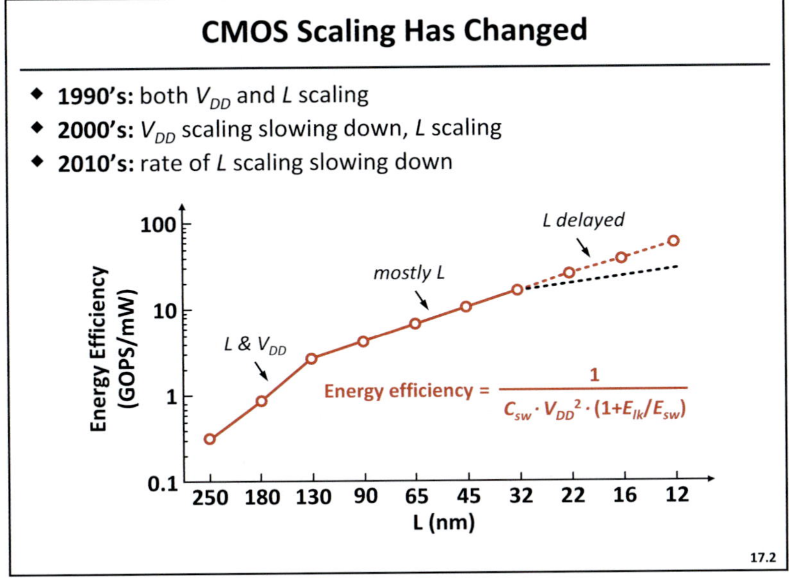

$$\text{Energy efficiency} = \frac{1}{C_{sw} \cdot V_{DD}^2 \cdot (1 + E_{lk}/E_{sw})}$$

Slide 17.2

We can no longer rely on technology scaling to improve energy efficiency. The plot on the slide shows energy efficiency in GOPS/mW, which is how many billion operations per second you can do in one milliwat of power, versus technology generation. Such energy efficiency represents the intrinsic computational capability of silicon technology. The numbers on the horizontal axis represent channel length (L).

In the past (e.g. 1990s), the number of digital computations per unit energy greatly improved with smaller transistors and lower operating voltage (V_{DD}). Things are now different. Energy efficiency is tapering off with scaling. Scaling of voltage has to slow down due to increased leakage and process variation, which results in reduced energy efficiency according to the formula. In the future, the rate of shrinking the channel length will be delayed due to increased development and manufacturing cost.

Technology scaling, overall, is no longer providing benefits in energy efficiency as in the past. This change in technology greatly emphasizes the need for energy-efficient design.

D. Marković and R.W. Brodersen, *DSP Architecture Design Essentials*, Electrical Engineering Essentials, DOI 10.1007/978-1-4419-9660-2, © Springer Science+Business Media New York 2012

Applications Have Changed

- **Signal processing content increasing**
  - Increasing functional diversity
    - Communications, multimedia, gaming, etc.
    - Hardware accelerators
  - Increasing complexity
    - High-speed WiFi: multi-antenna, cognitive radio, mm-wave, etc.
    - Healthcare, neuroscience

- **Apple iPad example**
  - Support for H.264 decoder was done in hardware
  - Software solution was too much power

- **Specialized hardware is not the answer for future problems**
  - Chips that are energy efficient and flexible are needed

17.3

Slide 17.3

Applications are getting more complex. Increased functional diversity has to be supported on the same device. The amount of data processing for adding new features is exploding. Let's take a recent (2010) example of an iPad. When Apple decided to put H.264 decoder on iPad, they had to use specialized hardware to meet the energy efficiency requirements. Software was not an option, because it would have consumed too much power. Future applications will be even more constraining. In applications such as high-speed wireless, multi-antenna and cognitive radio or mm-wave beamforming, software solutions wouldn't even meet real-time requirement regardless of power. Software solutions wouldn't even be an option. There are numerous other applications where real-time throughput and energy efficiency have to come from specialized hardware.

Adding too many pieces of specialized hardware, however, to support a diverse set of features would not be effective. Future designs must be energy efficient and flexible at the same time.

Ways to Provide Flexibility

Software	vs.	Hardware

Programmable DSP	FPGA (Flexible DSP)
Specialized μProc.	Reconfigurable hardware
Conditional ops, floating point	Repetitive operations
Multi-core programming is difficult	Implicitly parallel hardware
Low throughput (~10MS/s apps)	High throughput (10-100x ↑ GOPS)

17.4

Slide 17.4

There are two ways to provide flexibility. We can use a programmable DSP processor and develop application-specific software. People have been doing that for a long time. This approach works well for low-throughput and ad-hoc operations, but falls short on delivering the performance and energy efficiency required from high-throughput applications such as high-sped wireless.

Alternatively, we could use an FPGA, which is a reconfigurable hardware that you customize each time you need to execute a new algorithm. Unlike programmable DSP where your hardware is fixed, now you have uncommitted resources, which you can configure to support large degrees of parallelism and provide very high throughput.

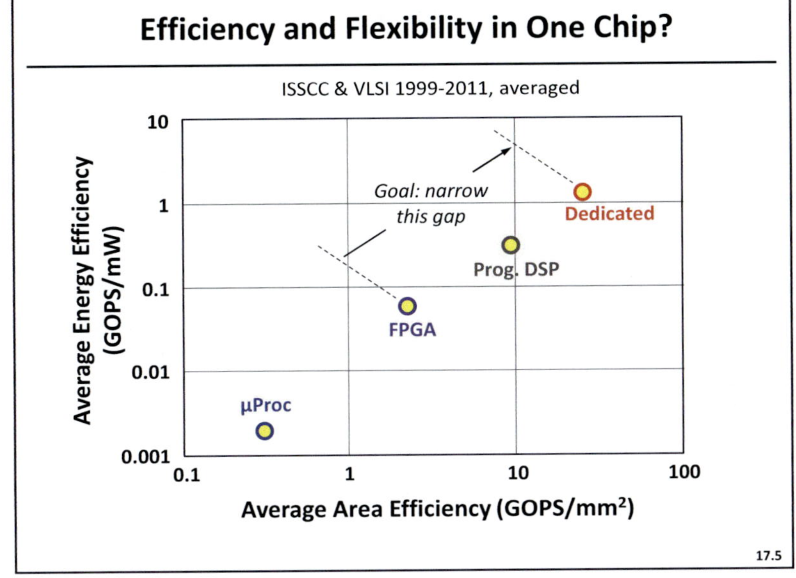

Slide 17.5

In both cases, the cost of flexibility is quite high. This slide shows energy and area efficiency comparisons for different types of processors. We took data from the top two international conferences in chip design over the past 12 years and averaged the numbers to observe general ranking. FPGA companies don't publish their data, but it is widely believed that FPGAs are at least 15x less efficient in both energy and area than dedicated chips.

Since programmable DSPs can't deliver the performance and efficiency for high-throughput applications, as we've discussed, we need to drive up efficiency of reconfigurable hardware towards the upper-right corner. It would be great if we could possibly eliminate the need for dedicated chips.

Rising Cost of ASIC Design

% Design Starts by Technology

Node (nm)	2002 (%)	2003 (%)	2004 (%)	2005 (%)	2006 (%)	2007 (%)	2008 (%)	2009 (%)	2010 (%)	2011 (%)	Cost ($M)
28	0	0	0	0	0	0	0	0	>0	>0	110
32	0	0	0	0	0	0	0	1	FPGA	FPGA	80
40	0	0	0	0	0	1	FPGA	FPGA	6	7	60
45	0	0	0	0	0	1	2	4	6	7	60
65	0	0	0	1	2	FPGA	8	10	13	15	55
90	0	1	8	FPGA	FPGA	23	23	24	24	24	30
130	FPGA	FPGA	FPGA	29	29	27	27	25	24	24	20
180	38	27	23	20	17	14	12	10	10	8	13
250	16	15	12	12	11	9	9	8	7	6	5
350	21	16	12	12	11	10	8	8	6	5	3
500	5	4	3	7	7	6	6	5	4	4	2
>500	1	1	0	6	6	6	5	5	5	5	<1
Total	100	100	100	100	100	100	100	100	100	100	-

Integration, Lower Cost, Performance · *Increased Risks & Development Cost*

#1 Design Technology: Dedicated   FPGA

Source: Altera & Gartner (2009)
M.C. Chian, FPGA 2009 (panel).

17.6

Slide 17.6

Apart from technical reasons, we need to take a quick look into the economic implications. Here, we see which technologies have been used by dedicated designs in the past 10 years. Entries in the table indicate percentage of new designs in respective technologies. Good news is on the left: scaling improves performance and lowers cost for high-volume chips. Bad news on the right is that the cost of design development is inversely proportional to feature size. We need over $100 million to develop a chip in 28-nm technology.

That's why dedicated chips still use 90 nm as the preferred technology. On the other hand, FPGA companies exploit the most advanced technology available. Due to their regularity, the development cost is not as high. So, if we retain the efficiency benefits of dedicated chips without giving up the flexibility benefits of FPGAs, that would be a revolutionary change.

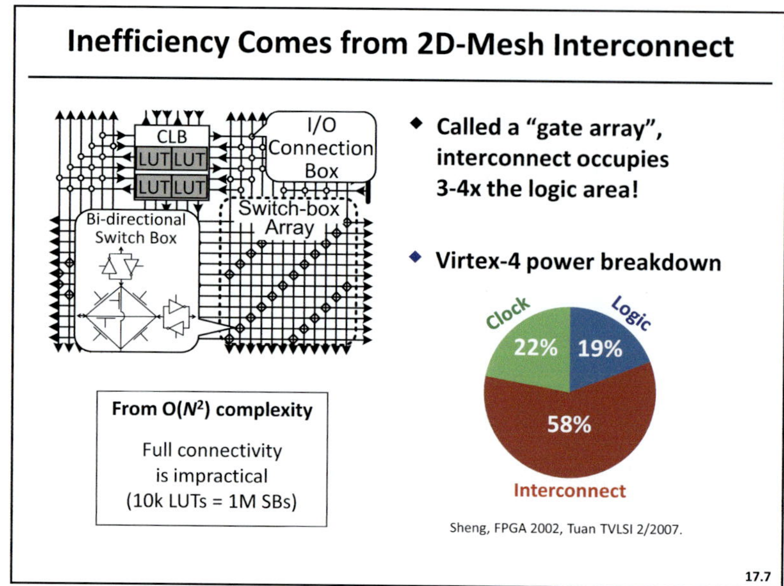

Slide 17.7

FPGAs are energy inefficient, because of their interconnect architecture. This slide shows a small section of an FPGA chip representing key FPGA building blocks. The configurable logic block (CLB) consists of look-up table (LUT) elements that can be configured to do arbitrary logic. The switch-box array consists of bi-directional switches that can be configured to establish connections between CLBs.

The architecture shown on the slide is derived from $O(N^2)$ complexity, where N represents the number of logic blocks. Clearly, full connectivity cannot be supported, because the number of switches would outpace Moore's law. In other words, if the number of logic elements N were to double, the number of switches N^2 would quadruple. This is why FPGAs never have full connectivity.

Depopulation and segmentation are two techniques that are used to manage connectivity. The switch-box array shown on the slide would have 12×12 switches for full connectivity, but only a few diagonal switches are provided. This is called depopulation. When two blocks that are physically close are connected, there is no reason to propagate electricity down the rest of the wire, so the wire is then split into segments. This technique is called segmentation. Both of the techniques are used heuristically to control connectivity. As a result, it is nearly impossible to fully utilize an FPGA chip without routing congestion and/or performance degradation.

Despite reduced connectivity, FPGA chips still have more than 75% of chip area allocated for the interconnect switches. The impact on power is also quite significant: interconnect takes up about 60% of the total chip power.

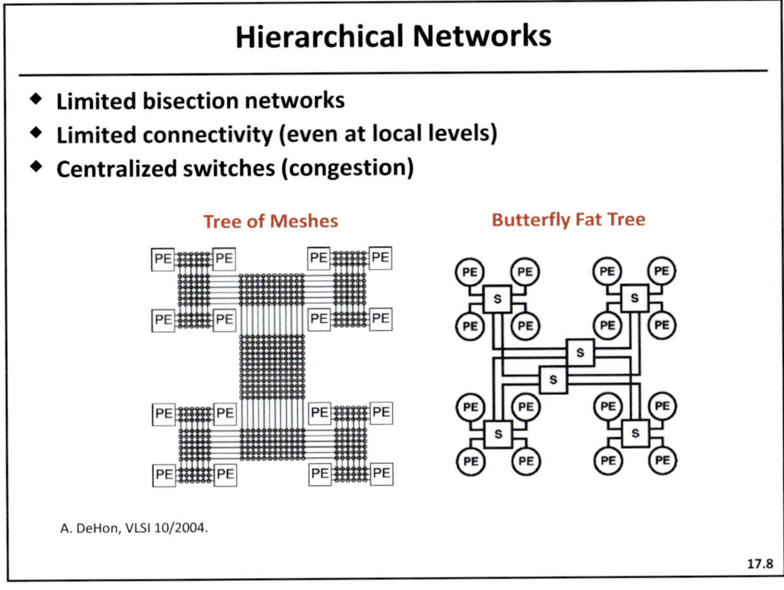

Slide 17.8

To improve energy efficiency, hierarchical networks have been considered. Two representative approaches, tree of meshes and butterfly fat tree, are shown on the slide. Both networks have limited connectivity even at local levels and also result in some form of a mesh.

Consider the tree of meshes, for example. Locally connected groups of 4 PEs are arranged in a network. As you can see, each PE has 3 wires. We would then need $4*3 = 12$ switches, while only 6 are available. This means 50% of connectivity even at the lowest level. Also, the complexity of the centralized mesh grows quickly.

Butterfly fat tree attempts to provide more dedicated connectivity at each level of hierarchy, but still results in a large central switch. We, again, have very similar problem as in 2D mesh: new levels of hierarchy use centralized global resources for routing. Dedicated resources would be more desirable for new levels of hierarchy. This is a critical problem to address in the future in order to provide energy efficiency without giving up the benefits of flexibility.

Slide 17.9

In summary, we are near the end of CMOS scaling for both technical and economic reasons. Energy efficiency is tapering off, design cost is going up. We must investigate architecture efficiency in light of these challenges.

Applications are getting more complex and the amount of digital signal processing is growing rapidly. Future applications, therefore, require energy efficient flexible hardware. Architecture of the interconnect network is crucial for providing energy efficiency.

Design problem of the future, therefore, is how to simultaneously achieve hardware flexibility and energy efficiency.

References

- M.C. Chian, Int. Symp. on FPGAs, 2009, Evening Panel: CMOS vs. NANO, Comrades or Rivals?

- T. Tuan, et al., "A 90-nm Low-Power FPGA for Battery-Powered Applications," *IEEE Trans. Computer-Aided Design of Integrated Circuits and Systems*, vol. 26, no. 2, pp. 296-300, Feb. 2007.

- L. Sheng, A.S. Kaviani, and K. Bathala, "Dynamic Power Consumption in Virtex™-II FPGA Family," in *Proc. FPGA 2002*, pp. 157-164.

- A. DeHon, "Unifying Mesh- and Tree-Based Programmable Interconnect," *IEEE Trans. Very Large Scale Integration Systems*, vol. 12, no. 10, pp. 1051-1065, Oct. 2004.

Index

The numbers indicate slide numbers (e.g. 12.15 indicates Slide 15 in Chapter 12).

D. Marković and R.W. Brodersen, *DSP Architecture Design Essentials*, Electrical Engineering Essentials,
DOI 10.1007/978-1-4419-9660-2, © Springer Science+Business Media New York 2012